Philosophy of mathematics

D1500356

Philosophy of mathematics

Selected readings
SECOND EDITION

Edited by
Paul Benacerraf

STUART PROFESSOR OF PHILOSOPHY
PRINCETON UNIVERSITY

Hilary Putnam

WALTER BEVERLY PEARSON PROFESSOR OF
MODERN MATHEMATICS AND MATHEMATICAL LOGIC
HARVARD UNIVERSITY

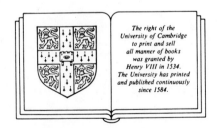

The right of the
University of Cambridge
to print and sell
all manner of books
was granted by
Henry VIII in 1534.
The University has printed
and published continuously
since 1584.

Cambridge University Press

Cambridge
New York   Port Chester   Melbourne   Sydney

Published by the Press Syndicate of the University of Cambridge
The Pitt Building, Trumpington Street, Cambridge CB2 1RP
40 West 20th Street, New York, NY 10011, USA
10 Stamford Road, Oakleigh, Melbourne 3166, Australia

First published by Prentice-Hall, Inc., 1964
Second edition published by Cambridge University Press 1983
Reprinted 1985, 1986, 1987, 1988, 1989

Printed in the United States of America

Library of Congress Cataloging in Publication Data
Main entry under title:
Philosophy of mathematics.
Bibliography: p.
I. Mathematics—Philosophy.   I. Benacerraf, Paul.
II. Putnam, Hilary.
QA8.4.P48   1983   510'.   82-25257
ISBN 0 521 22796 8 hard covers
ISBN 0 521 29648 X paperback

Contents

v

Contents

Preface to the second edition

Even a casual comparison of the table of contents of the present collection with that of its predecessor will reveal significant differences as well as much overlap. By and large, the present selection is the product of two forces: (a) comments from users of the first edition (and from potential users of the second) and (b) our own sense of the direction the field has taken during the past two decades.

We are grateful to our many friends and colleagues, too numerous to thank individually, who have commented on what they found useful and less than useful in our first effort, as well as on what they felt it would be good to have available in one volume. Their perspective has been invaluable, though the responsibility for our selections remains largely our own.

Needless to say, we would have liked in a way to reissue the first edition and simply add a second, companion, volume. But we are deterred by the prohibitive cost (to the user) of the two volumes. Hence the inevitable compromise: A selection was made, omitting several things to make room for new ones. In a number of cases (most notably the Wittgenstein material and "Two Dogmas of Empiricism"), the (present) availability of most of the material enabled us to omit it with less of a sense of loss. Not so with the rest. The selection of new material was even more difficult, as these years have been particularly fecund, both in relevant semi-technical results and in philosophical explorations.

As before, we limited our selections to those we felt would be accessible to the philosophically educated reader with enough background in logic to understand an exposition of some of the results of twentieth-century logic. (An important example is the independence of Cantor's Continuum Hypothesis.) In a similar vein, we tried also to narrow the range of *philosophical* issues discussed in the selection to ones that could most easily be recognized as concerning the philosophy of *mathematics*. Both of these admittedly loose principles served as guidelines only; but any attempt to observe them inevitably constrains the range of literature available for consideration. Except for these rules of thumb, in the end, we followed no overarching principle other than that of making a selection of items that, in our judgment, would make interesting reading when taken together.

Another major point of difference between this volume and its predecessor lies in the extended Bibliography that appears at the end of this edition. It was compiled by starting with the selected bibliography of the first edition, adding a number of items that we felt were missing from it, and closing it by including everything referred to in any of the selections included in the book. Inevitably, there have been many important omissions – as we go to press we must resist the urge to keep adding to it. We wish at this point to record our gratitude: once again to George Boolos, who compiled the initial bibliography; to Takashi Yagasawa, for his help in augmenting it; and to Ann Getson, James Cappio, and Ann Ivins for their invaluable contributions to its completion and to the preparation of the manuscript.

Finally, a word about format. Collecting all of the bibliographies into one comprehensive Bibliography enabled us to put all bibliographical references into a standard form for the book as a whole. This has meant the elimination of a large number of purely referential footnotes (for the convenience of including them in the text), as well as the shortening of many others. Also, we have identified as such any references to items that appear in this collection, inserting the present page number(s) where relevant. Wherever feasible, we retained the author's reference to the original source and simply added the locus of that reference in the present reprinting. We hope that the effort that went into this task is rewarded with a book that is, as a consequence, much more useful.

Princeton, N.J., and Cambridge, Mass.   PAUL BENACERRAF
November 1982   HILARY PUTNAM

Introduction

1. General remarks

It would be difficult to say just what comprises the philosophy of mathematics – what questions, views, and general areas should be covered in a book such as this. With that as our excuse, we have not tried to bring together a collection of selections that could be said to cover the field in a comprehensive way. We have tried rather to bring together selections that we felt were interesting in their own right, and that offered interesting comparisons when read together, all with the proviso that the issues discussed in them were in most cases central to the field. If we have succeeded, then we are certain that the reader has an adequate introduction to the philosophy of mathematics.

The divisions we have chosen are largely arbitrary, as the following remarks will indicate, and no importance should be attached to a particular article's being in this section rather than that. With this much said, we can state that the motivation behind the sections is roughly as follows: We included in Part I those selections that centered around three traditionally important views on the nature of mathematics: logicism, intuitionism, and formalism. This is not to say that other articles in the book do not bear on these views. For example, the article by Hempel in Part III is itself a very clear exposition of logicism. Like the other pieces in Part III, however, it discusses the view on a (mathematically) less technical plane and is therefore more readily accessible to people with no formal training in logic than the papers in Part I. The discussion of the three aforementioned views is thus the unifying thread that runs through the section called "The Foundations of Mathematics."

Questions concerning mathematical existence (we leave open here the question of whether mathematical existence is a different sort of existence or the existence of a different sort of thing, or both, or neither) are touched upon principally in Part II. But is is clear to anyone with the slightest familiarity with these matters that intuitionism, at least, is a view concerning mathematical existence, at least insofar as it includes conditions on what is to count as *proof* of the existence of certain mathematical structures, entities, and so on. Therefore, an adequate considera-

1

tion of these questions (adequate in that it takes in the leading points of view) would include items from Part I as well.

The first three sections overlap further in that the third, "Mathematical Truth," contains, besides the Hempel selection just mentioned, the quasi-intuitionistic piece by Poincaré. Quine ("Truth by Convention") discusses conventionalism, a view also expressed by Carnap in connection with mathematical existence and truth in his article in Part II, and Benacerraf's article in Part II, "What Numbers Could Not Be," is a discussion of an issue central to logicism: the nature of numbers.

We feel that such overlap is unavoidable. The division into problems is, at best, a guide for the reader. It is evident that one's view on the nature of mathematical truth (if there is indeed such a beast) will affect one's views on mathematical existence and will constitute a position on the "Foundations of Mathematics."

Despite this overlap, there is a further division to which we can point and which may prove helpful in organizing the array of views represented in this book: There is a suggestive distinction to be drawn between the items in Part I on the one hand, and those in Parts II and III on the other (Part IV, as we shall see, thoroughly straddles this distinction). Part I contains contributions belonging to what we should like to call the "epistemology of mathematics." With the possible exception of the selections by Frege and Russell, the authors of these pieces devote a good part of their attention to the question of what an acceptable mathematics *should* be like: what methods, practices, proofs, and so on, are *legitimate* and therefore justifiably used. They don't take existing mathematics and mathematical activity as sacrosanct and immune from criticism; according to them, there are justifiable and unjustifiable methods in mathematics, and acceptable results are those obtainable by justifiable methods. In fact, a good portion of the effort of the mathematician should be devoted to trying to recast intuitively desirable and acceptable results in forms that show them to be ultimately acceptable. If the author in question is an intuitionist, then it will be his view, e.g., that any part of real analysis that cannot be obtained by intuitionistic methods ought to be discarded. But in most cases, it is vitally important to carry on the search for intuitionistic proofs of as yet unobtained classical theorems.

And what we have seen to be true of the intuitionist is also true of formalists (Hilbert, von Neumann, Curry). Members of this latter group are concerned about the legitimacy of references to infinite collections, structures, and the like, in mathematics. More particularly, the concern takes the form of a concern that such reference, since it is to things so far from what we are capable of experiencing, might lead to contradictions (presumably because the candle of intuition casts but a dim light from such a distance). What then becomes important for the formalist is the search for a proof that these "infinitistic" methods form a consistent whole. The proof must, of course, be one that does not employ these questioned methods. So here again there appears a restriction on what methods should be countenanced in mathematics. And so on for the others. The other feature characterizing the members of this group is that they are predominantly mathematicians rather than philosophers. And we say this without in any way wishing to play down their philosophical contributions, any more than we would be inclined to deny the mathematical contributions made by members of the other group. The first group consists, therefore, largely of mathematicians who criticize the very foundations of their subject. These are the "epistemologists." (Poincaré and Bernays really belong in group one, if we are to judge from the bulk of their work, but the passages we have chosen assimilate them more to the second; Dummett, on the other hand, although primarily a philosopher, is writing to supply philosophical underpinnings to the intuitionist position and, as such, fits more naturally in group one.)

In contrast with the "epistemologists of mathematics," there are those who accept mathematics as, if not sacrosanct, then at least not their province to criticize. Their task is a different one: They do not want to *promulgate* certain mathematical methods as *acceptable*; they want to *describe* the *accepted* ones. Mathematics is something given and to be accounted for, explained, and accurately described. For them, epistemology is not a tool to help sort the good mathematics from the bad – it is a scheme within which mathematics as such must fit ("Mathematical propositions are analytic," "Mathematical statements are true by convention," and so on). One way of describing the difference between these two groups is to say that, for one group, the epistemological principles have a higher priority or centrality than most particular bits of mathematics, and hence can be used as a critical tool; whereas for the other group just the reverse is the case: Existing mathematics is used as a touchstone for the formation of an epistemology, one of whose conditions of adequacy will be its ability to put all of mathematics in the proper perspective. To put it somewhat crudely, if some piece of mathematics doesn't fit the scheme, then a writer in the first group will tend to throw out the mathematics, whereas one in the second will tend to throw out the scheme.

Of course, matters are not quite that neat. For both groups there is a constant interplay between epistemological principle and mathematical activity. Members of the first group will sometimes start with some paradigm cases of acceptable mathematical practice (e.g., intuitionists and formalists both start with parts of number theory) and then try to arrive at principles that will account for the validity of this starting point. These principles are then used either to criticize what fails to conform to them, or as a guide for the erection of standards that further proofs must meet, especially proofs of those "theorems" already "proved" but not "acceptably" so. Similarly, it would be an exaggeration to saddle the second group with absolutely everything a mathematician might produce. Their account of mathematics might very well force them to renounce and denounce some piece of mathematics as unacceptable. But by and large this is very unlikely. Ayer, Hempel, Boolos, Benacerraf, and Putnam are all prepared to take mathematics pretty much as is. Russell and the Carnap of "The Logicist Foundations of Mathematics" present a problem, depending on how seriously one takes their discussions of the vicious circle principle and impredicative definitions. There is no question that the Carnap of "Empiricism, Semantics, and Ontology" has abandoned any *critical* function for epistemology. Quine is a borderline case, almost professionally so. Insofar as he has abandoned his ontic qualms, however, he presents no problem. But leaving the borderline cases straddling the border line, Heyting, Brouwer, Hilbert, von Neumann, Curry, and Poincaré (in other passages than those we are presenting, alas) are quite clearly on the other side. Kreisel belongs, if anywhere, to this latter group, for he has strong constructivistic leanings. The "if anywhere" is inserted because Kreisel's function has been more or less to reconstruct and make mathematical sense of the philosophical pronouncements of the members of this latter group. And it is quite conceivable that he should wish to see just how much of extant mathematics one can obtain on this or that restriction without much caring whether or not one adopts the restriction. This would make his, as it were, a metatask.

The articles in Part IV weave both strands together into a hopeless tangle. This is hardly surprising, as the subject matter is set theory – a branch of mathematics with powerful ancestral ties to philosophy, and which has served as the battleground for a host of philosophic disputes ever since its explosive Cantorian(-Fregean) birth and its stormy Russellian adolescence. It has survived these traumas, as well as those inflicted by Gödel and Paul Cohen through their discovery of the independence results. Now, philosophers and mathematicians alike are scrutinizing its history and prehistory to tease out both the strengths responsible for its survival and the genetic weaknesses that almost caused it to perish. Inevitably, the reformers and apologists rub elbows.

But the distinction is a vague one and we should not try to make too much of it. Though vague, we hope it is nonetheless suggestive; and it

Introduction

should be of some help in understanding how the writings of the authors included in this collection are related. What becomes of interest, once one has seen the distinction, is the way in which one can view the discussions included in this book as continuous with one another. At first sight, it might appear that the two groups did not even discuss the same problems. But it should be seen that Hilbert is just as much concerned with the determinants of "mathematical truth" as, say, Ayer. The positions they adopt are rather different, but both should be read as writing on the same questions, or very nearly so. And similarly with Gödel. When Gödel discusses the continuum hypothesis, he is merely focusing on a particular mathematical proposition and the ways in which it might be shown to be true or false. Hempel's remarks on the nature of mathematical truth should, if cogent, be relevant to this discussion. And so on with the rest of the selections.

Consequently, it is our view that the questions we have chosen for study are intimately related. Furthermore, the authors whose discussions of these questions we have selected are, on the whole, concerned with very much the same questions, superficial differences to the contrary. These differences bespeak differences in point of view and differences in methods of attack, rather than simply different concerns. We believe that the discussion of all these problems benefits greatly from the interplay of these differences – but only when it becomes clear what unites them as discussions of the same problems. It is our hope that reading these selections together will make this clear.

So much by way of introduction. The remainder of this Introduction will consist of remarks on a number of problems, some of which are discussed rather fully, and others of which are merely touched upon by our authors. We hope that these remarks will make it easier to understand the selections and the issues involved.

A word of warning is in order. We have not attempted, in the sections that follow, to present a single, unified point of view on all of the problems and authors that we discuss. Instead we speak from many, different, and often incompatible viewpoints. For a couple of reasons. First, we are more concerned here to *raise* useful and interesting questions than to attempt to answer them. Second, it is unlikely that there is one on which we could agree, or even agree to agree for the nonce. Thus, the attentive reader, moving through this Introduction from section to section and even within individual sections, will discern shifts of focus and point of view. We hope that, rather than distract, they will prove helpful through the variety of perspectives they offer on the extremely complex problems under discussion.

5

2. The actual infinite and formalism

Although this collection does not contain a section titled "The Infinite in Mathematics," anyone who reads the selections from the writings of Brouwer, Heyting, and Hilbert that we have included under the general title "The Foundations of Mathematics" will quickly realize that the role played by infinite structures, collections, quantities, and the like in classical mathematics has a great deal to do with the controversy between the different "schools" in the philosophy of mathematics. By the same token, the measure of success attained by Cauchy and Weierstrass in eliminating "infinite quantities" from the calculus had a great deal to do with the ideal, shared by thinkers with views as mutually antagonistic and those of Hilbert and Brouwer, of eliminating the infinite from mathematics altogether.

But why should it be deemed desirable to avoid reference to the infinite in mathematics? Sometimes it is said – even by Hilbert – that references to the infinite are "meaningless." But why should one suppose that this is so? Classical philosophers – in particular Hume – had argued against the notion (in connection with infinite divisibility) on the basis of an identification of what is intelligible with what can be *visualized*; but the "image-in-the-head" theory of meaning no longer seems tenable, and attacks on the notion of the infinite must depend on something more reasonable than this if they are to be taken seriously.

In point of fact, it is very hard to find reasoned and even moderately detailed argument on this point. Opponents of the "actual infinite" tend to assume that the burden of proof lies on the other side. "Show us that the notion makes sense," they seem to say, where the criterion of making sense seems to be expressibility in *their* terms. We cannot discuss this issue here: Suffice it to say that readers who sympathize with the demand of the classical empiricists that all concepts be legitimized by being "derived from experience" will probably find themselves inclined to sympathize with those who doubt that any notion of an infinite structure is a clear one, whereas readers who are either of a more realistic or of a more pragmatist turn of mind may have difficulty in seeing "what the fuss is all about."

Suppose, however, that we assume that statements about infinite structures "make sense." Are there in fact any such structures to talk about? Hilbert argues convincingly that physics provides no clear evidence for the existence of such structures: In fact, the progress of physics has, as he points out, introduced finiteness and discontinuity in area after area in which the infinite and the continuous once reigned supreme. Today even the possibility of a beginning (and an end) to "physical time" is under discussion among physicists. Thus we must agree with Hilbert that if mathematics is to be independent of dubious empirical assumptions, it must not base assertions concerning the existence of infinite structures on physical considerations.

To this Russell replied, in a slightly different context, that mathematics is concerned not with (physical) existence, but only with the *possibility* of existence. Thus, in the second edition of *Principia Mathematica* (henceforth: *PM*), Russell and Whitehead chose not to assume the so-called Axiom of Infinity, which asserts that there are infinitely many objects in the universe of discourse, but rather explicitly included it among the hypotheses of each "theorem" in whose proof it was used. If T was the "theorem" in question, then Whitehead and Russell asserted only '*if* Inf. Ax., *then* T'.

But is it clear that an infinite totality could *possibly* exist? If the Axiom of Infinity leads to a contradiction, then the theorems that list it as hypothesis are certainly not very interesting. Since these form a large part of mathematics (at least as reconstructed by Russell and Whitehead), should there not be some proof of the *consistency* of the Axiom of Infinity, whether it is to be used as a postulate of the system or only as a hypothesis of a large number of important theorems?

Here we get a parting of the ways in the philosophy of mathematics. Russell and his followers apparently regard the *possible,* if not actual, existence of infinitely many objects as self-evident, whereas for Hilbert and the formalists the consistency of this assumption must be *proved.* Moreover it must be proved by "finitist" means – that is to say, the assumption itself must obviously not figure, even in a disguised way, among the assumptions of the consistency proof. The reader will observe that this kind of question is a bit like a political question – it is not a "purely theoretical" question, in the sense of making no difference to practice, but rather it affects one's standards in mathematics and one's program as a mathematician. Hilbert did not think it very likely that the system of *PM* was, in fact, inconsistent; he simply felt that to take its consistency, or even the consistency of elementary number theory, without proof, was to adopt too low a standard of mathematical exactness and to risk unpleasant surprises in the future.

Another perspective that might prove helpful in understanding Hilbert's desire for a consistency proof for infinitistic classical mathematics is the following.

Hilbert took certain kinds of mathematical assertions to be philosophically (i.e., epistemologically) unproblematic. These were assertions whose truth or falsity could be determined by combinatorial calculation – by the observation of combinatorial facts that could be ascertained by immediate perception. Call them "basic propositions" – for example, whether one (finite) string of symbols was longer than another. Assertions that made essential reference to infinite collections, in ways that deprived them of this property of finite verifiability/falsifiability in terms of observable combinatorial facts, were considered to be strictly meaningless. Hilbert recognized that their introduction into mathematics considerably simplifies the statement of a number of laws, thus making the theory considerably more elegant and appealing. In this regard he likened it to the introduction of ideal elements, such as points at infinity in projective geometry, or *i*.

But is it safe?

Hilbert allowed that they might nonetheless be admitted into mathematics if it could be shown that their admission would be harmless – that it would not enable one to prove falsehoods, that is, false basic propositions. A consistency proof for classical mathematics that made no essential reference to infinite collections would do just that. If the proof made use only of finitistic (i.e., sanitary) principles, the reference to infinite collections would have been justified on finitistic grounds. It could then be regarded as simply an elaborate *façon de parler* and indulged without risk.

3. The "potential infinite" and intuitionism

For the intuitionists, the position with respect to the infinite was different. Given a set of statements describing an infinite structure, there are two sorts of doubts that may arise. First, one may question the *consistency* of the statements: This was Hilbert's worry. Second, one may doubt that the statements "pick out" a *unique* and well-defined mathematical structure. Intuitionists sometimes write as if even the notion of an "arbitrary finite magnitude" is not completely fixed in advance.[1] We know, indeed, that 1, 2, 3 are integers. We know that certain operations applied to integers lead to integers – e.g., addition, multiplication, exponentiation. But it does not follow that we have a perfectly definite notion of "any integer" – because this involves the idea of iterating an operation (say, "adding 1") an *arbitrary finite number of times,* and we need not admit that we have a clear notion of what this means. The intuitionist does not, of course, propose to do without the concept "integer" – that would be to abandon mathematics altogether. The proposal (cf. Heyting,

---

[1] In what follows, we present an account of intuitionism directed at the nonintuitionist. It is possibly an account acceptable to no intuitionist, but we feel that such "falsification" is justified if it helps bridge the gulf that presently exists between the intuitionist and the "classical mathematician."

elaborating and formalizing Brouwer's ideas) is rather to develop a propositional calculus for dealing with concepts that do not necessarily correspond to a well-defined totality (and "statements" that do not necessarily have a truth-value). This attitude is often described as "countenancing the potential infinite but not the actual infinite." What it comes to is this:

1. A statement about an infinite structure – say, an infinite sequence of zeros and ones – may be regarded as *true* if proved and *false* if refuted, but in all other cases it is regarded as *neither true nor false.*
2. Since the structure is not thought of as well-defined, a statement about it can be proved only *if it is actually proved for a much larger class of structures.* In fact, to prove a statement about an infinite structure, we must prove the statement on the basis of verifiable statements either about some *finite part* of the structure (e.g., the first ten places of the sequence), or about the *rule* (if there is one) for successively producing the finite initial segments of the structure.

An example may help to make this position more clear. Consider the assertion that the sum of the first n odd number $(1+3+\cdots+(2n-1))$ is always a perfect square (in fact n^2). The sum of the first *one* odd numbers, that is, 1, is a perfect square, since $1 = 1^2$. And if the sum of the first n odd numbers is n^2, then the sum of the first $n+1$ must be $(n+1)^2$, or n^2+2n+1 (since the $n+1$st odd number is $(2n-1)+2$, and this is equal to $2n+1$). Thus we have proved the theorem for 1, and if we have proved the theorem for n, we can prove it for $n+1$. Accordingly, the intuitionist – like the classical mathematician – concludes that the theorem holds for every number. The philosophical difference is that the intuitionist does not assume that the numbers are a well-defined totality. But in this case it doesn't matter. (Although there are many cases in which intuitionists are led by their position to reject classically valid proofs; for example, proofs that assume that every statement about an infinite totality is either true or false – which amounts to assuming that the totality is well-defined – are rejected by intuitionists.) For, even if we extend the notion of an integer to cover a new "object," if all theorems proved in the preceding fashion hold for all the things previously counted as integers (notice that there need be only finitely many of these at any given time), and if the "new" integer is always "one plus" something previously counted as an integer, then the theorems in question will hold also for the "new" integer.

By way of contrast, consider the assertion that the number of "twin primes"[2] is infinite. For the classical mathematician this has a unique truth-value (even if he doesn't presently know it). But for the intuitionist it doesn't: For he has no proof of the statement, nor does he have a proof of its negation, and statements about an ill-defined totality don't have a truth-value unless they are proved (or disproved) from a *partial* determination of the totality.

A classical mathematician can get an approximate idea of what the intuitionist has in mind in the following way: (1) *Drop* the assumption that there is a well-defined "standard model" for number theory. (2) Don't assume that we can characterize by any finite number of axioms all of the things that we would intuitively recognize as correct methods of proof. (I.e., take "number theory" itself as a concept in the process of being created.) Then there will be three classes of statements in number theory: statements that are "true," that is, true in all models of number theory; statements that are "false," that is, false in all models; and statements that are "neither true nor false," that is, true in some models and false in others. Also, the "true" statements will all be provable – but not necessarily in any one formal system.

This explanation is, however, not intuitionistic, since an intuitionist would not accept the idea that the undecidable statements are "true in some models and false in others." Moreover, the intuitionist surely objects with reason here: for if "number theory" is not a closed concept, then the notion of a "model" of number theory is surely not a mathematically meaningful one even from the standpoint of classical mathematics. Yet one can still make sense of proving that a statement is "true in *all* models," namely, if one can prove that a statement is true in all models for some finite fragment of number theory, then, however the concept of "number theory" may be *enlarged* in the future, the statement in question must be, indeed, "true in all models of number theory" (since models of the whole must be models of each part). However, the idea of an infinite "model" – a *well-defined* infinite collection satisfying the axioms of a formal system – is not one acceptable to an intuitionist.

For the intuitionist, also, the problem of consistency does not arise because any statement about a "potential infinite" can be interpreted as a statement about a *finite* (but extendable) structure.[3] Thus the intuition-

---

[2] A *prime number* is one that cannot be divided without remainder except by itself and 1. *Twin primes* are primes whose difference is 2: e.g., (5, 7), (11, 13), (17, 19), etc. Whether the number of such pairs is finite or infinite is an unsolved (and apparently hopelessly difficult) problem in number theory.

[3] Strictly speaking, this is true only of free-variable statements. And even there, although the assumptions used in any one free-variable proof always have a finite model, the proof that *this* is so may require non-finitist methods. This is one problem with "finitist mathe-

ist and the formalist are in agreement as to the part of mathematics that is "safe"; that is, whose consistency may be taken as evident on the basis of an interpretation; namely, the part that may be interpreted as referring only to finite structures.[4]

4. Logicism

Logicism (Frege–Russell–Whitehead) arose out of a concern with a different problem: the nature of mathematical truth. Logicists hoped to show, as against Kant, that mathematics did not have any "subject matter," but dealt with pure relations among concepts,[5] and that these relations were "analytic," that is, of the same character as the principle of noncontradiction, or the rule of *modus ponens*. In contrast, Hilbert maintained that mathematics *did* have an extralogical subject matter, namely *expressions*[6] (e.g., series of strokes |, ||, |||, $\cdots$) and that its simplest truths (e.g., "|| added to ||| is |||||") were *anschaulich* (a German word that can mean both "visual" – in colloquial German – and "self-evident" or "intuitive" in philosophical German).

Logicism had one great and undeniable achievement – it succeeded in reducing all of classical mathematics (by any reasonable standard excluding completeness) to a single formal system. This achievement was much admired by the formalists, even if they did not agree that "mathematics has been reduced to logic." Formalists held that, as a result of the work of Whitehead and Russell, one had at last a clear formalization of what it was that had to be proved consistent.

Logicists, of course, thought they had done more than just axiomatize

matics": the consistency of "finitist" systems is not, in general, demonstrable by strictly finitist means.

[4]We are indebted to Georg Kreisel for the remark that the intuitionist notion of the "potential infinite" has *two* classical analogues: the "ill-defined infinite set" and the "finite (but unbounded) segment."

[5]For Frege, this is a complicated issue. Concepts for him are not objects or "entities" of any sort (what this pronouncement *means* is far from clear, and the subject of much controversy among Frege scholars). So, for logic (and hence mathematics) to deal with the relations among concepts is not for them to have a special subject matter – in the way, say, that living organisms constitute the subject matter of biology. Unfortunately, the issue is further complicated by Frege's view that concepts have *extensions,* which *are* objects (indeed, numbers, for Frege, are the extensions of concepts). Thus, to deny logic and mathematics any special subject matter, the logicist must argue that extensions of concepts (what some would simply call "sets") do not themselves constitute a special domain – a special subject matter.

[6]In place of "expressions" one might, of course, use other things: e.g., tables and chairs, or musical tunes. The important thing for Hilbert was not that finitist mathematics should be literally about series of *marks* (e.g., |, ||, |||, etc.) but that the subject matter, whatever it might be, should be wholly finite, discriminable, and *anschaulich* in all of its relevant parts and relations.

11

extant mathematics. They believed that they had *derived* all of mathematics from pure logic, without using any extralogical assumptions, and thus shown it all to be analytic. To assess this claim, we must ask at least whether what the logicist derives in his formal system is the mathematics he sets out to derive, and whether the premises of the derivation belong to logic.

Whether the logicist reduction should count as a derivation of mathematics depends, of course, on the character of the definitions employed – more specifically on what those definitions *preserve*. If they preserve meaning, at least sentence by sentence, then the answer is clearly yes. For he has shown that sentences with the same meaning as those of mathematics are logical consequences of the axioms of his formal system. To the extent that something less than meaning is preserved, the claim that it is the *propositions of mathematics* that have been derived must at least be questioned.

The issue of whether the derivation is from *logical* premises is regarded by many as largely a verbal issue (at least it need not be settled to see the bearing of the logicist reduction on Kant's claim that arithmetic was synthetic a priori): Logicists did not reduce all of mathematics to *elementary* logic, but they did reduce mathematics to elementary logic *plus* the theory of properties (or sets), properties of properties, properties of properties of properties, and so on. Thus if property theory (or set theory) may be counted as part of logic, mathematics is reducible to logic. But to what extent this refutes Kant's claim and establishes mathematics as analytic is something still open to question, for several objections may be raised. (It is not our purpose here to argue these points in detail, but we feel that it is particularly important to raise objections to logicism because it is a view that has received very little criticism in the literature: Of all the authors we reprint, only the intuitionists are really seriously critical of it. But they attack it from a very different point of view, as we will point out later.)

What Kant had denied was that the propositions of mathematics (arithmetic would be the relevant branch here, since he might have conceded the analyticity of algebra) were analytic. But 'analytic' for him meant either 'following from the law of noncontradiction' or being (a "logical truth") of the form 'All *A* are *B*', where "the idea of being a *B* is contained in the idea of being an *A*." A relevant example of such an analytic truth might be 'All spinsters are females'. Now, it is hardly the case that the logicist reduction of mathematics clearly shows the propositions of mathematics to be of either of these kinds. "Following from the law of noncontradiction" is itself at best a very unclear notion. The most likely (and most charitable) construal for it is something like 'whose negations are self-contradictory'. Thus construed, the question becomes that of deciding whether showing that on one set of plausible definitions arithmetic can be derived from set theory establishes that the negations of (presumably the true) arithmetic propositions are self-contradictory. It establishes, to be sure, that *if* these definitions are correct analyses of the meanings of the arithmetic terms, and *if* the set-theoretic axioms are themselves analytic in the relevant sense, and *if* being derivable in first-order logic from analytic propositions via definitions representing correct analyses constitutes "following from the law of noncontradiction," *then* indeed the logicists have shown that the propositions of arithmetic follow from the law of noncontradiction. But these are very big *if*s. Probably the biggest of them is the one concerning the analyticity of the set-theoretic axioms. In what sense would *they* be "analytic"? Many, even of those who don't doubt their consistency, would balk at their analyticity. But even should this *if* be granted, the other two loom large. The reader should refer to Quine's "Carnap and Logical Truth" for some objections to the first.

And, of course, it is just not the case that mathematical propositions have been shown to be analytic in the second of the two Kantian senses cited (i.e., reducible to "logical truths" that have the form 'All *A* are *B*'). It might be objected that to defend Kant on this basis is to trivialize him in the process, because he surely would have widened his notion of what constitutes logic if presented with, say, quantification theory. Therefore, the claim that ought to be examined is whether mathematics is reducible to quantification theory. To this, two replies might be offered. The first is, of course, that mathematics is not so reducible. Some set theory or its equivalent is needed as well. Hence this widening would not suffice. The second might be that Kant would not have agreed to a widening of the notion of logic beyond the *monadic* predicate calculus, so that the question does not even arise. And in either case, the problem of what constitutes a correct analysis of the meaning of a mathematical term is still with us and likely to remain for a while to come.

Yet it should not be forgotten that if today it seems somewhat arbitrary just where one draws the line between logic and mathematics, this is itself a victory for Frege, Russell, and Whitehead: Before their work, the gulf between the two subjects seemed absolute.

One difficulty with calling set theory "logic" concerns the axioms of set (or property) existence; e.g., $(\exists P)(x)(\sim P(x))$ (read: 'There is a P such that for all x, x does not have P' or more simply 'there is an empty property [or set]'). In his last years, Frege came to the conclusion that such assertions of existence were not part of *logic* at all and repudiated "logicism," which he had founded. Another difficulty is the need for an Axiom of Infinity in deriving mathematics: In order to meet this difficulty, Frege, having given up logicism, proposed to derive mathematics from *geometry* (where the Axiom of Infinity is true, since presumably there are infinitely many points) instead of from "logic."

Russell, as has already been mentioned, proposed in the second edition of *PM* not to take the Axiom of Infinity as a postulate of the system, but to list it as a hypothesis whenever it was needed to prove a theorem. But then it becomes puzzling how mathematics is *useful* (if a great many of its theorems have the form 'if Inf. Ax., then p', and "Inf. Ax." - the Axiom of Infinity - is, in fact, empirically false).

In connection with the first difficulty, it has been argued that '$(\exists P)(x)(\sim P(x))$' is a necessary truth, since there is indeed a proposition '$P(x)$' that is false for every x, namely '$x \neq x$' (or any other self-contradictory proposition). More generally, Russell has sometimes suggested that '$(\exists P)$' need not be interpreted as meaning that some extra-logical entity "exists," but may only be a way of indicating that there is a meaningful proposition '$P(x)$' with the specified characteristics. (Hilbert would reply: You still need the notion of the *existence* of *formulas*, i.e., *expressions*.) Sometimes Russell writes in this way: as if a property (or, as he says, "propositional function") were only a linguistic expression containing free variables (e.g., 'x', 'y',...) - or, perhaps, the meaning of such an expression. However, this interpretation of *PM* is, in fact, excluded if "impredicative definitions" are permitted. (For explanation, see the article by Carnap in Part I.) For, if 'P' ranges only over "properties nameable by formulas of *PM*" then this restriction - to objects nameable in *PM* - will appear in the definition of every set. In particular, 'real number' will only be able to mean 'real number nameable in *PM*'. However, under the intended interpretation of the version of *PM* that permits impredicative definitions, there is an expression that stands for the set of *all* real numbers, not just *nameable* real numbers.

Strangely, Russell never appreciated this difficulty, and called *PM* a "no-class theory" to the end, although his "propositional functions" are nothing but arbitrary sets (or "classes") under another name, if impredicative definitions are permitted.

Another achievement of Frege and Russell was the analysis of the concept "number." Since this analysis is presented in detail in several of our selections, we shall not review it here. However, it raises several points of disagreement between logicists and intuitionists.

According to the intuitionists, one cannot understand 'two', 'three', etc., unless one has the general notion of a *number*. On the other hand, the logicists maintain that 'two', for example, is (contextually) definable thus:

$$2(P) \equiv (\exists x)(\exists y)[P(x) \cdot P(y) \cdot x \neq y \cdot (z)(P(z) \supset . z = x \lor z = y)]$$

(read: 'there are two Ps if and only if there are x, y such that x is P and y is P and x is not the same thing as y and for all z, if z is P then z is the same as x or z is the same as y'.)

Here the intuitionists may perhaps be right. The logicist account, however, could easily be modified so as to take care of this criticism: namely, define "number" just as the logicists do[7] (roughly, a "number" is anything that can be obtained from zero – or the class of all empty classes – by repeatedly applying a certain "successor" operation), and then define 'zero' *not* as 'the class of all empty classes' but rather as 'the smallest *number*,' (defining 'smallest' in some suitable way, or as 'the *number* that is not a successor'), 'one' as 'the *number* that is the successor of zero', 'two' as 'the *number* that is the successor of one', etc. Then the notion 'number' will be part of the definition of *each* number.[8] Of course, the definition of 'two' will be *equivalent* to the one Russell employed, but not synonymous with it word-by-word (or rather symbol-by-symbol). So one who used the new definition could perfectly well agree with the intuitionists that the Russell definition does not express the customary meaning exactly.

Another disagreement between logicists and intuitionists is over the identification of numbers with sets of sets (e.g., of zero with the set of all empty sets). As a point separating these two camps, it has less importance than is customarily accorded to it. In the first place, it is questionable whether Frege[9] held that "zero," "one," "two," etc., *had* to be identified with any particular entities: the important thing was the analysis of 'there are two Ps', 'there are three Ps', etc. The intuitionists accept this analysis as mathematically correct. Perhaps the intuitionist would prefer to render 'there are two Ps' by 'the species of Ps can be put in one-to-one correspondence with the numbers one, two' – however, Frege would certainly have accepted this definition.

[7] We mean to suggest here that an intuitionist could accept the logicist definition of number in terms of the "ancestral" (a number is something that is either 0 or bears the ancestral of the successor relation to 0); not that the famous Frege–Russell definition of the ancestral would in turn be acceptable to an intuitionist.

[8] This procedure may sound circular, but clearly it is not, provided that the expression 'zero' does not appear in the definition of 'number'; i.e., that a number be defined, e.g., as either the set of all empty sets or something bearing the ancestral of the successor relation to the set of all empty sets. *Zero* could then be identified with the "smallest" number, etc.

[9] We are indebted to Michael Dummett for this and other points in connection with Frege.

A more important disagreement concerns the logicist claim that "mathematics can be reduced to logic." Intuitionists reject this claim on the following grounds:

1. Understanding any system of deduction involves already having the notion of *iterating an operation an arbitrary finite number of times*; and this the intuitionists regard as a fundamentally *mathematical* and not logical notion. (Recall that they also regard it as a "creative" or extendable notion – not one whose every application is completely clear and specifiable in advance.)

2. *The principle of mathematical induction* (which we used in our proof that n^2 is the sum of the first n odd numbers) is a fundamentally *mathematical* one (closely connected with the idea of the *iteration* of an operation), and not reducible to logic. Frege did indeed reduce mathematical induction to what *he* called logic – via the "definition of the ancestral" (see the Frege, Russell, and Hempel selections). This reduction, however, depends on the use of impredicative definitions, which are rejected by intuitionists, and also on the axioms of set existence, which would not be called "logic" by intuitionists even if they did accept them.

5. Tautologies and sets

In his *Tractatus Logico-Philosophicus,* Wittgenstein maintained, following Russell and Frege, that mathematics was reducible to logic. Logic, in turn, was reducible to propositional calculus, according to Wittgenstein. This is correct, for the system *PM,* whenever the number of individuals is a fixed finite number, but it is correct in the infinite case only if infinitely "big" expressions[10] are permitted. The idea is, briefly, to treat universal statements as infinite conjunctions: 'Everything is F' is treated as 'x_1 is F & x_2 is F & ... & ...' (where $x_1, x_2,...$ are all the individuals in the universe of discourse, in some order). Now, the truths of the propositional calculus are all "tautologies" – they come out true, combinatorially, under all possible assignments of 'true' and 'false' to the "elementary propositions." Thus was born the very popular philosophical slogan that "mathematics consists wholly of tautologies."

Of course, closer examination revealed serious difficulties with the *Tractatus* view. A quantifier over properties (i.e., such an expression as 'for all properties P') is expanded as a conjunction with one clause for

[10]One cannot even say "infinitely long," because some of the expressions that would arise at higher types if all quantifiers were "expanded" as truth-functions would be *nondenumerably* infinite, and could not be thought of as existing ("written out") in primitive notation.

16

each property of individuals. But this presupposes not only that the individuals form a well-defined totality, but that the properties (or *sets* of individuals) form a well-defined totality, and similarly for sets of sets, sets of sets of sets, and so on. This is already debatable for 'property' in the sense of the term in which each property corresponds to a possible "rule for selecting"; for 'property' (or rather 'set') in the sense of *arbitrary collection* (*any* collection, whether given by a rule or by "chance"), the situation is even worse. Consider, for example, the famous "continuum problem" of Cantor. This asks whether there exists some set (in the sense of *arbitrary* set) of real numbers (*arbitrary* sequences of integers) that can be put into one-to-one correspondence with neither the set of all integers nor the set of *all* real numbers. The answer 'no' has been proved by Gödel to be *consistent* with the axioms of set theory. And the answer 'yes' was proved by Paul J. Cohen *also* to be consistent with those axioms. In what sense then would it be true that "there really is" (or "really isn't") any such set? One might answer: 'in the sense that if you listed all the sets of real numbers, you would (or wouldn't) find one such that if you listed all the one-to-one correspondences (arbitrary sets of pairs consisting of a real number and an integer, or of two real numbers, satisfying the "one-to-one" condition) you would not find a correspondence mapping the set in question onto the integers, and you would similarly fail to find a correspondence mapping the set in question onto the real numbers.' This answer, however, is completely unhelpful for many reasons: e.g., the notion of "listing" all the sets of real numbers is absurd if taken literally; so is the notion of completely examining even one nondenumerably infinite and "random" collection of real numbers in detail; and then how much more absurd is the notion of examining *all* "one-to-one correspondences"!

Today, very few philosophers or mathematicians of any school would maintain that the notion of, say, an arbitrary set of sets of real numbers is a completely clear one, or that all the mathematical statements one can write down in terms of this notion have a truth-value that is well-defined in the sense of being fixed by a rule – even a non-constructive rule – that does not assume that the notion of an "arbitrary set" has already been made clear. The contention that, even in the absence of such rules, questions such as the continuum problem have a definite meaning and, having a definite meaning, have a definite answer, quite independently of the state of our knowledge, forms the core of what has variously been called "realism" or "platonism" in the philosophy of mathematics. (For a remarkably lucid and forceful statement of this position, see Gödel's article in Part IV on the continuum problem, especially the supplementary section.)

Nevertheless, there is a respect in which this is the natural position to take. We normally do not require an effective method of verification as the sine qua non of meaningfulness. This was a requirement made in quite another context (empirical science) by the Vienna Circle, and long since abandoned by most of its proponents. Why should it be different with mathematics? If we think we understand what is meant by a set, a one-to-one correspondence, and so on, why shouldn't we say that the continuum problem has a definite answer, no matter how far we may be from finding out what it is? What do the two have to do with one another? A split on this question normally reveals a split on the most fundamental issues in the philosophy of mathematics, on the very nature of mathematical activity.

In general, the platonists will be those who consider mathematics to be the *discovery* of truths about structures that exist independently of the activity or thought of mathematicians. For others not so platonistically minded, mathematics is an activity in which the mathematician plays a more creative role. To put it crudely, propositions are true at best insofar as they follow from assumptions and definitions we have made. If we can show that a proposition is *undecidable* from the assumptions we currently accept, the question of its "truth" or "falsity" vanishes in a puff of metaphysical smoke. Our assumptions, definitions, and methods of proof constitute the rules determining the truth or falsity of the propositions formulated in their terms. If a proposition is undecidable from our current assumptions, then its "truth" is not determined by the available rules. Since nothing else is relevant, the question of truth does not arise. The platonist does not agree because, for him, the truth of mathematical propositions is not determined by the rules we adopt, but rather by the correspondence or noncorrespondence between the propositions and the mathematical structures to which the terms in those propositions refer. In his view, mathematical terms and propositions have meaning above and beyond that conferred on them by the assumptions and methods of proof accepted at any one time.

To get an idea of the objections that might be raised to the platonistic way of looking at the continuum problem let us look briefly at the notion of an "arbitrary set," which is needed for the formulation of the problem. The reader may perhaps wonder what is wrong with our preceding explanation: "Arbitrary set" means "any set, whether given by a rule or by chance." The difficulty is that the notion of *chance* makes no sense in pure mathematics, except as a figure of speech. Suppose, however, we took this explanation literally: We might, for example, define an "arbitrary sequence of integers" as a sequence that *could* be generated by a "random device." One difficulty is then the word 'could'. 'Could' can only mean mathematical possibility here, since we do not want to let physical laws have any effect on mathematical truth. But "mathematical possibility" is itself a disputed notion, where infinite structures are concerned. And a further difficulty is that, according to classical mathematics, there are other infinite sets, for instance the set of all sets *of sets* of real numbers, which are so "big" that they cannot be put into one-to-one correspondence with the set of all integers or even with the set of all real numbers: Such sets could not be identical with the "output" of any possible physical process, even if we were to take the notion of a "possible (actually infinite) physical process" as itself a clear one.

Again, some people say: "Why worry about possible physical models at all? You know what a *collection* is (as in 'collection of oranges') and you know what an integer is; therefore you know what is meant by 'collection of integers', and by 'collection of collections of integers', etc." This "simple-minded" point of view hardly seems satisfying, however. In the first place, our ordinary notion of a "collection" is loaded with physical connotations. If we say that these are to be disregarded, and that the members of a "collection" need not be proximate in space and time, need not be "similar" in any particular respect, and so on, then we are left with the notion of something like a random listing of objects. And if we say that the members of a "collection" (a) need not be objects, numbers, and so on, but may themselves be "collections," and (b) need not even be capable of being listed (or for that matter, named in language), even by a random device working through an infinity of time, then what notion are we supposed to form at all?

Second, the presence of statements (such as the continuum hypothesis) corresponding to which there is no verification or refutation procedure (except looking for a proof – which is most certainly not going to do any good, if 'proof' means 'proof in present-day set theory') is, perhaps, a reason for at least suspecting an unclarity in our notion of a "set."

Here it is instructive to compare set theory with number theory. In number theory too there are statements that are neither provable nor refutable from the axioms of present-day mathematics. Intuitionists might agree that this shows (not by itself, of course, but together with other considerations) that we do not have a clear notion of "truth" in number theory, and that our notion of a "totality of all integers" is not precise. Most mathematicians would reject this conclusion. Yet most mathematicians feel that the notion of an "arbitrary set" is somewhat unclear. What is the reason for this difference in attitude?

Perhaps the reason is that a verification/refutation procedure is inconceivable for number theory only if we require that the procedure be effective.[11] If we take the stand that "nonconstructive" procedures – i.e., procedures that require us to perform infinitely many operations in a finite time – are conceivable,[12] though not *physically* possible (owing mainly to the existence of a limit to the velocity with which physical operations can be performed), then we can say that there does "in principle" exist a verification/refutation procedure for number theory. For instance, to "verify" that an equation $P(x, y, z) = 0$ has a solution using the "procedure," we *check each ordered triple x, y, z of integers.* (Of course, this requires working forever, or else completing an infinite series of operations in a finite time.) Similarly, to check a statement of the form $(x)(\exists y)P(x, y) = 0$ (read: 'For every x there is a y such that $P(x, y) = 0$') by the "procedure," we have to substitute 0 for x, and then check through $y = 0, 1, 2, \ldots$ until we find a y_0 such that $P(0, y_0) = 0$; then we substitute 1 for x, and look for a y_1 such that $P(1, y_1) = 0$; and so on (again this requires an infinite series of operations). What this shows is: The notion of "truth" in number theory is not a dubious one if the notion of a completed actually infinite series (of, say, definitely specifiable physical operations) is itself not dubious. Since many mathematicians do not share intuitionist doubts about the clarity of the actual infinite, it is understandable that such mathematicians are willing to take the notion of number-theoretic truth as precise. For instance, Carnap argued for the legitimacy of such "nonconstructive rules" in explaining the notion of number-theoretic truth in his famous book *The Logical Syntax of Language.*

By way of contrast, we recall that *no* physical structure (not even an infinite one) can serve as a "standard model" for set theory. In addition, even if we allow "working forever," "completing infinite series in a finite time," and so on, no precisely definable sequence of operations exists by means of which we could "in principle" tell (in the sense explained in connection with number theory) whether an arbitrary statement of set theory was true or false by the procedure of exhaustively checking all cases. In short, if you understand such notions as *counting, adding, multiplying,* and *seeing if two numbers are equal,* we can explain to you the notion of a "true statement of number theory" (though not if you consistently "boggle" at all quantifications over an infinite domain);

---

[11]An effective procedure is (roughly) one that a computing machine could be "programed" to employ. Gödel proved the impossibility of an effective decision procedure for number theory.

[12]E.g., if one has an infinite series of operations to perform, say $S_1, S_2, S_3, \ldots$ and one is able to perform S_1 in 1 minute, S_2 in 1/2 minute, S_3 in 1/4 minute, etc.; then in 2 minutes one will have completed the whole infinite series.

but to have explained to you the notion of a "true statement of set theory" or of an "arbitrary set," it would appear that you must already have some such notion in your conceptual vocabulary.

6. Mathematical truth

An admittedly naive view, which has certain attractions (perhaps its very innocence is among the greatest of them), runs like this:

Some propositions are true. Others aren't. Generally, a proposition is true if the reality it purports to describe is as that proposition depicts it – if the things referred to (if any) have the properties they are said to have or stand in the relations in which they are said to stand. All of this is quite independent of whether we know or have any reason to believe (or in the extreme – *could* ever have any reason to believe) that they are true. Our language can express (and we can understand) a whole range of questions, quite independently of whether we possess their answers, or ever will possess their answers, or even ever *could* possess their answers. Consider. On January 1, 1901, at 12 noon GMT, every molecule on Earth had an approximate location (never mind the rest of the universe). We don't now know that distribution. We never will know it. Perhaps there are reasons why in principle we never *could* know it. (Perhaps any representation of such a map is too big for us to comprehend, and comprehend we must if we are to know.) Still it seems plausible that such a distribution existed. We can frame the question. It has an answer. But the answer is too complicated.

And so too with certain mathematical questions.

Until recently it was not known whether every map can be colored with four colors, with no two regions that share a common boundary being colored the same color. Now we know. Mathematical research proceeds, at least in part, by answering questions previously put. At each stage there are meaningful questions that haven't been answered – and perhaps others that it is beyond us to answer despite the fact that it was we who framed them. In brief, every "well-formed" sentence in the language of science and mathematics expresses a meaningful proposition (one having a truth-value) about its subject matter. Whether a question is meaningful, whether it has an answer, does not depend on whether we know the answer, ever will know the answer, or even whether it is "in principle" possible for us to find it out.

Semantics is independent of epistemology.

Such an approach fits nicely with a conception of man and his place in nature in which man, like other animals, is a limited being. There are bounds on our epistemic powers – bounds dictated by the number of cells in our bodies, by the arrangement of those cells (the structure of our eyes is a good model: Things far too small for us to see, wave lengths too long to register, are not thereby deprived of existence). And so too there are things too complex for us to understand, questions just too hard for us to answer – though if we were differently constituted they might be within our reach. And might some of these questions not nevertheless be simple enough for us to frame? Any philosophy that adjusts the bounds of reality to those imposed by the existing (and in some sense necessary) limitations on our *epistemic* powers is bound to be shortchanging the world.

A referential semantics exhibits the propositions of physics as being "about" rigid bodies, fields, electrons; those of number theory as about numbers; set theory about sets. They are true if and only if the relevant entities have the properties ascribed to them.

In this form, our naive view is a kind of realism. Conjoin it with a platonist's perspective on the nature of the objects populating the domains of many mathematical theories (such as numbers, sets, functions, and spaces) – that these are abstract, exist outside of space-time, and independently of our conceptions. Plausible as such a perspective may seem (what and where *could* they be?), the result begins to threaten the pastoral calm of our opening scene. Because it now becomes unclear how we could know *any* of these mathematical propositions on their platonist construals. Our accounts of the truth-conditions of mathematical propositions and the nature of the objects that form their subject matter, independently plausible as these may have seemed, when joined clash with our most fundamental epistemological theories (for more on this general problem see Benacerraf "Mathematical Truth" and Section 8 of this Introduction).

Although not always explicitly stated, we feel that this is an important component of many of the questions that prompt the myriad of answers we call theories of the "foundations of mathematics," by mathematicians and philosophers alike:

 - Formalists deny, among other things, the associated platonism and attempt to supply mathematics with a more visible subject matter and truth conditions for its propositions whose presence or absence it is at least sometimes clearly within our power to ascertain. Indeed, Hilbert wanted his program to produce, in addition to a proof of the consistency of mathematics, a general method which, applied to any given mathematical question (framable in a particular formal language), would either answer it or show it to be independent of existing assumptions. Church's proof of the unsolvability of the decision problem for first-order logic dashed that hope, after Gödel had shown by his second incompleteness theorem that another principal aim of Hilbert's program – a finitist consistency proof for arithmetic – was already beyond reach.

- Nominalists traditionally have objected to the postulation of objects satisfying the platonist's description (although more recently the focus of the objection has shifted – see Part II, on the existence of mathematical objects, for more details), sometimes because they thought there simply weren't any such things, but more often for epistemological reasons – because they found the idea of an abstract object *unintelligible*.

- Conventionalists attempt to account for mathematical truth by bypassing the referential semantics (thus avoiding altogether the issues raised by platonism – mathematics has no "objects," or if it does, they simply have the properties we assign to them by convention). Mathematical truth thus reduces to the truth of certain conventions (by our fiat) plus the preservation of truth through logical consequence. Quine addresses these problems in "Truth by Convention" and in "Carnap and Logical Truth" far better than we could here. Although it is to some extent parasitic on Quine's reply, we should mention one reply to the conventionalist that is not often made but is worth considering:

  Everything "true by convention" is supposedly true. But conventions, however well-intentioned, can turn out to be inconsistent. First, their consistency or inconsistency is a mathematical fact of combinatorial mathematics (the fact that certain programs for computing do not lead to the "output" '1 = 0'), one that is itself hard to represent as a matter of convention. If this is right, then not *all* of mathematics can be true by convention. This suggests that our ability to make even axioms "true by convention" is already limited by the (nonconventional) fact of their logical consequences. But might we not make contradictions *themselves* true by convention? Few so far have been so devoted to conventionalism as to suggest that as a way out.

  And so on with a number of other views.

- Logicism is a special and difficult case – because it is really many cases – and we have dealt with it in Section 4 of this Introduction.

But there is another way in which epistemic considerations can entice one's account of mathematical truth away from the naive view described at the beginning of this section – other, that is, than the direct clash between epistemology and platonism. This is through the rejection of the

23

realist principles proclaiming the independence of semantics (the theories of meaning and of reference and truth) from epistemology. It represents semantical features of sentences (meaning, truth conditions) as parasitic on epistemological ones (conditions of warranted assertability). For the naive realist we depicted earlier, it was the *absence* of such a coupling that opened the floodgates to propositions (mathematical or other) that were meaningful, had determinate truth-values (and thus represented in some respects possible states of "the world"), but yet remained entirely beyond our epistemic grasp. A philosopher can hold such an epistemically determined view about both mathematics and empirical science, or about either of them separately. Many phenomenalistically inclined positivists were verificationists about empirical matters but not about the formal sciences (perhaps because, in the idiom of the day, these were already thought to be "empty of factual content"). Intuitionists, on the other hand, appear to be verificationists in *mathematics,* but not necessarily elsewhere.

Michael Dummett, in his article "The Philosophical Basis of Intuitionistic Logic," urges a more thoroughgoing abandonment of the naive realist position.

To simplify considerably, Dummett's intuitionist construes mathematical propositions as having truth-conditions that *coincide* with their verification conditions, thus very neatly bridging the gulf that, for the platonist, separates truth and knowledge. This is proposed in support of the claim that the canons of reasoning appropriate to mathematics are those of intuitionistic logic. Ingeniously, Dummett gives a *general* argument in favor of taking verification conditions as truth conditions – *general* because it does not depend on any special character of mathematical propositions and thus applies everywhere if it applies anywhere. Hence, in this respect, it satisfies our instinctive demand that our theories of meaning and truth (whether or not the latter is based on the former) be uniform across the language, a demand that is made all the more plausible by the fact that the logical vocabulary is the common property of *all* segments of our language: It could be a theoretical embarrassment to be obliged to assign the logical particles different meanings from one subject matter to the next. (What would we do in *mixed* contexts?)

So the generality of his account, and its attendant reconciliation of semantics and epistemology are achieved through a single fundamental move: the founding of the theory of meaning on epistemology – and the theory of truth on this already (for the realist) truncated theory of meaning. The natural result is a theory of truth for propositions, mathematical and nonmathematical alike, in which the *truth-conditions* are the *assertability conditions.* Sentences with no assertability conditions just do not express propositions – they do not describe possible but possibly unknowable states of the universe.

But one pays a price.

It is the abandonment of classical logic in favor of intuitionistic logic. Of course, to describe it as a *price* is to side with the platonist in the dispute: For the intuitionist there is no cost, simply gains on all fronts, as logic, mathematics, epistemology, semantics, and metaphysics are all finally brought into harmony.

Whatever one thinks of the details of these views, it is important to notice that a number of positions in the philosophy of mathematics (and other branches of philosophy that don't concern us here) are united by precisely such epistemic considerations.

An excellent example of a view that is similar in spirit is to be found in Putnam's "Models and Reality," a piece falling squarely in the pragmatist tradition. Certain authors, in the course of expressing similar concerns, have tended to legislate out of mathematics (and out of the realm of sense altogether as "speculative metaphysics") any proposition that couldn't "in principle" be decided by us.

The lines of demarcation vary from author to author – depending to a large extent on how "in principle" and "decided" are understood. On this spectrum, Putnam does not come off as a "hard-liner." His view amounts to this: Call our "theory of the world" some set of our beliefs, augmented by their logical consequences and indeed corrected and extended by any canons of reasoning, inductive or deductive, that might ever find favor with us. By a famous theorem due to Löwenheim and Skolem, such a theory – far vaster than any theory anyone has ever actually held or conceivably ever could hold – if it is consistent, has among its models (interpretations) ones of every cardinality from $\aleph_0$ on up, as well as others with even more horrifying pathologies. Which, if any, of these models is "the real world"? There is, *for us,* no distinguishing among them; for any distinctions that could be made on the basis of any principles we might hold or observations we might make *have already been taken into account* in constructing the theory (and therefore in selecting the set of models).

Putnam now asks if there is a fact of the matter as to *which* (if any) of the models of this theory (assuming it has some) is "the real world." (Whether the real world is even among the models may be in question for some. Consider the skeptic who feels there are ways in which our best efforts, as outlined above, might have been *mistaken,* thus possibly excluding "the real world" from the model set under consideration.)

His pragmatist answer is no.

Yet, one feels compelled to ask, might we not have been differently constructed – perhaps in such a way that our reconstructed selves made finer-grained distinctions that those we are now able to make or could ever conceivably make? Or simply different distinctions? If there is to be no fact of the matter about *which* of the models of our idealized theory is the real world, then there cannot *be* possible distinctions we cannot actually make – or could not make in Putnam's ideally extended theory.

The naive realist feels short-changed.

Given the way we are, some worlds are indistinguishable by us, and by the idealized possessors of our ideally extended idealized theories. But surely, had we been differently made, different theories might have evolved, and still different theories might have resulted from a yet different process of idealization. If such theories would be possible theories, surely worlds distinguishable in terms of them would be genuinely different, even if not distinguishable by Putnam's theory.

Or so a naive realist might want to reply.

Whatever the merits of such a reply, its bearing on Putnam's conclusions as applied to mathematics is more difficult to assess. One line of argument might be the following: Perhaps a way in which we might have been different involves the ability to "count" uncountable sets, or "compute" uncomputable functions, or the like. Whether, in Russell's famous phrase, such feats are at most "medically impossible" is certainly one of the questions at issue. If the impossibility *is* merely medical – if it resides in the accidents of our genetic makeup – then such feats might not have been impossible. And if they weren't impossible, we would be able to decide issues we are not able to decide. And if these are issues we *might* have been able to decide, they must be real issues, about which there must be a fact of the matter.

Or so the naive realist might urge.

In their curtailment of reality, some deal evenhandedly with all of it (Putnam). Others (intuitionists) address only mathematical reality – perhaps because they see mathematics as our *creation,* or because they see it as subjective in some important way, or... In each case, the result is a limitation of reality to what, in their view, could possibly be known. And in each case, too, it proceeds through an adjustment of the bounds of possible sense: through the theory of meaning.

Of course it is tendentious to describe their doings as a *curtailment.* This is a luxury we allow ourselves only because we have adopted, for expository purposes, the standpoint of the naive realist, the target of many of these views. In any event, whether such "curtailment" is legitimate depends on subtle questions in the philosophy of language – questions not likely to be resolved very soon, at least to the satisfaction of all combatants, for they are among the most profound that have plagued philosophers ever since the subject began.

To mention but one vexing question that runs through all the views discussed in this section: How may one express the "fact" that certain propositions cannot be expressed – because if they were expressible, we could not decide them and they would then have violated our *epistemic* constraints on sense. It seems that for us to know that they would violate such constraints *on meaning* we must know *on the basis of what they mean* that we could not know them to be true or false if they meant anything at all.

This problem is an ancient one: How can we circumscribe the bounds of sense without stepping outside of them?

7. The iterative conception of set

Consider the following view (which should be thought of as a sort of mathematico-philosophical metaphor, rather than as a serious suggestion): Sets are created by the mind. They are created by the mental act of "collecting" objects (or mental representations of the objects) together. This act can be performed only if the objects collected together are *already* in existence. So, as a result, the members of a set are always *prior* (in time) to the set.

Such a view has a number of attractive immediate consequences. For one thing, there cannot possibly be a collection of *all* sets. (It would have to contain itself, and this would mean that it would have to exist *prior* to itself.) Nor is there any reason why there *must* be a collection of *all* sets that are not members of themselves (this would, in fact, be the collection of all sets, since *no* set is a member of itself on this conception). So the Russell paradox is avoided. On the other hand, if a set has already been created, then, for any condition *P* that refers only to sets already created, there should be a set consisting of all members of the given set satisfying that condition *P*. For what can stop the mind from "collecting" all the members of the given set that satisfy *P*? So many plausible principles of set existence (e.g., existence of a union of any two sets, of an intersection, of an empty set), including the existence of a power set (a set of *all* subsets of a given set), seem to follow from this conception, while no obviously paradox-breeding principle of set existence seems to.

Although the metaphor of sets coming into existence (or being created) *in time* by the human *mind* cannot be taken seriously (even if the mind could create the set of all integers, separately creating each of its subsets would keep the mind rather busy, not to mention collecting these all together into the power set of the set of integers), these advantages of the metaphor can be retained if we suppose that there is *some* relation of "priority" that is transitive, irreflexive, and asymmetrical, and such that the members of any set are always *prior* to the set. Indeed, just such an intuition led Russell to the Vicious Circle Principles and ultimately to the theory of types. Any conception of set in which this figures as a prominent motivating force is today referred to as *the* iterative conception of set (despite the fact that there is more than one such conception). A number of papers in Part IV of this book discuss this conception, explain its great mathematical importance, and explore what the needed relation of "priority" might be.

One such conception is as follows: There is a well-defined totality of *all* sets (giving up the original metaphor of sets "coming into existence" or being constructed by us completely). This totality is not itself a set. Every set belongs to a set of a special kind, called a *rank* by some authors. (These ranks are a modified version of Russell's types, extended into the transfinite.) The ranks are indexed by numbers (including transfinite numbers, or Cantorian "ordinals"); the relation of priority is just the relation "belonging to a smaller rank" (i.e., a rank indexed by a smaller ordinal). The rank indexed by the number zero contains only the empty set (in the case of pure set theory; in a set theory with individuals ("*Urelemente*") we would take the collection of all the individuals as the rank zero). The rank indexed by the successor of any number (or any ordinal) is the power set of the rank indexed by the number. At those transfinite numbers that are not successors (the so-called "limit numbers," of which the least is ω, the ordinal number of the sequence of all integers) we take the union of all the ranks indexed by earlier ordinals to be the rank indexed by the limit ordinal. (For example, rank ω contains all the sets belonging to any *finite* rank. Thus rank ω is the union, in the set-theoretic sense, of ranks $0, 1, 2, \ldots$)

This explanation of the iterative conception makes free use of *numbers* (in fact, of *transfinite* numbers) and of mappings from sets to numbers, notwithstanding the fact that the numbers will eventually be identified with certain sets. This is not objectionable unless we think of the possibility of identifying mathematical objects with sets as of greater epistemological or metaphysical significance than is now customary among either philosophers or mathematicians. There is no reason why we cannot think of numbers, functions, and so on, as objects concerning which we have a certain amount of mathematical theory prior to doing set theory. This prior theorizing enables us to state certain assumptions about sets (e.g., that every set belongs to some rank); and when we see that these assumptions lead to an attractive theory we can adopt it and work out a formalization of it that seems attractive. Certain of the sets in such a formalization may *end up* being singled out as numbers (ordinals); but this feature has only "elegance" to recommend it, it cannot, on pain of circularity, be claimed to have epistemological significance – at least not by one who views the iterative conception *itself* as having some epistemological force. (Nor need it be viewed as having profound metaphysical significance either, but that is a more complicated matter.)

If we start with a weak (second-order) theory of ordinal numbers, then the axioms of set theory can actually be *derived* from the assumptions that the ranks exist and are related as described. Thus the iterative picture has the pleasing property of *unifying* the axioms of set theory; it gives a "model" such that we can *prove* the axioms to be true in that model (assuming, of course, the existence of the model), and this is clearly preferable to just assuming the various axioms without giving any intuitive picture of how or why they are supposed to be true. Although some logicians (notably Quine) claim that no intuitive justification can be given for the acceptance of set-theoretic axioms after the discovery of the Russell paradox, the fact is that almost all of our authors (Gödel, Wang, Boolos, and Parsons, among others) maintain that this iterative picture *is* an intuitive justification for the standard axioms in the sense that (1) the picture is natural and persuasive (i.e., it seems to these authors that there is *a* notion of set that we had all along on which the elements of a set had to be "prior" to the set, and on which a "set of all sets" was impossible); and (2) as just remarked, the assumption of a structure of "ranks" with the properties we mentioned enables us to derive almost all of the standard axioms of set theory. (The one doubtful case being the Axiom of Replacement – the image of a set under a function is a set – although that, too, seems natural to many people under this picture even if it is not deducible from the hypothesis that the system of ranks exists.)

The version of the iterative conception just described is "platonistic" in that it views the sets as all existing at once. The priority relation is simply an ordering defined in terms of the membership relation of set theory itself. Others espouse the iterative conception but seek to avoid assuming a well-defined totality of all sets. Saul Kripke has suggested that this might be made sense of by construing set-theoretical quantifiers intuitionistically rather than classically. On such a view it is not clear why a condition *P* that quantifies over *all* sets ("all" read intuitionistically) should single out a well-defined subset of a *given* set, however; so this construal might leave us with a problem in justifying or even stating the Axiom of Selection. In an interesting article in Part IV, Charles Parsons proposes a *modal* interpretation of the "priority" relation on which the iterative conception rests. (The idea being that the members of a set *could exist* – in some possible well-founded structure that is thought of as a "realization" of set theory in a possible world – without the set existing. The priority relation is a kind of *presupposition* relation; *x* is prior to *y* if *y*'s existence *presupposes* *x*'s existence, but not vice versa.)

There is no doubt that the iterative conception connects with a powerful and useful mathematical metaphor. On the other hand, the large number of statements to the effect that *the* iterative conception is a perfectly clear and consistent conception that shows there is no difficulty at all with our set-theoretic "intuition" might suggest to some readers that "the lady doth protest too much." The problem in either justifying or doing without the assumption that all sets form a well-defined totality on the iterative conception, and the *epistemological* unclarity of the "priority" relation, suggest that the iterative conception is not without its own problems.

8. The problem of "access"

At one extreme of the spectrum of views one might hold on the foundations of set theory, and of mathematics generally, stands a form of "platonism" that might be described as follows. Mathematics consists of a body of propositions about an independent reality composed of the familiar mathematical objects (such as sets, numbers, functions, and spaces). Mathematical *discovery* is the uncovering of truths about this independently existing reality by deduction from axioms that we see to be true by a special faculty of intuition distinct from sense experience (which gives us knowledge only of the empirical world). Mathematical objects are independent of our minds and, unlike physical objects, do not interact with our bodies to cause alterations in our brains that lead ultimately to knowledge of them. But they must be postulated to account for the existence and growth of mathematical knowledge and, to the extent to which other knowledge is dependent on mathematical knowledge, of other knowledge as well.

It is hard to pin this view in its pure form on anyone, although Gödel perhaps comes as close to it as anyone since Plato. For the time being, we will consider the view in the abstract, returning later to the details of the view that Gödel expresses in his various writings.

It is likely that most mathematicians would reject this extreme form of platonism; certainly few contemporary philosophers or psychologists would find the idea of a nonphysical power of surveying a realm of independently existing objects outside of space and time very congenial. But if one rejects the idea of returning to such a view (a view that is "platonist" in a more literal sense than that overused epithet frequently has in philosophical discussion), then one cannot escape the problems posed by the talk of "intuition" that one encounters in the writings of many philosophers and set theorists. What is "intuition," and how can there be such a faculty if platonist views such as those mentioned above are totally wrong?

We might describe the problem as a problem of *access: here* are we, evolving social organisms in space-time. Our sense organs are admirably suited to bringing us information about tables and chairs, trees, fruits and vegetables, other organisms, the sky, the weather, and so on. We have managed to devise electronic and optical extensions of these sense organs that enable us to observe objects as small as viruses (and even smaller) and as distant as remote galaxies. But none of these sense organs, natural or artificial, extended or unextended, ever causally interacts with, observes, or perceives a *set. There* are the sets; beautiful (at least to some), imperishable, multitudinous, intricately connected. They toil not, neither do they spin. Nor, and this is the rub, do they interact with us in any way. So how are we supposed to have *epistemological access* to them? To answer, "by intuition," is hardly satisfactory. We need some account of how we can have *knowledge* of these beasties, some account of our cognitive relationship to them.

(We referred earlier to worries about whether the notion of a set is "clear." It seems to us that many such worries are really grounded in precisely this problem of access. It is not, after all, that set language is "unclear" in any ordinary linguistic sense: too many ambiguities, too many possible paraphrases, and the like. What is "unclear" is whether such objects really exist and, if they do, how we can possibly know what we claim to know about them.)

One suggestion that a number of authors have advanced is that sets may exist not as platonic, extra-mental objects, but *in the mind,* as objects of our own making. If sets are, in some way, our own creation and in our minds, then the problem of accounting for our access to them should be easier. Or so it might seem.

But how can a nondenumerable infinity of sets exist as mental objects in our all-too-finite minds? One possible answer would be that they exist as *intentional objects,* that is, as objects whose existence is the *content* of certain thoughts of ours. These thoughts need not be supposed to be true of anything "external" any more than the play *Hamlet* is true of any actual prince of Denmark; truth in set theory, on such an account, would be no more than *truth in the story.* Two problems with such an account (which, along with Gödel's account and others, is discussed by Wang in Part IV) are:

31

(1) On such an account, how are we to choose among rival set theories? Or should we just be relativists? Should we say that within broad limits, one set theory is as good as another? And if so, what would establish even the broad limits? Wang points out that such an attitude is anathema to most working set theorists. He suggests (tentatively) that perhaps truth in set theory might be defined in terms of convergence in the intuitions of set theorists in the long run. But is such convergence, assuming it exists at all, epistemologically significant if it is founded on "aesthetic" preferences for certain stories as opposed to others? And how much of the existing agreement among set theorists is just the product of academic fashion?

(2) How, on such an account, can we explain the apparent truth or approximate truth of empirical laws (e.g., Newton's Law of Gravity, Quantum Mechanics, Relativity) that require higher mathematics to state? Such a law, as standardly formalized, makes reference both to physical entities (forces, masses, particles, fields) and to functions and sets. If the functions and sets are just *intentional* objects, objects "in the story," then the physical theories are to that extent also about fictional (or at least intentional) objects, too. This view would seem to require a radical adjustment in philosophy of science as a whole, not just a new philosophy of mathematics. If mathematics is fiction, should the Mathematics Department be renamed "Creative Writing"? And what about the Physics Department? (Perhaps all there is is Creative Writing.)

Parsons's paper, which is related to Putnam's "Mathematics without Foundations," explores the possibility of a modal interpretation of set theory. (Parsons is continuing to explore this possibility in subsequent work that is not yet published as of this writing.) The idea is that set theory should be interpreted as a theory of what sorts of structures *could* exist (in a special mathematical sense of "could") and, in particular, as a theory of what models for iterative set theory *could* exist.

One question of mathematical interest raised by this program is the following: Is there some way of making the Axiom of Replacement (the range of values of any function whose domain is a set is also a set) more evident by deriving it from suitable assumptions about possible existence? Are there axioms that are "evident" on the modal interpretation that imply the Axiom of Replacement?

At first blush, the modal logical interpretation seems attractive because it eschews the whole picture that makes the problem of "access" seem so terrifying. It avoids saying that there is a nondenumerable infinity of actual objects called sets *there,* thus bypassing the problem of how we *here* can know about those objects *there* if there are no telegraph (or other) wires running from *there* to *here.* (It also rejects Quine's view that we should reformulate all scientific theories in quantificational logic in order to determine their "ontological commitments." Recasting set theory in *modal* logic leaves it unclear what the "ontological commitments" of set theory are, since there is no presently agreed upon account of ontological commitment in the case of a logic with modal operators. But if one thinks that the whole picture is wrong, then one will not regard this as a defect of the reformulation.) The problem of accounting for our epistemic access to a nondenumerable infinity of *recherché* entities is replaced with the problem of accounting for our ability to know modal truths, truths about what is and is not possible. And perhaps this will prove in the long run to be more tractable; to some it sounds less frightening.

Be that as it may, at present it is not clear just *how* the modal logical interpretation *can* help with the epistemological problem. This is particularly so if modalities are themselves understood in terms of "possible world" semantics: *S* is mathematically necessary if it is true in every mathematically possible world; *S* is mathematically possible if *S* is true in at least one mathematically possible world. The problem then shifts to that of explaining how we can know what we appear to know about this new breed of platonic entity – the (mathematically) possible world. After all, the theories that naturally come to mind as candidate accounts of our ability to know modal truths are just the ones that come to mind on the standard "Mathematics as the theory of Mathematical Objects" picture: for example, that modal truths are "analytic" or "true by convention"; that there are special Faculties of the Mind that enable one to know (and perhaps actually *define*) what is and is not "possible" in the logical/ mathematical sense; that our theory of what is and is not *mathematically* possible is, like our theory of what is and is not *physically* possible, a part of total scientific theory and to be accepted, modified, or rejected on grounds similar to those on which one accepts, modifies, or rejects empirical theories. It may be useful to have an alternative to the Heaven of Mathematical Objects picture; but we haven't been shown (yet) *how* it is useful.

9. Quine and Gödel

In line with what we said at the outset about the importance of seeing connections between issues raised by authors who are philosophers and authors who are mathematicians, and the importance of seeing connections between essays in different sections of this anthology, we recommend reading Quine's "Carnap and Logical Truth" in connection with the papers in Part IV as well as in connection with the other papers in

Part III. In particular, the view just mentioned, that the grounds for accepting or rejecting mathematical theories are analogous to the grounds for accepting or rejecting physical theories, has long been urged by Quine.

It is not, of course, that Quine is unaware of the fact that there are *experiments* in physics and no experiments in mathematics (or, at least, none in the same sense). But Quine emphasizes the idea that mathematics has to be viewed not by itself but rather as a part of an all-embracing conceptual scheme, and he claims that the necessity for quantification over mathematical objects (e.g., *sets*) is all the reason one needs for making the "posit" of the existence of sets, numbers, and so on. Sets and electrons are alike for Quine in being objects we need to assume to do science.

While this sort of holistic pragmatism is attractive in that it recognizes what Russell and the logicists were for a long time alone in emphasizing – that we must account for the use of mathematical locutions in empirical statements, and not only in the context of pure set theory, number theory, and so on – and in that it provides a reason to believe in the existence of sets without postulating mysterious Faculties of the Mind, it too runs into serious difficulties. (So does *every* view; that is why the philosophy of mathematics is so fascinating.) Quine *seems* to be saying (on one reading; despite the deceptive clarity of his style, he is not always an easy philosopher to interpret) that science as a whole is to be viewed as a single explanatory theory and that the theory is to be justified *as a whole* by its ability to explain *sensations*. But is is not clear what the acceptance or nonacceptance of the Axiom of Choice or the Continuum Hypothesis has to do with explaining sensations.

The idea that there is something *analogous* to empirical reasoning in pure mathematics has also been advanced by Putnam ("Mathematical Truth," not in this volume) and even by Gödel ("What is Cantor's Continuum Problem?" in Part IV). In spite of his platonism, Gödel is much too sophisticated to think acts of "intuition" are *all* that is involved in mathematical "self-evidence," "plausibility," and the like. The fact that two philosophers as radically different as Quine and Gödel both recognize the presence of an element of something like "hypothetico-deductive" reasoning in pure mathematics is certainly striking. (Such an element was also pointed out by Russell, in the preface to *PM* and in earlier publications.)

Quine recognizes that even in empirical science there are considerations other than predicting sensations that play a crucial role in theory selection. He speaks most often of "conservatism" – the desire to preserve principles that have long been regarded as "central" or "obvious" or both, and of "simplicity" – a desire for elegance, which occasionally

34

makes us fly in the face of conservatism when a radical change of a central (or even a "self-evident") principle turns out to lead to far-reaching simplifications of the whole system.

But why should the simplest and most conservative system (or rather, the system that best balances simplicity and conservatism, by our lights) have any tendency to be *true*? Quine, good pragmatist that he is, tends to pooh-pooh this sort of question; but more realistically minded philosophers are sure to be bothered. It is hard enough to believe that the natural world is so nicely arranged that what is simplest, etc., by *our* lights is always the same as what is *true* (or, at least, *generally* the same as what is true); why should one believe that the universe of sets (or the totality of modal truths) is so nicely arranged that there is a preestablished harmony between *our* feelings of simplicity, etc., and *truth*?

It might be rewarding at this point to go into Gödel's view in more detail and particularly to compare it to Quine's; for despite superficial differences, they agree in a surprising number of respects. And what is perhaps of greater interest, the least satisfactory portion of each account comes at precisely the same spot: at the place where it must be explained how the criteria of truth that are advanced by each are related to the truth of the propositions for whose truths they are criteria.

For Gödel, as for Quine, objects exist in exactly the same sense as the objects of physical theory. And our reasons for believing in them are every bit as good as our reasons for believing in the existence of fields, protons, and so on. Both Gödel and Quine declare our experience to be in some sense the touchstone of our theorizing; they differ principally in what they admit as constituting that experience. Quine insists on an almost thoroughgoing holism, broken only by the independence of observation sentences (they record our experience) and (sometimes) logical truths, whereas Gödel adopts a modified Kantian position about experience: Our experience of physical objects goes beyond mere sensation – our concept of physical object contains an admixture of elements not conceivably derived from sensation. But the "added ingredients" belong to our perceptual faculty and, unlike for Kant, they are not subjective, not contributed by the perceiving subject ("... by our thinking we cannot create any qualitatively new elements, but only reproduce and combine those that are given"; p. 484). This faculty of grouping sensations into sensations *of objects* is, for Gödel, very closely related to the faculty of intuition, in virtue of which we have the (iterative) concept of set ("... the function of both is 'synthesis,' i.e., the generating of unities out of manifolds...."; p. 484, fn. 26). He emphasizes that it (intuition) need not be conceived of as giving us *immediate* knowledge of mathematics. But this should not exclude it from being recognized as part of experience,

35

particularly since even our experience of physical objects transcends sensation. He concludes: "Evidently the 'given' underlying mathematics is closely related to the abstract elements contained in our empirical ideas" (p. 484).

Despite this difference (and who can say how much of a difference it really is, since Quine much more readily discusses the kinds of sentences we may admit into our theories than the kinds of evidence appropriate to each?), both agree that "empirical" considerations might figure among the criteria for the truth of mathematical axioms. According to Quine's view, this is axiomatic, since mathematical axioms are not observation sentences, while Gödel concedes that "besides mathematical intuition, there exists another (though only probable) criterion of the truth of mathematical axioms, namely their fruitfulness in mathematics and, one may add, possibly also in physics" (p. 485). Someone wishing to minimize the difference would emphasize that Gödel's account of the connection between mathematical experience and our acceptance of a mathematical theory plays a similar *explanatory* role in his philosophy to that played by the concept of simplicity, conservatism, and explanatory power in the philosophies of pragmatists and empiricists: In neither case is it made crystal clear why theories that (1) accord with intuition or (2) appear simpler and appear to explain better are more likely to be true than the others. Far from being insensitive to the problem of "access," as we have described it in this section, Gödel rather founds his philosophy on the view that it must be taken very seriously indeed. Believing that we *have* peculiarly mathematical knowledge, he tries to explain our possession of it by (1) noting that there are objective elements of experience that do not derive from sensation and (2) proposing to account for our mathematical knowledge at least in part in terms of such nonsensational aspects of experience. To be sure, we are left without an account of the *mechanism* by which these experiential elements reflect aspects of the alleged reality they allegedly betray. But it would be hard to argue that such a view is in a worse position for giving an account of our overall knowledge than are its holist, empiricist, or pragmatist competitors. In both cases, the philosophical lacuna occurs in exactly the same place: when it must be explained why the criteria that are advocated for selecting a theory (fit with intuition – simplicity) should pick out a theory that is more likely to be true than one of its competitors.

An interesting and perhaps undesired consequence of Gödel's view, and one that makes the *rapprochement* with Quine even closer, is that mathematics no longer appears to be a priori, unless that is reinterpreted as "independent of *sensation*." For now, our only mathematical knowledge is what we have derived from our experience. And Gödel dispenses with the very feature of the Kantian view that was designed to guarantee the a priori character of mathematics when he says:

It by no means follows, however, that the data of this second kind [what we have called the nonsensational component of experience. Eds.], because they cannot be associated with actions of certain things upon our sense organs, are something purely subjective, as Kant asserted. Rather they, too, may represent an aspect of objective reality, but, as opposed to the sensations, their presence may be due to another kind of relationship between ourselves and reality. (p. 484)

The very question "Why should one believe that there is a preestablished harmony between our feelings of simplicity, or intuition, and *truth*?" presupposes a notion of truth that is independent of our standards of assertability. In Dummett's terminology, it assumes a "realist" notion of truth; and it is at this point that the discussion of issues in the foundations of set theory leads back to the discussion of the issues that we grouped together in Section 6 under the heading "Mathematical Truth." It is possible that significant further progress on these issues cannot be made until we have a more satisfactory account of the nature of truth and of the ways in which truth and reference are and are not linked to assertability. In any case, there is certainly a close connection between discussions in the philosophy of mathematics and discussions in general philosophy concerning the metaphysical issue of realism. But this is not to say that philosophy of mathematics ought simply to wait for improved views in general philosophy. Quite the contrary, for it is even more likely that one way in which theories of truth and knowledge in general philosophy will be shown to be adequate (or inadequate) is by their ability (or inability) to account for mathematical knowledge; and it is only in philosophy of mathematics that one finds searching attempts to apply theories of truth and knowledge to the special case of mathematics. For this reason, it is our conviction that philosophers interested in fundamental questions in epistemology, in theory of reference and truth, in philosophy of language, and in metaphysics should pay closer attention to the case of mathematics than they generally have. General philosophical theories that appear to account adequately for our intercourse with protons and pachyderms but fail on polynomials are, for that very reason, inadequate treatments of the very cases they seem to fit so well. The world of mathematics is not a world apart. We will not have an adequate account of the physical world and our knowledge of it until we understand better than we presently do the role played by mathematics in our accounts of physical phenomena. And it is not likely that we will have satisfied ourselves on that score until we have produced accounts of knowledge, truth, and reality that deal adequately with pure mathematics as well.

PART I
The foundations of mathematics

Symposium on the foundations
of mathematics

1. The logicist foundations of mathematics

RUDOLF CARNAP

The problem of the logical and epistemological foundations of mathematics has not yet been completely solved. This problem vitally concerns both mathematicians and philosophers, for any uncertainty in the foundations of the "most certain of all the sciences" is extremely disconcerting. Of the various attempts already made to solve the problem none can be said to have resolved every difficulty. These efforts, the leading ideas of which will be presented in these three papers, have taken essentially three directions: *Logicism,* the chief proponent of which is Russell; *Intuitionism,* advocated by Brouwer; and Hilbert's *Formalism.*

Since I wish to draw you a rough sketch of the salient features of the logicist construction of mathematics, I think I should not only point out those areas in which the logicist program has been completely or at least partly successful but also call attention to the difficulties peculiar to this approach. One of the most important questions for the foundations of mathematics is that of the relation between mathematics and logic. *Logicism* is the thesis that mathematics is reducible to logic, hence nothing but a part of logic. Frege was the first to espouse this view (1884). In their great work, *Principia Mathematica,* the English mathematicians A. N. Whitehead and B. Russell produced a systematization of logic from which they constructed mathematics.

We will split the logicist thesis into two parts for separate discussion:

1. The *concepts* of mathematics can be derived from logical concepts through explicit definitions.
2. The *theorems* of mathematics can be derived from logical axioms through purely logical deduction.

I. The derivation of mathematical concepts

To make precise the thesis that the concepts of mathematics are derivable from logical concepts, we must specify the logical concepts to be employed

The first three essays in this chapter form part of a symposium on the foundations of mathematics which appeared in Erkenntnis (1931), pp. 91–121. They were translated by Erna Putnam and Gerald J. Massey and appear here with the kind permission of Rudolf Carnap, Arend Heyting, and Klara von-Neumann Eckart. The last of these appears in A. H. Taub, ed., *John von Neumann Collected Works,* Vol. 2 (New York: Pergamon Press, 1961).

in the derivation. They are the following: In propositional calculus, which deals with the relations between unanalyzed sentences, the most important concepts are: the negation of a sentence p, 'not-p' (symbolized '$\sim p$'); the disjunction of two sentences, 'p or q' ('$p \vee q$'); the conjunction, 'p and q' ('$p \cdot q$'); and the implication, 'if p, then q' ('$p \supset q$'). The concepts of functional calculus are given in the form of functions, e.g., '$f(a)$' (read 'f of a') signifies that the property f belongs to the object a. The most important concepts of functional calculus are universality and existence: '$(x)f(x)$' (read 'for every x, f of x') means that the property f belongs to every object; '$(\exists x)f(x)$' (read 'there is an x such that f of x') means that f belongs to at least one object. Finally there is the concept of identity: '$a = b$' means that 'a' and 'b' are names of the same object.

Not all these concepts need be taken as undefined or primitive, for some of them are reducible to others. For example, '$p \vee q$' can be defined as '$\sim (\sim p \cdot \sim q)$' and '$(\exists x)f(x)$' as '$\sim (x) \sim f(x)$'. It is the logicist thesis, then, that the logical concepts just given suffice to define all mathematical concepts, that over and above them no specifically mathematical concepts are required for the construction of mathematics.

Already before Frege, mathematicians in their investigations of the interdependence of mathematical concepts had shown, though often without being able to provide precise definitions, that all the concepts of arithmetic are reducible to the natural numbers (i.e., the numbers $1, 2, 3, \ldots$ which are used in ordinary counting). Accordingly, the *main problem* which remained for logicism was to derive the natural numbers from logical concepts. Although Frege had already found a solution to this problem, Russell and Whitehead reached the same results independently of him and were subsequently the first to recognize the agreement of their work with Frege's. The crux of this solution is the correct recognition of the logical status of the natural numbers; they are logical attributes which belong, not to things, but to concepts. That a certain number, say 3, is the number of a concept means that three objects fall under it. We can express the very same thing with the help of the logical concepts previously given. For example, let '$2_m (f)$' mean that at least two objects fall under the concept f. Then we can define this concept as follows (where '$=_{Df}$' is the symbol for definition, read as "means by definition"):

$$2_m (f) =_{Df} (\exists x)(\exists y)[\sim (x = y) \cdot f(x) \cdot f(y)]$$

or in words: there is an x and there is a y such that x is not identical with y and f belongs to x and f belongs to y. In like manner, we define 3_m, 4_m, and so on. Then we define the number two itself thus:

$$2(f) =_{Df} 2_m (f) \cdot \sim 3_m (f)$$

42

or in words: at least two, but not at least three, objects fall under f. We can also define arithmetical operations quite easily. For example, we can define addition with the help of the disjunction of two mutually exclusive concepts. Furthermore, we can define the concept of natural number itself.

The derivation of the other kinds of number – i.e., the positive and negative numbers, the fractions, the real and the complex numbers – is accomplished, not in the usual way by adding to the domain of the natural numbers, but by the construction of a completely new domain. The natural numbers do not constitute a subset of the fractions but are merely correlated in obvious fashion with certain fractions. Thus the natural number 3 and the fraction 3/1 are not identical but merely correlated with one another. Similarly we must distinguish the fraction 1/2 from the real number correlated with it. In this paper, we will treat only the definition of the real numbers. Unlike the derivations of the other kinds of numbers which encounter no great difficulties, the derivation of the real numbers presents problems which, it must be admitted, neither logicism, intuitionism, nor formalism has altogether overcome.

Let us assume that we have already constructed the series of fractions (ordered according to magnitude). Our task, then, is to supply definitions of the real numbers based on this series. Some of the real numbers, the rationals, correspond in obvious fashion to fractions; the rest, the irrationals, correspond as Dedekind showed (1872) to "gaps" in the series of fractions. Suppose, for example, that we divide the (positive) fractions into two classes, the class of all whose square is less than 2, and the class comprising all the rest of the fractions. This division forms a "cut" in the series of fractions which corresponds to the irrational real number $\sqrt{2}$. This cut is called a "gap" since there is no fraction correlated with it. As there is no fraction whose square is two, the first or "lower" class contains no greatest member, and the second or "upper" class contains no least member. Hence, to every real number there corresponds a cut in the series of fractions, each irrational real number being correlated with a gap.

Russell developed further Dedekind's line of thought. Since a cut is uniquely determined by its "lower" class, Russell defined a real number as the lower class of the corresponding cut in the series of fractions. For example, $\sqrt{2}$ is defined as the class (or property) of those fractions whose square is less than two, and the rational real number 1/3 is defined as the class of all fractions smaller than the fraction 1/3. On the basis of these definitions, the entire arithmetic of the real numbers can be developed. This development, however, runs up against certain difficulties connected with so-called "impredicative definition," which we will discuss shortly.

The essential point of this method of introducing the real numbers is that they are *not postulated but constructed*. The logicist does not establish the existence of structures which have the properties of the real numbers by laying down axioms or postulates; rather, through explicit definitions, he produces logical constructions that have, by virtue of these definitions, the usual properties of the real numbers. As there are no "creative definitions," definition is not creation but only name-giving to something whose existence has already been established.

In similarly constructivistic fashion, the logicist introduces the rest of the concepts of mathematics, those of analysis (e.g., convergence, limit, continuity, differential, quotient, integral, etc.) and also those of set theory (notably the concepts of the transfinite cardinal and ordinal numbers). This "constructivist" method forms part of the very texture of logicism.

II. The derivation of the theorems of mathematics

The second thesis of logicism is that the *theorems of mathematics* are derivable from logical axioms through logical deduction. The requisite system of logical axioms, obtained by simplifying Russell's system, contains four axioms of propositional calculus and two of functional calculus. The rules of inference are a rule of substitution and a rule of implication (the *modus ponens* of ancient logic). Hilbert and Ackermann have used these same axioms and rules of inference in their system.

Mathematical predicates are introduced by explicit definitions. Since an explicit definition is nothing but a convention to employ a new, usually much shorter, way of writing something, the *definiens* or the new way of writing it can always be eliminated. Therefore, as every sentence of mathematics can be translated into a sentence which contains only the primitive logical predicates already mentioned, this second thesis can be restated thus: Every provable mathematical sentence is translatable into a sentence which contains only primitive logical symbols and which is provable in logic.

But the derivation of the theorems of mathematics poses certain difficulties for logicism. In the first place it turns out that some theorems of arithmetic and set theory, if interpreted in the usual way, require for their proof besides the logical axioms still other special axioms known as the *axiom of infinity* and the *axiom of choice* (or multiplicative axiom). The axiom of infinity states that for every natural number there is a greater one. The axiom of choice states that for every set of disjoint non-empty sets, there is (at least) one selection-set, i.e., a set that has exactly one member in common with each of the member sets. But we are not concerned here with the content of these axioms but with their logical character. Both are existential sentences. Hence, Russell was right in hesitating to present them as logical axioms, for logic deals only with possible entities and cannot make assertions about whether something does or does not exist. Russell found a way out of this difficulty. He reasoned that since mathematics was also a purely formal science, it too could make only conditional, not categorical, statements about existence: if certain structures exist, then there also exist certain other structures whose existence follows logically from the existence of the former. For this reason he transformed a mathematical sentence, say S, the proof of which required the axiom of infinity, I, or the axiom of choice, C, into a conditional sentence; hence S is taken to assert not S, but $I \supset S$ or $C \supset S$, respectively. This conditional sentence is then derivable from the axioms of logic.

A greater difficulty, perhaps the greatest difficulty, in the construction of mathematics has to do with another axiom posited by Russell, the so-called *axiom of reducibility,* which has justly become the main bone of contention for the critics of the system of *Principia Mathematica.* We agree with the opponents of logicism that it is inadmissible to take it as an axiom. As we will discuss more fully later, the gap created by the removal of this axiom has certainly not yet been filled in an entirely satisfactory way. This difficulty is bound up with Russell's *theory of types* which we shall now briefly discuss.

We must distinguish between a "simple theory of types" and a "ramified theory of types." The latter was developed by Russell but later recognized by Ramsey to be an unnecessary complication of the former. If, for the sake of simplicity, we restrict our attention to one-place functions (properties) and abstract from many-place functions (relations), then type theory consists in the following classification of expressions into different "types": To type 0 belong the names of the objects ("individuals") of the domain of discourse (e.g., $a, b, \ldots$). To type 1 belong the properties of these objects (e.g., $f(a), g(a), \ldots$). To type 2 belong the properties of these properties (e.g., $F(f), G(f), \ldots$); for example, the concept $2(f)$ defined above belongs to this type. To type 3 belong the properties of properties of properties, and so on. The basic rule of type theory is that every predicate belongs to a determinate type and can be meaningfully applied only to expressions of the next lower type. Accordingly, sentences of the form $f(a), F(f), 2(f)$ are always meaningful, i.e., either true or false; on the other hand combinations like $f(g)$ and $f(F)$ are neither true nor false but meaningless. In particular, expressions like $f(f)$ or $\sim f(f)$ are meaningless, i.e., we cannot meaningfully say of a property either that it belongs to itself or that it does not. As we shall see, this last result is important for the elimination of the antinomies.

This completes our outline of the simple theory of types, which most proponents of modern logic consider legitimate and necessary. In his system, Russell introduced the ramified theory of types, which has not found much acceptance. In this theory the properties of each type are further subdivided into "orders." This division is based, not on the kind of objects to which the property belongs, but on the form of the definition which introduces it. Later we shall consider the reasons why Russell believed this further ramification necessary. Because of the introduction of the ramified theory of types, certain difficulties arose in the construction of mathematics, especially in the theory of real numbers. Many fundamental theorems not only could not be proved but could not even be expressed. To overcome this difficulty, Russell had to use brute force; i.e., he introduced the axiom of reducibility by means of which the different orders of a type could be reduced in certain respects to the lowest order of the type. The sole justification for this axiom was the fact that there seemed to be no other way out of this particular difficulty engendered by the ramified theory of types. Later Russell himself, influenced by Wittgenstein's sharp criticism, abandoned the axiom of reducibility in the second edition of *Principia Mathematica* (1925). But, as he still believed that one could not get along without the ramified theory of types, he despaired of the situation. Thus we see how important it would be, not only for logicism but for any attempt to solve the problems of the foundations of mathematics, to show that the simple theory of types is sufficient for the construction of mathematics out of logic. A young English mathematician and pupil of Russell, Ramsey (who unfortunately died this year, i.e., 1930), in 1926 made some efforts in this direction which we will discuss later.

III. The problem of impredicative definition

To ascertain whether the simple theory of types is sufficient or must be further ramified, we must first of all examine the reasons which induced Russell to adopt this ramification in spite of its most undesirable consequences. There were two closely connected reasons: the necessity of eliminating the logical antinomies and the so-called "vicious circle" principle. We call "logical antinomies" the contradictions which first appeared in set theory (as so-called "paradoxes") but which Russell showed to be common to all logic. It can be shown that these contradictions arise in logic if the theory of types is not presupposed. The simplest antinomy is that of the concept "impredicable." By definition a property is "impredicable" if it does not belong to itself. Now is the property "impredicable" itself impredicable? If we assume that it is, then since it belongs to itself it would be, according to the definition of "impredicable," not impredicable. If we assume that it is not impredicable, then it does not belong to itself and hence, according to the definition of "impredicable," is impredicable. According to the law of excluded middle, it is either impredicable or not, but both alternatives lead to a contradiction. Another example is Grelling's antinomy of the concept "heterological." Except that it concerns predicates rather than properties, this antinomy is completely analogous to the one just described. By definition, a predicate is "heterological" if the property designated by the predicate does not belong to the predicate itself. (For example, the word 'monosyllabic' is heterological, for the word itself is not monosyllabic.) Obviously both the assumption that the word 'heterological' is itself heterological as well as the opposite assumption lead to a contradiction. Russell and other logicians have constructed numerous antinomies of this kind.

Ramsey has shown that there are two completely different kinds of antinomies. Those belonging to the first kind can be expressed in logical symbols and are called "logical antinomies" (in the narrower sense). The "impredicable" antinomy is of this kind. Ramsey has shown that this kind of antinomy is eliminated by the simple theory of types. The concept "impredicable," for example, cannot even be defined if the simple theory of types is presupposed, for an expression of the form, a property does not belong to itself ($\sim f(f)$), is not well-formed, and meaningless according to that theory.

Antinomies of the second kind are known as "semantical" or "epistemological" antinomies. They include our previous example, "heterological," as well as the antinomy, well-known to mathematicians, of the smallest natural number which cannot be defined in German with fewer than 100 letters. Ramsey has shown that antinomies of this second kind cannot be constructed in the symbolic language of logic and therefore need not be taken into account in the construction of mathematics from logic. The fact that they appear in word languages led Russell to impose certain restrictions on logic in order to eliminate them, viz., the ramified theory of types. But perhaps their appearance is due to some defect of our ordinary word language.

Since antinomies of the first kind are already eliminated by the simple theory of types and those of the second kind do not appear in logic, Ramsey declared that the ramified theory of types and hence also the axiom of reducibility were superfluous.

Now what about Russell's second reason for ramifying the theory of types, viz., the vicious circle principle? This principle, that "no whole may contain parts which are definable only in terms of that whole", may also be called an "injunction against impredicative definition." A definition is said to be "impredicative" if it defines a concept in terms of a totality to which the concept belongs. (The concept "impredicative" has nothing to do with the aforementioned pseudo concept "impredicable.") Russell's main reason for laying down this injunction was his belief that antinomies arise when it is violated. From a somewhat different standpoint Poincaré before, and Weyl after, Russell also rejected impredicative definition. They pointed out that an impredicatively defined concept was meaningless because of the circularity in its definition. An example will perhaps make the matter clearer:

We can define the concept "inductive number" (which corresponds to the concept of natural number including zero) as follows: A number is said to be "inductive" if it possesses all the hereditary properties of zero. A property is said to be "hereditary" if it always belongs to the number $n+1$ whenever it belongs to the number n. In symbols,

$$\text{Ind}(x) =_{\text{Df}} (f)[(\text{Her}(f) \cdot f(0)) \supset f(x)]$$

To show that this definition is circular and useless, one usually argues as follows: In the *definiens* the expression '(f)' occurs, i.e., "for all properties (of numbers)". But since the property "inductive" belongs to the class of all properties, the very property to be defined already occurs in a hidden way in the *definiens* and thus is to be defined in terms of itself, an obviously inadmissible procedure. It is sometimes claimed that the meaninglessness of an impredicatively defined concept is seen most clearly if one tries to establish whether the concept holds in an individual case. For example, to ascertain whether the number three is inductive, we must, according to the definition, investigate whether every property which is hereditary and belongs to zero also belongs to three. But if we must do this for every property, we must also do it for the property "inductive" which is also a property of numbers. Therefore, in order to determine whether the number three is inductive, we must determine among other things whether the property "inductive" is hereditary, whether it belongs to zero, and finally – this is the crucial point – whether it belongs to three. But this means that it would be impossible to determine whether three is an inductive number.

Before we consider how Ramsey tried to refute this line of thought, we must get clear about how these considerations led Russell to the ramified theory of types. Russell reasoned in this way: Since it is inadmissible to define a property in terms of an expression which refers to "all properties," we must subdivide the properties (of type 1): To the "first order" belong those properties in whose definition the expression 'all properties' does not occur; to the "second order" those in whose definition the expression 'all properties of the first order' occurs; to the "third order" those in whose definition the expression 'all properties of the second order' occurs, and so on. Since the expression 'all properties' without reference to a determinate order is held to be inadmissible, there never occurs in the definition of a property a totality to which it itself belongs. The property "inductive," for example, is defined in this no longer impredicative way: A number is said to be "inductive" if it possess all the hereditary properties of the first order which belong to zero.

But the ramified theory of types gives rise to formidable difficulties in the treatment of the real numbers. As we have already seen, a real number is defined as a class, or what comes to the same thing, as a property of fractions. For example, we say that $\sqrt2$ is defined as the class or property of those fractions whose square is less than two. But since the expression 'for all properties' without reference to a determinate order is inadmissible under the ramified theory of types, the expression 'for all real numbers' cannot refer to all real numbers without qualification but only to the real numbers of a determinate order. To the first order belong those real numbers in whose definition an expression of the form 'for all real numbers' does not occur; to the second order belong those in whose definition such an expression occurs, but this expression must be restricted to "all real numbers of the first order," and so on. Thus there can be neither an admissible definition nor an admissible sentence which refers to all real numbers without qualification.

But as a consequence of this ramification, many of the most important definitions and theorems of real number theory are lost. Once Russell had recognized that his earlier attempt to overcome it, viz., the introduction of the axiom of reducibility, was itself inadmissible, he saw no way out of this difficulty. The *most difficult problem* confronting contemporary studies in the foundations of mathematics is this: How can we develop logic if, on the one hand, we are to avoid the danger of the meaninglessness of impredicative definitions and, on the other hand, are to reconstruct satisfactorily the theory of real numbers?

IV. Attempt at a solution

Ramsey (1926a) outlined a construction of mathematics in which he courageously tried to resolve this difficulty by declaring the forbidden impredicative definitions to be perfectly admissible. They contain, he contended, a circle but the circle is harmless, not vicious. Consider, he said, the description 'the tallest man in this room'. Here we describe something in terms of a totality to which it itself belongs. Still no one thinks this description inadmissible since the person described already exists and is only singled out, not created, by the description. Ramsey believed that the same considerations applied to properties. The totality of properties already exists in itself. That we men are finite beings who cannot name individually each of infinitely many properties but can describe some of them only with reference to the totality of all properties is an empirical fact that has nothing to do with logic. For these reasons Ramsey allows impredicative definition. Consequently, he can both get along with the simple theory of types and still retain all the requisite mathematical definitions, particularly those needed for the theory of the real numbers.

Although this happy result is certainly tempting, I think we should not let ourselves be seduced by it into accepting Ramsey's basic premise; viz., that the totality of properties already exists before their characterization by definition. Such a conception, I believe, is not far removed from a belief in a platonic realm of ideas which exist in themselves, independently of *if* and *how* finite human beings are able to think them. I think we ought to hold fast to Frege's dictum that, in mathematics, only that may be taken to exist whose existence has been proved (and he meant proved in finitely many steps). I agree with the intuitionists that the finiteness of every logical-mathematical operation, proof, and definition is not required because of some accidental empirical fact about man but is required by the very nature of the subject. Because of this attitude, intuitionist mathematics has been called "anthropological mathematics." It seems to me that, by analogy, we should call Ramsey's mathematics "theological mathematics," for when he speaks of the totality of properties he elevates himself above the actually knowable and definable and in certain respects reasons from the standpoint of an infinite mind which is not bound by the wretched necessity of building every structure step by step.

We may now rephrase our crucial question thus: Can we have Ramsey's result without retaining his absolutist conceptions? His result was this: Limitation to the simple theory of types and retention of the possibility of definitions for mathematical concepts, particularly in real number theory. We can reach this result if, like Ramsey, we allow impredicative definition, but can we do this without falling into his conceptual absolutism? I will try to give an affirmative answer to this question.

Let us go back to the example of the property "inductive" for which we gave an impredicative definition:

$$\text{Ind}(x) =_{\text{Df}} (f)[(\text{Her}(f) \cdot f(0)) \supset f(x)]$$

Let us examine once again whether the use of this definition, i.e., establishing whether the concept holds in an individual case or not, really leads to circularity and is therefore impossible. According to this definition, that the number two is inductive means:

$$(f)[(\text{Her}(f) \cdot f(0)) \supset f(2)]$$

in words: Every property f which is hereditary and belongs to zero belongs also to two. How can we verify a universal statement of this kind? If we had to examine every single property, an unbreakable circle would indeed result, for then we would run headlong against the property "inductive." Establishing whether something had it would then be impossible in principle, and the concept would therefore be meaningless. But the verification of a universal logical or mathematical sentence does not consist in running through a series of individual cases, for impredicative definitions usually refer to infinite totalities. The belief that we must run through all the individual cases rests on a confusion of "numerical" generality, which refers to objects already given, with "specific" generality (cf. Kaufmann 1930). We do not establish specific generality by running through individual cases but by logically deriving certain properties from certain others. In our example, that the number two is inductive means that the property "belonging to two" follows logically from the property "being hereditary and belonging to zero." In symbols, '$f(2)$' can be derived for an arbitrary f from '$\text{Her}(f) \cdot f(0)$' by logical operations. This is indeed the case. First, the derivation of '$f(0)$' from '$\text{Her}(f) \cdot f(0)$' is trivial and proves the inductiveness of the number zero. The remaining steps are based on the definition of the concept "hereditary":

$$\text{Her}(f) =_{\text{Df}} (n)[f(n) \supset f(n+1)]$$

Using this definition, we can easily show that '$f(0+1)$' and hence '$f(1)$' are derivable from '$\text{Her}(f) \cdot f(0)$' and thereby prove that the number one is inductive. Using this result and our definition, we can derive '$f(1+1)$' and hence '$f(2)$' from '$\text{Her}(f) \cdot f(0)$', thereby showing that the number two is inductive. We see then that the definition of inductiveness, although impredicative, does not hinder its utility. That proofs that the defined property obtains (or does not obtain) in individual cases can be given shows that the definition is meaningful. If we reject the belief that it is necessary to run through individual cases and rather make it clear to ourselves that the complete verification of a statement about an arbitrary property means nothing more than its logical (more exactly, tautological) validity for an arbitrary property, we will come to the conclusion that impredicative definitions are logically admissible. If a property is defined impredicatively, then establishing whether or not it obtains in an individual case may, under certain circumstances, be difficult, or it may even be impossible if there is no solution to the decision problem for that logical system. But in no way does impredicativeness make such decisions impossible in principle for all cases. If the theory just sketched proves feasible, logicism will have been helped over its greatest difficulty, which consists in steering a safe course between the Scylla of the axiom of reducibility and the Charybdis of the allocation of the real numbers to different orders.

Logicism as here described has several features in common both with intuitionism and with formalism. It shares with intuitionism a constructivistic tendency with respect to definition, a tendency which Frege also emphatically endorsed. A concept may not be introduced axiomatically but must be constructed from undefined, primitive concepts step by step through explicit definitions. The admission of impredicative definitions seems at first glance to run counter to this tendency, but this is only true for constructions of the form proposed by Ramsey. Like the intuitionists, we recognize as properties only those expressions (more precisely, expressions of the form of a sentence containing one free variable) which are constructed in finitely many steps from undefined primitive properties of the appropriate domain according to determinate rules of construction. The difference between us lies in the fact that we recognize as valid not only the rules of construction which the intuitionists use (the rules of the so-called "strict functional calculus"), but in addition, permit the use of the expression 'for all properties' (the operations of the so-called "extended functional calculus").

Further, logicism has a methodological affinity with formalism. Logicism proposes to construct the logical-mathematical system in such a way that, although the axioms and rules of inference are chosen with an interpretation of the primitive symbols in mind, nevertheless, *inside the system* the chains of deductions and of definitions are carried through formally as in a pure calculus, i.e., without reference to the meaning of the primitive symbols.

2. The intuitionist foundations of mathematics

[Die intuitionistische Grundlegung der Mathematik]
AREND HEYTING

The intuitionist mathematician proposes to do mathematics as a natural function of his intellect, as a free, vital activity of thought. For him, mathematics is a production of the human mind. He uses language, both natural and formalized, only for communicating thoughts, i.e., to get others or himself to follow his own mathematical ideas. Such a linguistic accompaniment is not a representation of mathematics; still less is it mathematics itself.

It would be most in keeping with the active attitude of the intuitionist to deal at once with the construction of mathematics. The most important building block of this construction is the concept of unity which is the architectonic principle on which the series of integers depends. The integers must be treated as units which differ from one another only by their place in this series. Since in his *Logischen Grundlagen der exakten Wissenschaften* Natorp has already carried out such an analysis, which in the main conforms tolerably well to the intuitionist way of thinking, I will forego any further analysis of these concepts. But I must still make one remark which is essential for a correct understanding of our intuitionist position: we do not attribute an existence independent of our thought, i.e., a transcendental existence, to the integers or to any other mathematical objects. Even though it might be true that every thought refers to an object conceived to exist independently of it, we can nevertheless let this remain an open question. In any event, such an object need not be completely independent of human thought. Even if they should be independent of individual acts of thought, mathematical objects are by their very nature dependent on human thought. Their existence is guaranteed only insofar as they can be determined by thought. They have properties only insofar as these can be discerned in them by thought. But this possibility of knowledge is revealed to us only by the act of knowing itself. Faith in transcendental existence, unsupported by concepts, must be rejected as a means of mathematical proof. As I will shortly illustrate more fully by an example, this is the reason for doubting the law of excluded middle.

Oskar Becker has dealt thoroughly with the problems of mathematical existence in his book on that subject. He has also uncovered many connections between these questions and the most profound philosophical problems.

We return now to the construction of mathematics. Although the introduction of the fractions as pairs of integers does not lead to any basic difficulties, the definition of the irrational numbers is another story. A real number is defined according to Dedekind by assigning to every rational number either the predicate 'Left' or the predicate 'Right' in such a way that the natural order of the rational numbers is preserved. But if we were to transfer this definition into intuitionist mathematics in exactly this form, we would have no guarantee that Euler's constant C is a real number. We do not need the definition of C. It suffices to know that this definition amounts to an algorithm which permits us to enclose C within an arbitrarily small rational interval. (A rational interval is an interval whose end points are rational numbers. But, as absolutely no ordering relations have been defined between C and the rational numbers, the word 'enclose' is obviously vague for practical purposes. The practical question is that of computing a series of rational intervals each of which is contained in the preceding one in such a way that the computation can always be continued far enough so that the last interval is smaller than an arbitrarily given limit.) But this algorithm still provides us with no way of deciding for an arbitrary rational number A whether it lies left or right of C or is perhaps equal to C. But such a method is just what Dedekind's definition, interpreted intuitionistically, would require.

The usual objection against this argument is that it does not matter whether or not this question can be decided, for, if it is not the case that $A = C$, then either $A < C$ or $A > C$, and this last alternative is decided after a finite, though perhaps unknown, number of steps N in the computation of C. I need only reformulate this objection to refute it. It can mean only this: either there exists a natural number N such that after N steps in the computation of C it turns out that $A < C$ or $A > C$; or there is no such N and hence, of course, $A = C$. But, as we have seen, the existence of N signifies nothing but the possibility of actually producing a number with the requisite property, and the non-existence of N signifies the possibility of deriving a contradiction from this property. Since we do not know whether or not one of these possibilities exists, we may not assert that N either exists or does not exist. In this sense, we can say that the law of excluded middle may not be used here.

In its original form, then, Dedekind's definition cannot be used in intuitionist mathematics. Brouwer, however, has improved it in the following way: Think of the rational numbers enumerated in some way. For the sake of simplicity, we restrict ourselves to the numbers in the closed unit interval and take always as our basis the following enumeration:

$$(A) \quad 0, 1, \frac{1}{2}, \frac{1}{3}, \frac{2}{3}, \frac{1}{4}, \frac{3}{4}, \frac{1}{5}, \frac{2}{5}, \frac{3}{5}, \frac{4}{5}, \cdots$$

A real number is determined by a cut in the series (A); i.e., by a rule which assigns to each rational number in the series either the predicate 'Left' or the predicate 'Right' in such a way that the natural order of the rational numbers is preserved. At each step, however, we permit one individual number to be left out of this mapping. For example, let the rule be so formed that the series of predicates begins this way:

$$0,\ 1,\ \frac{1}{2},\ \frac{1}{3},\ \frac{2}{3},\ \frac{1}{4},\ \frac{3}{4},\ \frac{1}{5},\ \frac{2}{5},\ \frac{3}{5},\ \frac{4}{5},\cdots$$

L, R, L, L, ?, L,

Here 2/3 is temporarily left out of the mapping. We need not know whether or not the predicate for 2/3 is ever determined. But it is also a possibility that 3/4 should become a new excluded number and hence that 2/3 would receive the predicate 'Left'.

It is easy to give a cut for Euler's constant. Let d_n be the smallest difference between two successive numbers in the first n numbers of (A). Now if we compute C far enough to get a rational interval i which is smaller than d_n, then at most one of these n numbers can fall within i. If there is such a number, it becomes the excluded number for the cut. Thus, we can see how closely Brouwer's definition is related to the actual computation of a real number.

We can now take an important step forward. We can drop the requirement that the series of predicates be determined to infinity by a rule. It suffices if the series is determined step by step in some way, e.g., by free choices. I call such sequences "infinitely proceeding." Thus the definition of real numbers is extended to allow infinitely proceeding sequences in addition to rule-determined sequences. Before discussing this new definition in detail, we will give a simple example. We begin with this "Left-Right" choice-sequence:

$$0,\ 1,\ \frac{1}{2},\ \frac{1}{3},\ \frac{2}{3},\ \frac{1}{4},\ \frac{3}{4},\ \frac{1}{5},\ \frac{2}{5},\ \frac{3}{5},\ \frac{4}{5},$$

L, R, L, L, R, L, R, L, L,

Here the question about which predicate 3/5 receives cannot be answered yet, for it must still be decided which predicate to give it. The question about the predicate which 4/5 receives, on the other hand, can be answered now by 'Right,' since that choice would hold for every possible continuation of the sequence. In general, only those questions about an infinitely proceeding sequence which refer to every possible continuation of the sequence are susceptible of a determinate answer. Other questions, like the foregoing about the predicate for 3/5, must therefore be regarded as meaningless. Thus choice-sequences supplant, not so much the individual rule-determined sequences, but rather the totality of all possible rules. A "Left-Right" choice-sequence, the freedom of choice for which is limited only by the conditions which result from the natural order of the rational numbers, determines not just one real number but the spread of all real numbers or the continuum. Whereas we ordinarily think of each real number as individually defined and only afterwards think of them all together, we here define the continuum as a totality. If we restrict this freedom of choice by rules given in advance, we obtain spreads of real numbers. For example, if we prescribe that the sequence begin in the way we have just written it, we define the spread of real numbers between 1/2 and 2/3. An infinitely proceeding sequence gradually becomes a rule-determined sequence when more and more restrictions are placed on the freedom of choice.

We have used the word 'spread' exactly in Brouwer's sense. His definition of a spread is a generalization of this notion. In addition to choice-sequences, Brouwer treats sequences which are formed from choice-sequences by mapping rules. A spread involves two rules. The first rule states which choices of natural numbers are allowed after a determinate finite series of permitted choices has been made. The rule must be so drawn that at least one new permissible choice is known after each finite series of permitted choices has been made. The natural order of the rational numbers is an example of such a rule for our "Left-Right" sequence previously given. The second rule involved in a spread assigns a mathematical object to each permissible choice. The mathematical object may, of course, depend also on choices previously made. Thus it is permissible to terminate the mapping at some particular number and to assign nothing to subsequent choices. A sequence which results from a permissible choice-sequence by a mapping-rule is called an "element" of the spread.

To bring our previous example of the spread of real numbers between 1/2 and 2/3 under this general definition, we will replace the predicates 'Left', 'Right', and 'temporarily undetermined', by 1, 2, and 3; and we will derive the rule for permissible choices from the natural order of the rational numbers and from the requirement that the sequence begin in a particular way; and we will take identity for the mapping-rule.

A spread is not the sum of its elements (this statement is meaningless unless spreads are regarded as existing in themselves). Rather, a spread is identified with its defining rules. Two elements of a spread are said to be equal if equal objects exist at the nth place in both for every n. Equality of elements of a spread, therefore, does not mean that they are the same element. To be the same, they would have to be assigned to the same spread by the same choice-sequence. It would be impractical to call two mathematical objects equal only if they are the same object. Rather, every kind of object must receive its own definition of equality.

Brouwer calls "species" those spreads which are defined, in classical terminology, by a characteristic property of their members. A species, like a spread, is not regarded as the sum of its members but is rather identified with its defining property. Impredicative definitions are made impossible by the fact, which intuitionists consider self-evident, that only previously defined objects may occur as members of a species. There results, consequently, a step-by-step introduction of species. The first level is made up of those spread-species whose defining property is identity with an element of a particular spread. Hence, to every spread M there corresponds the spread-species of those spread-elements which are identical with some element of M.[1] A species of the first order can contain spread-elements and spread-species. In addition, a species of the second order contains species of the first order as members, and so on.

The introduction of infinitely proceeding sequences is not a necessary consequence of the intuitionist approach. Intuitionist mathematics could be constructed without choice-sequences. But the following set-theoretic theorem about the continuum shows how much mathematics would thereby be impoverished. This theorem will also serve as an example of an intuitionist reasoning process.

Let there be a rule assigning to each real number a natural number as its correlate. Assume that the real numbers a and b have different correlates, e.g., 1 and 2. Then, by a simple construction, we can determine a third number c which has the following property: in every neighborhood of c, no matter how small, there is a mapped number other than c; i.e., every finite initial segment of the cut which defines c can be continued so as to get a mapped number other than c. We define the number d by a choice-sequence thus: we begin as with c but we reserve the freedom to continue at an arbitrary choice in a way different from that for c. Obviously the correlate of d is not determined after any previously known finite number of choices. Accordingly, no definite correlate is assigned to d. But this conclusion contradicts our premise that every real number has a correlate. Our assumption that the two numbers a and b have different correlates is thus shown to be contradictory. And, since two natural numbers which cannot be distinguished are the same number, we have the following theorem: if every real number is assigned a correlate, then all the real numbers have the same correlate.

As a special result, we have: if a continuum is divided into two subspecies in such a way that every member belongs to one and only one of these subspecies, then one of the subspecies is empty and other other is identical with the continuum.

The unit continuum, for example, cannot be subdivided into the species of numbers between 0 and 1/2 and the species of numbers between

---

[1] This definition of spread-species is taken from a communication of Professor Brouwer.

1/2 and 1, for the preceding construction produces a number for which one need never decide whether it is larger or smaller than 1/2. The theorems about the continuity of a function determined in an interval are also connected with the foregoing theorem. But Brouwer's theorem about the uniform continuity of all full functions goes far beyond these results.

But what becomes of the theorem we have just proved if no infinitely proceeding sequences are allowed in mathematics? In that event, the species of numbers defined by rule-determined sequences would have to take the place of the continuum. This definition is admissible if we take it to mean that a number belongs to this species only if there is a rule which permits us actually to determine all the predicates of the sequence successively.

In this event, the foregoing proof continues to hold only if we succeed in defining the number d by a rule-determined sequence rather than by a choice-sequence. We can probably do it if we make use of certain unresolved problems; e.g., whether or not the sequence 0123456789 occurs in the decimal expansion of π. We can let the question – whether or not to deviate from the predicate series for c, at the nth predicate in the predicate sequence for d – depend on the occurrence of the preceding sequence at the nth digit after the decimal point in π. This proof obviously is weakened as soon as the question about the sequence is answered. But, in the event that it is answered, we can replace this question by a similar unanswered question, if there are any left. We can prove our theorem for rule-determined sequences only on the condition that there always remain unsolved problems. More precisely, the theorem is true if there are two numbers, determined by rule-determined sequences, such that the question about whether they are the same or different poses a demonstrably unsolvable problem. It is false if the existence of two such numbers is contradictory. But the problem raised by these conditions is insuperable. Even here choice-sequences prove to be superior to rule-determined sequences in that the former make mathematics independent of the question of the existence of unsolvable problems.

We conclude our treatment of the construction of mathematics in order to say something about the intuitionist propositional calculus. We here distinguish between propositions and assertions. An assertion is the affirmation of a proposition. A mathematical proposition expresses a certain expectation. For example, the proposition, 'Euler's constant C is rational', expresses the expectation that we could find two integers a and b such that $C = a/b$. Perhaps the word 'intention', coined by the phenomenologists, expresses even better what is meant here. We also use the word 'proposition' for the intention which is linguistically expressed by

58

the proposition. The intention, as already emphasized above, refers not only to a state of affairs thought to exist independently of us but also to an experience thought to be possible, as the preceding example clearly brings out.

The affirmation of a proposition means the fulfillment of an intention. The assertion '*C* is rational', for example, would mean that one has in fact found the desired integers. We distinguish an assertion from its corresponding proposition by the assertion sign '⊢' that Frege introduced and which Russell and Whitehead also used for this purpose. The affirmation of a proposition is not itself a proposition; it is the determination of an empirical fact, viz., the fulfillment of the intention expressed by the proposition.

A logical function is a process for forming another proposition from a given proposition. Negation is such a function. Becker, following Husserl, has described its meaning very clearly. For him negation is something thoroughly positive, viz., the intention of a contradiction contained in the original intention. The proposition '*C* is not rational', therefore, signifies the expectation that one can derive a contradiction from the assumption that *C* is rational. It is important to note that the negation of a proposition always refers to a proof procedure which leads to the contradiction, even if the original proposition mentions no proof procedure. We use ¬ as the symbol for negation.

For the law of excluded middle we need the logical function "either-or". '$p \vee q$' signifies that intention which is fulfilled if and only if at least one of the intentions p and q is fulfilled. The formula for the law of excluded middle would be '⊢$p \vee \neg p$'. One can assert this law for a particular proposition p only if p either has been proved or reduced to a contradiction. Thus, a proof that the law of excluded middle is a general law must consist in giving a method by which, when given an arbitrary proposition, one could always prove either the proposition itself or its negation. Thus the formula '$p \vee \neg p$' signifies the expectation of a mathematical construction (method of proof) which satisfies the aforementioned requirement. Or, in other words, this formula is a mathematical proposition; the question of its validity is a mathematical problem which, when the law is stated generally, is unsolvable by mathematical means. In this sense, logic is dependent on mathematics.

We conclude with some remarks on the question of the solvability of mathematical problems. A problem is posed by an intention whose fulfillment is sought. It is solved either if the intention is fulfilled by a construction or if it is proved that the intention leads to a contradiction. The question of solvability can, therefore, be reduced to that of provability.

A proof of a proposition is a mathematical construction which can itself be treated mathematically. The intention of such a proof thus yields a new proposition. If we symbolize the proposition 'the proposition p is provable' by '$+p$', then '$+$' is a logical function, viz., "provability." The assertions '$\vdash p$' and '$\vdash +p$' have exactly the same meaning. For, if p is proved, the provability of p is also proved, and if $+p$ is proved, then the intention of a proof of p has been fulfilled, i.e., p has been proved. Nevertheless, the propositions p and $+p$ are not identical, as can best be made clear by an example. In the computation of Euler's constant C, it can happen that a particular rational value, say A, is contained for an unusually long time within the interval within which we keep more narrowly enclosing C so that we finally suspect that $C=A$; i.e., we expect that, if we continued the computation of C, we would keep on finding A within this interval. But such a suspicion is by no means a proof that it will always happen. The proposition $+(C=A)$, therefore, contains more than the proposition $(C=A)$.

If we apply negation to both of these propositions, then we get not only two different propositions, '$\neg p$' and '$\neg +p$', but also the assertions, '$\vdash \neg p$' and '$\vdash \neg +p$', are different. '$\vdash \neg +p$' means that the assumption of such a construction as $+p$ requires is contradictory. The simple expectation p, however, need not lead to a contradiction. Here is how this works in our example just cited. Assume that we have proved the contradictoriness of the assumption that there is a construction which proves that A lies within every interval that contains C ($\vdash \neg +p$). But still the assumption that in the actual computation of C we will always in fact find A within our interval need not lead to a contradiction. It is even conceivable that we might prove that the latter assumption could never be proved to be contradictory, and hence that we could assert at the same time both '$\vdash \neg +p$' and '$\vdash \neg \neg p$'. In such an event, the problem whether $C=A$ would be essentially unsolvable.

The distinction between p and $+p$ vanishes as soon as a construction is intended in p itself, for the possibility of a construction can be proved only by its actual execution. If we limit ourselves to those propositions which require a construction, the logical function of provability generally does not arise. We can impose this restriction by treating only propositions of the form 'p is provable' or, to put it another way, by regarding every intention as having the intention of a construction for its fulfillment added to it. It is in this sense that intuitionist logic, insofar as it has been developed up to now without using the function $+$, must be understood. The introduction of provability would lead to serious complications. Yet its minimal practical value would hardly make it worthwhile to deal with those complications in detail.[2] But here this notion has given us an insight into how to conceive of essentially unsolvable problems.

We will have accomplished our purpose if we have shown you that intuitionism contains no arbitrary assumptions. Still less does it contain artificial prohibitions, such as those used to avoid the logical paradoxes. Rather, once its basic attitude has been adopted, intuitionism is the only possible way to construct mathematics.

3. The formalist foundations of mathematics
JOHANN VON NEUMANN

I

Critical studies of the foundations of mathematics during the past few decades, in particular Brouwer's system of "intuitionism," have re-opened the question of the origins of the generally supposed absolute validity of classical mathematics. Noteworthy is the fact that this question, in and of itself philosophico-epistemological, is turning into a logico-mathematical one. As a result of three important advances in the field of mathematical logic (namely: Brouwer's sharp formulation of the defects of classical mathematics; Russell's thorough and exact description of its methods – both the good and the bad; and Hilbert's contributions to the mathematical-combinatorial investigation of these methods and their relations), more and more it is unambiguous mathematical questions, not matters of taste, that are being investigated in the foundation of mathematics. As the other papers have dealt extensively both with the domain (delimited by Brouwer) of unconditionally valid (i.e., needing no justification) "intuitionist" or "finitistic" definitions and methods of proof and with Russell's formal characterization (which has been further developed by his school) of the nature of classical mathematics, we need not dwell on these topics any longer. An understanding of them is, of course, a necessary prerequisite for an understanding of the utility, tendency, and *modus procedendi* of Hilbert's theory of proof. We turn instead directly to the theory of proof.

The leading idea of Hilbert's theory of proof is that, even if the statements of classical mathematics should turn out to be false as to content, nevertheless, classical mathematics involves an internally closed procedure which operates according to fixed rules known to all mathematicians

[2]The question dealt with in this paragraph was fully clarified only in a discussion with H. Freudenthal after the conference. The results of this discussion are reproduced in the text.

and which consists basically in constructing successively certain combinations of primitive symbols which are considered "correct" or "proved." This construction-procedure, moreover, is "finitary" and directly constructive. To see clearly the essential difference between the occasionally non-constructive handling of the "content" of mathematics (real numbers and the like) and the always constructive linking of the steps in a proof, consider this example: Assume that there exists a classical proof of the existence of a real number x with a certain very complicated and deep-seated property $E(x)$. Then it may well happen that, from this proof, we can in no way derive a procedure for constructing an x such that $E(x)$. (We shall give an example of such a proof in a moment.) On the other hand, if the proof somehow violated the conventions of mathematical inference, i.e., if it contained an error, we could, of course, find this error by a finitary process of checking. In other words, although the content of a classical mathematical sentence cannot always (i.e., generally) be finitely verified, the formal way in which we arrive at the sentence can be. Consequently, if we wish to prove the validity of classical mathematics, which is possible in principle only by reducing it to the *a priori* valid finitistic system (i.e., Brouwer's system), then we should investigate, not statements, but methods of proof. We must regard classical mathematics as a combinatorial game played with the primitive symbols, and we must determine in a finitary combinatorial way to which combinations of primitive symbols the construction methods or "proofs" lead.

As we promised, we now produce an example of a non-constructive existence proof. Let $f(x)$ be a function which is linear from 0 to 1/3, from 1/3 to 2/3, from 2/3 to 1, and so on. Let

$$f(0) = -1; \quad f\left(\frac{1}{3}\right) = -\sum_{n=1}^{n=\infty} \frac{\epsilon_{2n}}{2^n}; \quad f\left(\frac{2}{3}\right) = \sum_{n=1}^{n=\infty} \frac{\epsilon_{2n}}{2^n}; \quad \text{and} \quad f(1) = 1$$

ϵ_n is defined as follows: if $2k$ is the sum of two prime numbers, then $\epsilon_k = 0$; otherwise $\epsilon_k = 1$. Obviously $f(x)$ is continuous and calculable with arbitrary accuracy at any point x. Since $f(0) < 0$ and $f(1) > 0$, there exists an x, where $0 \leqslant x \leqslant 1$, such that $f(x) = 0$. (In fact we readily see that $1/3 \leqslant x \leqslant 2/3$.) However the task of finding a root with an accuracy greater than $\pm 1/6$ encounters formidable difficulties. Given the present state of mathematics, these difficulties are insuperable, for if we could find such a root, then we could predict with certitude the existence of a root $< 2/3$ or $> 1/3$, according as its approximate value were $\leqslant 1/2$ or $\geqslant 1/2$, respectively. The former case (where the approximate value of the root is $\leqslant 1/2$) excludes both that $f(1/3) < 0$ and that $f(2/3) = 0$; the latter case (where the approximate value of the root $\geqslant 1/2$) excludes both that $f(1/3)=0$ and that $f(2/3)>0$. In other words, in the former case the value of ϵ_n must be 0 for all even n but not for all odd n; in the latter case the value of ϵ_n must be 0 for all odd n but not for all even n. Hence we would have proved that Goldbach's famous conjecture (that $2n$ is always the sum of two prime numbers), instead of holding universally, must already fail to hold for odd n in the former case and for even n in the latter. But no mathematician today can supply a proof for either case, since no one can find the solution of $f(x)=0$ more accurately than with an error of $1/6$. (With an error of $1/6$, $1/2$ would be an approximate value of the root, for the root lies between $1/3$ and $2/3$, i.e., between $1/2-1/6$ and $1/2+1/6$.)

II

Accordingly, the tasks which Hilbert's theory of proof must accomplish are these:

1. To enumerate all the symbols used in mathematics and logic. These symbols, called "primitive symbols," include the symbols '~' and ' → ' (which stand for "negation" and "implication" respectively).

2. To characterize unambiguously all the combinations of these symbols which represent statements classified as "meaningful" in classical mathematics. These combinations are called "formulas." (Note that we said only "meaningful," not necessarily "true." '$1+1=2$' is meaningful but so is '$1+1=1$', independently of the fact that one is true and the other false. On the other hand, combinations like '$1+ \rightarrow =1$' and '$++1= \rightarrow$' are meaningless.)

3. To supply a construction procedure which enables us to construct successively all the formulas which correspond to the "provable" statements of classical mathematics. This procedure, accordingly, is called "proving."

4. To show (in a finitary combinatorial way) that those formulas which correspond to statements of classical mathematics which can be checked by finitary arithmetical methods can be proved (i.e., constructed) by the process described in (3) if and only if the check of the corresponding statement shows it to be true.

To accomplish tasks 1–4 would be to establish the validity of classical mathematics as a short-cut method for validating arithmetical statements whose elementary validation would be much too tedious. But since this is in fact the way we use mathematics, we would at the same time sufficiently establish the empirical validity of classical mathematics.

We should remark that Russell and his school have almost completely accomplished tasks 1–3. In fact, the formalization of logic and mathematics suggested by tasks 1–3 can be carried out in many different ways. The real problem, then, is (4).

In connection with (4) we should note the following: If the "effective check" of a numerical formula shows it to be false, then from that formula we can derive a relation $p=q$ where p and q are two different, effectively given numbers. Hence (according to task 3) this would give us a formal proof of $p=q$ from which we could obviously get a proof of $1=2$. Therefore, the sole thing we must show to establish (4) is the formal unprovability of $1=2$; i.e., we need to investigate only this one particular false numerical relation. The unprovability of the formula $1=2$ by the methods described in (3) is called "consistency." The real problem, then, is that of finding a finitary combinatorial proof of consistency.

III

To be able to indicate the direction which a proof of consistency takes, we must consider formal proof procedure – as in (3) – a little more closely. It is defined as follows:

3_1. Certain formulas, characterized in an unambiguous and finitary way, are called "axioms." Every axiom is considered proved.

3_2. If a and b are two meaningful formulas, and if a and $a \rightarrow b$ have both been proved, then b also has been proved.

Note that, although (3_1) and (3_2) do indeed enable us to write down successively all provable formulas, still this process can never be finished. Further, (3_1) and (3_2) contain no procedure for deciding whether a given formula e is provable. As we cannot tell in advance which formulas must be proved successively in order ultimately to prove e, some of them might turn out to be far more complicated and structurally quite different from e itself. (Anyone who is acquainted, for example, with analytic number theory knows just how likely this possibility is, especially in the most interesting parts of mathematics.) But the problem of deciding the provability of an arbitrarily given formula by means of a (naturally finitary) general procedure, i.e., the so-called decision problem for mathematics, is much more difficult and complex than the problem discussed here.

As it would take us too far afield to give the axioms which are used in classical mathematics, the following remarks must suffice to characterize them. Although infinitely many formulas are regarded as axioms (for example, by our definition each of the formulas $1=1, 2=2, 3=3,\ldots$ is an axiom), they are nevertheless constructed from finitely many schemata by substitution in this manner: 'If a, b, and c are formulas, then $(a \rightarrow b) \rightarrow ((b \rightarrow c) \rightarrow (a \rightarrow c))$ is an axiom', and the like.

Now if we could succeed in producing a class R of formulas such that

(α) Every axiom belongs to R,
(β) If a and $a \rightarrow b$ belong to R, then b also belongs to R,
(γ) '$1=2$' does not belong to R,

then we would have proved consistency, for according to (α) and (β) every proved formula obviously must belong to R, and according to (γ), $1=2$ must therefore be unprovable. The actual production of such a class at this time is unthinkable, however, for it poses difficulties comparable to those raised by the decision problem. But the following remark leads from this problem to a much simpler one: If our system were inconsistent, then there would exist a proof of $1=2$ in which only a finite number of axioms are used. Let the set of these axioms be called M. Then the axiom system M is already inconsistent. Hence the axiom system of classical mathematics is certainly consistent if every finite subsystem thereof is consistent. And this is surely the case if, for every finite set of axioms M, we can give a class of formulas R_M which has the following properties:

(α) Every axiom of M belongs to R_M.
(β) If a and $a \rightarrow b$ belong to R_M, then b also belongs to R_M.
(γ) $1=2$ does not belong to R_M.

This problem is not connected with the (much too difficult) decision problem, for R_M depends only on M and plainly says nothing about provability (with the help of all the axioms). It goes without saying that we must have an effective, finitary procedure for constructing R_M (for every effectively given finite set of axioms M) and that the proofs of (α), (β), and (γ) must also be finitary.

Although the consistency of classical mathematics has not yet been proved, such a proof has been found for a somewhat narrower mathematical system. This system is closely related to a system which Weyl proposed before the conception of the intuitionist system. It is substantially more extensive than the intuitionist system but narrower than classical mathematics (for bibliographical material, see Weyl 1927).

Thus Hilbert's system has passed the first test of strength: the validity of a non-finitary, not purely constructive mathematical system has been established through finitary constructive means. Whether someone will succeed in extending this validation to the more difficult and more important system of classical mathematics, only the future will tell.

Disputation

AREND HEYTING

Persons of the dialogue: *Class, Form, Int, Letter, Prag, Sign*

Class: How do you do, Mr. Int? Did you not flee the town on this fine summer day?

Int: I had some ideas and worked them out at the library.

Class: Industrious bee! How are you getting along?

Int: Quite well. Shall we have a drink?

Class: Thank you. I bet you worked on that hobby of yours, rejection of the excluded middle, and the rest. I never understood why logic should be reliable everywhere else, but not in mathematics.

Int: We have spoken about that subject before. The idea that for the description of some kinds of objects another logic may be more adequate than the customary one has sometimes been discussed. But it was Brouwer who first discovered an object which actually requires a different form of logic, namely the mental mathematical construction [L. E. J. Brouwer 1908]. The reason is that in mathematics from the very beginning we deal with the infinite, whereas ordinary logic is made for reasoning about finite collections.

Class: I know, but in my eyes logic is universal and applies to the infinite as well as to the finite.

Int: You ought to consider what Brouwer's program was [L. E. J. Brouwer 1907]. It consisted in the investigation of mental mathematical construction as such, without reference to questions regarding the nature of the constructed objects, such as whether these objects exist independently of our knowledge of them. That this point of view leads immediately to the rejection of the principle of excluded middle, I can best demonstrate by an example.

Let us compare two definitions of natural numbers, say k and l.

Excerpted by kind permission of the author and publisher from *Intuitionism: an Introduction,* Arend Heyting, North-Holland, 1956 (3rd ed., 1971).

66

I. k is the greatest prime such that $k-1$ is also a prime, or $k=1$ if such a number does not exist.

II. l is the greatest prime such that $l-2$ is also a prime, or $l=1$ if such a number does not exist.

Classical mathematics neglects altogether the obvious difference in character between these definitions. k can actually be calculated ($k=3$), whereas we possess no method for calculating l, as it is not known whether the sequence of pairs of twin primes $p, p+2$ is finite or not. Therefore intuitionists reject II as a definition of an integer; they consider an integer to be well-defined only if a method for calculating it is given. Now this line of thought leads to the rejection of the principle of excluded middle, for if the sequence of twin primes were either finite or not finite, II would define an integer.

Class: One may object that the extent of our knowledge about the existence or non-existence of a last pair of twin primes is purely contingent and entirely irrelevant in questions of mathematical truth. Either an infinity of such pairs exist, in which case $l=1$; or their number is finite, in which case l equals the greatest prime such that $l-2$ is also a prime. In every conceivable case l is defined; what does it matter whether or not we can actually calculate the number?

Int: Your argument is metaphysical in nature. If "to exist" does not mean "to be constructed", it must have some metaphysical meaning. It cannot be the task of mathematics to investigate this meaning or to decide whether it is tenable or not. We have no objection against a mathematician privately admitting any metaphysical theory he likes, but Brouwer's program entails that we study mathematics as something simpler, more immediate than metaphysics. In the study of mental mathematical constructions "to exist" must be synonymous with "to be constructed".

Class: That is to say, as long as we do not know if there exists a last pair of twin primes, II is not a definition of an integer, but as soon as this problem is solved, it suddenly becomes such a definition. Suppose on January 1, 1970 it is proved that an infinity of twin primes exists; from that moment $l=1$. Was $l=1$ before that date or not? [Menger 1930].

Int: A mathematical assertion affirms the fact that a certain mathematical construction has been effected. It is clear that before the construction was made, it had not been made. Applying this remark to your example, we see that before Jan. 1, 1970 it had not been proved that $l=1$. But this is not what you mean. It seems to me that in order to clarify the sense of your question you must again refer to metaphysical concepts: to some world of mathematical things existing independently of our knowledge, where "$l=1$" is true in some absolute sense. But I repeat that mathematics ought not to depend upon such notions as these. In fact all mathematicians and even intuitionists are convinced that in some sense mathematics bear upon eternal truths, but when trying to define precisely this sense, one gets entangled in a maze of metaphysical difficulties. The only way to avoid them is to banish them from mathematics. This is what I meant by saying that we study mathematical constructions as such and that for this study classical logic is inadequate.

Class: Here come our friends Form and Letter. Boys, we are having a most interesting discussion on intuitionism.

Letter: Could you speak about anything else with good old Int? He is completely submerged in it.

Int: Once you have been struck with the beauty of a subject, devote your life to it!

Form: Quite so! Only I wonder how there can be beauty in so indefinite a thing as intuitionism. None of your terms are well-defined, nor do you give exact rules of derivation. Thus one for ever remains in doubt as to which reasonings are correct and which are not [R. Carnap 1934b, p. 41; 1937, p. 46; W. Dubislav 1932, pp. 57, 75]. In daily speech no word has a perfectly fixed meaning; there is always some amount of free play, the greater, the more abstract the notion is. This makes people miss each other's point, also in non-formalized mathematical reasonings. The only way to achieve absolute rigour is to abstract all meaning from the mathematical statements and to consider them for their own sake, as sequences of signs, neglecting the sense they may convey. Then it is possible to formulate definite rules for deducing new statements from those already known and to avoid the uncertainty resulting from the ambiguity of language.

Int: I see the difference between formalists and intuitionists mainly as one of taste. You also use meaningful reasoning in what Hilbert called metamathematics, but your purpose is to separate these reasonings from purely formal mathematics and to confine yourself to the most simple reasonings possible. We, on the contrary, are interested not in the formal side of mathematics, but exactly in that type of reasoning which appears in metamathematics; we try to develop it to its farthest consequences. This preference arises from the conviction that we find here one of the most fundamental faculties of the human mind.

Form: If you will not quarrel with formalism, neither will I with intuitionism. Formalists are among the most pacific of mankind. Any theory may be formalized and then becomes subject to our methods. Also intuitionistic mathematics may and will be thus treated [R. Carnap 1934b, p. 44; 1937, p. 51].

Class: That is to say, intuitionistic mathematics ought to be studied as a part of mathematics. In mathematics we investigate the consequences of given assumptions; the intuitionistic assumptions may be interesting, but they have no right to a monopoly.

Int: Nor do we claim that; we are content if you admit the good right of our conception. But I must protest against the assertion that intuitionism starts from definite, more or less arbitrary assumptions. Its subject, constructive mathematical thought, determines uniquely its premises and places it beside, not interior to, classical mathematics, which studies another subject, whatever subject that may be. For this reason an agreement between formalism and intuitionism by means of the formalization of intuitionistic mathematics is also impossible. It is true that even in intuitionistic mathematics the finished part of a theory may be formalized. It will be useful to reflect for a moment upon the meaning of such a formalization. We may consider the formal system as the linguistic expression, in a particularly suitable language, of mathematical thought.

If we adopt this point of view, we clash against the obstacle of the fundamental ambiguousness of language. As the meaning of a word can never be fixed precisely enough to exclude every possibility of misunderstanding, we can never be mathematically sure that the formal system expresses correctly our mathematical thoughts.

However, let us take another point of view. We may consider the formal system itself as an extremely simple mathematical structure; its entities (the signs of the system) are associated with other, often very complicated, mathematical structures. In this way formalizations may be carried out inside mathematics, and it becomes a powerful mathematical tool. Of course, one is never sure that the formal system represents fully any domain of mathematical thought; at any moment the discovering of new methods of reasoning may force us to extend the formal system.

Form: For several years we have been familiar with this situation. Gödel's incompleteness theorem showed us that any consistent formal system of number-theory may be extended consistently in different ways.

Int: The difference is that intuitionism proceeds independently of the formalization, which can but follow after the mathematical construction.

Class: What puzzles me most is that you both seem to start from nothing at all. You seem to be building castles in the air. How can you know if your reasoning is sound if you do not have at your disposal the infallible criterion given by logic? Yesterday I talked with Sign, who is still more of a relativist than either of you. He is so slippery that no argument gets hold of him, and he never comes to any somewhat solid conclusion. I fear this fate for anybody who discards the support of logic, that is, of common sense.

Sign: Speak of the devil and his imp appears. Were you speaking ill of me?

Class: I alluded to yesterday's discussion. To-day I am attacking these other two damned relativists.

Sign: I should like to join you in that job, but first let us hear the reply of your opponents. Please meet my friend Prag; he will be interested in the discussion.

Form: How do you do? Are you also a philosopher of science?

Prag: I hate metaphysics.

Int: Welcome, brother!

Form: Why, I would rather not defend my own position at the moment, as our discussion has dealt mainly with intuitionism and we might easily confuse it. But I fear that you are wrong as to intuitionistic logic. It has indeed been formalized and valuable work in this field has been done by a score of authors. This seems to prove that intuitionists esteem logic more highly than you think, though it is another logic than you are accustomed to.

Int: I regret to disappoint you. Logic is not the ground upon which I stand. How could it be? It would in turn need a foundation, which would involve principles much more intricate and less direct than those of mathematics itself. A mathematical construction ought to be so immediate to the mind and its result so clear that it needs no foundation whatsoever. One may very well know whether a reasoning is sound without using any logic; a clear scientific conscience suffices. Yet it is true that intuitionistic logic has been developed. To indicate what its significance is, let me give you an illustration. Let A designate the property of an integer of being divisible by 8, B the same by 4, C the same by 2. For $8a$ we may write $4 \times 2a$; by this mathematical construction P we see that the property A entails B ($A \rightarrow B$). A similar construction Q shows $B \rightarrow C$. By effecting first P, then Q (juxtaposition of P and Q) we obtain $8a = 2 \times (2 \times 2a)$ showing $A \rightarrow C$. This process remains valid if for A, B, C we substitute arbitrary properties: If the construction P shows that $A \rightarrow B$ and Q shows that $B \rightarrow C$, then the juxtaposition of P and Q shows that $A \rightarrow C$. We have obtained a logical theorem. The process by which it is deduced shows us that it does not differ essentially from mathematical theorems; it is only more general, e.g., in the same sense that "addition of integers

70

is commutative" is a more general statement than "$2 + 3 = 3 + 2$". This is the case for every logical theorem: it is but a mathematical theorem of extreme generality; that is to say, logic is a part of mathematics, and can by no means serve as a foundation for it. At least, this is the conception of logic to which I am naturally led; it may be possible and desirable to develop other forms of logic for other purposes.

It is the mathematical logic which I just described that has been formalized. The resulting formal system proves to have peculiar properties, very interesting when compared to those of other systems of formal logic. This fact has led to the investigations to which Mr. Form alluded, but, however interesting, they are tied but very loosely to intuitionistic mathematics.

Letter: In my opinion all these difficulties are imaginary or artificial. Mathematics is quite a simple thing. I define some signs and I give some rules for combining them; that is all.

Form: You want some modes of reasoning to prove the consistency of your formal system.

Letter: Why should I want to prove it? You must not forget that our formal systems are constructed with the aim towards applications and that in general they prove useful; this fact would be difficult to explain if every formula were deducible in them. Thereby we get a practical conviction of consistency which suffices for our work. What I contest in intuitionism is the opinion that mathematics has anything to do with the infinite. I can write down a sign, say α, and call it the cardinal number of the integers. After that I can fix rules for its manipulation in agreement with those which Mr. Class uses for this notion; but in doing this I operate entirely in the finite. As soon as the notion of infinity plays a part, obscurity and confusion penetrate into the reasoning. Thus all the intuitionistic assertions about the infinite seem to me highly ambiguous, and it is even questionable whether such a sign as $10^{10^{10}}$ has any other meaning than as a figure on paper with which we operate according to certain rules [J. Dieudonné 1951].

Int: Of course your extreme finitism grants the maximum of security against misunderstanding, but in our eyes it implies a denial of understanding which it is difficult to accept. Children in the elementary school understand what the natural numbers are and they accept the fact that the sequence of natural numbers can be indefinitely continued.

Letter: It is suggested to them that they understand.

Int: That is no objection, for every communication by means of language may be interpreted as suggestion. Also Euclid in the 20th proposition of Book IX, where he proved that the set of prime numbers is infinite, knew what he spoke about. This elementary notion of natural numbers, familiar to every thinking creature, is fundamental in intuitionistic mathematics. We do not claim for it any form of certainty or definiteness in an absolute sense, which would be unrealizable, but we content that it is sufficiently clear to build mathematics upon.

Letter: My objection is that you do not suppose too little, as Mr. Class thinks, but far too much. You start from certain principles which you take as intuitively clear without any explanation and you reject other modes of reasoning without giving any grounds for that discrimination. For instance, to most people the principle of the excluded middle seems at least as evident as that of complete induction. Why do you reject the former and accept the latter? Such an unmotivated choice of first principles gives to your system a strongly dogmatic character.

Int: Indeed intuitionistic assertions must seem dogmatic to those who read them as assertions about facts, but they are not meant in this sense. Intuitionistic mathematics consists, as I have explained already to Mr. Class, in mental constructions; a mathematical theorem expresses a purely empirical fact, namely the success of a certain construction. "$2 + 2 = 3 + 1$" must be read as an abbreviation for the statement: "I have effected the mental constructions indicated by "$2 + 2$" and by "$3 + 1$" and I have found that they lead to the same result." Now tell me where the dogmatic element can come in; not in the mental construction itself, as is clear by its very nature as an activity, but no more in the statements made about the constructions, for they express purely empirical results.

Letter: Yet you contend that these mental constructions lead to some sort of truth; they are not a game of solitaire, but in some sense must be of value for mankind, or you would be wrong in annoying others with them. It is in this pretence that I see the dogmatic element. The mathematical intuition inspires you with objective and eternal truths; in this sense your point of view is not only dogmatic, but even theological [H. B. Curry 1951, p. 6].

Int: In the first instance, my mathematical thoughts belong to my individual intellectual life and are confined to my personal mind, as is the case for other thoughts as well. We are generally convinced that other people have thoughts analogous to our own and that they can understand us when we express our thoughts in words, but we also know that we are never quite sure of being faultlessly understood. In this respect, mathematics does not essentially differ from other subjects; if for this reason you consider mathematics to be dogmatic, you ought to call any human reasoning dogmatic. The characteristic of mathematical thought is, that it does not convey truth about the external world, but is only concerned with mental constructions. Now we must distinguish between the simple practice of mathematics and its valuation. In order to construct mathematical theories no philosophical preliminaries are needed, but the value we attribute to this activity will depend upon our philosophical ideas.

Sign: In the way you treat language you put the clock back. Primitive language has this floating, unsteady character you describe, and the language of daily life is still in the main of the same sort, but as soon as scientific thought begins, the formalization of language sets in. In the last decades significists have studied this process. It has not yet come to an end, for more strictly formalized languages are still being formed.

Int: If really the formalization of language is the trend of science, then intuitionistic mathematics does not belong to science in this sense of the word. It is rather a phenomenon of life, a natural activity of man, which itself is open to study by scientific methods; it has actually been studied by such methods, namely that of formalizing intuitionistic reasoning and the signific method, but it is obvious that this study does not belong to intuitionistic mathematics, nor do its results. That such a scientific examination of intuitionistic mathematics will never produce a complete and definite description of it, no more than a complete theory of other phenomena is attainable, is clearly to be seen. Helpful and interesting as these metaintuitionistic considerations may be, they cannot be incorporated into intuitionistic mathematics itself. Of course, these remarks do not apply to formalization inside mathematics, as I described it a few moments ago.

Prag: Allow me to underline what Mr. Sign said just now. Science proceeds by formalization of language; it uses this method because it is efficient. In particular the modern completely formalized languages have appeared to be most useful. The ideal of the modern scientist is to prepare an arsenal of formal systems ready for use from which he can choose, for any theory, that system which correctly represents the experimental results. Formal systems ought to be judged by this criterion of usefulness and not by a vague and arbitrary interpretation, which is preferred for dogmatic or metaphysical reasons.

Int: It seems quite reasonable to judge a mathematical system by its usefulness. I admit that from this point of view intuitionism has as yet little chance of being accepted, for it would be premature to stress the few weak indications that it might be of some use in physics [J. L. Destouches 1951]; in my eyes its chances of being useful for philosophy, history and the social sciences are better. In fact, mathematics, from the intuitionistic point of view, is a study of certain functions of the human mind, and as such it is akin to these sciences. But is usefulness really the only measure of value? It is easy to mention of score of valuable activities which in no way support science, such as the arts, sports, and light entertainment. We claim for intuitionism a value of this sort, which it is difficult to define beforehand, but which is clearly felt in dealing with the matter. You know how philosophers struggle with the problem of defining the concept of value in art; yet every educated person feels this value. The case is analogous for the value of intuitionistic mathematics.

Form: For most mathematicians this value is affected fatally by the fact that you destroy the most precious mathematical results; a valuable method for the foundation of mathematics ought to save as much as possible of its results [D. Hilbert 1922]. This might even succeed by constructive methods; for definitions of constructiveness other than that advocated by the intuitionists are conceivable. For that matter, even the small number of actual intuitionists do not completely agree about the delimitation of the constructive. The most striking example is the rejection by Griss of the notion of negation, which other intuitionists accept as perfectly clear [H. Freudenthal 1936; G. F. C. Griss 1946a, p. 24; 1946b]. It seems probable, on the other hand, that a somewhat more liberal conception of the constructive might lead to the saving of the vital parts of classical mathematics.

Int: As intuitionists speak a non-formalized language, slight divergences of opinion between them can be expected. Though they have arisen sooner and in more acute forms than we could foresee, they are in no way alarming, for they all concern minor points and do not affect the fundamental ideas, about which there is complete agreement. Thus it is most unlikely that a wider conception of constructiveness could obtain the support of intuitionists. As to the mutilation of mathematics of which you accuse me, it must be taken as an inevitable consequence of our standpoint. It can also be seen as the excision of noxious ornaments, beautiful in form, but hollow in substance, and it is at least partly compensated for by the charm of subtle distinctions and witty methods by which intuitionists have enriched mathematical thought.

Form: Our discussion has assumed the form of a discussion of values. I gather from your words that you are ready to acknowledge the value of other conceptions of mathematics, but that you claim for your conception a value of its own. Is that right?

Int: Indeed, the only positive contention in the foundation of mathematics which I oppose is that classical mathematics has a clear sense; I must confess that I do not understand that. But even those who maintain that they do understand it might still be able to grasp our point of view and to value our work.

Letter: It is shown by the paradoxes that classical mathematics is not perfectly clear.

Form: Yes, but intuitionistic criticism goes much farther than is necessary to avoid the paradoxes; Mr. Int has not even mentioned them as an argument for his conception, and no doubt in his eyes consistency is but a welcome by-product of intuitionism.

Sign: You describe your activity as mental construction, Mr. Int, but mental processes are only observable through the acts to which they lead, in your case through the words you speak and the formulas you write. Does not this mean that the only way to study intuitionism is to study the formal system which it constructs?

Int: When looking at the tree over there, I am convinced I see a tree, and it costs considerable training to replace this conviction by the knowledge that in reality lightwaves reach my eyes, leading me to the construction of an image of the tree. In the same way, in speaking to you I am convinced that I press my opinions upon you, but you instruct me that in reality I produce vibrations in the air, which cause you to perform some action, e.g. to produce other vibrations. In both cases the first view is the natural one, the second is a theoretical construction. It is too often forgotten that the truth of such constructions depends upon the present state of science and that the words "in reality" ought to be translated into "according to the contemporary view of scientists". Therefore I prefer to adhere to the idea that, when describing intuitionistic mathematics, I convey thoughts to my hearers; these words ought to be taken not in the sense of some philosophical system, but in the sense of every-day life.

Sign: Then intuitionism, as a form of interaction between men, is a social phenomenon and its study belongs to the history of civilization.

Int: Its study, not its practice. Here I agree with Mr. Prag: *primum vivere, deinde philosophari,* and if we like we can leave the latter to others. Let those who come after me wonder why I built up these mental constructions and how they can be interpreted in some philosophy; I am content to build them in the conviction that in some way they will contribute to the clarification of human thought.

Prag: It is a common fault of philosophers to speak about things they know but imperfectly and we are near to being caught in that trap. Is Mr. Int willing to give us some samples of intuitionistic reasoning, in order that we may better be able to judge the quality of the stuff?

Int: Certainly, and even I am convinced that a few lessons will give you a better insight into it than lengthy discussions. May I beg those gentlemen who are interested in my explanations, to follow me to my classroom?

Intuitionism and formalism

L. E. J. BROUWER

The subject for which I am asking your attention deals with the foundations of mathematics. To understand the development of the opposing theories existing in this field one must first gain a clear understanding of the concept "science"; for it is as a part of science that mathematics originally took its place in human thought.

By science we mean the systematic cataloguing by means of laws of nature of causal sequences of phenomena, i.e., sequences of phenomena which for individual or social purposes it is convenient to consider as repeating themselves identically, – and more particularly of such causal sequences as are of importance in social relations.

That science lends such great power to man in his action upon nature is due to the fact that the steadily improving cataloguing of ever more causal sequences of phenomena gives greater and greater possibility of bringing about desired phenomena, difficult or impossible to evoke directly, by evoking other phenomena connected with the first by causal sequences. And that man always and everywhere creates order in nature is due to the fact that he not only isolates the causal sequences of phenomena (i.e., he strives to keep them free from disturbing secondary phenomena) but also supplements them with phenomena caused by his own activity, thus making them of wider applicability. Among the latter phenomena the results of counting and measuring take so important a place, that a large number of the natural laws introduced by science treat only of the mutual relations between the results of counting and measuring. It is well to notice in this connection that a natural law in the statement of which measurable magnitudes occur can only be understood to hold in nature with a certain degree of approximation; indeed natural laws as a rule are not proof against sufficient refinement of the measuring tools.

The exceptions to this rule have from ancient times been practical arithmetic and geometry on the one hand, and the dynamics of rigid bodies and celestial mechanics on the other hand. Both these groups have so far resisted all improvements in the tools of observation. But while

Inaugural address at the University of Amsterdam, read October 14, 1912. Translated by Professor Arnold Dresden. Reprinted by the kind permission of the author and the editor from the *Bulletin of the American Mathematical Society,* 20 (November, 1913), 81–96.

this has usually been looked upon as something accidental and temporal for the latter group, and while one has always been prepared to see these sciences descend to the rank of approximate theories, until comparatively recent times there has been absolute confidence that no experiment could ever disturb the exactness of the laws of arithmetic and geometry; this confidence is expressed in the statement that mathematics is "the" exact science.

On what grounds the conviction of the unassailable exactness of mathematical laws is based has for centuries been an object of philosophical research, and two points of view may here be distinguished, *intuitionism* (largely French) and *formalism* (largely German). In many respects these two viewpoints have become more and more definitely opposed to each other; but during recent years they have reached agreement as to this, that the exact validity of mathematical laws as laws of nature is out of the question. The question where mathematical exactness does exist, is answered differently by the two sides; the intuitionist says: in the human intellect, the formalist says: on paper.

In Kant we find an old form of intuitionism, now almost completely abandoned, in which time and space are taken to be forms of conception inherent in human reason. For Kant the axioms of arithmetic and geometry were synthetic a priori judgments, i.e., judgments independent of experience and not capable of analytical demonstration; and this explained their apodictic exactness in the world of experience as well as in abstracto. For Kant, therefore, the possibility of disproving arithmetical and geometrical laws experimentally was not only excluded by a firm belief, but it was entirely unthinkable.

Diametrically opposed to this is the view of formalism, which maintains that human reason does not have at its disposal exact images either of straight lines or of numbers larger than ten, for example, and that therefore these mathematical entities do not have existence in our conception of nature any more than in nature itself. It is true that from certain relations among mathematical entities, which we assume as axioms, we deduce other relations according to fixed laws, in the conviction that in this way we derive truths from truths by logical reasoning, but this non-mathematical conviction of truth or legitimacy has no exactness whatever and is nothing but a vague sensation of delight arising from the knowledge of the efficacy of the projection into nature of these relations and laws of reasoning. For the formalist therefore mathematical exactness consists merely in the method of developing the series of relations, and is independent of the significance one might want to give to the relations or the entities which they relate. And for the consistent formalist these meaningless series of relations to which mathematics are reduced have mathematical existence only when they have been represented in spoken or written language together with the mathematical-logical laws upon which their development depends, thus forming what is called symbolic logic.

Because the usual spoken or written languages do not in the least satisfy the requirements of consistency demanded of this symbolic logic, formalists try to avoid the use of ordinary language in mathematics. How far this may be carried is shown by the modern Italian school of formalists, whose leader, Peano, published one of his most important discoveries concerning the existence of integrals of real differential equations in the *Mathematische Annalen* in the language of symbolic logic; the result was that it could only be read by a few of the initiated and that it did not become generally available until one of these had translated the article into German.

The viewpoint of the formalist must lead to the conviction that if other symbolic formulas should be substituted for the ones that now represent the fundamental mathematical relations and the mathematical-logical laws, the absence of the sensation of delight, called "consciousness of legitimacy," which might be the result of such substitution would not in the least invalidate its mathematical exactness. To the philosopher or to the anthropologist, but not to the mathematician, belongs the task of investigating why certain systems of symbolic logic rather than others may be effectively projected upon nature. Not to the mathematician, but to the psychologist, belongs the task of explaining why we believe in certain systems of symbolic logic and not in others, in particular why we are averse to the so-called contradictory systems in which the negative as well as the positive of certain propositions are valid (Mannoury 1909: 149–54).

As long as the intuitionists adhered to the theory of Kant it seemed that the development of mathematics in the nineteenth century put them in an ever weaker position with regard to the formalists. For in the first place this development showed repeatedly how complete theories could be carried over from one domain of mathematics to another: projective geometry, for example, remained unchanged under the interchange of the rôles of point and straight line, an important part of the arithmetic of real numbers remained valid for various complex number fields and nearly all the theorems of elementary geometry remained true for non-archimedian geometry, in which there exists for every straight line segment another such segment, infinitesimal with respect to the first. These discoveries seemed to indicate indeed that of a mathematical theory only the logical form was of importance and that one need no more be concerned with the material than it is necessary to think of the significance of the digit groups with which one operates, for the correct solution of a problem in arithmetic.

But the most serious blow for the Kantian theory was the discovery of non-euclidean geometry, a consistent theory developed from a set of axioms differing from that of elementary geometry only in this respect that the parallel axiom was replaced by its negative. For this showed that the phenomena usually described in the language of elementary geometry may be described with equal exactness, though frequently less compactly in the language of non-euclidean geometry; hence, it is not only impossible to hold that the space of our experience has the properties of elementary geometry but it has no significance to ask for *the* geometry which would be true for the space of our experience. It is true that elementary geometry is better suited than any other to the description of the laws of kinematics of rigid bodies and hence of a large number of natural phenomena, but with some patience it would be possible to make objects for which the kinematics would be more easily interpretable in terms of non-euclidean than in terms of euclidean geometry (Poincaré 1903: 104).

However weak the position of intuitionism seemed to be after this period of mathematical development, it has recovered by abandoning Kant's apriority of space but adhering the more resolutely to the apriority of time. This neo-intuitionism considers the falling apart of moments of life into qualitatively different parts, to be reunited only while remaining separated by time, as the fundamental phenomenon of the human intellect, passing by abstracting from its emotional content into the fundamental phenomenon of mathematical thinking, the intuition of the bare two-oneness. This intuition of two-oneness, the basal intuition of mathematics, creates not only the numbers one and two, but also all finite ordinal numbers, inasmuch as one of the elements of the two-oneness may be thought of as a new two-oneness, which process may be repeated indefinitely; this gives rise still further to the smallest infinite ordinal number ω. Finally this basal intuition of mathematics, in which the connected and the separate, the continuous and the discrete are united, gives rise immediately to the intuition of the linear continuum, i.e., of the "between," which is not exhaustible by the interposition of new units and which therefore can never be thought of as a mere collection of units.

In this way the apriority of time does not only qualify the properties of arithmetic as synthetic a priori judgments, but it does the same for those of geometry, and not only for elementary two- and three-dimensional geometry, but for non-euclidean and *n*-dimensional geometries as well. For since Descartes we have learned to reduce all these geometries to arithmetic by means of the calculus of coordinates.

From the present point of view of intuitionism therefore all mathematical sets of units which are entitled to that name can be developed out of the basal intuition, and this can only be done by combining a finite number of times the two operations: "to create a finite ordinal number" and "to create the infinite ordinal number ω"; here it is to be understood that for the latter purpose any previously constructed set or any previously performed constructive operation may be taken as a unit. Consequently the intuitionist recognizes only the existence of denumerable sets, i.e., sets whose elements may be brought into one-to-one correspondence either with the elements of a finite ordinal number or with those of the infinite ordinal number ω. And in the construction of these sets neither the ordinary language nor any symbolic language can have any other rôle than that of serving as a nonmathematical auxiliary, to assist the mathematical memory or to enable different individuals to build up the same set.

For this reason the intuitionist can never feel assured of the exactness of a mathematical theory by such guarantees as the proof of its being noncontradictory, the possibility of defining its concepts by a finite number of words (Poincaré 1908a: 6), or the practical certainty that it will never lead to a misunderstanding in human relations (Borel 1912: 221).

As has been stated above, the formalist wishes to leave to the psychologist the task of selecting the "truly-mathematical" language from among the many symbolic languages that may be consistently developed. Inasmuch as psychology has not yet begun in this task, formalism is compelled to mark off, at least temporarily, the domain that it wishes to consider as "true mathematics" and to lay down for that purpose a definite system of axioms and laws of reasoning, if it does not wish to see its work doomed to sterility. The various ways in which this attempt has actually been made all follow the same leading idea, viz., the presupposition of the existence of a world of mathematical objects, a world independent of the thinking individual, obeying the laws of classical logic and whose objects may possess with respect to each other the "relation of a set to its elements." With reference to this relation various axioms are postulated, suggested by the practice with natural finite sets; the principal of these are: *"a set is determined by its elements"; "for any two mathematical objects it is decided whether or not one of them is contained in the other one as an element"; "to every set belongs another set containing as its elements nothing but the subsets of the given set";* the axiom of selection: *"a set which is split into subsets contains at least one subset which contains one and not more than one element of each of the first subsets";* the axiom of inclusion: *"if for any mathematical object it is decided whether a certain property is valid for it or not, then there exists a set containing nothing but those objects for which the property does hold";* the axiom of composition: *"the elements of all sets that belong to a set of sets form a new set."*

On the basis of such a set of axioms the formalist develops now in the first place the theory of "finite sets." A set is called finite if its elements cannot be brought into one-to-one correspondence with the elements of one of its subsets; by means of relatively complicated reasoning the principle of complete induction is proved to be a fundamental property of these sets (Zermelo 1909: 185–93); this principle states that a property will be true for all finite sets if, first, it is true for all sets containing a single element, and, second, its validity for an arbitrary finite set follows from its validity for this same set reduced by a single one of its elements. That the formalist must give an explicit proof of this principle, which is self-evident for the finite numbers of the intuitionist on account of their construction, shows at the same time that the former will never be able to justify his choice of axioms by replacing the unsatisfactory appeal to inexact practice or to intuition equally inexact for him by a proof of the non-contradictoriness of his theory. For in order to prove that a contradiction can never arise among the infinitude of conclusions that can be drawn from the axioms he is using, he would first have to show that if no contradiction had as yet arisen with the nth conclusion then none could arise with the $(n+1)$th conclusion, and secondly, he would have to apply the principle of complete induction intuitively. But it is this last step which the formalist may not take, even though he should have proved the principle of complete induction; for this would require mathematical certainty that the set of properties obtained after the nth conclusion had been reached, would satisfy for an arbitrary n his definition for finite sets (Poincaré 1905: 834), and in order to secure this certainty he would have to have recourse not only to the unpermissible application of a symbolic criterion to a concrete example but also to another intuitive application of the principle of complete induction; this would lead him to a vicious circle reasoning.

In the domain of finite sets in which the formalist axioms have an interpretation perfectly clear to the intuitionists, unreservedly agreed to by them, the two tendencies differ solely in their method, not in their results; this becomes quite different however in the domain of infinite or transfinite sets, where, mainly by the application of the axiom of inclusion, quoted above, the formalist introduces various concepts, entirely meaningless to the intuitionist, such as for instance *"the set whose elements are the points of space," "the set whose elements are the continuous functions of a variable," "the set whose elements are the discontinuous functions of a variable,"* and so forth. In the course of these formalistic developments it turns out that the consistent application of the axiom of inclusion leads inevitably to contradictions. A clear illustration of this fact is furnished by the so-called paradox of Burali-Forti (1897). To exhibit it we have to lay down a few definitions.

A set is called ordered if there exists between any two of its elements a relation of "higher than" or "lower than," with this understanding that if the element *a* is higher than the element *b*, then the element *b* is lower than the element *a*, and if the element *b* is higher than *a* and *c* is higher than *b*, then *c* is higher than *a*.

A well-ordered set (in the formalistic sense) is an ordered set, such that every subset contains an element lower than all others.

Two well-ordered sets that may be brought into one-to-one correspondence under invariance of the relations of "higher than" and "lower than" are said to have the same ordinal number.

If two ordinal numbers *A* and *B* are not equal, then one of them is greater than the other one, let us say *A* is greater than *B*; this means that *B* may be brought into one-to-one correspondence with an initial segment of *A* under invariance of the relations of "higher than" and "lower than." We have introduced above, from the intuitionist viewpoint, the smallest infinite ordinal number ω, i.e., the ordinal number of the set of all finite ordinal numbers arranged in order of magnitude.[1] Well-ordered sets having the ordinal number ω are called elementary series.

It is proved without difficulty by the formalist that an arbitrary subset of a well-ordered set is also a well-ordered set, whose ordinal number is less than or equal to that of the original set; also, that if to a well-ordered set that does not contain all mathematical objects a new element be added that is defined to be higher than all elements of the original set, a new well-ordered set arises whose ordinal number is greater than that of the first set.

We construct now on the basis of the axiom of inclusion the *set s which contains as elements all the ordinal numbers arranged in order of magnitude;* then we can prove without difficulty, on the one hand that *s* is a well-ordered set whose ordinal number can not be exceeded by any other ordinal number in magnitude, and on the other hand that it is possible, since not all mathematical objects are ordinal numbers, to create an ordinal number greater than that of *s* by adding a new element to *s*, – a contradiction.[2]

[1] The more general ordinal numbers of the intuitionist are the numbers constructed by means of Cantor's two principles of generation (cf. Cantor 1895-7, 49: 226).

[2] It is without justice that the paradox of Burali-Forti is sometimes classed with that of Richard, which in a somewhat simplified form reads as follows: "Does there exist a *least integer, that can not be defined by a sentence of at most twenty words?* On the one hand *yes,* for the number of sentences of at most twenty words is of course finite; on the other

---

L. E. J. BROUWER

Although the formalists must admit contradictory results as mathematical if they want to be consistent, there is something disagreeable for them in a paradox like that of Burali-Forti because at the same time the progress of their arguments is guided by the principium contradictionis, i.e., by the rejection of the simultaneous validity of two contradictory properties. For this reason the axiom of inclusion has been modified to read as follows: *"If for all elements of a set it is decided whether a certain property is valid for them or not, then the set contains a subset containing nothing but those elements for which the property does hold"* (Zermelo 1908: 263).

In this form the axiom permits only the introduction of such sets as are subsets of sets previously introduced; if one wishes to operate with other sets, their existence must be explicitly postulated. Since however in order to accomplish anything at all the existence of a certain collection of sets will have to be postulated at the outset, the only valid argument that can be brought against the introduction of a new set is that it leads to contradictions; indeed the only modifications that the discovery of paradoxes has brought about in the practice of formalism has been the abolition of those sets that had given rise to these paradoxes. One continues to operate without hesitation with other sets introduced on the basis of the old axiom of inclusion; the result of this is that extended fields of research, which are without significance for the intuitionist are still of considerable interest to the formalist. An example of this is found in the theory of potencies, of which I shall sketch the principal features here, because it illustrates so clearly the impassable chasm which separates the two sides.

Two sets are said to possess the same potency, or power, if their elements can be brought into one-to-one correspondence. The power of set A is said to be greater than that of B, and the power of B less than that of A, if it is possible to establish a one-to-one correspondence between B and a part of A, but impossible to establish such a correspondence between A and a part of B. The power of a set which has the same power as one of its subsets, is called infinite, other powers are called finite. Sets that have the same power as the oridinal number ω are called denumerably infinite and the power of such sets is called aleph-null: it proves to be the smallest infinite power. According to the statements previously made, this power aleph-null is the only infinite power of which the intuitionists recognize the existence.

hand *no,* for if it should exist, it would be defined by the sentence of fifteen words formed by the words italicized above."

The origin of this paradox does not lie in the axiom of inclusion but in the variable meaning of the word *"defined"* in the italicized sentence, which makes it possible to define by means of this sentence an infinite number of integers in succession.

84

Let us now consider the concept: "denumerably infinite ordinal number." From the fact that this concept has a clear and well-defined meaning for both formalist and intuitionist, the former infers the right to create the "set of all denumerably infinite ordinal numbers," the power of which he calls aleph-one, a right not recognized by the intuitionist. Because it is possible to argue to the satisfaction of both formalist and intuitionist, first, that denumerably infinite sets of denumerably infinite ordinal numbers can be built up in various ways, and second, that for every such set it is possible to assign a denumerably infinite ordinal number, *not* belonging to this set, the formalist concludes: "aleph-one is greater than aleph-null," a proposition that has no meaning for the intuitionist. Because it is possible to argue to the satisfaction of both formalist and intuitionist that it is impossible to construct[3] a set of denumerably infinite ordinal numbers, which could be proved to have a power less than that of aleph-one, but greater than that of aleph-null, the formalist concludes: "aleph-one is the second smallest infinite ordinal number," a proposition that has no meaning for the intuitionist.

Let us consider the concept: "real number between 0 and 1." For the formalist this concept is equivalent to "elementary series of digits after the decimal point,"[4] for the intuitionist it means "law for the construction of an elementary series of digits after the decimal point, built up by means of a finite number of operations." And when the formalist creates the "set of all real numbers between 0 and 1," these words are without meaning for the intuitionist, even whether one thinks of the real numbers of the formalist, determined by elementary series of freely selected digits, or of the real numbers of the intuitionist, determined by finite laws of construction. Because it is possible to prove to the satisfaction of both formalist and intuitionist, first, that denumerably infinite sets of real numbers between 0 and 1 can be constructed in various ways, and second that for every such set it is possible to assign a real number between 0 and 1, not belonging to the set, the formalist concludes: "the power of the continuum, i.e., the power of the set of real numbers between 0 and 1, is greater than aleph-null," a proposition that is without meaning for the intuitionist; the formalist further raises the question, whether there exist sets of real numbers between 0 and 1, whose power is less than that of the continuum, but greater than aleph-null, in other words, "whether the power of the continuum is the second smallest infinite power," and this

---

[3]If "construct" were here replaced by "define" (in the formalistic sense), the proof would *not* be satisfactory to the intuitionist. For, in Cantor's argument it is not allowed to replace the words "können wir bestimmen" (1895-7, 49: 214, line 17 from top) by the words "muss es geben."

[4]Here as everywhere else in this paper, the assumption is tacitly made that there are an infinite number of digits different from 9.

question, which is still waiting for an answer, he considers to be one of the most difficult and most fundamental of mathematical problems.

For the intuitionist, however, the question as stated is without meaning; and as soon as it has been so interpreted as to get a meaning, it can easily be answered.

If we restate the question in this form: "Is it impossible to construct[5] infinite sets of real numbers between 0 and 1, whose power is less than that of the continuum, but greater than aleph-null?" then the answer must be in the affirmative; for the intuitionist can only construct denumerable sets of mathematical objects and if, on the basis of the intuition of the linear continuum, he admits elementary series of free selections as elements of construction, then each non-denumerable set constructed by means of it contains a subset of the power of the continuum.

If we restate the question in the form: "Is it possible to establish a one-to-one correspondence between the elements of a set of denumerably infinite ordinal numbers on the one hand, and a set of real numbers between 0 and 1 on the other hand, both sets being indefinitely extended by the construction of new elements, of such a character that the correspondence shall not be disturbed by any continuation of the construction of both sets?" then the answer must also be in the affirmative, for the extension of both sets can be divided into phases in such a way as to add a denumerably infinite number of elements during each phase.[6]

If however we put the question in the following form: "Is it possible to construct a law which will assign a denumerably infinite ordinal number to every elementary series of digits and which will give certainty a priori that two different elementary series will never have the same denumerably infinite ordinal number corresponding to them?" then the answer must be in the negative; for this law of correspondence must prescribe in some way a construction of certain denumerably infinite ordinal numbers at each of the successive places of the elementary series; hence there is for each place c_ν a well-defined largest denumerably infinite number α_ν, the construction of which is suggested by that particular place; there is then also a well-defined denumerably infinite ordinal number α_ω,

---

[5]If "construct" were here replaced by "define" (in the formalistic sense), and if we suppose that the problem concerning the pairs of digits in the decimal fraction development of π, discussed on p. 88, *can not be solved,* then the question of the text must be answered negatively. For, let us denote by Z the set of those infinite binary fractions, whose nth digit is 1, if the nth pair of digits in the decimal fraction development of π consists of unequal digits; let us further denote by X the set of all finite binary fractions. Then the power of $Z + X$ is greater than aleph-null, but less than that of the continuum.

[6]Calling *denumerably unfinished* all sets of which the elements can be individually realized, and in which for every denumerably infinite subset there exists an element not belonging to this subset, we can say in general, in accordance with the definitions of the text: *"All denumerably unfinished sets have the same power."*

greater than all α_ν's and that can not therefore be exceeded by any of the ordinal numbers involved by the law of correspondence; hence the power of that set of ordinal numbers can not exceed aleph-null.

As a means for obtaining ever greater powers, the formalists define with every power μ a "set of all the different ways in which a number of selections of power μ may be made," and they prove that the power of this set is greater than μ. In particular, when it has been proved to the satisfaction of both formalist and intuitionist that it is possible in various ways to construct laws according to which functions of a real variable different from each other are made to correspond to all elementary series of digits, but that it is impossible to construct a law according to which an elementary series of digits is made to correspond to every function of a real variable and in which there is certainty a priori that two different functions will never have the same elementary series corresponding to them, the formalist concludes: "the power c' of the set of all functions of a real variable is greater than the power c of the continuum," a proposition without meaning to the intuitionist; and in the same way in which he was led from c to c', he comes from c' to a still greater power c''.

A second method used by the formalists for obtaining ever greater powers is to define for every power μ, which can serve as a power of ordinal numbers, "the set of all ordinal numbers of power μ," and then to prove that the power of this set is greater than μ. In particular they denote by aleph-two the power of the set of all ordinal numbers of power aleph-one and they prove that aleph-two is greater than aleph-one and that it follows in magnitude immediately after aleph-one. If it should be possible to interpret this result in a way in which it would have meaning for the intuitionist, such interpretation would not be as simple in this case as it was in the preceding cases.

What has been treated so far must be considered to be the negative part of the theory of potencies; for the formalist there also exists a positive part however, founded on the theorem of Bernstein: "If the set A has the same power as a subset of B and B has the same power as a subset of A, then A and B have the same power" or, in an equivalent form: "If the set $A = A_1 + B_1 + C_1$, has the same power as the set A_1, then it also has the same power as the set $A_1 + B_1$.

This theorem is self-evident for denumerable sets. If it is to have any meaning at all for sets of higher power for the intuitionist, it will have to be interpretable as follows: "If it is possible, *first* to construct a law determining a one-to-one correspondence between the mathematical entities of type A and those of type A_1, and *second* to construct a law determining a one-to-one correspondence between the mathematical entities of type A and those of types A_1, B_1, and C_1, then it is possible to determine from these two laws by means of a finite number of operations a third law, determining a one-to-one correspondence between the mathematical entities of type A and those of types A_1 and B_1."

In order to investigate the validity of this interpretation, we quote the proof:

"From the division of A into $A_1 + B_1 + C_1$, we secure by means of the correspondence γ_1 between A and A_1 a division of A_1 into $A_2 + B_2 + C_2$, as well as a one-to-one correspondence γ_2 between A_1 and A_2. From the division of A_1 into $A_2 + B_2 + C_2$, we secure by means of the correspondence between A_1 and A_2 a division of A_2 into $A_3 + B_3 + C_3$, as well as a one-to-one correspondence γ_3 between A_2 and A_3. Indefinite repetition of this procedure will divide the set A into an elementary series of subsets $C_1, C_2, C_3, \ldots$, an elementary series of subsets $B_1, B_2, B_3, \ldots$, and a remainder set D. The correspondence γ_C between A and $A_1 + B_1$ which is desired is secured by assigning to every element of C_ν the corresponding element of $C_{\nu+1}$ and by assigning every other element of A to itself."

In order to test this proof on a definite example, let us take for A the set of all real numbers between 0 and 1, represented by infinite decimal fractions, for A_1 the set of those decimal fractions in which the $(2n-1)$th digit is equal to the $2n$th digit; further a decimal fraction that does not belong to A_1 will be counted to belong to B_1 or to C_1 according as the above-mentioned equality of digits occurs an infinite or a finite number of times. By replacing successively each digit of an arbitrary element of A by a pair of digits equal to it, we secure at once a law determining a one-to-one correspondence γ_1 between A and A_1. For of the element of A_1 that corresponds to an arbitrary well-defined element of A, such as, e.g., $\pi - 3$, we can determine successively as many digits as we please; it must therefore be considered as being well-defined.

In order to determine the element corresponding to $\pi - 3$ according to the correspondence γ_C, it is now necessary to decide first whether it happens an infinite or a finite number of times in the decimal fraction development of $\pi - 3$ that a digit in an odd-numbered place is equal to the digit in the following even-numbered place; for this purpose we should either have to invent a process for constructing an elementary series of such pairs of equal digits, or to deduce a contradiction from the assumption of the existence of such an elementary series. There is, however, no ground for believing that either of these problems can be solved.[7]

[7]Such belief could be based only on an appeal to the principium tertii exclusi, i.e., to the axiom of the existence of the "set of all mathematical properties," an axiom of far wider range even than the axioms of inclusion, quoted above. Compare in this connection Brouwer 1908: 152–8.

Hence it has become evident that also the theorem of Bernstein, and with it the positive part of the theory of potencies, does not allow an intuitionistic interpretation.

So far my exposition of the fundamental issue, which divides the mathematical world. There are eminent scholars on both sides and the chance of reaching an agreement within a finite period is practically excluded. To speak with Poincaré: "Les hommes ne s'entendent pas, parce qu'ils ne parlent pas la même langue et qu'il y a des langues qui ne s'apprennent pas."

Consciousness, philosophy, and mathematics

L. E. J. BROUWER

The ... point of view that there are no non-experienced truths and that logic is not an absolutely reliable instrument to discover truths has found acceptance with regard to mathematics much later than with regard to practical life and to science. Mathematics rigorously treated from this point of view, including deducing theorems exclusively by means of introspective construction, is called intuitionistic mathematics. In many respects it deviates from classical mathematics. In the first place because classical mathematics uses logic to generate theorems, believes in the existence of unknown truths, and in particular applies the *principle of the excluded third* expressing that every mathematical assertion (i.e. every assignment of a mathematical property to a mathematical entity) either is a truth or cannot be a truth. In the second place because classical mathematics confines itself to *predeterminate* infinite sequences for which from the beginning the nth element is fixed for each n. Owing to this confinement classical mathematics, to define real numbers, has only predeterminate convergent infinite sequences of rational numbers at its disposal. Out of real numbers defined in this way, only subspecies of "ever unfinished denumerable" species of real numbers can be composed by means of introspective construction. Such ever unfinished denumerable species all being of measure zero, classical mathematics, to create the continuum out of points, needs some logical process starting from one or more axioms. Consequently we may say that classical analysis, however appropriate it be for technique and science, has less mathematical truth than intuitionistic analysis performing the said composition of the continuum by considering the species of freely proceeding convergent infinite sequences of rational numbers, without having recourse to language or logic.

As a matter of course also the languages of the two mathematical schools diverge. And even in those mathematical theories which are covered by a neutral language, i.e. by a language understandable on both sides, either school operates with mathematical entities not recognized

Excerpted by kind permission of the publisher from 10th International Congress of Philosophy, Amsterdam, 1948, *Proceedings I,* Fascicule II (Amsterdam: North-Holland Publishing Company, 1949), pp. 1243–9.

by the other one: there are intuitionist structures which cannot be fitted into any classical logical frame, and there are classical arguments not applying to any introspective image. Likewise, in the theories mentioned, mathematical entities recognized by both parties on each side are found satisfying theorems which for the other school are either false, or senseless, or even in a way contradictory. In particular, theorems holding in intuitionism, but not in classical mathematics, often originate from the circumstance that for mathematical entities belonging to a certain species, the possession of a certain property imposes a special character on their way of development from the basic intuition, and that from this special character of their way of development from the basic intuition, properties ensue which for classical mathematics are false. A striking example is the intuitionist theorem that a full function of the unity continuum, i.e. a function assigning a real number to every non-negative real number not exceeding unity, is necessarily uniformly continuous.

To elucidate the consequences of the rejection of the principle of the excluded third as an instrument to discover truths, we shall put the wording of this principle into the following slightly modified, intuitionistically more adequate form, called the *simple principle of the excluded third:*

> *Every assignment τ of a property to a mathematical entity can be judged, i.e. either proved or reduced to absurdity.*

Then for a single such assertion τ the enunciation of this principle is non-contradictory in intuitionistic as well as in classical mathematics. For, if it were contradictory, then the absurdity of τ would be true and absurd at the same time, which is impossible. Moreover, as can easily be proved, for a *finite* number of such assertions τ the simultaneous enunciation of the principle is non-contradictory likewise. However, for the simultaneous enunciation of the principle for all elements of an *arbitrary* species of such assertions τ this non-contradictority cannot be maintained.

E.g. from the supposition, for a definite real number c_1, that the assertion: c_1 *is rational,* has been proved to be either true or contradictory, no contradiction can be deduced. Furthermore, $c_1, c_2, \ldots c_m$ being real numbers, neither the simultaneous supposition, for each of the values $1, 2, \ldots m$ of ν, that the assertion: c_ν *is rational,* has been proved to be either true or contradictory, can lead to a contradiction. However, the simultaneous supposition for *all* real numbers c that the assertion: c *is rational,* has been proved to be either true or contradictory, does lead to a contradiction.

Consequently if we formulate the *complete principle of the excluded third* as follows:

> *If a, b, and c are species of mathematical entities, if further both a and b form part of c, and if b consists of those elements of c which cannot belong to a, then c is identical with the union of a and b,*

the latter principle is contradictory.

A corollary of the *simple* principle of the excluded third says that:

> *If for an assignment τ of a property to a mathematical entity the non-contradictority, i.e. the absurdity of the absurdity, has been established, the truth of τ can be demonstrated likewise.*

The analogous corollary of the *complete* principle of the excluded third is the *principle of reciprocity of complementarity,* running as follows:

> *If a, b, and c are species of mathematical entities, if further a and b form part of c, and if b consists of the elements of c which cannot belong to a, then a consists of the elements of c which cannot belong to b.*

Another corollary of the *simple* principle of the excluded third is the *simple principle of testability* saying that

> *every assignment τ of a property to a mathematical entity can be tested, i.e. proved to be either non-contradictory or absurd.*

The analogous corollary of the *complete* principle of the excluded third is the following *complete principle of testability:*

> *If a, b, d, and c are species of mathematical entities, if each of the species a, b, and d forms part of c, if b consists of the elements of c which cannot belong to a, and d of the elements of c which cannot belong to b, then c is identical with the union of b and d.*

For intuitionism the principle of the excluded third and its corollaries are assertions σ about assertions τ, and these assertions σ only then are "realized", i.e. only then convey truths, if these truths have been experienced.

Each assertion τ of the possibility of a construction of bounded finite character in a finite mathematical system furnishes a case of realization of the principle of the excluded third. For every such construction can be attempted only in a finite number of particular ways, and each attempt proves successful or abortive in a finite number of steps.

If the assertion of an absurdity is called a *negative assertion,* then each negative assertion furnishes a case of realization of the principle of reciprocity of complementarity. For, let α be a negative assertion, indicating the absurdity of the assertion β. As, on the one hand, the implication of the truth of an assertion a by the truth of an assertion b implies the implication of the absurdity of b by the absurdity of a, whilst, on the other hand, the truth of β implies the absurdity of the absurdity of β, we conclude that the absurdity of the absurdity of the absurdity of β, i.e. the non-contradictority of α, implies the absurdity of β, i.e. implies α.

In consequence of this realization of the principle of reciprocity of complementarity the principles of testability and of the excluded third are equivalent in the domain of negative assertions. For, if for α the principle of testability holds, this means that either the absurdity of the absurdity of β or the non-contradictority of the absurdity of β, i.e. by the preceding paragraph, that either the absurdity of the absurdity of β or the absurdity of β, i.e. either the absurdity of α or α can be proved, so that α satisfies the principle of the excluded third.

To give some examples refuting the principle of the excluded third and its corollaries, we introduce the notion of a *drift*. By a drift we understand the union γ of a convergent fundamental sequence of real numbers $c_1(\gamma), c_2(\gamma), \ldots$, called the *counting-numbers* of the drift, and the limiting-number $c(\gamma)$ of this sequence, called the *kernel* of the drift, all counting-numbers lying apart[1] from each other and from the kernel. If $c_\nu(\gamma) < \circ c(\gamma)$ for each ν, the drift will be called *left-winged*. If $c_\nu(\gamma) \circ > c(\gamma)$ for each ν, the drift will be called *right-winged*. If the fundamental sequence $c_1(\gamma), c_2(\gamma), \ldots$ is the union of a fundamental sequence of *left counting-numbers* $l_1(\gamma), l_2(\gamma), \ldots$ such that $l_\nu(\gamma) < \circ c(\gamma)$ for each ν, and a fundamental sequence of *right counting-numbers* $d_1(\gamma), d_2(\gamma), \ldots$ such that $d_\nu(\gamma) \circ > c(\gamma)$ for each ν, the drift will be called *two-winged*.

Let α be a mathematical assertion so far neither tested nor recognized as testable. Then in connection with this assertion α and with a drift γ the creating subject can generate an infinitely proceeding sequence $R(\gamma, \alpha)$ of real numbers $c_1(\gamma, \alpha), c_2(\gamma, \alpha), \ldots$ according to the following direction: As long as during the choice of the $c_n(\gamma, \alpha)$ the creating subject has experienced neither the truth, nor the absurdity of α, each $c_n(\gamma, \alpha)$ is chosen equal to $c(\gamma)$. But as soon as between the choice of $c_{r-1}(\gamma, \alpha)$ and that of $c_r(\gamma, \alpha)$ the creating subject has experienced either the truth or the absurdity of α, $c_r(\gamma, \alpha)$, and likewise $c_{r+\nu}(\gamma, \alpha)$ for each natural

---

[1] If for two real numbers a and b defined by convergent infinite sequences of rational numbers $a_1, a_2, \ldots$ and $b_1, b_2, \ldots$ respectively, two such natural numbers m and n can be calculated that $b_\nu - a_\nu > 2^{-n}$ for $\nu \geqslant m$, we write $b \circ > a$ and $a < \circ b$, and a and b are said to lie *apart* from each other. If $a = b$ is absurd, we write $a \neq b$. If $a < \circ b$ is absurd, we write $a \geqslant b$. If both $a = b$ and $a < \circ b$ are absurd, we write $a > b$. The absurdities of $a < \circ b$ and $a < b$ prove to be mutually equivalent, and the absurdity of $a \geqslant b$ proves to be equivalent to $a < b$.

number ν, is chosen equal to $c_r(\gamma)$. This sequence $R(\gamma, \alpha)$ converges to a real number $D(\gamma, \alpha)$ which will be called a *direct checking-number of γ through α*.

Again, in connection with α and with a two-winged drift γ the creating subject can generate an infinitely proceeding sequence $S(\gamma, \alpha)$ of real numbers $\omega_1(\gamma, \alpha), \omega_2(\gamma, \alpha), \ldots$ according to the following direction: As long as during the choice of the $\omega_n(\gamma, \alpha)$ the creating subject has experienced neither the truth, nor the absurdity of α, each $\omega_n(\gamma, \alpha)$ is chosen equal to $c(\gamma)$. But as soon as between the choice of $\omega_{r-1}(\gamma, \alpha)$ and that of $\omega_r(\gamma, \alpha)$ the creating subject has experienced the truth of α, $\omega_r(\gamma, \alpha)$, and likewise $\omega_{r+\nu}(\gamma, \alpha)$ for each natural number ν, is chosen equal to $d_r(\gamma)$. And as soon as between the choice of $\omega_{s-1}(\gamma, \alpha)$ and that of $\omega_s(\gamma, \alpha)$ the creating subject has experienced the absurdity of α, $\omega_s(\gamma, \alpha)$, and likewise $\omega_{s+\nu}(\gamma, \alpha)$ for each natural number ν, is chosen equal to $l_s(\gamma)$. This sequence $S(\gamma, \alpha)$ converges to a real number $E(\gamma, \alpha)$ which will be called an *oscillatory checking-number of γ through α*.

Let γ be a right-winged drift whose counting-numbers are rational. Then the assertion of the rationality of $D(\gamma, \alpha)$ is testable, but not judgable, and its non-contradictoriness is not equivalent to its truth. Furthermore we have $D(\gamma, \alpha) > c(\gamma)$, but not $D(\gamma, \alpha) \circ > c(\gamma)$.

Let γ be a two-winged drift whose right counting-numbers are rational, and whose left counting-numbers are irrational. Then the assertion of the rationality of $E(\gamma, \alpha)$ is neither judgeable, nor is it testable, nor is its non-contradictoriness equivalent to its truth. Furthermore $E(\gamma, \alpha)$ is neither $\geqslant c(\gamma)$, nor $\leqslant c(\gamma)$.

The long belief in the universal validity of the principle of the excluded third in mathematics is considered by intuitionism as a phenomenon of history of civilization of the same kind as the old-time belief in the rationality of π or in the rotation of the firmament on an axis passing through the earth. And intuitionism tries to explain the long persistence of this dogma by two facts: firstly the obvious non-contradictoriness of the principle for an arbitrary single assertion; secondly the practical validity of the whole of classical logic for an extensive group of *simple everyday phenomena*. The latter fact apparently made such a strong impression that the play of thought that classical logic originally was, became a deep-rooted habit of thought which was considered not only as useful but even as aprioristic.

Obviously the field of validity of the principle of the excluded third is identical with the intersection of the fields of validity of the principle of testability and the principle of reciprocity of complementarity. Furthermore the former field of validity is a *proper* subfield of each of the latter ones, as is shown by the following examples:

Let A be the species of the direct checking-numbers of drifts with rational counting-numbers, B the species of the irrational real numbers, C the union of A and B. Then all assertions of rationality of an element of C satisfy the principle of testability, whilst there are assertions of rationality of an element of C not satisfying the principle of the excluded third. Again, all assertions of equality of two real numbers satisfy the principle of reciprocity of complementarity, whereas there are assertions of equality of two real numbers not satisfying the principle of the excluded third.

In the domain of mathematical assertions the property of absurdity, just as the property of truth, is a *universally additive property,* that is to say, if it holds for each element α of a species of assertions, it also holds for the assertion which is the union of the assertions α. *This property of universal additivity does not obtain for the property of non-contradictority.* However, non-contradictority does possess the weaker property of *finite additivity,* that is to say, if the assertions ρ and σ are non-contradictory, the assertion τ which is the union of ρ and σ, is also non-contradictory. For, let us start for a moment from the supposition ω that τ is contradictory. Then the truth of ρ would entail the contradictority of σ, which would clash with the data, so that the truth of ρ is absurd, i.e. ρ is absurd. This consequence of the supposition ω clashing with the data, the supposition ω is contradictory, i.e. τ is non-contradictory.

Application of this theorem to the special non-contradictory assertions that are the enunciations of the principle of the excluded third for a single assertion, establishes the above-mentioned non-contradictority of the simultaneous enunciation of this principle for a finite number of assertions.

Within some species of mathematical entities the absurdities of two non-equivalent[2] assertions may be equivalent. E.g. each of the following three pairs of non-equivalent assertions relative to a real number a:

I 1. $a=a$;	I 2. either $a \leqslant 0$ or $a \geqslant 0$
II 1. $a \geqslant 0$;	II 2. either $a=0$ or $a \circ > 0$
III 1. $a > 0$;	III 2. $a \circ > 0$

furnishes a pair of equivalent absurdities.

It occurs that within some species of mathematical entities some absurdities of constructive properties can be given a constructive form. E.g. for a natural number a the absurdity of the existence of two natural numbers different from a and from 1 and having a as their product is equivalent to the existence, whenever a is divided by a natural number dif-

---

[2] By non-equivalence we understand absurdity of equivalence, just as by noncontradictority we understand absurdity of contradictority.

95

ferent from a and from 1, of a remainder. Likewise, for two real numbers a and b the relation $a \geqslant b$ introduced above as an absurdity of a constructive property can be formulated constructively as follows: Let $a_1, a_2, \ldots$ and $b_1, b_2, \ldots$ be convergent infinite sequences of rational numbers defining a and b respectively. Then, for any natural number n, a natural number m can be calculated such that $a_\nu - b_\nu \circ > -2^{-n}$ for $\nu \geqslant m$.

On the other hand there seems to be little hope for reducing irrationality of a real number a, or one of the relations $a \neq b$ and $a > b$ for real numbers a and b, to a constructive property, if we remark that a direct checking-number of a drift whose kernel is rational and whose counting-numbers are irrational, is irrational without lying apart from the species of rational numbers; further that a direct checking-number of an arbitrary drift differs from the kernel of the drift without lying apart from it, and that a direct checking-number of a right-winged drift lies to the right of the kernel of the drift without lying apart from it.

It occurs that within some species of mathematical entities some non-contradictorities of constructive properties ζ can be given either a constructive form (possibly, but not necessarily, in consequence of reciprocity of complementarity holding for ζ) or the form of an absurdity of a constructive property. E.g. for real numbers a and b the non-contradictory of $a = b$ is equivalent to $a = b$, and the non-contradictory of: *either $a = b$ or $a \circ > b$*, is equivalent to $a \geqslant b$; further the non-contradictory of $a \circ > b$ is equivalent to the absurdity of $a \leqslant b$ as well as to the absurdity of: *either $a = b$ or $a < \circ b$*.

On the other hand, if we think of the property of non-contradictority of rationality existing for all direct checking-numbers of drifts whose counting-numbers are rational, there seems to be little hope for reducing non-contradictority of rationality of a real number to a constructive property or to an absurdity of a constructive property.

If we understand by the *simple absurdity* of the property η the absurdity of η, and by the $(n+1)$-*fold absurdity* of η the absurdity of the n-fold absurdity of η, then a theorem established above expresses that *threefold absurdity is equivalent to simple absurdity*. And a corollary of this theorem is that *n-fold absurdity is equivalent to simple or to double absurdity according as n is odd or even*.

I should like to terminate here. I hope I have made clear that intuitionism on the one hand subtilizes logic, on the other hand denounces logic as a source of truth. Further that intuitionistic mathematics is inner architecture, and that research in foundations of mathematics is inner inquiry with revealing and liberating consequences, also in non-mathematical domains of thought.

The philosophical basis of intuitionistic logic

MICHAEL DUMMETT

The question with which I am here concerned is: What plausible rationale can there be for repudiating, within mathematical reasoning, the canons of classical logic in favour of those of intuitionistic logic? I am, thus, not concerned with justifications of intuitionistic mathematics from an eclectic point of view, that is, from one which would admit intuitionistic mathematics as a legitimate and interesting form of mathematics alongside classical mathematics: I am concerned only with the standpoint of the intuitionists themselves, namely that classical mathematics employs forms of reasoning which are not valid on any legitimate way of construing mathematical statements (save, occasionally, by accident, as it were, under a quite unintended reinterpretation). Nor am I concerned with exegesis of the writings of Brouwer or of Heyting: the question is what forms of justification of intuitionistic mathematics will stand up, not what particular writers, however eminent, had in mind. And, finally, I am concerned only with the most fundamental feature of intuitionistic mathematics, its underlying logic, and not with the other respects (such as the theory of free choice sequences) in which it differs from classical mathematics. It will therefore be possible to conduct the discussion wholly at the level of elementary number theory. Since we are, in effect, solely concerned with the logical constants – with the sentential operators and the first-order quantifiers – our interest lies only with the most general features of the notion of a mathematical construction, although it will be seen that we need to consider these in a somewhat delicate way.

Any justification for adopting one logic rather than another as the logic for mathematics must turn on questions of *meaning*. It would be impossible to contrive such a justification which took meaning for granted, and represented the question as turning on knowledge or certainty. We are certain of the truth of a statement when we have conclusive grounds for it and are certain that the grounds which we have *are* valid grounds for it and *are* conclusive. If classical arguments for mathematical statements are called in question, this cannot possibly be because

Reprinted with the kind permission of the author, the editors, and the publisher from *Proceedings of the Logic Colloquium, Bristol, July 1973,* H. E. Rose and J. C. Shepherdson, eds., North-Holland 1975, pp. 5–40.

it is thought that we are, in general, unable to tell with certainty whether an argument is classically valid, unless it is also intuitionistically valid: rather, it must be that what is being put in doubt is whether arguments which are valid by classical but not by intuitionistic criteria are absolutely valid, that is, whether they really do conclusively establish their conclusions as true. Even if it were held that classical arguments, while not in general absolutely valid, nevertheless always conferred a high probability on their conclusions, it would be wrong to characterise the motive for employing only intuitionistic arguments as lying in a desire to attain knowledge in place of mere probable opinion in mathematics, since the very thesis that the use of classical arguments did not lead to knowledge would represent the crucial departure from the classical conception, beside which the question of whether or not one continued to make use of classical arguments as mere probabilistic reasoning is comparatively insignificant. (In any case, within standard intuitionistic mathematics, there is no reason whatever why the existence of a classical proof of it should render a statement probable, since if, e.g., it is a statement of analysis, its being a classical theorem does not prevent it from being intuitionistically disprovable.)

So far as I am able to see, there are just two lines of argument for repudiating classical reasoning in mathematics in favour of intuitionistic reasoning. The first runs along the following lines. The meaning of a mathematical statement determines and is exhaustively determined by its *use*. The meaning of such a statement cannot be, or contain as an ingredient, anything which is not manifest in the use made of it, lying solely in the mind of the individual who apprehends that meaning: if two individuals agree completely about the use to be made of the statement, then they agree about its meaning. The reason is that the meaning of a statement consists solely in its rôle as an instrument of communication between individuals, just as the powers of a chess-piece consist solely in its rôle in the game according to the rules. An individual cannot communicate what he cannot be observed to communicate: if one individual associated with a mathematical symbol or formula some mental content, where the association did not lie in the use he made of the symbol or formula, then he could not convey that content by means of the symbol or formula, for his audience would be unaware of the association and would have no means of becoming aware of it.

The argument may be expressed in terms of the *knowledge* of meaning, i.e. of understanding. A model of meaning is a model of understanding, i.e. a representation of what it is that is known when an individual knows the meaning of a particular symbol or expression is frequently verbalisable knowledge, that is, knowledge which consists in the ability to state the rules in accordance with which the expression or symbol is used or the way in which it may be replaced by an equivalent expression or sequence of symbols. But to suppose that, in general, a knowledge of meaning consisted in verbalisable knowledge would involve an infinite regress: if a grasp of the meaning of an expression consisted, in general, in the ability to *state* its meaning, then it would be impossible for anyone to learn a language who was not already equipped with a fairly extensive language. Hence that knowledge which, in general, constitutes the understanding of the language of mathematics must be implicit knowledge. Implicit knowledge cannot, however, meaningfully be ascribed to someone unless it is possible to say in what the manifestation of that knowledge consists: there must be an observable difference between the behaviour or capacities of someone who is said to have that knowledge and someone who is said to lack it. Hence it follows, once more, that a grasp of the meaning of a mathematical statement must, in general, consist of a capacity to use that statement in a certain way, or to respond in a certain way to its use by others.

Another approach is via the idea of learning mathematics. When we learn a mathematical notation, or mathematical expressions, or, more generally, the language of a mathematical theory, what we learn to do is to make use of the statements of that language: we learn when they may be established by computation, and how to carry out the relevant computations, we learn from what they may be inferred and what may be inferred from them, that is, what rôle they play in mathematical proofs and how they can be applied in extra-mathematical contexts, and perhaps we learn also what plausible arguments can render them probable. These things are all that we are shown when we are learning the meanings of the expressions of the language of the mathematical theory in question, because they are all that we can be shown: and, likewise, our proficiency in making the correct use of the statements and expressions of the language is all that others have from which to judge whether or not we have acquired a grasp of their meanings. Hence it can only be in the capacity to make a correct use of the statements of the language that a grasp of their meanings, and those of the symbols and expressions which they contain, can consist. To suppose that there is an ingredient of meaning which transcends the use that is made of that which carries the meaning is to suppose that someone might have learned all that is directly taught when the language of a mathematical theory is taught to him, and might then behave in every way like someone who understood the language, and yet not actually understand it, or understand it only incorrectly. But to suppose this is to make meaning ineffable, that is, in principle incommunicable. If this is possible, then no one individual ever has a guarantee that he is understood by any other individual; for all he knows, or can ever know, everyone else may attach to his words or to the symbols which he employs a meaning quite different fom that which he attaches to them. A notion of meaning so private to the individual is one that has become completely irrelevant to mathematics as it is actually practised, namely as a body of theory on which many individuals are corporately engaged, an enquiry within which each can communicate his results to others.

It might seem that an approach to meaning which regarded it as exhaustively determined by use would rule out any form of revisionism. If use constitutes meaning, then, it might seem, use is beyond criticism: there can be no place for rejecting any established mathematical practice, such as the use of certain forms of argument or modes of proof, since that practice, together with all others which are generally accepted, is simply constitutive of the meanings of our mathematical statements, and we surely have the right to make our statements mean whatever we choose that they shall mean. Such an attitude is one possible development of the thesis that use exhaustively determines meaning: it is, however, one which can, ultimately, be supported only by the adoption of a holistic view of language. On such a view, it is illegitimate to ask after the content of any single statement, or even after that of any one theory, say a mathematical or a physical theory; the significance of each statement or of each deductively systematised body of statements is modified by the multiple connections which it has, direct and remote, with other statements in other areas of our language taken as a whole, and so there is no adequate way of understanding the statement short of knowing the entire language. Or, rather, even this image is false to the facts: it is not that a statement or even a theory has, as it were, a primal meaning which then gets modified by the interconnections that are established with other statements and other theories; rather, its meaning simply consists in the place which it occupies in the complicated network which constitutes the totality of our linguistic practices. The only thing to which a definite content may be attributed is the totality of all that we are, at a given time, prepared to assert; and there can be no simple model of the content which that totality of assertions embodies; nothing short of a complete knowledge of the language can reveal it.

Frequently such a holistic view is modified to the extent of admitting a class of observation statements which can be regarded as more or less directly registering our immediate experience, and hence as each carrying a determinate individual content. These observation statements lie, in Quine's famous image of language, at the periphery of the articulated structure formed by all the sentences of our language, where alone experience impinges. To these peripheral sentences, meanings may be ascribed in a more or less straightforward manner, in terms of the observational stimuli which prompt assent to and dissent from them. No comparable model of meaning is available for the sentences which lie further towards the interior of the structure: an understanding of them consists solely in a grasp of their place in the structure as a whole and their interaction with its other constituent sentences. Thus, on such a view, we may accept a mathematical theory, and admit its theorems as true, only because we find in practice that it serves as a convenient substructure deep in the interior of the complex structure which forms the total theory: there can be no question of giving a representation of the truth-conditions of the statements of the mathematical theory under which they may be judged individually as acceptable, or otherwise, in isolation from the rest of language.

Such a conception bears an evident analogy with Hilbert's view of classical mathematics; or, more accurately, with Boole's view of his logical calculus. For Hilbert, a definite individual content, according to which they may be individually judged as correct or incorrect, may legitimately be ascribed only to a very narrow range of statements of elementary number theory: these correspond to the observation statements of the holistic conception of language. All other statements of mathematics are devoid of such a content, and serve only as auxiliaries, though psychologically indispensable auxiliaries, to the recognition as correct of the finitistic statements which alone are individually meaningful. The other mathematical statements are not, on such a view, devoid of significance: but their significance lies wholly in the rôle which they play within the mathematical theories to which they belong, and which are themselves significant precisely because they enable us to establish the correctness of finitistic statements. Boole likewise distinguished, amongst the formulas of his logical calculus, those which were interpretable from those which were uninterpretable: a deduction might lead from some interpretable formulas as premisses, via uninterpretable formulas as intermediate steps, to a conclusion which was once more interpretable.

The immediately obvious difficulty about such a manner of construing a mathematical, or any other, theory is to know how it can be justified. How can we be sure that the statements or formulas to which we ascribe a content, and which are derived by such a means, are true? The difference between Hilbert and Boole, in this respect, was that Hilbert took the demand for justification seriously, and saw the business of answering it as the prime task for his philosophy of mathematics, while Boole simply ignored the question. Of course, the most obvious way to find a justification is to extend the interpretation to all the statements or formulas with which we are concerned, and, in the case of Boole's calculus, this is very readily done, and indeed yields a great simplification of the calculus. Even in Hilbert's case, the consistency proof, once found, does yield an interpretation of the infinitistic statements, though one which is relative to the particular proof in which they occur, not one uniform for all contexts. Without such a justification, the operation of the mechanism of the theory or the language remains quite opaque to us; and it is because the holist is oblivious of the demand for justification, or of the unease which the lack of one causes us, that I said that he is to be compared to Boole rather than to Hilbert. In his case, the question would become: With what right do we feel an assurance that the observation statements deduced with the help of the complex theories, mathematical, scientific and otherwise, embedded in the interior of the total linguistic structure, are true, when these observation statements are interpreted in terms of their stimulus meanings? To this the holist attempts no answer, save a generalised appeal to induction: these theories have 'worked' in the past, in the sense of having for the most part yielded true observation statements, and so we have confidence that they will continue to work in the future.

The path of thought which leads from the thesis that use exhaustively determines meaning to an acceptance of intuitionistic logic as the correct logic for mathematics is one which rejects a holistic view of mathematics and insists that each statement of any mathematical theory must have a determinate individual content. A grasp of this content cannot, in general, consist of a piece of verbalisable knowledge, but must be capable of being fully manifested by the use of the statement: but that does not imply that every aspect of its existing use is sacrosanct. An existing practice in the use of a certain fragment of language is capable of being subjected to criticism if it is impossible to systematise it, that is, to frame a model whereby each sentence carries a determinate content which can, in turn, be explained in terms of the use of that sentence. What makes it possible that such a practice may prove to be incoherent and therefore in need of revision is that there are different aspects to the use of a sentence; if the whole practice is to be capable of systematisation in the present sense, there must be a certain harmony between these different aspects. This is already apparent from the holistic examples already cited. One aspect of the use of observation statements lies in the propensities we have acquired to assent to and dissent from them under certain types of stimuli; another lies in the possibility of deducing them by means of non-observational statements, including highly theoretical ones. If the linguistic system as a whole is to be coherent, there must be harmony between these two aspects: it must not be possible to deduce observation statements from which the perceptual stimuli require dissent. Indeed, if the observation statements are to retain their status as observation statements, a stronger demand must be made: of an observation statement deduced by means of theory, it must hold that we can place ourselves in a situation in which stimuli occur which require assent to it. This condition is thus a demand that, in a certain sense, the language as a whole be a conservative extension of that fragment of the language containing only observation statements. In just the same way, Hilbert's philosophy of mathematics requires that classical number theory, or even classical analysis, be a conservative extension of finitistic number theory.

For utterances considered quite generally, the bifurcation between the two aspects of their use lies in the distinction between the conventions governing the occasions on which the utterance is appropriately made and those governing both the responses of the hearer and what the speaker commits himself to by making the utterance: schematically, between the *conditions for* the utterance and the *consequences of* it. Where, as in mathematics, the utterances with which we are concerned are *statements,* that is, utterances by means of which assertions can be effected, this becomes the distinction between the grounds on which the statement can be asserted and its inferential consequences, the conclusions that can be inferred from it. Plainly, the requirement of harmony between these in respect of some type of statement is the requirement that the addition of statements of that type to the language produces a conservative extension of the language; i.e., that it is not possible, by going via statements of this type as intermediaries, to deduce from premisses not of that type conclusions, also not of that type, which could not have been deduced before. In the case of the logical constants, a loose way of putting the requirement is to say that there must be a harmony between the introduction and elimination rules; but, of course, this is not accurate, since the whole system has to be considered (in classical logic, for example, it is possible to infer a disjunctive statement, say by double negation elimination, without appeal to the rule of disjunction introduction). An alternative way of viewing the dichotomy between the two principal aspects of the use of statements is as a contrast between *direct* and *indirect* means of establishing them. So far as a logically complex statement is concerned, the introduction rules governing the logical constants occurring in the statement display the most direct means of establishing the statement, step by step in accordance with its logical structure; but the statement may be accepted on the basis of a complicated deduction which relies also on elimination rules, and we require a harmony which obtains only if a statement that has been indirectly established always could (in some sense of 'could') have been established directly.

103

Here again the demand is that the admission of the more complex infer-ences yield a conservative extension of the language. When only intro-duction rules are used, the inference involves only statements of logical complexity no greater than that of the conclusion: we require that the derivation of a statement by inferences involving statements of greater logical complexity shall be possible only when its derivation by the more direct means is in some sense already possible.

On any molecular view of language – any view on which individual sentences carry a content which belongs to them in accordance with the way they are compounded out of their own constituents, independently of other sentences of the language not involving those constituents – there must be some demand for harmony between the various aspects of the use of sentences, and hence some possibility of criticising or rejecting existing practice when it does not display the required harmony. Exactly what the harmony is which is demanded depends upon the theory of meaning accepted for the language, that is, the general model of that in which the content of an individual sentence consists; that is why I ren-dered the above remarks vague by the insertion of phrases like 'in some sense'. It will always be legitimate to demand, of any expression or form of sentence belonging to the language, that its addition to the language should yield a conservative extension; but, in order to make the notion of a conservative extension precise, we need to appeal to some concept such as that of truth or that of being assertible or capable in principle of being established, or the like; and just which concept is to be selected, and how it is to be explained, will depend upon the theory of meaning that is adopted.

A theory of meaning, at least of the kind with which we are mostly familiar, seizes upon some one general feature of sentences (at least of assertoric sentences, which is all we need be concerned with when con-sidering the language of mathematics) as central: the notion of the con-tent of an individual sentence is then to be explained in terms of this central feature. The selection of some one such feature of sentences as central to the theory of meaning is what is registered by philosophical dicta of the form, 'Meaning is ...' – e.g., 'The meaning of a sentence is the method of its verification', 'The meaning of a sentence is determined by its truth-conditions', etc. (The slogan 'Meaning is use' is, however, of a different character: the 'use' of a sentence is not, in this sense, a *single* feature; the slogan simply restricts the *kind* of feature that may legiti-mately be appealed to as constituting or determining meaning.) The justification for thus selecting some one single feature of sentences as central – as being that in which their individual meanings consist – is that it is hoped that every other feature of the use of sentences can be derived, in a uniform manner, from this central one. If, e.g., the notion of truth is taken as central to the theory of meaning, then the meanings of individual expressions will consist in the manner in which they contribute to determining the truth-conditions of sentences in which they occur; but this conception of meaning will be justified only if it is possible, for an arbitrary assertoric sentence whose truth-conditions are taken as known, to describe, in terms of the notion of truth, our actual practice in the use of such a sentence; that is, to give a general characterisation of the linguistic practice of making assertions, of the conditions under which they are made and the responses which they elicit. Obviously, we are very far from being able to construct such a general theory of the use of sentences, of the practice of speaking a language; equally obviously, it is likely that, if we ever do attain such an account, it will involve a considerable modification of the ideal pattern under which the account will take a quite general form, irrespective of the individual content of the sentence as given in terms of whatever is taken as the central notion of the theory of meaning. But it is only to the extent that we shall eventually be able to approximate to such a pattern that it is possible to give substance to the claim that it is in terms of some *one* feature, such as truth or verification, that the individual meanings of sentences and of their component expressions are to be given.

It is the multiplicity of the different features of the use of sentences, and the consequent legitimacy of the demand, given a molecular view of language, for harmony between them, that makes it possible to criticise existing practice, to call in question uses that are actually made of sentences of the language. The thesis with which we started, that use exhaustively determines meaning, does not, therefore, conflict with a revisionary attitude to some aspect of language: what it does do is to restrict the selection of the feature of sentences which is to be treated as central to the theory of meaning. On a platonistic interpretation of a mathematical theory, the central notion is that of truth: a grasp of the meaning of a sentence belonging to the language of the theory consists in a knowledge of what it is for that sentence to be true. Since, in general, the sentences of the language will not be ones whose truth-value we are capable of effectively deciding, the condition for the truth of such a sentence will be one which we are not, in general, capable of recognising as obtaining whenever it obtains, or of getting ourselves into a position in which we can so recognise it. Nevertheless, on the theory of meaning which underlies platonism, an individual's grasp of the meaning of such a sentence consists in his knowledge of what the condition is which has to obtain for the sentence to be true, even though the condition is one which he cannot, in general, recognise as obtaining when it does obtain.

This conception violates the principle that use exhaustively determines meaning; or, at least, if it does not, a strong case can be put up that it does, and it is this case which constitutes the first type of ground which appears to exist for repudiating classical in favour of intuitionistic logic for mathematics. For, if the knowledge that constitutes a grasp of the meaning of a sentence has to be capable of being manifested in actual linguistic practice, it is quite obscure in what the knowledge of the condition under which a sentence is true can consist, when that condition is not one which is always capable of being recognised as obtaining. In particular cases, of course, there may be no problem, namely when the knowledge in question may be taken as verbalisable knowledge, i.e. when the speaker is able to *state,* in other words, what the condition is for the truth of the sentence; but, as we have already noted, this cannot be the general case. An ability to state the condition for the truth of a sentence is, in effect, no more than an ability to express the content of the sentence in other words. We accept such a capacity as evidence of a grasp of the meaning of the original sentence on the presumption that the speaker understands the words in which he is stating its truth-condition; but at some point it must be possible to break out of the circle: even if it were always possible to find an equivalent, understanding plainly cannot in general consist in the ability to find a synonymous expression. Thus the knowledge in which, on the platonistic view, a grasp of the meaning of a mathematical statement consists must, in general, be implicit knowledge, knowledge which does not reside in the capacity to state that which is known. But, at least on the thesis that use exhaustively determines meaning, and perhaps on any view whatever, the ascription of implicit knowledge to someone is meaningful only if he is capable, in suitable circumstances, of fully manifesting that knowledge. (Compare Wittgenstein's question why a dog cannot be said to expect that his master will come home next week.) When the sentence is one which we have a method for effectively deciding, there is again no problem: a grasp of the condition under which the sentence is true may be said to be manifested by a mastery of the decision procedure, for the individual may, by that means, get himself into a position in which he can recognise that the condition for the truth of the sentence obtains or does not obtain, and we may reasonably suppose that, in this position, he displays by his linguistic behaviour his recognition that the sentence is, respectively, true or false. But, when the sentence is one which is not in this way effectively decidable, as is the case with the vast majority of sentences of any interesting mathematical theory, the situation is different. Since the sentence is, by hypothesis, effectively undecidable, the condition which must, in general, obtain for it to be true is not one which we are capable of recog-

106

nising whenever it obtains, or of getting ourselves in a position to do so. Hence any behaviour which displays a capacity for acknowledging the sentence as being true in all cases in which the condition for its truth can be recognised as obtaining will fall short of being a full manifestation of the knowledge of the condition for its truth: it shows only that the condition can be recognised in certain cases, not that we have a grasp of what, in general, it is for that condition to obtain even in those cases when we are incapable of recognising that it does. It is, in fact, plain that the knowledge which is being ascribed to one who is said to understand the sentence is knowledge which transcends the capacity to manifest that knowledge by the way in which the sentence is used. The platonistic theory of meaning cannot be a theory in which meaning is fully determined by use.

If to know the meaning of a mathematical statement is to grasp its use; if we learn the meaning by learning the use, and our knowledge of its meaning is a knowledge which we must be capable of manifesting by the use we make of it: then the notion of *truth,* considered as a feature which each mathematical statement either determinately possesses or determinately lacks, independently of our means of recognising its truth-value, cannot be the central notion for a theory of the meanings of mathematical statements. Rather, we have to look at those things which are actually features of the use which we learn to make of mathematical statements. What we actually learn to do, when we learn some part of the language of mathematics, is to recognise, for each statement, what counts as establishing that statement as true or as false. In the case of very simple statements, we learn some computation procedure which decides their truth or falsity: for more complex statements, we learn to recognise what is to be counted as a proof or a disproof of them. That is the practice of which we acquire a mastery: and it is in the mastery of that practice that our grasp of the meanings of the statements must consist. We must, therefore, replace the notion of truth, as the central notion of the theory of meaning for mathematical statements, by the notion of *proof:* a grasp of the meaning of a statement consists in a capacity to recognise a proof of it when one is presented to us, and a grasp of the meaning of any expression smaller than a sentence must consist in a knowledge of the way in which its presence in a sentence contributes to determining what is to count as a proof of that sentence. This does not mean that we are obliged uncritically to accept the canons of proof as conventionally acknowledged. On the contrary, as soon as we construe the logical constants in terms of this conception of meaning, we become aware that certain forms of reasoning which are conventionally accepted are devoid of justification. Just because the conception of meaning in terms of proof is as much a molecular, as opposed to holistic, theory of meaning as that of meaning in terms of truth-conditions, forms of inference stand in need of justification, and are open to being rejected as unjustified. Our mathematical practice has been disfigured by a false conception of what our understanding of mathematical theories consisted in.

This sketch of one possible route to an account of why, within mathematics, classical logic must be abandoned in favour of intuitionistic logic obviously leans heavily upon Wittgensteinian ideas about language. Precisely because it rests upon taking with full seriousness the view of language as an instrument of social communication, it looks very unlike traditional intuitionist accounts, which, notoriously, accord a minimum of importance to language or to symbolism as a means of transmitting thought, and are constantly disposed to slide in the direction of solipsism. However, I said at the outset that my concern in this paper was not in the least with the exegesis of actual intuitionist writings: however little it may jibe with the view of the intuitionists themselves, the considerations that I have sketched appear to me to form one possible type of argument in favour of adopting an intuitionistic version of mathematics in place of a classical one (at least as far as the logic employed is concerned), and, moreover, an argument of considerable power. I shall not take the time here to attempt an evaluation of the argument, which would necessitate enquiring how the platonist might reply to it, and how the debate between them would then proceed: my interest lies, rather, in asking whether this is the only legitimate route to the adoption of an intuitionistic logic for mathematics.

Now the first thing that ought to strike us about the form of argument which I have sketched is that it is virtually independent of any considerations relating specifically to the *mathematical* character of the statements under discussion. The argument involved only certain considerations within the theory of meaning of a high degree of generality, and could, therefore, just as well have been applied to any statements whatever, in whatever area of language. The argument told in favour of replacing, as the central notion for the theory of meaning, the condition under which a statement is true, whether we know or can know when that condition obtains, by the condition under which we acknowledge the statement as conclusively established, a condition which we must, by the nature of the case, be capable of effectively recognising whenever it obtains. Since we were concerned with mathematical statements, which we recognise as true by means of a proof (or, in simple cases, a computation), this meant replacing the notion of truth by that of proof: evidently, the appropriate generalisation of this, for statements of an arbitrary kind, would be the replacement of the notion of truth, as the central notion of the theory of meaning, by that of verification; to know the meaning of a statement is, on such a view, to be capable of recognising whatever counts as verifying the statement, i.e. as conclusively establishing it as true. Here, of course, the verification would not ordinarily consist in the bare occurrence of some sequence of sense-experiences, as on the positivist conception of the verification of a statement. In the mathematical case, that which establishes a statement as true is the production of a deductive argument terminating in that statement as conclusion; in the general case, a statement will, in general, also be established as true by a process of reasoning, though here the reasoning will not usually be purely deductive in character, and the premisses of the argument will be based on observation; only for a restricted class of statements – the observation statements – will their verification be of a purely observational kind, without the mediation of any chain of reasoning or any other mental, linguistic or symbolic process.

It follows that, in so far as an intuitionist position in the philosophy of mathematics (or, at least, the acceptance of an intuitionistic logic for mathematics) is supported by an argument of this first type, similar, though not necessarily identical, revisions must be made in the logic accepted for statements of other kinds. What is involved is a thesis in the theory of meaning of the highest possible level of generality. Such a thesis is vulnerable in many places: if it should prove that it cannot be coherently applied to any one region of discourse, to any one class of statements, then the thesis cannot be generally true, and the general argument in favour of it must be fallacious. Construed in this way, therefore, a position in the philosophy of mathematics will be capable of being undermined by considerations which have nothing directly to do with mathematics at all.

Is there, then, any alternative defence of the rejection, for mathematics, of classical in favour of intuitionistic logic? Is there any such defence which turns on the fact that we are dealing with *mathematical* statements in particular, and leaves it entirely open whether or not we wish to extend the argument to statements of any other general class?

Such a defence must start from some thesis about mathematical statements the analogue of which we are free to reject for statements of other kinds. It is plain what this thesis must be: namely the celebrated thesis that mathematical statements do not relate to an objective mathematical reality existing independently of us. The adoption of such a view apparently leaves us free either to reject or to adopt an analogous view for statements of any other kind. For instance, if we are realists about the physical universe, then we may contrast mathematical statements with

statements ascribing physical properties to material objects: on this combination of views, material-object statements do relate to an objective
reality existing independently of ourselves, and are rendered true or
false, independently of our knowledge of their truth-values or of our
ability to attain such knowledge or the particular means, if any, by which
we do so, by that independently existing reality; the assertion that mathematical statements relate to no such external reality gains its substance by
contrast with the physical case. Unlike material objects, mathematical
objects are, on this thesis, creations of the human mind: they are objects
of thought, not merely in the sense that they can be thought about, but in
the sense that their being is to be thought of; for them, *esse est concipi.*

On such a view, a conception of meaning as determined by truth-
conditions is available for any statements which do relate to an independently existing reality, for then we may legitimately assume, of each such
statement, that it possesses a determinate truth-value, true or false, independently of our knowledge, according as it does or does not agree with
the constitution of that external reality which it is about. But, when the
statements of some class do not relate to such an external reality, the
supposition that each of them possesses such a determinate truth-value is
empty, and we therefore cannot regard them as being given meanings by
associating truth-conditions with them; we have, in such a case, *faute de
mieux,* to take them as having been given meaning in a different way,
namely by associating with them conditions of a different kind – conditions that we are capable of recognising when they obtain – namely,
those conditions under which we take their assertion or their denial as
being conclusively justified.

The first type of justification of intuitionistic logic which we considered conformed to Kreisel's dictum, 'The point is not the existence of
mathematical objects, but the objectivity of mathematical truth': it bore
directly upon the claim that mathematical statements possess objective
truth-values, without raising the question of the ontological status of
mathematical objects or the metaphysical character of mathematical
reality. But a justification of the second type violates the dictum: it
makes the question whether mathematical statements possess objective
truth-values depend upon a prior decision as to the being of mathematical objects. And the difficulty about it lies in knowing on what we
are to base the premiss that mathematical objects are the creations of
human thought in advance of deciding what is the correct model for the
meanings of mathematical statements or what is the correct conception
of truth as relating to them. It appears that, on this view, before deciding
whether a grasp of the meaning of a mathematical statement is to be considered as consisting in a knowledge of what has to be the case for it to be

110

true or in a capacity to recognise a proof of it when one is presented, we have first to resolve the metaphysical question whether mathematical objects – natural numbers, for example – are, as on the constructivist view, creations of the human mind, or, as on the platonist view, independently existing abstract objects. And the puzzle is to know on what basis we could possibly resolve this metaphysical question, at a stage at which we do not even know what model to use for our understanding of mathematical statements. We are, after all, being asked to choose between two metaphors, two pictures. The platonist metaphor assimilates mathematical enquiry to the investigations of the astronomer: mathematical structures, like galaxies, exist, independently of us, in a realm of reality which we do not inhabit but which those of us who have the skill are capable of observing and reporting on. The constructivist metaphor assimilates mathematical activity to that of the artificer fashioning objects in accordance with the creative power of his imagination. Neither metaphor seems, at first sight, especially apt, nor one more apt than the other: the activities of the mathematician seem strikingly unlike those either of the astronomer or of the artist. What basis can exist for deciding which metaphor is to be preferred? How are we to know in which respects the metaphors are to be taken seriously, how the pictures are to be used?

Preliminary reflection suggests that the metaphysical question ought not to be answered first: we cannot, as the second type of approach would have us do, *first* decide the ontological status of mathematical objects, and then, with that as premiss, deduce the character of mathematical truth or the correct model of meaning for mathematical statements. Rather, we have first to decide on the correct model of meaning – either an intuitionistic one, on the basis of an argument of the first type, or a platonistic one, on the basis of some rebuttal of it; and then one or other picture of the metaphysical character of mathematical reality will force itself on us. If we have decided upon a model of the meanings of mathematical statements according to which we have to repudiate a notion of truth considered as determinately attaching, or failing to attach, to such statements independently of whether we can now, or ever will be able to, prove or disprove them, then we shall be unable to use the picture of mathematical reality as external to us and waiting to be discovered. Instead, we shall inevitably adopt the picture of that reality as being the product of our thought, or, at least, as coming into existence only as it is thought. Conversely, if we admit a notion of truth as attaching objectively to our mathematical statements independently of our knowledge, then, likewise, the picture of mathematical reality as existing, like the galaxies, independently of our observation of it will force itself on us in an equally irresistible manner. But, when we approach the matter in this way, there is no puzzle over the interpretation of these metaphors: psychologically inescapable as they may be, their non-metaphorical content will consist entirely in the two contrasting models of the meanings of mathematical statements, and the issue between them will become simply the issue as to which of these two models is correct. If, however, a view as to the ontological status of mathematical objects is to be treated as a *premiss* for deciding between the two models of meaning, then the metaphors cannot without circularity be explained solely by reference to those models; and it is obscure how else they are to be explained.

These considerations appear, at first sight, to be reinforced by reflection upon Frege's dictum, 'Only in the context of a sentence does a name stand for anything'. We cannot refer to an object save in the course of saying something about it. Hence, any thesis concerning the ontological status of objects of a given kind must be, at the same time, a thesis about what makes a statement involving reference to such objects true, in other words, a thesis about what properties an object of that kind can have. Thus, to say that fictional characters are the creations of the imagination is to say that a statement about a fictional character can be true only if it is imagined as being true, that a fictional character can have only those properties which it is part of the story that he has; to say that something is an object of sense – that for it *esse est percipi* – is to say that it has only those properties it is perceived as having: in both cases, the ontological thesis is a ground for rejecting the law of excluded middle as applied to statements about those objects. Thus we cannot separate the question of the ontological status of a class of objects from the question of the correct notion of truth for statements about those objects, i.e. of the kind of thing in virtue of which such statements are true, when they are true. This conclusion corroborates the idea that an answer to the former question cannot serve as a premiss for an answer to the latter one.

Nevertheless, the position is not so straightforward as all this would make it appear. From the possibility of an argument of the first type for the use of intuitionistic logic in mathematics, it is evident that a model of the meanings of mathematical statements in terms of proof rather than of truth need not rest upon any particular view about the ontological character of mathematical objects. There is no substantial disagreement between the two models of meaning so long as we are dealing only with decidable statements: the crucial divergence occurs when we consider ones which are not effectively decidable, and the linguistic operation which first enables us to frame effectively undecidable mathematical statements is that of quantification over infinite totalities, in the first place over the totality of natural numbers. Now suppose someone who has, on whatever grounds, been convinced by the platonist claim that we do not create the natural numbers, and yet that reference to natural numbers is not a mere *façon de parler,* but is a genuine instance of reference to objects: he believes, with the platonist, that natural numbers are abstract objects, existing timelessly and independently of our knowledge of them. Such a person may, nevertheless, when he comes to consider the meaning of existential and universal quantification over the natural numbers, be convinced by a line of reasoning such as that which I sketched as constituting the first type of justification for replacing classical by intuitionistic logic. He may come to the conclusion that quantification over a denumerable totality cannot be construed in terms of our grasp of the conditions under which a quantified statement is true, but must, rather, be understood in terms of our ability to recognise a proof or disproof of such a statement. He will therefore reject a classical logic for number-theoretic statements in general, admitting only intuitionistically valid arguments involving them. Such a person would be accepting a platonistic view of the existence of mathematical objects (at least the objects of number theory), but rejecting a platonistic view of the objectivity of mathematical statements.

Our question is, rather, whether the opposite combination of views is possible: whether one may consistently hold that natural numbers are the creations of human thought, but yet believe that there is a notion of truth under which each number-theoretic statement is determinately either true or false, and that it is in terms of our grasp of their truth-conditions that our understanding of number-theoretic statements is to be explained. If such a combination is possible, then, it appears, there can be no route from the ontological thesis that mathematical objects are the creations of our thought to the model of the meanings of mathematical statements which underlies the adoption of an intuitionistic logic.

This is not the only question before us: for, even if these two views cannot be consistently combined, it would not follow that the ontological thesis could serve as a premiss for the constructivist view of the meanings of mathematical statements; our difficulty was to understand how the ontological thesis could have any substance if it were not merely a picture encapsulating that conception of meaning. The answer is surely this: that, while it is surely correct that a thesis about the ontological status of objects of a given kind, e.g. natural numbers, must be understood as a thesis about that in which the truth of certain statements about those objects consists, it need not be taken as, in the first place, a thesis about the entire class of such statements; it may, instead, be understood as a thesis only about some restricted subclass of such statements, those which are basic to the very possibility of making reference to those objects. Thus, for example, the thesis that natural numbers are creations of human thought may be taken as a thesis about the sort of thing which makes a numerical equation or inequality true, or, more generally, a statement formed from such equations by the sentential operators and bounded quantification. To say that the only notion of truth we can have for number-theoretic statements generally is that which equates truth with our capacity to prove a statement is to prejudge the issue about the correct model of meaning for such statements, and therefore cannot serve as a premiss for the constructivist view of meaning. But to say that, for decidable number-theoretic statements, truth consists in provability, is not in itself to prejudge the question in what the truth of undecidable statements, involving unbounded quantification, consists: and hence the possibility is open that a view about the one might serve as a premiss for a view about the other. Our problem is to discover whether it can do so in fact: whether there is any legitimate route from the thesis that natural numbers are creations of human thought, construed as a thesis about the sort of thing which makes decidable number-theoretic statements true, to a view of the meanings of number-theoretic statements generally which would require the adoption for them of an intuitionistic rather than a classical logic.

In order to resolve this question, it is necessary for us to take a rather closer look at the notion of truth for mathematical statements, as under-stood intuitionistically. The most obvious suggestion that comes to mind in this connection is that the intuitionistic notion of truth conforms, just as does the classical notion, to Tarski's schema:

(T) S is true iff A,

where an instance of the schema is to be formed by replacing 'A' by some number-theoretic statement and 'S' by a canonical name of that sentence, as, e.g., in:

'There are infinitely many twin primes' is true iff there are infinitely many twin primes.

It is necessary to admit counter-examples to the schema (T) in any case in which we wish to hold that there exist sentences which are neither true nor false: for if we replace 'A' by such a sentence, the left-hand side of the biconditional becomes false (on the assumption that, if the negation of a sentence is true, that sentence is false), although, by hypothesis, the right-hand side is not false. But, in intuitionistic logic, that semantic principle holds good which stands to the double negation of the law of excluded middle as the law of bivalence stands to the law of excluded middle itself: it is inconsistent to assert of any statement that it is neither true nor false; and hence there seems no obstacle to admitting the correctness of the schema (T). Of course, in doing so, we must construe the statement which appears on the right-hand side of any instance of the schema in an intuitionistic manner. Provided we do this, a truth-definition for the sentences of an intuitionistic language, say that of Heyting arithmetic, may be constructed precisely on Tarski's lines, and will yield, as a consequence, each instance of the schema (T).

However, notoriously, such an approach leaves many philosophical problems unresolved. The truth-definition tells us, for example, that

$$\text{`}598017 + 246532 = 844549\text{' is true}$$

just in the case in which $598017 + 246532 = 844549$. We may perform the computation, and discover that $598017 + 246532$ does indeed equal 844549: but does that mean that the equation was already true before the computation was performed, or that it would have been true even if the computation had never been performed? The truth-definition leaves such questions quite unanswered, because it does not provide for inflections of tense or mood of the predicate 'is true': it has been introduced only as a predicate as devoid of tense as are all ordinary mathematical predicates; but its rôle in our language does not reveal why such inflections of tense or even of mood should be forbidden.

These difficulties raise their heads as soon as we make the attempt to introduce tense into mathematics, as intuitionism provides us with some inclination to do; this can be seen from the problems surrounding the theory of the creative subject. These problems are well brought out in Troelstra's discussion of the topic. It is evident that we ought to admit as an axiom

(α) $\qquad\qquad\qquad (\vdash_n A) \rightarrow A;$

if we know that, at any stage, A has been (or will be) proved, then we are certainly entitled to assert A. But ought we to admit the converse in the form

(β) $\qquad\qquad\qquad A \rightarrow \exists n (\vdash_n A)?$

Its double negation

(γ) $\qquad\qquad\qquad A \rightarrow \neg\neg\exists n (\vdash_n A)$

is certainly acceptable: if we know that A is true, then we shall certainly never be able to assert, at least on purely mathematical grounds, that it will never be proved. But can we equate truth with the obtaining of a proof at some stage, in the past or in the future, as the equivalence:

(δ) $$A \leftrightarrow \exists n(\vdash_n A)$$

requires us to do? (To speak of 'truth' here seems legitimate, since, while Tarski's truth-predicate is a predicate of sentences, the sentential operator to which it corresponds is a redundant one, which can be inserted before or deleted from in front of any clause without change of truth-value.)

If we accept the axiom (β), and hence the equivalence (δ), we run into certain difficulties, on which Troelstra comments. The operator '$\exists n(\vdash_n \ldots)$' becomes a redundant truth-operator, and hence may be distributed across any logical constant, as in

(ε) $$(\vdash_k \forall m A(m)) \rightarrow \forall m \exists n(\vdash_n A(m)).$$

As Troelstra observes, this appears to have the consequence that, if we have once proved a universally quantified statement, we are in some way committed to producing, at some time in the future, individual proofs of all its instances, whereas, palpably, we are under no such constraint. The solution to which he inclines is that proposed by Kreisel, namely that the operator '$\vdash_n$' must be so construed that a proof, at stage n, of a universally quantified statement counts as being, at the same time, a proof of each instance, so that we could assert the stronger thesis

(ζ) $$(\vdash_k \forall m A(m)) \rightarrow \forall m(\vdash_k A(m)).$$

(Troelstra in fact recommends this interpretation on separate grounds, as enabling us to escape a paradox about constructive functions; he himself points out, however, that this paradox can alternatively be avoided by introducing distinctions of level which seem intrinsically plausible.) The difficulty about this solution is that it must be extended to every recognised logical consequence. From

(η) $$(m \leqslant n \ \& \ (\vdash_m A)) \rightarrow (\vdash_n A)$$

we have

(θ) $$(n = \max(m, k) \ \& \ (\vdash_m A) \ \& \ (\vdash_k C)) \rightarrow ((\vdash_n A) \ \& \ (\vdash_n C)),$$

while from (δ) we obtain

(ι) $$(\vdash_m A) \ \& \ (\vdash_k (A \rightarrow B)) \rightarrow \exists n(\vdash_n B).$$

We could in the same way complain that this committed us, whenever we had proved a statement A and had recognised some other statement B as being a consequence of A, to actually drawing that consequence some time in the future; and, if our interpretation of the operator '$\vdash_n$' is to be capable of dealing with this difficulty in the same way as with the special case of instances of a universally quantified statement, we should have to allow that a proof that a theorem had a certain consequence was, at the same time, a proof of that consequence, and, likewise, that a proof of a statement already known to have a certain consequence was, at the same time, a proof of that consequence; we should, that is, have to accept the law

(κ) $(n = \max(m, k) \,\&\, (\vdash_m A) \,\&\, (\vdash_k (A \to B))) \to (\vdash_n B)$.

We should thus have so to construe the notion of proof that a proof of a statement is taken as simultaneously constituting a proof of anything that has already been recognised as a consequence of that statement. We can, no doubt, escape having to say that it is simultaneously a proof of whatever, in a platonistic sense, is as a matter of fact an intuitionistic consequence of the statement: but when are we to be said to have recognised that one statement is a consequence of another? If a proof of a universally quantified statement is simultaneously a proof of all its instances, it is difficult to see how we can avoid conceding that a demonstration of the validity of a schema of first-order predicate logic is simultaneously a demonstration of the truth of all its instances, or an acceptance of the induction schema simultaneously an acceptance of all cases of induction. The resulting notion of proof would be far removed indeed from actual mathematical experience, and could not be explained as no more than an idealisation of it.

The trouble with all this is that, as a representation of actual mathematical experience, we are operating with too simplified a notion of proof. The axiom (η) is acceptable in the sense that, prescinding from the occasional accident, once a theorem has been proved, it always remains *available* to be subsequently appealed to: but the idea that, having acknowledged the two premisses of a modus ponens, we have *thereby* recognised the truth of the conclusion, is plausible only in a case in which we are simultaneously bearing in mind the truth of the two premisses. To have once proved a statement is not thereafter to be continuously aware of its truth: if it were, then we should indeed always know the logical consequences of everything which we know, and should have no need of proof.

Acceptance of axiom (β) leads to the conclusion that we shall eventually prove every logical consequence of everything we prove. This, as a representation of the intuitionist notion of proof, is an improvement upon Beth trees, as normally presented: for these are set up in such a way that, at any stage (node), every logical consequence of statements true at that stage is already true; the Beth trees are adapted only to situations, such as those involving free choice sequences, where new information is coming in that is not derived from the information we have at earlier stages. But the idea that we shall eventually establish every logical consequence of everything we know is implausible and arbitrary: and it cannot be rescued by construing each proof as, implicitly, a proof also of the consequences of the statement proved, save at the cost of perverting the whole conception. If we wish to do so, there seems no reason why we should not take the stages represented by the numerical subscripts as punctuated by proofs, however short the stages thereby become, and the notion of proof as relating only to what is quite explicitly proved, so that, at each stage, one and only new statement is proved, and consider what axioms hold under the resulting interpretation of the symbol '$\vdash_n$'. It thus appears that, under this interpretation, the axiom (β) must be rejected in favour of the weaker axiom (γ).

Looked at in another way, however, the stronger axiom (β) seems entirely acceptable. If, that is, we interpret the implication sign in its intuitionistic sense, the axiom merely says that, given a proof of A, we can effectively find a proof that A was proved at some stage; and this seems totally innocuous and banal. But, if axiom (β) is innocuous, how did we arrive at our earlier difficulties? The only possibility seems to be that our logical laws are themselves at fault. For instance, the law

$$(\lambda) \qquad \forall x A(x) \rightarrow A(m)$$

leads, via axiom (β), to the conclusion

$$(\mu) \qquad \forall x A(x) \rightarrow \exists n (\vdash_n A(m)),$$

which appears, on the present interpretation of '$\vdash_n$', to say that we shall explicitly prove every instance of every universally quantified statement which we prove; so perhaps the error lies in the law (λ) itself. A law such as (λ) is ordinarily justified by saying that, given a proof of $\forall x A(x)$, we can, for each m, effectively find a proof of $A(m)$. If this is to remain a sufficient justification of (μ), then (μ) must be construed as saying that, given a proof of $\forall x A(x)$, we can effectively find a proof that $A(m)$ will be proved at some stage. How can we do this, for given m? Obviously, by proving $A(m)$ and noting the stage at which we do so. This means, then, that the existentially quantified statement

$$(\nu) \qquad \exists n (\vdash_n A(m))$$

is to be so understood that its assertion does not amount to a claim that we shall, as a matter of fact, prove $A(m)$ at some stage n, but only that we are capable of bringing it about that $A(m)$ is proved at some stage.

118

Our difficulties thus appear to have arisen from understanding the existential quantifier in (β) in an excessively classical or realistic manner, namely as meaning that there will in fact be a stage n at which the statement is proved, rather than as meaning that we have an effective means, if we choose to apply it, of making it the case that there is such a stage. The point here is that it is not merely a question of interpreting the existential quantifier intuitionistically rather than classically in the sense that we can assert that there is a stage n at which a statement will be proved only if we have an effective means for identifying a particular such stage. Rather, if quantification over temporal stages is to be introduced into mathematical statements, then it must be treated like quantification over mathematical objects and mathematical constructions: the assertion that there is a stage n at which such-and-such will hold is justified provided that we possess an enduring capability of bringing about such a stage, regardless of whether we ever exercise this capability or not.

The confusions concerning the theory of the creative subject which we have been engaged in disentangling arose in part from a perfectly legitimate desire, to relate the intuitionistic truth of a mathematical statement with a use of the logical constants which is alien to intuitionistic mathematics. Troelstra's difficulties sprang from his desire to construe the expression '$\exists n(\vdash_n A)$' as meaning that A would in fact be proved at some stage: but, whether we interpret the existential quantifier classically or constructively, such a way of construing it fails to jibe with the way it and the other logical constants are construed within ordinary mathematical statements, and hence, however we try to modify our notion of a statement's being proved, we shall not obtain anything equivalent to the mathematical statement A itself. Nevertheless, the desire to express the condition for the intuitionistic truth of a mathematical statement in terms which do not presuppose an understanding of the intuitionistic logical constants as used within mathematical statements is entirely licit. Indeed, if it were impossible to do so, intuitionists would have no way of conveying to platonist mathematicians what it was that they were about: we should have a situation quite different from that which in fact obtains, namely one in which some people found it natural to extend basic computational mathematics in a classical direction, and others found it natural to extend it in an intuitionistic direction, and neither could gain a glimmering of what the other was at. That we are not in this situation is because intuitionists and platonists can find a common ground, namely statements, both mathematical and non-mathematical, which are, in the view of both, decidable, and about whose meaning there is therefore no serious dispute and which both sides agree obey a classical logic. Each party can, accordingly, by use of and reference to

119

these unproblematic statements, explain to the other what his conception of meaning is for those mathematical statements which are in dispute. Such an explanation may not be accepted as legitimate by the other side (the whole point of the intuitionist position is that undecidable mathematical statements cannot legitimately be given a meaning by laying down truth-conditions for them in the platonistic manner): but at least the conception of meaning held by each party is not wholly opaque to the other.

This dispute between platonists and intuitionists is a dispute over whether or not a realist interpretation is legitimate for mathematical statements: and the situation I have just indicated is quite characteristic for disputes concerning the legitimacy of a realist interpretation of some class of statements, and is what allows a *dispute* to take place at all. Typically, in such a dispute there is some auxiliary class of statements about which both sides agree that a realist interpretation is possible (depending upon the grounds offered by the anti-realists for rejecting a realist interpretation for statements of the disputed class, this auxiliary class may or may not consist of statements agreed to be effectively decidable); and, typically, it is in terms of the truth-conditions of statements of this auxiliary class that the anti-realist frames his conception of meaning, his non-classical notion of truth, for statements of the disputed class, while the realist very often appeals to statements of the auxiliary class as providing an analogy for his conception of meaning for statements of the disputed class. Thus, when the dispute concerns statements about the future, statements about the present will form the auxiliary class; when it concerns statements about material objects, the auxiliary class will consist of sense-data statements; when the dispute concerns statements about character-traits, the auxiliary class will consist of statements about actual or hypothetical behaviour; and so on.

If the intuitionistic notion of truth for mathematical statements can be explained only by a Tarski-type truth-definition which takes for granted the meanings of the intuitionistic logical constants, then the intuitionist notion of truth, and hence of meaning, cannot be so much as conveyed to anyone who does not accept it already, and no debate between intuitionists and platonists is possible, because they cannot communicate with one another. It is therefore wholly legitimate, and, indeed, essential, to frame the condition for the intuitionistic truth of a mathematical statement in terms which are intelligible to a platonist and do not beg any questions, because they employ only notions which are not in dispute.

The obvious way to do this is to say that a mathematical statement is intuitionistically true if there exists an (intuitionistic) proof of it, where the existence of a proof does not consist in its platonic existence in a realm outside space and time, but in our actual possession of it. Such a notion of truth, obvious as it is, already departs at once from that supplied by the analogue of the Tarski-type truth-definition, since the predicate 'is true', thus explained, is significantly tensed: a statement not now true may later become true. For this reason, when 'true' is so construed, the schema (T) is incorrect: for the negation of the right-hand side of any instance will be a mathematical statement, while the negation of the left-hand side will be a non-mathematical statement, to the effect that we do not as yet possess a proof of a certain mathematical statement, and hence the two sides cannot be equivalent. We might, indeed, seek to restore the equivalence by replacing 'is true' on the left-hand side by 'is or will be true': but this would lead us back into the difficulties we encountered with the theory of the creative subject, and I shall not further explore it.

What does require exploration is the notion of proof being appealed to, and that also of the existence of a proof. It has often, and, I think, correctly, been held that the notion of proof needs to be specialised if it is to supply a non-circular account of the meanings of the intuitionistic logical constants. It is possible to see this by considering disjunction and existential quantification. The standard explanation of disjunction is that a construction is a proof of $A \vee B$ just in case it is a proof either of A or of B. Despite this, it is not normally considered legitimate to assert a disjunction, say in the course of a proof, only when we actually have a proof of one or other disjunct. For instance, it would be quite in order to assert that

$$10^{10^{10}} + 1 \text{ is either prime or composite}$$

without being able to say which alternative held good, and to derive some theorem by means of an argument by cases. What makes this legitimate, on the standard intuitionist view, is that we have a method which is in principle effective for deciding which of the two alternatives is correct: if we were to take the trouble to apply this method, the appeal to an argument by cases could be dispensed with. Generally speaking, therefore, if we take a statement as being true only when we actually possess a proof of it, an assertion of a disjunctive statement will not amount to a claim that it is true, but only to a claim that we have a means, effective in principle, for obtaining a proof of it. This means, however, that we have to distinguish between a proof proper, a proof in the sense of 'proof' used in the explanations of the logical constants, and a cogent argument. In the course of a cogent argument for the assertibility of a mathematical statement, a disjunction of which we do not possess, an actual proof may be asserted, and an argument by cases based upon this disjunction. This argument will not itself be a proof, since any initial segment of a proof

121

must again be a proof: it merely indicates an effective method by which we might obtain a proof of the theorem if we cared to apply it. We thus appear to require a distinction between a proof proper – a canonical proof – and the sort of argument which will normally appear in a mathematical article or textbook, an argument which we may call a 'demonstration'. A demonstration is just as cogent a ground for the assertion of its conclusion as is a canonical proof, and is related to it in this way: that a demonstration of a proposition provides an effective means for finding a canonical proof. But it is in terms of the notion of a canonical proof that the meanings of the logical constants are given. Exactly similar remarks apply to the existential quantifier.

There is some awkwardness about this way of looking at disjunction and existential quantification, namely in the divorce between the notions of truth and of assertibility. It might be replied that the significance of the act of assertion is not, in general, uniquely determined by the notion of truth: for instance, even when we take the notion of truth for mathematical statements as given, it still needs to be stipulated whether the assertion of a mathematical statement amounts to a claim to have a proof of it, or whether it may legitimately be based on what Polya calls a 'plausible argument' of a non-apodictic kind. (We can imagine people whose mathematics wholly resembles ours, save that they do not construe an assertion as embodying a claim to have more than a plausible argument.) It nevertheless remains that, if the truth of a mathematical statement consists in our possession of a canonical proof of it, while its assertion need be based on possession of no more than a demonstration, we are forced to embrace the awkward conclusion that it may be legitimate to assert a statement even though it is *known* not to be true. Still, if the sign of disjunction and the existential quantifier were the only logical constants whose explanation appeared to call for a distinction between canonical proofs and demonstrations, the distinction might be avoided altogether by modifying their explanations, to allow that a proof of a disjunction consisted in any construction of which we could recognise that it would effectively yield a proof of one or other disjunct, and similarly for existential quantification: we should then be able to say that a statement could be asserted only when it was (known to be) true.

However, the distinction is unavoidable if the explanations of universal quantification, implication and negation are to escape circularity. The standard explanation of implication is that a proof of $A \rightarrow B$ is a construction of which we can recognise that, applied to any proof of A, it would yield a proof of B. It is plain that the notion of proof being used here cannot be one which admits unrestricted use of modus ponens: for, if it did, the explanation would be quite empty. We could admit anything we liked as constituting a proof of $A \rightarrow B$, and it would remain the case that, given such a proof, we had an effective method of converting any proof of A into a proof of B, namely by adding the proof of $A \rightarrow B$ and performing a single inference by modus ponens. Obviously, this is not what is intended: what is intended is that the proof of $A \rightarrow B$ should supply a means of converting a proof of A into a proof of B without appeal to modus ponens, at least, without appeal to any modus ponens containing $A \rightarrow B$ as a premiss. The kind of proof in terms of which the explanation of implication is being given is, therefore, one of a restricted kind. On the assumption that we have, or can effectively obtain, a proof of $A \rightarrow B$ of this restricted kind, an inference from $A \rightarrow B$ by modus ponens is justified, because it is in principle unnecessary. The same must, by parity of reasoning, hold good for any other application of modus ponens in the main (though not in any subordinate) deduction of any proof. Thus, if the intuitionistic explanation of implication is to escape, not merely circularity, but total vacuousness, there must be a restricted type of proof – canonical proof – in terms of which the explanation is given, and which does not admit modus ponens save in subordinate deductions. Arguments employing modus ponens will be perfectly valid and compelling, but they will, again, not be proofs in this restricted sense: they will be demonstrations, related to canonical proofs as supplying a means effective in principle for finding canonical proofs. Exactly similar remarks apply to universal quantification *vis-à-vis* universal instantiation and to negation *vis-à-vis* the rule ex falso quodlibet: the explanations of these operators presuppose a restricted type of proof in which the corresponding elimination rules do not occur within the main deduction.

What exactly the notion of a canonical proof amounts to is obscure. The deletion of elimination rules from a canonical proof suggests a comparison with the notion of a normalised deduction. On the other hand, Brouwer's celebrated remarks about fully analysed proofs in connection with the bar theorem do not suggest that such a proof is one from which unnecessary detours have been cut out – the proof of the bar theorem consists in great part in cutting out such detours from a proof taken already to be in 'fully analysed' form. Rather, Brouwer's idea appears to be that, in a fully analysed proof, all operations on which the proof depends will actually have been carried out. That is why such a proof may be an infinite structure: a proof of a universally quantified statement will be an operation which, applied to each natural number, will yield a proof of the corresponding instance; and, if this operation is carried out for each natural number, we shall have proofs of denumerably many statements. The conception of the mental construction which is the fully analysed proof as being an infinite structure must, of course, be interpreted in the light of the intuitionist view that all infinity is potential infinity: the mental construction consists of a grasp of general principles according to which any finite segment of the proof could be explicitly constructed. The direction of analysis runs counter to the direction of deduction; while one could not be convinced by an actually infinite proof-structure (because one would never reach the conclusion), one may be convinced by a potentially infinite one, because its infinity consists in our grasp of the principles governing its analysis. Indeed, it might reasonably be said that the standard intuition-istic meanings of the universal and conditional quantifiers involve that a proof is such a potentially infinite structure. Nevertheless, the notion of a fully analysed proof, that is, of the result of applying every operation involved in the proof, is far from clear, because it is obscure what the effect of the analysis would be on conditionals and negative statements. We can systematically display the results of applying the operation which constitutes a proof of a statement involving universal quantification over the natural numbers, because we can generate each natural number in sequence. But the corresponding application of the operation which con-stitutes the proof of a statement of the form $A \rightarrow B$ would consist in running through all putative canonical proofs of A and either showing, in each case, that it was not a proof of A, or transforming it into a proof of B; and, at least without a firm grasp upon the notion of a canonical proof, we have no idea how to generate all the possible candidates for being a proof of A.

The notion of canonical proof thus lies in some obscurity; and this state of affairs is not indefinitely tolerable, because, unless it is possible to find a coherent and relatively sharp explanation of the notion, the via-bility of the intuitionist explanations of the logical constants must remain in doubt. But, for present purposes, it does not matter just how the notion of canonical proof is to be explained; all that matters is that we require some distinction between canonical proofs and demonstrations, related to one another in the way that has been stated. Granted that such a distinction is necessary, there is no motivation for refusing to apply it to the case of disjunctions and existential statements.

Let us now ask whether we want the intuitionistic truth of a mathe-matical statement to consist in the existence of a canonical proof or of a demonstration. If by the 'existence' of a proof or demonstration we mean that we have actually explicitly carried one out, then either choice leaves us with certain counter-intuitive consequences. On either view, naturally, a valid rule of inference will not always lead from true premisses to a true conclusion, namely if we have not explicitly drawn the inference: this will always be so on any view which equates truth with our actual possession of some kind of proof. If we take the stricter line, and hold a statement to be true only when we possess a canonical proof of it, then, as we have seen, we shall have to allow that a statement may be asserted even though it is known not to be true. If, on the other hand, we allow that a statement is true when we possess merely a demonstration of it, then truth will not distribute over disjunction: we may possess a demonstration of $A \vee B$ without having a demonstration either of A or of B. Now, admittedly, once we have admitted a significant tense for the predicate 'is true', then, as we have noted, the schema (T) cannot be maintained as in all cases correct: but our instinct is to permit as little divergence from it as possible, and it is for this reason that we are uneasy about a notion of truth which is not distributive over disjunction or existential quantification.

A natural emendation is to relax slightly the requirement that a proof or demonstration should have been explicitly given. The question is how far we may consistently go along this path. If we say merely that a mathematical statement is true just in case we are aware that we have an effective means of obtaining a canonical proof of it, this will not be significantly different from equating truth with our actual possession of a demonstration. It might be allowed that there would be some cases when we had demonstrated the premisses of, say, an inference by modus ponens in which we were aware that we could draw the conclusion, though we had not quite explicitly done so; but there will naturally be others in which we were not aware of this, i.e. had not noticed it; if it were not so, we could never discover new demonstrations. It is therefore tempting to go one step further, and say that a statement is true provided that we are in fact in possession of a means of obtaining a canonical proof of it, whether or not we are aware of the fact. Would such a step be a betrayal of intuitionist principles?

In which cases would it be correct to say that we possess an effective means of finding a canonical proof of a statement, although we do not know that we have such a means? Unless we are to suppose that we can attain so sharp a notion of a canonical proof that it would be possible to enumerate effectively all putative such proofs of a given statement (the supposition whose implausibility causes our difficulty over the notion of a fully analysed proof), there is only one such case: that in which we possess a demonstration of a disjunctive or existential statement. Such a demonstration provides us with what we recognise as an effective means (in principle) for finding a canonical proof of the disjunctive or existential statement demonstrated. Such a canonical proof, when found, will be a proof of one or other disjunct, or of one instance of the existentially quantified statement: but we cannot, in general, tell which. For example, when $A(x)$ is a decidable predicate, the decision procedure constitutes a demonstration of the disjunction '$A(\bar{n}) \vee \neg A(\bar{n})$', for specific n; but, until we apply the procedure, we do not know which of the two disjuncts we can prove. It is very difficult for us to resist the temptation to suppose that there is already, unknown to us, a determinate answer to the question which of the two disjuncts we should obtain a proof of, were we to apply the decision procedure; that, for example, that it is already the case either that, if we were to test it out, we should find that $10^{10^{10}}+1$ is prime, or that, if we were to test it out, we should find that it was composite. What is involved here is the passage from a subjunctive conditional of the form:

$$A \rightarrow (B \vee C)$$

to a disjunction of subjunctive conditionals of the form

$$(A \rightarrow B) \vee (A \rightarrow C).$$

Where the conditional is interpreted intuitionistically, this transition is, of course, invalid: but the subjunctive conditional of natural language does not coincide with the conditional of intuitionistic mathematics. It is, indeed, the case that the transition is not in general valid for the subjunctive conditional of natural language either: but, when we reflect on the cases in which the inference fails, it is difficult to avoid thinking that the present case is not one of them.

There are two obvious kinds of counter-example to this form of inference for ordinary subjunctive conditionals: perhaps they are really two sub-varieties of a single type. One is the case in which the antecedent A requires supplementation before it will yield a determinate one of the disjuncts B and C. For instance, we may safely agree that, if Fidel Castro were to meet President Carter, he would either insult him or speak politely to him; but it might not be determinately true, of either of those things, that he would do it, since it might depend upon some so far unspecified further condition, such as whether the meeting took place in Cuba or outside. Schematically, this kind of case is one in which we can assert:

$$A \rightarrow (B \vee C),$$
$$(A \,\&\, Q) \rightarrow B,$$
$$(A \,\&\, \neg Q) \rightarrow C,$$

but in which the subjunctive antecedent A neither implies nor presupposes either Q or its negation; in such a case, we cannot assert either

126

$A \to B$ or $A \to C$. The other kind of counter-example is that in which we do not consider the disjuncts to be determined by anything at all: no supplementation of the antecedent would be sufficient to decide between them in advance. If that light-beam were to fall upon an atom, either it would assume a higher energy level, or it would remain in its ground state; but nothing can determine for certain in advance which would happen. Similar cases will arise, for those who believe in free will in the traditional sense, in respect of human actions.

If we were to carry out the decision procedure for determining the primality or otherwise of some specific large number N, we should either obtain the result that N is prime or obtain the result that N is composite. Is this, or is it not, a case in which we may conclude that it either holds good that, if we were to carry out the procedure, we should find that N is prime, or that, if we were to carry out the procedure, we should find that N is composite? The difficulty of resisting the conclusion that it is such a case stems from the fact that it does not display either of the characteristics found in the two readily admitted types of counter-example to the form of inference we are considering. No further circumstance could be relevant to the result of the procedure – this is part of what is meant by calling it a computation; and, since at each step the outcome of the procedure is determined, how can we deny that the overall outcome is determinate also?

If we yield to this line of thought, then we must hold that every statement formed by applying a decidable predicate to a specific natural number already has a definite truth-value, true or false, although we may not know it. And, if we hold this, it makes no difference whether we chose at the outset to say that natural numbers are creations of the human mind or that they are eternally existing abstract objects. Whichever we say, our decision how to interpret undecidable statements of number theory, and, in the first place, statements of the forms $\forall x A(x)$ and $\exists x A(x)$, where $A(x)$ is decidable, will be independent of our view about the ontological status of natural numbers. For, on this view of the truth of mathematical statements, each decidable number-theoretic statement will already be determinately true or false, independently of our knowledge, just as it is on a platonistic view; any thesis about the ontological character of natural numbers will then be quite irrelevant to the interpretation of the quantifiers. As we noted, it would be possible for someone to be prepared to regard natural numbers as timeless abstract objects, and to regard decidable predicates as being determinately true or false of them, and yet to be convinced by an argument of the first type, based on quite general considerations concerning meaning, that unbounded quantification over natural numbers was not an operation which in all cases preserved the property of possessing a determinate truth-value, and therefore to fall back upon a constructivist interpretation of it. Conversely, if someone who thought of the natural numbers as creations of human thought also believed, for the reasons just indicated, that each decidable predicate was determinately true or false of each of them, he might accept a classical interpretation of the quantifiers. He would do so if he was unconvinced by the general considerations about meaning which we reviewed, i.e., by the first type of argument for the adoption of an intuitionistic logic for mathematics: the fact that he was prepared to concede that the natural numbers come into existence only in virtue of our thinking about them would play no part in his reflections on the meanings of the quantifiers. Dedekind, who declared that mathematical structures are free creations of the human mind, but nevertheless appears to have construed statements about them in a wholly platonistic manner, may perhaps be an instance of just such a combination of ideas.

One who rejects the idea that there is already a determinate outcome for the application, to any specific case, of an effective procedure is, however, in a completely different position. If someone holds that the only acceptable sense in which a mathematical statement, even one that is effectively decidable, can be said to be true is that in which this means that we presently possess an actual proof or demonstration of it, then a classical interpretation of unbounded quantification over the natural numbers is simply unavailable to him. As is frequently remarked, the classical or platonistic conception is that such quantification represents an infinite conjunction or disjunction: the truth-value of the quantified statement is determined as the infinite sum or product of the truth-values of the denumerably many instances. Whether nor not this be regarded as an acceptable means of determining the meaning of these operators, the explanation presupposes that all the instances of the quantified statement themselves already possess determinate truth-values: if they do not, it is impossible to take the infinite sum or product of these. But if, for example, we do not hold that such a predicate as 'x is odd $\rightarrow x$ is not perfect' already has a determinate application to each natural number, though we do not know it, then it is just not open to us to think that, by attaching a quantifier to this predicate, we obtain a statement that is determinately true or false.

One question which we asked earlier was this: Can the thesis that natural numbers are creations of human thought be taken as a premiss for the adoption of an intuitionistic logic for number-theoretic statements? And another question was: What content can be given to the thesis that natural numbers are creations of human thought that does not prejudge the question what is the correct notion of truth for number-theoretic statements in general? The tentative answer which we gave to this latter question was that the thesis might be taken as relating to the appropriate notion of truth for a restricted class of number-theoretic statements, say numerical equations, or, more generally, decidable statements. From what we have said about the intuitionistic notion of truth for mathematical statements, it has now become apparent that there is one way in which the thesis that natural numbers are creations of the human mind might be taken, namely as relating precisely to the appropriate notion of truth for decidable statements of arithmetic, which would provide a ground for rejecting a platonistic interpretation of number-theoretic statements generally, without appeal to any general thesis concerning the notion of meaning. This way of taking the thesis would amount to holding that there is no notion of truth applicable even to numerical equations save that in which a statement is true when we have actually performed a computation (or effected a proof) which justifies that statement. Such a claim must rest, as we have seen, on the most resolute scepticism concerning subjunctive conditionals: it must deny that there exists any proposition which is now true about what the result of a computation which has not yet been performed would be if it were to be performed. Anyone who can hang on to a view as hard-headed as this has no temptation at all to accept a platonistic view of number-theoretic statements involving unbounded quantification: he has a rationale for an intuitionistic interpretation of them which rests upon considerations relating solely to mathematics, and demanding no extension to other realms of discourse (save in so far as the subjunctive conditional is involved in explanations of the meanings of statements in these other realms). But, for anyone who is not prepared to be quite as hard-headed as that, the route to a defence of an intuitionistic interpretation of mathematical statements which begins from the ontological status of mathematical objects is closed; the only path that he can take to this goal is that which I sketched at the outset: one turning on the answers given to general questions in the theory of meaning.

The concept of number

GOTTLOB FREGE

Each individual number is an independent object

55. Having recognized that a statement of number is an assertion about a concept, we can attempt to supplement the leibnizian definitions of the individual numbers by means of the definitions of 0 and of 1.

Right away we might say: the number 0 applies to a concept, if no object falls under that concept. Here, however, "no" appears to have been substituted for 0, with which it is synonymous. Therefore the following wording is preferable: the number 0 applies to a concept if, no matter what *a* might be, the statement always holds that *a* does not fall under this concept.

Similarly we could say: the number 1 applies to a concept *F* if it is not the case that no matter what *a* is, *a* does not fall under *F*, and if from the statement

'*a* falls under *F*' and '*b* falls under *F*'

it always follows that *a* and *b* are the same.

We must still define in general the transition from one number to the next. We will try the following formulation: the number $(n+1)$ applies to the concept *F* if there is an object *a* which falls under *F* and such that the number *n* applies to the concept "falling under *F* but not [identical with] *a*."

56. These definitions appear so natural, following our previous results, that an explanation is called for to show why they cannot satisfy us.

The last definition will most quickly arouse hesitation, for, strictly speaking, the sense of the expression 'the number *n* applies to the concept *G*' is just as unknown to us as that of the expression 'the number $(n+1)$ applies to the concept *F*'. To be sure, we can say by means of this and the next-to-last definition what

'the number $1+1$ applies to the concept *F*'

Translated by Michael S. Mahoney from Gottlob Frege, *Die Grundlagen der Arithmetik* (Breslau: 1884), pp. 67–104, 115–19.

means, and then, using this, indicate the sense of the expression

'the number $1+1+1$ applies to the concept F', etc.

But, to give a crude example, we can never decide by means of our definitions, whether the number *Julius Caesar* applies to a concept, whether this well-known conqueror of Gaul is a number or not. Furthermore, we cannot prove with the help of our attempted definitions that a must equal b if a applies to the concept F and b applies to the same concept. The expression '*the* number which applies to the concept F' would, therefore, not be justifiable, and it would consequently be completely impossible to prove a numerical equality because we could never isolate a definite number. It is only apparent that we have defined 0 and 1; as a matter of fact, we have only determined the sense of the expressions

'the number 0 applies to'

and

'the number 1 applies to';

but it is not permissible to isolate in these 0 and 1 as independent, recognizable objects.

57. Here is the place to examine somewhat more closely our statement that a statement of number involves an assertion about a concept. In the sentence 'the number 0 applies to the concept F', 0 is only a part of the predicate, if we consider the concept F as the actual subject. Therefore I have avoided calling numbers like 0, 1, 2 properties of concepts. The individual number appears as a separate independent object for the very reason that it forms only a part of the assertion. I have already called attention above to the fact that we say 'the [number] 1' and, by means of the definite article, set up 1 as an object.

This independence appears everywhere in arithmetic, e.g., in the equation '$1+1=2$'. Since the important thing here is to grasp the concept of number in such a way that it is useful for science, it needn't disturb us that in everyday usage the number appears attributively. This may always be avoided. E.g., the sentence 'Jupiter has four moons' may be rearranged to form 'The number of Jupiter's moons is four'. Here the 'is' is not to be considered merely a copula, as in the sentence 'the sky is blue'. This is shown by the fact that one can say 'the number of Jupiter's moons is four' or 'is the number four'. Here 'is' has the sense of 'is equal to', 'is the same as'. We have, therefore, an equation which asserts that the expression 'the number of Jupiter's moons' denotes the same object as the word 'four'. And the form of the equation is the reigning one in

131

arithmetic. The fact that nothing about Jupiter or about a moon is contained in the word 'four' is no objection to this interpretation. Neither is there anything in the name 'Columbus' to suggest discovery or America, and nonetheless the same man is called both Columbus and the discoverer of America.

58. One could object that we cannot at all represent[1] to ourselves the object which we call four or the number of Jupiter's moons as something separate and indepenent. However, it is not the separateness which we have given the number that is at fault. To be sure, one would like to believe that in picturing the four spots of a die something appears which corresponds to the word 'four' – but that is an illusion. Think of a green meadow and see whether the picture changes when the indefinite article is replaced by the number 'one'. Nothing is added, but there is certainly something in the picture corresponding to the word 'green'.

If one pictures for himself the printed word 'gold', one will not at first associate any number with it. Were one now to ask himself how many letters the word has, the result would be the number 4; the picture, however, will be in no way more definite, but can remain wholly unchanged. The added concept "letter of the word 'gold'" is the very thing in which we discover the number. In the case of the four spots of a die the situation is somewhat less obvious because the concept is forced upon us so directly by the similarity of the spots that we hardly notice its intrusion. The number can be *pictured* [translator's italics] neither as a separate object nor as a property of an outward thing, because it is neither something sensible nor the property of an outward thing. The situation is probably most clear in the sense of the number 0. One will try in vain to picture 0 visible stars. To be sure, one can think of the sky completely covered up by clouds; but there is nothing in this picture which might correspond to the word 'star' or to the 0. One is only imagining a situation in which one may conclude: now no star may be seen.

59. Perhaps each word awakens some sort of picture for us, even a word like 'only'. The picture, however, need not correspond to the content of the word; it can be an entirely different one for different men. One will then probably imagine a situation which evokes a sentence in which the word occurs; or the spoken word might call forth the written word in one's memory.

This does not occur only in the case of particles. There can be no doubt that we lack any idea [picture] of our distance from the sun. For, even if

[1]In the sense of 'picture'.

we know the rule about the number of times we must multiply a unit of measure, nevertheless any attempt by this rule to sketch a picture which even slightly approaches the one desired is doomed to fail. This is, however, no reason to doubt the correctness of the computation by which the distance has been found, and it in no way hinders us in basing further conclusions on this being the distance.

60. Even such a concrete thing as the earth we cannot picture in the way that we have learned it actually to be, but rather we are satisfied with a sphere of medium size, which serves us as a symbol for the earth, knowing nevertheless that the two are very different from one another. Now although our picture often does not at all meet the requirements, still we make judgments with great certainty about an object like the earth, even where its size is concerned.

Thought often leads us far beyond the imaginable without thereby depriving us of the basis for our conclusions. Even if, as it appears, thought without mental pictures is impossible for us men, still their connection with the object of thought can be wholly superficial, arbitrary, and conventional.

The unimaginability of the content of a word is no reason, then, to deny it any meaning or to exclude it from usage. That we are nevertheless inclined to do so is probably owing to the fact that we consider words individually and ask about their meaning [in isolation], for which we then adopt a mental picture. Thus a word for which we are lacking a corresponding inner picture will seem to have no content. However, we must always consider a complete sentence. Only in [the context of] the latter do the words really have a meaning. The inner pictures which somehow sway before us (in reading the sentence) need not correspond to the logical components of the judgment. It is enough if the sentence as a whole has a sense; by means of this its parts also receive their content.

This observation seems to me to be useful in throwing light on several difficult concepts, such as that of the infinitesimal,[2] and its scope is probably not limited to mathematics.

The separateness [independence] which I require for the number is not intended to mean that a number-word used outside of the context of a sentence shall denote anything, but rather I want only to exclude its use as a predicate or attribute, for such a use somewhat alters its meaning.

---

[2] What is in question here is defining the sense of an equation like

$$df(x) = g(x)dx$$

rather than finding an interval bounded by two distinct points and of length dx.

61. But, one might object, even if the earth is really unimaginable, still it is an external thing having a definite place. Where, however, is the number 4? It is neither outside of us nor inside of us. Taken in spatial terms, this is correct. A determination of the place of the number 4 makes no sense. But, from this it follows only that the number 4 is not a spatial object, not that it is no object at all. Not every object is somewhere. Even our mental pictures[3] are in this sense not in us (subcutaneously). In us there are ganglia cells, blood particles, etc., but no mental pictures. Spatial predicates are not applicable to them: the one is neither right nor left of the other. Mental pictures have no distances between them which may be stated in millimeters. When nevertheless we refer to them as in us, we mean that they are subjective.

Even if the subjective has no spatial location, however, how is it possible for the number 4, which is objective, to be nowhere? Now I maintain that there is no contradiction here. The number 4 is, as a matter of fact, exactly the same for everyone who works with it; but this has nothing to do with spatiality. Not every objective object has a place.

In order to obtain concept of number, one must determine the sense of a numerical equation

62. How shall we have a number, then, if we can have no idea or picture of it? Only in the context of a sentence do words have meaning. We must, therefore, define the sense of a sentence in which a number-word occurs. This seems at first to leave a lot of latitude, but we have already determined that number-words are to be understood as standing for independent objects. This already specifies a class of sentences which must have a sense, the class of those sentences which express the recognition [of a number as the same number]. If for us the symbol *a* is to denote an object, then we must have a criterion which determines in every case whether *b* is the same as *a*, even if it is not always within our power to apply this criterion. In our present case, we must explain the sense of the statement:

'the number which applies to the concept *F* is the same number as that which applies to the concept *G*',

i.e., we must reproduce the content of this statement in another way without using the expression

'the number which applies to the concept *F*'.

In doing this, we give a general criterion for the equality of numbers.

[3]This word is understood purely psychologically, not psychophysically.

Once we have obtained such a means of grasping a definite number and recognizing it as such, we can assign it a number-word as its proper name.

63. Hume (Baumann 1868–9, 2: 565) has already mentioned such a means: "If two numbers are so combined that the one always has a unit which corresponds to each unit of the other, then we claim they are equal." In more recent times, the opinion seems to have found much sympathy among mathematicians, that the equality of numbers must be defined in terms of a one-to-one correspondence. Immediately, however, there arise certain logical hesitations and difficulties, which we must not pass by without examination.

The relationship of equality does not hold only among numbers. It seems to follow from this that the relationship should be defined especially for numbers. One would think it possible to derive a criterion of when numbers are identical with one another from a previously determined concept of identity together with the concept of number, without its being necessary, for this purpose, to define a special concept of numerical identity.

Contrary to this, it should be noted that, for us, the concept of number has not yet been defined, but rather is to be determined by means of our definition of numerical identity. We intend to reconstruct the content of judgments interpretable as expressing identities each side of which is a number. We do not, therefore, want to define equality especially for this instance, but we wish rather, by means of the already familiar concept of equality, to determine that which is to be considered equal. This seems indeed to be a very unusual type of definition, which has probably not yet received sufficient attention from the logicians. Nevertheless, that it is not entirely unheard of may be shown by a few examples:

64. The judgment: 'the [straight] line *a* is parallel to the [straight] line *b*', or, symbolically:

$$a \parallel b,$$

can be interpreted as an equation. If we do this, we obtain the concept of direction and say: 'the direction of line *a* is the same as the direction of line *b*'. Hence, we replace the symbol '$\parallel$' by the more general '$=$', by distributing the particular content of the former to *a* and *b*. We split up the content in some way other than the original way and thus obtain a new concept. Often the situation is interpreted conversely, and several teachers define: parallel lines are those having the same direction. The theorem "if two straight lines are parallel to a third, then they are parallel to one another" can then be very easily proved on the basis of the similarly worded equality theorem. Unfortunately, this method reverses the natural order of things. For everything geometric must indeed be intuitive, at least originally. Now I ask whether anyone has ever had an intuition of the direction of a straight line? Of the straight line, yes, but can one also distinguish intuitively this line from its direction? Rather difficult! This concept is found only by means of a mental activity connected with intuition. On the other hand, one has a picture of parallel lines. That proof comes about only through a trick in which what is to be proved is covertly presupposed in the use of the word 'direction'; for, were the statement: 'if two straight lines are parallel to a third, then they are parallel to one another' false, then one could not change '$a \parallel b$' into an equation.

Thus one can obtain from the parallelism of planes a concept which corresponds to that of direction among straight lines. I have seen the name 'orientation' used for this concept. From geometric similarity there arises the concept of shape, so that, e.g., instead of 'the two triangles are similar', one says: 'the two triangles have the same shape' or 'the shape of the one triangle is equal to the shape of the other.' Similarly one can also obtain from the collinear relationship of geometric figures a concept for which a name is probably still lacking.

65. Now, in order to move, e.g., from parallelism[4] to the concept of direction, let us try the following definition: the sentence

'line a is parallel to line b'

is to be synonymous with

'the direction of line a is the same as the direction of line b'.

This definition departs from common practice insofar as it apparently defines the already familiar relation of equality, while it should in actuality introduce the expression 'the direction of line a', which occurs only incidentally. From this there arises a second hesitation; viz., whether, through such a stipulation, we could not become involved in contradictions with the familiar laws of equality. What are these? They will be developed as analytic truths from the concept itself. Now, Leibniz defines:[5]

---

[4]In order to be able to express myself more comfortably and to be more easily understood, I speak here of parallelism. The essential parts of these discussions are very easily carried over to the case of numerical equality.

[5]*Non inelegans specimen demonstrandi in abstractis* (Erdmann 1840: 94).

"Eadem sunt, quorum unum potest substitui alteri salva veritate."
["Things are equal which may be substituted for one another without change of truth [value]."]

I will adopt this definition. Whether, like Leibniz, one says 'the same'
or 'equal', is of little import. 'The same' does seem to express complete
agreement, 'equal' only agreement in this respect or that. One can, however, assume a manner of speaking in which this difference is eliminated,
e.g., by saying instead of 'the lines are equal in length' that 'the length of
the lines is equal' or 'the same'; instead of saying 'the surfaces are equal
in color' one might say 'the color of the surfaces is equal [identical]'.

And this is the way we used the word in the foregoing examples. In
fact, all the laws of equality are contained in the principle of universal
substitutivity.

In order to justify our proposed definition of the direction of a straight
line, we would have to show, then, that

'the direction of *a*'

can be everywhere replaced by

'the direction of *b*',

if line *a* is parallel to line *b*. This is simplified by the fact that, at first, we
know no assertion about the direction of a straight line other than its
agreement with the direction of another straight line. We would therefore
need to demonstrate only the substitutivity in such an equation or in contexts which would contain such equations as component parts.[6] All other
statements about directions would have to be defined first, and for these
definitions we can adopt the rule that the substitutivity of the direction of
a straight line for that of one parallel to it must be preserved.

66. Still a third hesitation arises, however, concerning our proposed definition. In the sentence

'the direction of *a* is equal to the direction of *b*',

the direction of *a* appears as an object,[7] and we have in our definition a
means of recognizing this object, should it appear in some other guise,

[6]For example, in a hypothetical judgment an equality of directions could occur either as
antecedent or as consequent.

[7]The definite article points to this. A concept is for me a possible predicate in a singular
thought content, an object a possible subject of the latter. [Although the terminology of
"thought contents" has been adopted, Frege must not be taken to mean anything psychological by 'thought'. For Frege a "thought content" is what is asserted in a statement,
asked in a question, etc....] If, in the sentence 'the direction of the axis of the telescope is
equal to the direction of the earth's axis', we consider the direction of the telescope's axis to

137

such as the direction of *b*. However, this method is not sufficient for all cases. One cannot use it to decide whether England is the same as the direction of the earth's axis. Please excuse this apparently nonsensical example! Naturally, no one is going to confuse England with the direction of the earth's axis; but this is not owing to our definition. The latter says nothing about whether the statement

'the direction of *a* is equal to *q*'

is to be affirmed or denied, if *q* itself is not given in the form 'the direction of *b*'. We lack the concept of direction; for, if we had this, then we could stipulate that, if *q* is not a direction, then our statement is to be denied; if *q* is a direction, then the earlier definition decides. It is now but a step away to define:

q is a direction if there is a straight line *b* whose direction is *q*.

However, it is clear that we have now come around in a circle. In order to apply this definition, we would first have to know in each case whether the statement

'*q* is equal to the direction of *b*'

was to be affirmed or denied.

67. If we were to say: *q* is a direction if it is introduced by means of the foregoing definitions, then we would be treating the manner by which the object *q* is introduced as a property of it, which it is not. The definition of an object, as such, really says nothing about that object; rather it stipulates the meaning of a symbol. Once that has happened, the definition becomes a judgment which treats of the object: it now no longer introduces the object but stands on equal footing with other statements about it. To choose this way out is to presuppose that an object could be given in one way only; otherwise it would not follow from the fact that *q* is not introduced by means of our definition that it could not be so introduced. The import of any equation would then be that what is given us in the same way should be recognized as the same. But this principle is so obvious and so unfruitful that there is little to be gained by stating it. As a matter of fact, no conclusion could be drawn from it which would not be the same as some premise. The many-sided and broad applicability of equations is based rather on the fact that something is recognizable again even though it is given in a different way.

be the subject, then the predicate is 'equal to the direction of the earth's axis'. This is a concept. But the direction of the earth's axis is only a part of the predicate; the direction is an object, since it can also be made the subject.

68. Since this method fails to yield a sharply delimited concept of direction and, for the same reason, would yield no such concept of number, let us try a different tack. If line *a* is parallel to line *b*, then the extension of the concept "line parallel to line *a*" is the same as the extension of the concept "line parallel to line *b*"; and conversely: if the extensions of the aforementioned concepts are equal, then *a* is parallel to *b*. Let us try, then, to define:

> the direction of line *a* is the extension of the concept "parallel to line *a*"
>
> the shape of triangle *d* is the extension of the concept "similar to triangle *d*."

If we want to apply this to our case, then we must substitute concepts for the lines or the triangles and, for parallelism or similarity, the possibility of correlating in one-to-one fashion the objects falling under the one concept with those falling under the other. As an abbreviation, I will call the concept *F* equinumerous[8] with the concept *G*, if this possibility exists; I must, however, request that this word be considered an arbitrarily chosen notational device whose meaning is not to be taken from its linguistic composition, but rather from the foregoing definition.

I define accordingly:

> the number which applies to the concept *F* is the extension[9] of the concept "equinumerous with the concept *F*."

69. That this definition is correct will, at first perhaps, not be so clear. Don't we mean something other than [different from] a number by the extension of a concept? What we do mean becomes clear from the basic statements that can be made about extensions of concepts. They are the following:

[8][Frege coined 'gleichzählig' for this. In his translation, J. L. Austin (Frege 1950) uses 'equal' and adds the following footnote: "*Gleichzählig* – an invented word, literally 'identinumerate' or 'tautarithmic'; but these are too clumsy for constant use. Other translators have used 'equinumerous'; 'equinumerate' would be better. Later writers have used 'similar' in this connection (but as a predicate of 'class' not of 'concept')." – Tr.]

[9]I think we could say for 'extension of the concept' simply 'concept'. However, there might be two objections:

1. This stands in contradiction to my earlier assertion that the individual number is an object, the latter being indicated by the use of the article in expressions like "the 2," by the impossibility of speaking about ones, twos, etc. in the plural, and by the fact that the number makes up only a part of the predicate of a statement of number.

2. Concepts can have the same extension without coinciding.

Now I am of the opinion that both these objections can be met, but doing this would lead us too far astray. I presuppose that one knows what the extension of a concept is.

1. that they are equal,
2. that the one encompasses more than the other.

Now the statement

> 'the extension of the concept "equinumerous with the concept F" is the same as the extension of the concept "equinumerous with the concept G"'

is true if and only if the statement

> 'the same number applies to the concept F as to the concept G'

is also true. Hence, there is complete agreement here.

To be sure, one does not say that one number encompasses more than another in the same sense that the extension of one concept encompasses more than does another; however, so is it impossible that

> the extension of the concept "equinumerous with the concept F"

should encompass more than

> the extension of the concept "equinumerous with the concept G"

Rather, if all concepts which are equinumerous with G are also equinumerous with F, then conversely, all concepts which are equinumerous with F are also equinumerous with G. This term 'more encompassing' should not, of course, be confused with the term 'greater', which occurs among numbers.

Certainly, it is also imaginable that the extension of the concept "equinumerous with the concept F" might encompass more or less than the extension of another concept; the latter, then, could not be a number according to our definition. Furthermore, it is not usual to call a number more or less encompassing than the extension of a concept. Nonetheless, there is nothing in the way of so speaking should the occasion arise.

Completion and confirmation of our definition

70. Definitions are confirmed by their fruitfulness. Those definitions which could just as easily be left out without invalidating proofs should be discarded as wholly worthless.

Let us see, then, whether some of the familiar properties of numbers can be derived from our definition of the number which applies to the concept F. We will be satisfied here by the most simple properties.

In order to do this, it is necessary to specify somewhat more exactly the meaning of equinumerosity. We defined it in terms of one-to-one correlation; just how I want to understand this expression must now be explained, since one might easily suspect a connection with intuition.

Let us consider the following example: If a waiter wants to be sure that he is placing just as many knives as plates on the table, he need count neither of them if he places a knife immediately to the right of each plate so that each knife on the table is located to the immediate right of a plate. The plates and knives are thus correlated in one-to-one fashion with one another, in this case through the same positional relationship. If, in the sentence

'α lies immediately to the right of A'

we imagine all sorts of objects substituted for α and A, then the part of the content which remains unchanged through all this forms the essence of the relation. Let us generalize this:

When, from a thought content which concerns an object a and an object b, we remove a and b, we retain the concept of a relation, which, accordingly, requires supplementation in two places. If, in the statement

'the earth has more mass than the moon',

we remove "the earth," then we obtain the concept "having more mass than the moon." If, on the other hand, we remove the object, "the moon," we gain the concept "having less mass than the earth." Removing both at once leaves a relational concept, which has in itself no more meaning than a simple concept, and which must be supplemented to become a thought content. But this supplementation can come about in various ways: instead of the earth and moon, I can take, e.g., the sun and earth, thus also effecting a removal of the earth and moon [and disclosing the relational nature of the concept].

The individual pairs of associated objects are related – one might say as subjects – to the relational concept in a manner similar to that of the individual object and the concept under which it falls. The subject here is a composite. At times, when the relation is a reversible one [symmetric in two argument places], this is also expressed linguistically, as in the sentence 'Peleus and Thetis were the parents of Achilles'.[10]

On the other hand, it would not be possible to reformulate the statement 'the earth is greater than the moon' so as to make 'the earth and the moon' appear as a compound subject, because the 'and' always indicates a certain equality of rank. This, however, does not affect the matter at hand.

The concept of relation, like the simple concept, belongs, then, to pure

[10]Do not confuse this with the case where the 'and' only seemingly connects the subjects, but in reality, however, connects two sentences.

logic. The particular content of the relation does not concern us here, but only its logical form. And [the truth of] whatever can be asserted about this form is analytic and is known *a priori*. This holds for the relational concepts as well as for the others.

Just as

'a falls under the concept F'

is the general form of a thought content concerning the object a, so can

'a stands in the relation ϕ to b'

be taken as the general form of a thought content concerning objects a and b.

71. Now if each object which falls under the concept F stands in the relation ϕ to an object falling under the concept G, and if, for each object which falls under G, there is an object falling under F which stands in the relation ϕ to it, then the objects falling under F and G are correlated with one another by means of the relation ϕ.

We may still ask what the expression

'each object which falls under F stands in the relation ϕ to an object falling under G'

means, if no object at all falls under F. By this I mean that the two statements

'a falls under F'

and

'a does not stand in the relation ϕ to any object falling under G'

cannot stand together, no matter what a denotes, so that either the first or the second or both are false. From this it follows that if there is no object falling under F, then "each object which falls under F stands in the relation ϕ to an object falling under G," because the first statement

'a falls under F'

is always to be denied, no matter what a might be.

Thus

'for each object which falls under G, there is an object falling under F which stands in the relation ϕ to it'

means that the two statements

'*a* falls under *G*'

and

'no object falling under *F* stands in the relation ϕ to *a*'

cannot stand together, whatever *a* may be.

72. We have now seen when the objects falling under the concepts *F* and *G* are correlated with one another by means of the relation ϕ. This correlation is here supposed to be one-to-one. By that I mean that the following two statements must hold:

1. If *d* stands in the relation ϕ to *a*, and if *d* stands in the relation ϕ to *e*, then, no matter what *d*, *a*, and *e* may be, *a* is always the same as *e*.
2. If *d* stands in the relation ϕ to *a*, and if *b* stands in the relation ϕ to *a*, then, whatever *d*, *b*, and *a* may be, *d* is always the same as *b*.

By these statements we have reduced one-to-one correlations to purely logical terms and can now offer the following definition:

the expression
    'the concept *F* is equinumerous with the concept *G*'
is to be synonymous with the expression
    'there is a relation ϕ which correlates in one-to-one fashion the objects falling under *F* with the objects falling under *G*'.

I [now] repeat [our original definition]:

the number which applies to the concept *F* is the extension of the concept "equinumerous with the concept *F*,"

and add to it:

the expression:
    '*n* is a number'
is to be synonymous with the expression
    'there is a concept to which the number *n* applies'.

Thus the concept of number is defined, apparently by means of itself, nevertheless without fallacy, because 'the number which applies to the concept *F*' has already been defined.

73. We want to show next, then, that the number which applies to the concept F is equal to the number which applies to the concept G, if the concept F is equinumerous with the concept G. This sounds like a tautology, but it is not, since the meaning of the word 'equinumerous' does not follow from its (linguistic) composition, but rather from the foregoing definition.

According to our definition, we must show that the extension of the concept "equinumerous with the concept F" is the same as that of the concept "equinumerous with the concept of G," if the concept F is equinumerous with the concept G. In other words, it must be shown that, under this hypothesis, the following statements always hold:

'if the concept H is equinumerous with the concept F, then it is also equinumerous with the concept G';

and

'if the concept H is equinumerous with the concept G, then it is also equinumerous with the concept F'.

The upshot of the first statement is that there is a relation which correlates in one-to-one fashion the objects falling under the concept H with those falling under the concept G, if there is a relation ϕ which correlates one-to-one the objects falling under the concept F with those falling under the concept G, and if there is a relation ψ which correlates one-to-one the objects falling under the concept H with those falling under the concept F. The following arrangement of the letters will make this easier to see

$$H\psi F\phi G.$$

Such a relation can in fact be given: it is [that] part of the thought content:

"there is an object to which c stands in the relation ψ and which stands in the relation ϕ to b"

[which remains] if we remove from it c and b (considering them as the things related). It can be shown that this relation is one-to-one and that it correlates the objects falling under the concept H with those falling under the concept G.

In a similar manner, the other theorem can also be proved.[11] Hopefully, these outlines will suffice to demonstrate that we need not borrow

[11]Similarly for its converse: If the number which applies to the concept F is the same as that which applies to the concept G, then the concept F is equinumerous with the concept G.

here any evidence from intuition, and that something may be done with our definitions.

74. We can now go on to the definitions of the individual numbers.
Because nothing falls under the concept "unequal to itself," I define:

0 is the number which applies to the concept "unequal to itself."

Perhaps someone will take exception to my speaking about a concept here. He will perhaps object that a contradiction is contained therein and will recall the old stand-bys, wooden iron and the square circle. To my mind, these are not at all as bad as they are made out to be. Of course, they are not exactly useful, but they can't do any harm, either, as long as one doesn't require that something fall under them; and *that* one does not yet do through the mere usage of the concepts. That a concept contains a contradiction is not always obvious without some examination; but to do that, one must have [the concept] and treat it logically just like any other. All that can be demanded of a concept from the point of view of logic and for rigor in proof procedure is its precise delineation; that, for each object, it be determined whether or not it falls under the concept. This requirement is fully satisfied, then, by concepts containing a contradiction, such as "unequal to itself," for it is known of every object that it does not fall under such a concept.[12]

I use the word 'concept' in such a way that

'*a* falls under the concept *F*'

is the general form of a thought content, which concerns an object *a* and which remains decidable, whatever one may put for *a*. And in this sense,

'*a* falls under the concept "unequal to itself"'

is synonymous with

'*a* is unequal to itself'

or

'*a* is unequal to *a*'.

In defining 0, I could have taken any other concept under which nothing

---

[12]Completely different from this is the definition of an object in terms of a concept under which it falls. The expression 'the greatest proper fraction' has, for example, no content, because the definite article carries with it the requirement that it refer to a definite object. On the other hand, the concept, "fraction which is less than 1 and has the property that no fraction which is less than 1 exceeds it in magnitude," is wholly unobjectionable. In fact, in order to prove that there is no such fraction, one even needs this concept, even though it contains a contradiction.

falls. It was up to me, however, to choose one of which this could be purely logically proved, and for this purpose "unequal to itself" presented itself most comfortably, whereby I let the previously presented definition of Leibniz hold, which is also purely logical.

75. We must now be able to prove, by means of what has already been said, that every concept under which nothing falls is equinumerous with any other concept under which nothing falls, and only with such a concept; from which it follows that 0 is the number which applies to such a concept and that no object falls under a concept if the number which applies to that concept is 0.

If we assume that no object falls either under the concept F or under the concept G, then, in order to prove that they are equinumerous, we need a relation ϕ about which the following statements hold:

'each object which falls under F stands in the relation ϕ to an object which falls under G; for each object which falls under G there is one falling under F which stands in the relation ϕ to it'.

According to what was said earlier about the meaning of these expressions, every relation fulfills these conditions under our hypotheses; hence also equality, which is, moreover, one-to-one. For, both the foregoing statements required of it hold.

If, on the other hand, an object falls under G, e.g., a, whereas none falls under F, then the two statements

'a falls under G'

and

'no object falling under F stands in the relation ϕ to a'

hold for every relation ϕ; for, the first holds true according to the first assumption, and the second, according to the second. That is, if there is no object falling under F, then there is also none which would stand in any sort of relation to a. There is, therefore, no relation which would, according to our definition, correlate the objects falling under F with those falling under G; accordingly, the concepts F and G are not equinumerous.

76. I want now to define the relation in which any two adjoining members of the series of natural numbers stand to one another. The statement

'there is a concept F and an object x falling under it such that the number which applies to the concept F is n, and that the number which applies to the concept "falling under F but not identical with x" is m',

is to be synonymous with

'n immediately follows m in the series of natural numbers'.

I am avoiding the expression 'n is *the* number immediately following m', because two theorems would first have to be proved in order to justify the use of the definite article.[13] For the same reason, I am not yet saying here '$n = m + 1$'; for, by means of the equals sign, $(m + 1)$ is also designated as an object.

77. Now in order to arrive at the number 1, we must first show, that there is something which immediately follows 0 in the series of natural numbers.

Let us consider the concept – or, if you prefer – the predicate 'equal to 0'. 0 falls under this. On the other hand, no object falls under the concept "equal to 0 but not equal to 0," so that 0 is the number which applies to this concept. We have therefore, a concept "equal to 0" and an object 0 falling under it, for which it holds that:

the number which applies to the concept "equal to 0" is equal to the number which applies to the concept "equal to 0";

the number which applies to the concept "equal to 0 but not equal to 0" is 0.

Therefore, according to our definition, the number which applies to the concept "equal to 0" follows immediately after 0 in the series of natural numbers.

If we define, then,

1 is the number which applies to the concept "equal to 0,"

then we can express the last statement so:

1 immediately follows 0 in the series of natural numbers.

Perhaps it is not superfluous to note that the definition of 1 does not presuppose any observed fact[14] for its objective legitimacy, for one can easily be confused by the fact that certain subjective conditions must be fulfilled in order to enable us to give the definition, and that sense impressions cause us to do so (cf. Erdmann 1877: 164). This can, nevertheless, be the case without the derived theorems ceasing to be *a priori*. To such conditions belongs the requirement, for example, that blood

[13]See footnote 12.
[14]A proposition that is not general.

flow through the brain in sufficient quantity and of the right concentration – at least as far as we know; however, the truth of our last proposition is independent of that; it continues to hold even if this flow no longer takes place. And even if all reasonable creatures should at some time simultaneously slip into hibernation, the truth of the statement would not, as it were, be suspended for the duration of this sleep, but would remain undisturbed. The truth of a statement is not its being thought.

78. I list here several theorems to be proved by means of our definitions. The reader will easily see how this may be done.

I. If a immediately follows 0 in the series of natural numbers, then $a=1$.

II. If 1 is the number which applies to a concept, then there is an object which falls under that concept.

III. If 1 is the number which applies to a concept F; if the object x falls under the concept F, and if y falls under the concept F, then $x=y$; i.e., x is the same as y.

IV. If an object falls under a concept F and if, from the fact that x falls under the concept F and that y falls under the concept F, it may always be inferred that $x=y$, then 1 is the number which applies to the concept F.

V. The relation that m bears to n, if and only if

"n immediately follows m in the series of natural numbers",

is a one-one relation.

Thus far it has not yet been said that for every number there is another which immediately follows it or is immediately followed by it in the series of natural numbers.

VI. Every number except 0 immediately follows another number in the series of natural numbers.

79. Now in order to be able to prove that every number (n) in the series of natural numbers is immediately followed by a number, one must come up with a concept to which this latter number applies. We choose for this:

"belonging to the series of natural numbers ending with n,"

but we must first define it.

To begin with I shall repeat, in somewhat different words, the definition I gave in my *Begriffsschrift* of following in a series:

148

The statement

'if every object to which x stands in the relation ϕ falls under the concept F, and if, from the fact that d falls under the concept F, it always follows, no matter what d may be, that every object to which d stands in the relation ϕ falls under the concept F, then y falls under the concept F, no matter what concept F might be',

is to be synonymous with

'y follows x in the ϕ-series'

and with

'x precedes y in the ϕ-series'.

80. Several remarks concerning this definition will not be superfluous here. Since the relation ϕ is left indeterminate, the series is not necessarily to be thought of in the form of a spatial or temporal arrangement, although these cases are not excluded.

Some other definition might be considered more natural, e.g., if, in proceeding from x, we always turn our attention from one object to another, to which it stands in the relation ϕ, and if, in this way, we can finally reach y, then we say that y follows x in the ϕ-series.

This is a way of looking at the matter, not a definition. Whether we reach y in the wanderings of our attention can depend on many subjective incidental circumstances; e.g., on the time we have available or on our knowledge of the things. Whether y follows x in the ϕ-series has, in general, nothing at all to do with our attention and the conditions of its progress, but rather it is a matter of objective fact: just as a green leaf reflects certain light rays whether or not they should meet my eye and summon up a sensation; just as a grain of salt is soluble in water whether or not I put it in water and observe the process; and just as it remains soluble even if it is not possible for me to experiment on it.

By means of my definition, the matter is elevated from the realm of the subjectively possible to that of the objectively definite. Indeed, the fact that from certain statements another statement follows is something objective, something independent of whatever laws may govern the wanderings of our attention; and it makes no difference whether we really make the inference or not. Here we have a criterion which decides the question, wherever it can be asked, even though we might be hindered by external difficulties from judging in individual cases whether it is applicable. That makes no difference to the issue itself.

We need not always run through all the intermediate members, from the initial member up to an object, in order to be sure that the latter follows the former. If, e.g., it is given that, in the ϕ-series, b follows a and c follows b, then we can conclude on the basis of our definition that c follows a, without even knowing the intermediate members.

Only by means of this definition of following in a series does it become possible to reduce the rule of inference from n to $(n+1)$, which apparently is peculiar to mathematics, to general logical laws.

81. Now if we have as our relation ϕ the one in which m is related to n by the statement

'n immediately follows m in the series of natural numbers',

then we say instead of 'ϕ-series', 'series of natural numbers'.
I define further:

the statement

'y follows x in the ϕ-series or y is the same as x',

is to be synonymous with

'y belongs to the ϕ-series starting with x'

and with

'x belongs to the ϕ-series ending with y'.

According to this, a belongs to the series of natural numbers ending with n if n either follows a in the series of natural numbers or is equal to a.[15]

82. We must now show that, under a condition still to be stated, the number which applies to the concept

"belonging to the series of natural numbers ending with n"

immediately follows n in the series of natural numbers. Having this result, we will have proved that there is a number which immediately follows n in the series of natural numbers; i.e., that there is no last member of this series. Obviously, this statement cannot be established empirically or by means of induction.

It would take us too far afield to give the proof itself. We can only give a brief sketch of it here. We must prove:

1.   If a immediately follows d in the series of natural numbers, and if the number which applies to the concept

"belonging to the series of natural numbers ending with d"

[15] If n is not a number, then only n itself belongs to the series of natural numbers ending with n. One should not object to this expression.

immediately follows *d* in the series of natural numbers, then the number which applies to the concept

"belonging to the series of natural numbers ending with *a*"

immediately follows *a* in the series of natural numbers.

2. We must prove that what has been asserted about *d* and *a* in the foregoing statements holds for 0, and then show that it also holds for *n*, if *n* belongs to the series of natural numbers beginning with 0. This will result from an application of my definition of

'*y* follows *x* in the series of natural numbers',

taking as the concept *F* the relation asserted above to hold between *d* and *a*, and substituting 0 and *n* for *d* and *a*.

83. In order to prove Theorem 1 of the last paragraph, we must show that *a* is the number which applies to the concept "belonging to the series of natural numbers ending with *a*, but not equal to *a*." And to this end, we must prove that this concept has the same extension as the concept "belonging to the series of natural numbers ending with *d*." For this, we need the theorem that no object which belongs to the series of natural numbers beginning with 0 can follow itself in the series of natural numbers. The latter must likewise be proved by means of our definition of following in a series, as it is outlined above.[16]

For this reason, we must add the condition that *n* belong to the series of natural numbers beginning with 0 to the statement that the number which applies to the concept

"belonging to the series of natural numbers ending with *n*,"

immediately follows *n* in the series of natural numbers. There is a shorter way of putting this, which I shall now define:

the statement

'*n* belongs to the series of natural numbers beginning with 0'

is to be synonymous with

'*n* is a finite number'.

We can now express the last theorem thus: no finite number follows itself in the series of natural numbers.

[16]E. Schröder (1873: 63) seems to look upon this theorem as the consequence of an ambiguous terminology. The difficulty which infects his whole presentation of the matter emerges here too; i.e., it is never quite clear whether the number is a symbol and, if so, what its meaning is, or whether it *is* this very meaning. From the fact that one sets up different symbols, so that the same one never recurs, it does not follow that these symbols mean different things.

Infinite numbers

84. In contrast to the finite numbers there are the infinite ones. The number which applies to the concept "finite number" is an infinite one. Let us denote it, say, by $\aleph_0$.[17] Were it a finite number, it could not follow itself in the series of natural numbers. One can show, however, that $\aleph_0$ does just this.

There is nothing somehow mysterious or marvellous about the infinite number $\aleph_0$ when so defined. 'The number which applies to the concept F is $\aleph_0$' says nothing more nor less than: there is a relation which establishes a one-to-one correlation between the objects falling under the concept F and the finite numbers. This has, according to our definitions, a completely clear and unambiguous sense, and that suffices to justify the use of the symbol $\aleph_0$ and to guarantee it a meaning. That we can form no mental picture of an infinite number is wholly irrelevant and would hold true of finite numbers as well. In this way, our number $\aleph_0$ is something just as determinate as any finite number: it can be recognized without a doubt as the same and differentiated from any other.

85. Recently, in a noteworthy paper (1883b), G. Cantor introduced infinite numbers. I agree with him completely in his evaluation of the view which would have only the finite numbers qualify as real. Neither these nor the fractions are sensibly perceptible and spatial, nor are the negative, irrational, and complex numbers. And if one calls real [only] that which affects the senses, or at least can have sense impressions as an immediate or distant consequence, then certainly none of these numbers is real. But we don't need such sense impressions as evidence for our theorems. A name or a symbol, which is introduced in a logically unobjectionable way, may be used by us without hesitation in our investigations, and thus our number $\aleph_0$ is just as firmly grounded as 2 or 3.

Although I believe I agree with Cantor in this matter, I do, however, deviate from him in terminology. He calls my numbers 'powers', whereas his concept[18] of number is based on ordering. To be sure, finite numbers end up being independent of order; however, this does not hold for infinite numbers. Now the linguistic usage of the word 'number' and of the question 'how many?' contains no indication of a definite order. Cantor's number answers rather the question: 'the last member is the how-manyth member of the sequence?' Therefore my terminology seems to me to

[17][Frege used '∞_1', but we adopt the aleph notation as being more in keeping with current practice. – Tr.]

[18]This expression may appear to contradict [my earlier remarks emphasizing] the objectivity of concepts; however, only the *terminology* is subjective here.

agree better with linguistic usage. If one extends the meaning of a word, then one must take care that as many general statements as possible retain their validity, and particularly statements as basic as, for instance, [the one asserting] for numbers their independence of the sequence. We have needed no extension at all, because our concept of number immediately embraces infinite numbers as well.

86. In order to obtain his infinite numbers, Cantor introduces the relational concept of following in a sequence, which differs from my "following in a series." According to him, for instance, a sequence would result if one were so to order the finite positive whole numbers that the odd numbers followed one another in their own natural order, and similarly the even numbers in theirs, and it were further stipulated that all the even numbers should come after all the odd numbers. In this sequence, e.g., 0 would follow 13. There would, however, be no number immediately preceding 0. Now this case cannot occur within my definition of following in a series. It may be strictly proved, without using intuition, that, if y follows x in the ϕ-series, there is an object which immediately precedes y in this series. It seems to me, then, that exact definitions of following in a sequence and of number [in Cantor's sense] are still lacking. Thus Cantor bases himself on a somewhat mysterious "inner intuition" where a proof from definitions should be striven for and would probably be found. For I think I can foresee how those concepts could be defined. In any case, I in no way wish these comments to be taken as an attack on the justifiability or fruitfulness of these concepts. On the contrary, I welcome these investigations as an extension of the science, especially because they strike a purely arithmetic path to higher infinite numbers (powers).

Conclusion

87. I hope in this monograph to have made it probable that arithmetic laws are analytic judgments, and therefore *a priori*. According to this, arithmetic would be only a further developed logic, every arithmetic theorem a logical law, albeit a derived one. The applications of arithmetic to the explanation of natural phenomena would be logical processing of observed facts;[19] computation would be inference. Numerical laws will not need, as Baumann (1868–9, 2: 670) contends, a practical confirmation in order to be applicable in the external world; for, in the external world, the totality of space and its contents, there are no concepts, no properties of concepts, no numbers. Therefore, the numerical laws are

[19]Observation itself already includes a logical activity.

153

really not applicable to the external world: they are not laws of nature. They are, however, applicable to judgments, which are true of things in the external world: they are laws of the laws of nature. They assert connections not between natural phenomena, but rather between judgments; and it is to the latter that the laws of nature belong.

88. Kant (1867–8, 3: 39ff) evidently underestimated the value of analytic judgments – probably as the result of having too narrow a definition of the concept – although he apparently also had in mind the broader concept used here.[20] Taking his definition as a basis, the division of judgments into the analytic and the synthetic is not exhaustive. He is thinking of universal affirmative judgments. In such cases, one can speak of a concept of the subject and inquire whether the concept of the predicate – as would result from *his* definition – is contained in it. How can we do this, however, when the subject is a single object? Or when the judgment is existential? In such cases there can be, in Kant's sense, no talk of a concept of the subject. Kant seems to have thought of the concept as determined by subordinate characteristics; that, however, is one of the least fruitful notions of concept. If one surveys the foregoing definitions, one will hardly find one of this kind. The same is true of the really fruitful definitions in mathematics, e.g., of the continuity of a function. There we don't have a series of subordinate characteristics but rather a more intimate, I should say more organic, connection between the [elements of the] definitions. The difference can be illustrated by means of a geometrical analogy. If the concepts (or their extensions) are represented by regions of a plane, then the concept defined by means of subordinate characteristics corresponds to the region which is the overlap of all the individual regions corresponding to these characteristics; it is enclosed by parts of their boundaries. Pictorially speaking, in such a definition, we delimit a region by using in a new way lines already given. In doing this, however, nothing essentially new comes out. The more fruitful definitions draw border lines which had not previously been given. What can be inferred from them cannot be seen in advance; one does not simply withdraw again from the box what one has put into it. These inferences expand our knowledge and one should, therefore, following Kant, consider them synthetic. Nevertheless, they can be proved purely logically and hence are analytic. They are in fact contained in the definitions, but like the plant in the seed, not like the rafter in the house. One often needs several definitions to prove a theorem, which consequently is contained in no single

---

[20][Kant] says that a synthetic statement can be understood according to the Theorem of Contradiction only if another synthetic statement is presupposed (1867–8, 3: 43).

definition, but nevertheless follows in a purely logical way from all of them together.

89. I must also contradict the generality of Kant's assertion (1867–8, 3: 82) that without sensible perception no object would be given us. Zero and 1 are objects that cannot be given us sensibly. And those who hold the smaller numbers to be intuitive will surely have to concede that none of the numbers greater than $1000^{1000^{1000}}$ can be given them intuitively, and that we nevertheless know a good deal about them. Perhaps Kant was using the word 'object' in a somewhat different sense; but then zero, 1, and our $\aleph_0$ disappear completely from his considerations; for, they are not concepts either, and Kant demands even of concepts that their objects be appended to them in intuition.

In order not to open myself to the criticism of carrying on a picayune search for faults in the work of a genius whom we look up to only with thankful awe, I believe I should also emphasize our areas of agreement, which are far more extensive than those of our disagreement. To touch on only the immediate points, I see a great service in Kant's having distinguished between synthetic and analytic judgments. In terming geometric truths synthetic and *a priori,* he uncovered their true essence. And this is still worth repeating today, because it is still often not recognized. If Kant erred with respect to arithmetic, this does not detract essentially, I think, from his merit. It was important for him that there should be synthetic judgments *a priori;* whether they occur only in geometry or also in arithmetic is of little importance.

90. I do not claim to have made the analytic nature of arithmetic theorems more than probable, because one can always still doubt whether their proof can be carried out completely from purely logical laws, whether evidence of another sort has not crept in unnoticed somewhere. This doubt is also not entirely relieved by the outlines which I have given of the proofs of a few theorems; it can only be alleviated by an airtight chain of reasoning, such that no step is made which is not in conformity with one of a few rules of inference recognized as purely logical. Thus until now, hardly a single [real] proof has ever been offered, because the mathematician is satisfied if every transition to a new judgment appears to him to be correct, without asking whether this appearance is logical or intuitive. A step in such a proof is often quite complex and involves several simple inferences, in addition to which intuitive considerations can creep in. One proceeds in jumps, and from this there arises the impression of an over-rich variety of rules of inference used in mathematics. For, the

155

greater the jumps, the more complex are the combinations of simple inferences and intuitive axioms which they can represent. Nevertheless, such a transition often occurs to us directly, without our being conscious of the intermediate steps, and since it does not present itself as one of the recognized logical rules of inference, we are immediately ready to consider this manifest transition as an intuitive one and the inferred truth as a synthetic one, even when the range of its validity extends far beyond intuition.

Proceeding in this way, it is not possible clearly to separate the synthetic, based on intuition, from the analytic. Nor will it be possible to compile with completeness and certainty the axioms of intuition needed to make every mathematical proof capable of proceeding from these axioms alone, according to logical laws.

91. The requirement of avoiding all jumps in a proof must, therefore, be imposed. That it is so difficult to satisfy is owing to the tediousness of a step-by-step procedure. Every proof, which is even slightly involved, threatens to become enormously long. In addition to this, the superfluity of logical forms expressed in language makes it difficult to extract a group of rules of inference sufficient for all cases and yet easy to survey.

In order to minimize the effects of these drawbacks, I have devised my concept writing. It strives for greater brevity and comprehensibility of expression and is manipulated in a few standard ways, as in a computation, so that no transition is permitted which does not conform to rules set up once for all.[21] No assumption can then slip in unnoticed. I have thus proved a theorem,[22] borrowing no axioms from intuition, which one would consider at first glance to be synthetic and which I shall state here as follows:

If the relation of each member of a series to its successor is one-to-one, and if *m* and *y* follow *x* in this series, then *y* precedes *m* in this series, or coincides with it, or follows *m*.

From this proof, one can see that theorems which expand our knowledge can contain analytic judgments.[23]

[21]It is, however, supposed to be able to express not only the logical form of a statement, as does the Boolean notation, but also its content.

[22]*Begriffsschrift,* 1879, p. 86, Formula 133.

[23]This proof will be found to be still much too lengthy, a disadvantage which may seem to more than balance out the almost unconditional guarantee against a mistake or a loophole. My purpose at that time was to reduce everything to the smallest possible number of the simplest possible logical laws. As a result of this, I applied only one rule of inference. I pointed out even then, in the foreword (p. vii) that, for further application, it would be recommended to admit more rules of inference. This can be done without impairing the validity of the chain of reasoning, and an important abbreviation could thereby be achieved.

[Recapitulation]

106. Let us cast a quick glance backward on the course of our investigation. After determining that a number was not a collection of things nor a property of such a collection, nor, furthermore, the subjective product of mental processes, we decided that a statement of number asserts something objective about a concept. We defined first the individual numbers 0, 1 etc., and then following in the number series. Our first attempt failed, because in it we stated the meaning of only whole assertions about concepts, and not of 0 and 1 separately, although these entered into those assertions. As a result of this, we could not prove the equality of numbers. It was shown that the numbers with which arithmetic concerns itself must be understood not as dependent attributes, but rather substantivally.[24] Thus numbers appeared to us as recognizable objects, although not physical ones or even merely spatial ones, nor ones which we could picture in imagination. We then established the basic theorem: that the meaning of a word is not to be defined separately, but rather in the context of a statement; only by following this theorem can we, I think, avoid the physical interpretation of number, without slipping into psychological interpretation. There is only one type of statement which must have a sense for every object; that is the recognition sentences, called equations in the case of numbers. We saw that statements of number are also to be interpreted as equations. It became a question, then, of determining the sense of a numerical equation and of expressing this sense without making use of the number-words or the word 'number'. The possibility of establishing a one-to-one correspondence between the objects falling under a concept F and those falling under a concept G was found to be the content of a recognition judgment about numbers. Our definition, therefore, had to posit that possibility as synonymous with a numerical equation. We recalled similar instances: the definition of direction from parallelism, of shape from similarity, etc.

107. The question then arose: when are we justified in interpreting a content to be that of a recognition judgment? For this, the condition must be fulfilled that in every judgment the left side of the tentatively assumed equation can be replaced by the right, without altering the truth of the judgment. Now, at first and without resorting to further definitions, no further assertion about the left or right side of such an equation is known to us beyond the assertion of their equality. Substitutivity had therefore to be proved only for equations.

---

[24]The difference corresponds to that between 'blue' and 'the color of the sky'.

A doubt still remained, however. A recognition statement must always have a sense. If we interpreted the possibility of correlating in one-to-one fashion the objects falling under the concept F with those falling under the concept G as an equation, by saying for it: 'the number which applies to the concept F is equal to the number which applies to the concept G' and thereby introducing the expression 'the number which applies is the concept F', then we have a sense for the equation only if both sides have the form just mentioned. We would not be able to judge according to such a definition whether an equation only one side of which had this form was true or false. That caused us to make the following definition:

The number which applies to the concept F is the extension of the concept "concept equinumerous with the concept F,"

by which we called a concept F equinumerous with a concept G, if there exists the possibility of correlating them one-to-one.

In doing this, we presuppose that the sense of the expression 'extension of a concept' is familiar. This method of overcoming the difficulty will probably not be everywhere applauded, and some will prefer to set aside this doubt in another way. I, too, place no decisive weight on the introduction of the extension of a concept.

108. We still had to define one-to-one correspondences; we reduced them to purely logical terms. After we had outlined the proof of the theorem that the number which applies to the concept F is equal to that which applies to the concept G, if the concept F is equinumerous with the concept G, we defined 0, the expression 'n immediately follows m in the series of natural numbers', and the number 1, and we showed that 1 immediately follows 0 in the series of natural numbers. We presented a few theorems which could be easily proved at this point and then went somewhat more deeply into the following, which demonstrates the infinity of the number series:

Every number in the series of natural numbers is followed by a number.

We were thereby led to the concept "belonging to the series of natural numbers ending with n," of which we wanted to show that the number applying to it immediately follows n in the series of natural numbers. We defined it at first by means of an object y following an object x in a general ϕ-series. The sense of this expression was also reduced to purely logical terms. And thereby we succeeded in proving that the rule of inference from n to $(n+1)$, which is usually considered a peculiarly mathematical one, is based on the general logical rules of inference.

For the proof of the infinity of the number series, we needed the theorem that no finite number follows itself in the series of natural numbers. We thus arrived at the concepts of finite and infinite numbers. We showed that the latter is basically no less justified logically than is the former. For the purposes of comparison, we drew upon Cantor's infinite numbers and his "following in a sequence," where we pointed out the difference in terminology.

109. We thus rendered the analytic and *a priori* character of arithmetic truths highly probable, arriving at an improvement on Kant's point of view. We saw further what was still lacking in order to elevate that probability to certainty and we indicated the path that must lead to this.

Selections from
Introduction to Mathematical Philosophy

BERTRAND RUSSELL

I. The series of natural numbers

Mathematics is a study which, when we start from its most familiar portions, may be pursued in either of two opposite directions. The more familiar direction is constructive, towards gradually increasing complexity: from integers to fractions, real numbers, complex numbers; from addition and multiplication to differentiation and integration, and on to higher mathematics. The other direction, which is less familiar, proceeds, by analysing, to greater and greater abstractness and logical simplicity; instead of asking what can be defined and deduced from what is assumed to begin with, we ask instead what more general ideas and principles can be found, in terms of which what was our starting-point can be defined or deduced. It is the fact of pursuing this opposite direction that characterises mathematical philosophy as opposed to ordinary mathematics. But it should be understood that the distinction is one, not in the subject matter, but in the state of mind of the investigator. Early Greek geometers, passing from the empirical rules of Egyptian land-surveying to the general propositions by which those rules were found to be justifiable, and thence to Euclid's axioms and postulates, were engaged in mathematical philosophy, according to the above definition; but when once the axioms and postulates had been reached, their deductive employment, as we find it in Euclid, belonged to mathematics in the ordinary sense. The distinction between mathematics and mathematical philosophy is one which depends upon the interest inspiring the research, and upon the stage which the research has reached; not upon the propositions with which the research is concerned.

We may state the same distinction in another way. The most obvious and easy things in mathematics are not those that come logically at the beginning; they are things that, from the point of view of logical deduction, come somewhere in the middle. Just as the easiest bodies to see are those that are neither very near nor very far, neither very small nor very

Reprinted by kind permission of the publishers from Bertrand Russell, *Introduction to Mathematical Philosophy* (New York: The Macmillan Company; London: George Allen & Unwin Ltd., 1919), pp. 1–19, 194–206.

great, so the easiest conceptions to grasp are those that are neither very complex nor very simple (using "simple" in a *logical* sense). And as we need two sorts of instruments, the telescope and the microscope, for the enlargement of our visual powers, so we need two sorts of instruments for the enlargement of our logical powers, one to take us forward to the higher mathematics, the other to take us backward to the logical foundations of the things that we are inclined to take for granted in mathematics. We shall find that by analysing our ordinary mathematical notions we acquire fresh insight, new powers, and the means of reaching whole new mathematical subjects by adopting fresh lines of advance after our backward journey. It is the purpose of this book to explain mathematical philosophy simply and untechnically, without enlarging upon those portions which are so doubtful or difficult that an elementary treatment is scarcely possible. A full treatment will be found in *Principia Mathematica* (1910–13); the treatment in the present volume is intended as an introduction.

To the average educated person of the present day, the obvious starting-point of mathematics would be the series of whole numbers,

$$1, 2, 3, 4, \ldots, \text{etc.}$$

Probably only a person with some mathematical knowledge would think of beginning with 0 instead of with 1, but we will presume this degree of knowledge; we will take as our starting-point the series:

$$0, 1, 2, 3, \ldots n, n+1, \ldots$$

and it is this series that we shall mean when we speak of the "series of natural numbers."

It is only at a high stage of civilisation that we could take this series as our starting-point. It must have required many ages to discover that a brace of pheasants and a couple of days were both instances of the number 2: the degree of abstraction involved is far from easy. And the discovery that 1 is a number must have been difficult. As for 0, it is a very recent addition; the Greeks and Romans had no such digit. If we had been embarking upon mathematical philosophy in earlier days, we should have had to start with something less abstract than the series of natural numbers, which we should reach as a stage on our backward journey. When the logical foundations of mathematics have grown more familiar, we shall be able to start further back, at what is now a late stage in our analysis. But for the moment the natural numbers seem to represent what is easiest and most familiar in mathematics.

But though familiar, they are not understood. Very few people are prepared with a definition of what is meant by "number," or "0," or "1."

It is not very difficult to see that, starting from 0, any other of the natural numbers can be reached by repeated additions of 1, but we shall have to define what we mean by "adding 1," and what we mean by "repeated." These questions are by no means easy. It was believed until recently that some, at least, of these first notions of arithmetic must be accepted as too simple and primitive to be defined. Since all terms that are defined are defined by means of other terms, it is clear that human knowledge must always be content to accept some terms as intelligible without definition, in order to have a starting-point for its definitions. It is not clear that there must be terms which are *incapable* of definition: it is possible that, however far back we go in defining, we always *might* go further still. On the other hand, it is also possible that, when analysis has been pushed far enough, we can reach terms that really are simple, and therefore logically incapable of the sort of definition that consists in analysing. This is a question which it is not necessary for us to decide; for our purposes it is sufficient to observe that, since human powers are finite, the definitions known to us must always begin somewhere, with terms undefined for the moment, though perhaps not permanently.

All traditional pure mathematics, including analytical geometry, may be regarded as consisting wholly of propositions about the natural numbers. That is to say, the terms which occur can be defined by means of the natural numbers, and the propositions can be deduced from the properties of the natural numbers – with the addition, in each case, of the ideas and propositions of pure logic.

That all traditional pure mathematics can be derived from the natural numbers is a fairly recent discovery, though it had long been suspected. Pythagoras, who believed that not only mathematics, but everything else, could be deduced from numbers, was the discoverer of the most serious obstacle in the way of what is called the "arithmetising" of mathematics. It was Pythagoras who discovered the existence of incommensurables, and, in particular, the incommensurability of the side of a square and the diagonal. If the length of the side is 1 inch, the number of inches in the diagonal is the square root of 2, which appeared not to be a number at all. The problem thus raised was solved only in our own day, and was only solved *completely* by the help of the reduction of arithmetic to logic, which will be explained in following chapters. For the present, we shall take for granted the arithmetisation of mathematics, though this was a feat of the very greatest importance.

Having reduced all traditional pure mathematics to the theory of the natural numbers, the next step in logical analysis was to reduce this theory itself to the smallest set of premisses and undefined terms from which it could be derived. This work was accomplished by Peano. He showed that the entire theory of the natural numbers could be derived from three primitive ideas and five primitive propositions in addition to those of pure logic. These three ideas and five propositions thus became, as it were, hostages for the whole of traditional pure mathematics. If they could be defined and proved in terms of others, so could all pure mathematics. Their logical "weight," if one may use such an expression, is equal to that of the whole series of sciences that have been deduced from the theory of the natural numbers; the truth of this whole series is assured if the truth of the five primitive propositions is guaranteed, provided, of course, that there is nothing erroneous in the purely logical apparatus which is also involved. The work of analysing mathematics is extraordinarily facilitated by this work of Peano's.

The three primitive ideas in Peano's arithmetic are:

0, number, successor.

By "successor" he means the next number in the natural order. That is to say, the successor of 0 is 1, the successor of 1 is 2, and so on. By "number" he means, in this connection, the class of natural numbers.[1] He is not assuming that we know all the members of this class, but only that we know what we mean when we say that this or that is a number, just as we know what we mean when we say "Jones is a man," though we do not know all men individually.

(1) 0 is a number.
(2) The successor of any number is a number.
(3) No two numbers have the same successor.
(4) 0 is not the successor of any number.
(5) Any property which belongs to 0, and also to the successor of every number which has the property, belongs to all numbers.

The last of these is the principle of mathematical induction. We shall have much to say concerning mathematical induction in the sequel; for the present, we are concerned with it only as it occurs in Peano's analysis of arithmetic.

Let us consider briefly the kind of way in which the theory of the natural numbers results from these three ideas and five propositions. To begin with, we define 1 as "the successor of 0," 2 as "the successor of 1," and so on. We can obviously go on as long as we like with these definitions, since, in virtue of (2), every number that we reach will have a successor, and, in virtue of (3), this cannot be any of the numbers already defined, because, if it were, two different numbers would have the same

[1] We shall use "number" in this sense in the present chapter. Afterwards the word will be used in a more general sense.

163

successor; and in virtue of (4) none of the numbers we reach in the series of successors can be 0. Thus the series of successors gives us an endless series of continually new numbers. In virtue of (5) all numbers come in this series, which begins with 0 and travels on through successive successors: for (a) 0 belongs to this series, and (b) if a number n belongs to it, so does its successor, whence, by mathematical induction, every number belongs to the series.

Suppose we wish to define the sum of two numbers. Taking any number m, we define $m + 0$ as m, and $m + (n + 1)$ as the successor of $m + n$. In virtue of (5) this gives a definition of the sum of m and n, whatever number n may be. Similarly we can define the product of any two numbers. The reader can easily convince himself that any ordinary elementary proposition of arithmetic can be proved by means of our five premises, and if he has any difficulty he can find the proof in Peano.

It is time now to turn to the considerations which make it necessary to advance beyond the standpoint of Peano, who represents the last perfection of the "arithmetisation" of mathematics, to that of Frege, who first succeeded in "logicising" mathematics, *i.e.* in reducing to logic the arithmetical notions which his predecessors had shown to be sufficient for mathematics. We shall not, in this chapter, actually give Frege's definition of number and of particular numbers, but we shall give some of the reasons why Peano's treatment is less final than it appears to be.

In the first place, Peano's three primitive ideas – namely, "0," "number," and "successor" – are capable of an infinite number of different interpretations, all of which will satisfy the five primitive propositions. We will give some examples.

(1) Let "0" be taken to mean 100, and let "number" be taken to mean the numbers from 100 onward in the series of natural numbers. Then all our primitive propositions are satisfied, even the fourth, for, though 100 is the successor of 99, 99 is not a "number" in the sense which we are now giving to the word "number." It is obvious that any number may be substituted for 100 in this example.

(2) Let "0" have its usual meaning, but let "number" mean what we usually call "even numbers," and let the "successor" of a number be what results from adding two to it. Then "1" will stand for the number two, "2" will stand for the number four, and so on; the series of "numbers" now will be

$$0, \text{ two, four, six, eight} \ldots$$

All Peano's five premises are satisfied still.

(3) Let "0" mean the number one, let "number" mean the set

$$1, \frac{1}{2}, \frac{1}{4}, \frac{1}{8}, \frac{1}{16}, \ldots$$

and let "successor" mean "half". Then all Peano's five axioms will be true of this set.

It is clear that such examples might be multiplied indefinitely. In fact, given any series

$$x_0, x_1, x_2, x_3, \ldots x_n, \ldots$$

which is endless, contains no repetitions, has a beginning, and has no terms that cannot be reached from the beginning in a finite number of steps, we have a set of terms verifying Peano's axioms. This is easily seen, though the formal proof is somewhat long. Let "0" mean x_0, let "number" mean the whole set of terms, and let the "successor" of x_n mean x_{n+1}. Then

(1) "0" is a number," i.e. x_0 is a member of the set.
(2) "The successor of any number is a number," i.e. taking any term x_n in the set, x_{n+1} is also in the set.
(3) "No two numbers have the same successor," i.e. if x_m and x_n are two different members of the set, x_{m+1} and x_{n+1} are different; this results from the fact that (by hypothesis) there are no repetitions in the set.
(4) "0 is not the successor of any number," i.e. no term in the set comes before x_0.
(5) This becomes: Any property which belongs to x_0, and belongs to x_{n+1} provided it belongs to x_n, belongs to all the x's.

This follows from the corresponding property for numbers.
A series of the form

$$x_0, x_1, x_2, \ldots x_n, \ldots$$

in which there is a first term, a successor to each term (so that there is no last term), no repetitions, and every term can be reached from the start in a finite number of steps, is called a *progression*. Progressions are of great importance in the principles of mathematics. As we have just seen, every progression verifies Peano's five axioms. It can be proved, conversely, that every series which verifies Peano's five axioms is a progression. Hence these five axioms may be used to define the class of progressions: "progressions" are "those series which verify these five axioms." Any progression may be taken as the basis of pure mathematics: we may give the name "0" to its first term, the name "number" to the whole set of its terms, and the name "successor" to the next in the progression. The progression need not be composed of numbers: it may be composed of points in space, or moments of time, or any other terms of which there is an infinite supply. Each different progression will give rise to a different interpretation of all the propositions of traditional pure mathematics; all these possible interpretations will be equally true.

In Peano's system there is nothing to enable us to distinguish between these different interpretations of his primitive ideas. It is assumed that we know what is meant by "0," and that we shall not suppose that this symbol means 100 or Cleopatra's Needle or any of the other things that it might mean.

This point, that "0" and "number" and "successor" cannot be defined by means of Peano's five axioms, but must be independently understood, is important. We want our numbers not merely to verify mathematical formulae, but to apply in the right way to common objects. We want to have ten fingers and two eyes and one nose. A system in which "1" meant 100, and "2" meant 101, and so on, might be all right for pure mathematics, but would not suit daily life. We want "0" and "number" and "successor" to have meanings which will give us the right allowance of fingers and eyes and noses. We have already some knowledge (though not sufficiently articulate or analytic) of what we mean by "1" and "2" and so on, and our use of numbers in arithmetic must conform to this knowledge. We cannot secure that this shall be the case by Peano's method; all that we can do, if we adopt his method, is to say "we know what we mean by '0' and 'number' and 'successor,' though we cannot explain what we mean in terms of other simpler concepts." It is quite legitimate to say this when we must, and at *some* point we all must; but it is the object of mathematical philosophy to put off saying it as long as possible. By the logical theory of arithmetic we are able to put it off for a very long time.

It might be suggested that, instead of setting up "0" and "number" and "successor" as terms of which we know the meaning although we cannot define them, we might let them stand for *any* three terms that verify Peano's five axioms. They will then no longer be terms which have a meaning that is definite though undefined: they will be "variables," terms concerning which we make certain hypotheses, namely, those stated in the five axioms, but which are otherwise undetermined. If we adopt this plan, our theorems will not be proved concerning an ascertained set of terms called "the natural numbers," but concerning all sets of terms having certain properties. Such a procedure is not fallacious; indeed for certain purposes it represents a valuable generalisation. But from two points of view it fails to give an adequate basis for arith-

166

metic. In the first place, it does not enable us to know whether there are any sets of terms verifying Peano's axioms; it does not even give the faintest suggestion of any way of discovering whether there are such sets. In the second place, as already observed, we want our numbers to be such as can be used for counting common objects, and this requires that our numbers should have a *definite* meaning, not merely that they should have certain formal properties. This definite meaning is defined by the logical theory of arithmetic.

II. Definition of number

The question "What is a number?" is one which has often been asked but has only been correctly answered in our own time. The answer was given by Frege in 1884, in his *Grundlagen der Arithmetik*.[2] Although this book is quite short, not difficult, and of the very highest importance, it attracted almost no attention, and the definition of number which it contains remained practically unknown until it was rediscovered by the present author in 1901.

In seeking a definition of number, the first thing to be clear about is what we may call the grammar of our inquiry. Many philosophers, when attempting to define number, are really setting to work to define plurality, which is quite a different thing. *Number* is what is characteristic of numbers, as *man* is what is characteristic of men. A plurality is not an instance of number, but of some particular number. A trio of men, for example, is an instance of the number 3, and the number 3 is an instance of number; but the trio is not an instance of number. This point may seem elementary and scarcely worth mentioning; yet it has proved too subtle for the philosophers, with few exceptions.

A particular number is not identical with any collection of terms having that number: the number 3 is not identical with the trio consisting of Brown, Jones, and Robinson. The number 3 is something which all trios have in common, and which distinguishes them from other collections. A number is something that characterises certain collections, namely, those that have that number.

Instead of speaking of a "collection," we shall as a rule speak of a "class," or sometimes a "set." Other words used in mathematics for the same thing are "aggregate" and "manifold." We shall have much to say later on about classes. For the present, we will say as little as possible. But there are some remarks that must be made immediately.

A class or collection may be defined in two ways that at first sight seem

[2]The same answer is given more fully and with more development in his *Grundgesetze der Arithmetik,* vol. 1, 1893.

quite distinct. We may enumerate its members, as when we say, "The collection I mean is Brown, Jones, and Robinson." Or we may mention a defining property, as when we speak of "mankind" or "the inhabitants of London." The definition which enumerates is called a definition by "extension," and the one which mentions a defining property is called a definition by "intension." Of these two kinds of definition, the one by intension is logically more fundamental. This is shown by two considerations: (1) that the extensional definition can always be reduced to an intensional one; (2) that the intensional one often cannot even theoretically be reduced to the extensional one. Each of these points needs a word of explanation.

(1) Brown, Jones, and Robinson all of them possess a certain property which is possessed by nothing else in the whole universe, namely, the property of being either Brown or Jones or Robinson. This property can be used to give a definition by intension of the class consisting of Brown and Jones and Robinson. Consider such a formula as "x is Brown or x is Jones or x is Robinson." This formula will be true for just three x's, namely, Brown and Jones and Robinson. In this respect it resembles a cubic equation with its three roots. It may be taken as assigning a property common to the members of the class consisting of these three men, and peculiar to them. A similar treatment can obviously be applied to any other class given in extension.

(2) It is obvious that in practice we can often know a great deal about a class without being able to enumerate its members. No one man could actually enumerate all men, or even all the inhabitants of London, yet a great deal is known about each of these classes. This is enough to show that definition by extension is not *necessary* to knowledge about a class. But when we come to consider infinite classes, we find that enumeration is not even theoretically possible for beings who only live for a finite time. We cannot enumerate all the natural numbers: they are 0, 1, 2, 3, *and so on.* At some point we must content ourselves with "and so on." We cannot enumerate all fractions or all irrational numbers, or all of any other infinite collection. Thus our knowledge in regard to all such collections can only be derived from a definition by intension.

These remarks are relevant, when we are seeking the definition of number, in three different ways. In the first place, numbers themselves form an infinite collection, and cannot therefore be defined by enumeration. In the second place, the collections having a given number of terms themselves presumably form an infinite collection: it is to be presumed, for example, that there are an infinite collection of trios in the world, for if this were not the case the total number of things in the world would be finite, which, though possible, seems unlikely. In the third place, we wish

168

to define "number" in such a way that infinite numbers may be possible; thus we must be able to speak of the number of terms in an infinite collection, and such a collection must be defined by intension, i.e. by a property common to all its members and peculiar to them.

For many purposes, a class and a defining characteristic of it are practically interchangeable. The vital difference between the two consists in the fact that there is only one class having a given set of members, whereas there are always many different characteristics by which a given class may be defined. Men may be defined as featherless bipeds, or as rational animals, or (more correctly) by the traits by which Swift delineates the Yahoos. It is this fact that a defining characteristic is never unique which makes classes useful; otherwise we could be content with the properties common and peculiar to their members.[3] Any one of these properties can be used in the place of the class whenever uniqueness is not important.

Returning now to the definition of number, it is clear that number is a way of bringing together certain collections, namely, those that have a given number of terms. We can suppose all couples in one bundle, all trios in another, and so on. In this way we obtain various bundles of collections, each bundle consisting of all the collections that have a certain number of terms. Each bundle is a class whose members are collections, i.e. classes; thus each is a class of classes. The bundle consisting of all couples, for example, is a class of classes: each couple is a class with two members, and the whole bundle of couples is a class with an infinite number of members, each of which is a class of two members.

How shall we decide whether two collections are to belong to the same bundle? The answer that suggests itself is: "Find out how many members each has, and put them in the same bundle if they have the same number of members." But this presupposes that we have defined numbers, and that we know how to discover how many terms a collection has. We are so used to the operation of counting that such a presupposition might easily pass unnoticed. In fact, however, counting, though familiar, is logically a very complex operation; moreover it is only available, as a means of discovering how many terms a collection has, when the collection is finite. Our definition of number must not assume in advance that all numbers are finite; and we cannot in any case, without a vicious circle, use counting to define numbers, because numbers are used in counting. We need, therefore, some other method of deciding when two collections have the same number of terms.

---

[3] As will be explained later, classes may be regarded as logical fictions, manufactured out of defining characteristics. But for the present it will simplify our exposition to treat classes as if they were real.

In actual fact, it is simpler logically to find out whether two collections have the same number of terms than it is to define what that number is. An illustration will make this clear. If there were no polygamy or polyandry anywhere in the world, it is clear that the number of husbands living at any moment would be exactly the same as the number of wives. We do not need a census to assure us of this, nor do we need to know what is the actual number of husbands and of wives. We know the number must be the same in both collections, because each husband has one wife and each wife has one husband. The relation of husband and wife is what is called "one-one."

A relation is said to be "one-one" when, if x has the relation in question to y, no other term x' has the same relation to y, and x does not have the same relation to any term y' other than y. When only the first of these two conditions is fulfilled, the relation is called "one-many"; when only the second is fulfilled, it is called "many-one." It should be observed that the number 1 is not used in these definitions.

In Christian countries, the relation of husband to wife is one-one; in Mahometan countries it is one-many; in Tibet it is many-one. The relation of father to son is one-many; that of son to father is many-one, but that of eldest son to father is one-one. If n is any number, the relation of n to $n+1$ is one-one; so is the relation of n to $2n$ or to $3n$. When we are considering only positive numbers, the relation of n to n^2 is one-one; but when negative numbers are admitted, it becomes two-one, since n and $-n$ have the same square. These instances should suffice to make clear the notions of one-one, one-many, and many-one relations, which play a great part in the principles of mathematics, not only in relation to the definition of numbers, but in many other connections.

Two classes are said to be "similar" when there is a one-one relation which correlates the terms of the one class each with one term of the other class, in the same manner in which the relation of marriage correlates husbands with wives. A few preliminary definitions will help us to state this definition more precisely. The class of those terms that have a given relation to something or other is called the *domain* of that relation: thus fathers are the domain of the relation of father to child, husbands are the domain of the relation of husband to wife, wives are the domain of the relation of wife to husband, and husbands and wives together are the domain of the relation of marriage. The relation of wife to husband is called the *converse* of the relation of husband to wife. Similarly *less* is the converse of *greater, later* is the converse of *earlier,* and so on. Generally, the converse of a given relation is that relation which holds between y and x whenever the given relation holds between x and y. The *converse domain* of a relation is the domain of its converse: thus the class of wives is the converse domain of the relation of husband to wife. We may now state our definition of similarity as follows: –

> *One class is said to be "similar" to another when there is a one-one relation of which the one class is the domain, while the other is the converse domain.*

It is easy to prove (1) that every class is similar to itself, (2) that if a class α is similar to a class β, then β is similar to α, (3) that if α is similar to β and β to γ, then α is similar to γ. A relation is said to be *reflexive* when it possesses the first of these properties, *symmetrical* when it possesses the second, and *transitive* when it possesses the third. It is obvious that a relation which is symmetrical and transitive must be reflexive throughout its domain. Relations which possess these properties are an important kind, and it is worth while to note that similarity is one of this kind of relations.

It is obvious to common sense that two finite classes have the same number of terms if they are similar, but not otherwise. The act of counting consists in establishing a one-one correlation between the set of objects counted and the natural numbers (excluding 0) that are used up in the process. Accordingly common sense concludes that there are as many objects in the set to be counted as there are numbers up to the last number used in the counting. And we also know that, so long as we confine ourselves to finite numbers, there are just n numbers from 1 up to n. Hence it follows that the last number used in counting a collection is the number of terms in the collection, provided the collection is finite. But this result, besides being only applicable to finite collections, depends upon and assumes the fact that two classes which are similar have the same number of terms; for what we do when we count (say) 10 objects is to show that the set of these objects is similar to the set of numbers 1 to 10. The notion of similarity is logically presupposed in the operation of counting, and is logically simpler though less familiar. In counting, it is necessary to take the objects counted in a certain order, as first, second, third, etc., but order is not of the essence of number: it is an irrelevant addition, an unnecessary complication from the logical point of view. The notion of similarity does not demand an order: for example, we saw that the number of husbands is the same as the number of wives, without having to establish an order of precedence among them. The notion of similarity also does not require that the classes which are similar should be finite. Take, for example, the natural numbers (excluding 0) on the one hand, and the fractions which have 1 for their numerator on the other hand: it is obvious that we can correlate 2 with 1/2, 3 with 1/3, and so on, thus proving that the two classes are similar.

We may thus use the notion of "similarity" to decide when two collections are to belong to the same bundle, in the sense in which we were asking this question earlier in this chapter. We want to make one bundle containing the class that has no members: this will be for the number 0. Then we want a bundle of all the classes that have one member: this will be for the number 1. Then, for the number 2, we want a bundle consisting of all couples; then one of all trios; and so on. Given any collection, we can define the bundle it is to belong to as being the class of all those collections that are "similar" to it. It is very easy to see that if (for example) a collection has three members, the class of all those collections that are similar to it will be the class of trios. And whatever number of terms a collection may have, those collections that are "similar" to it will have the same number of terms. We may take this as a *definition* of "having the same number of terms." It is obvious that it gives results conformable to usage so long as we confine ourselves to finite collections.

So far we have not suggested anything in the slightest degree paradoxical. But when we come to the actual definition of numbers we cannot avoid what must at first sight seem a paradox, though this impression will soon wear off. We naturally think that the class of couples (for example) is something different from the number 2. But there is no doubt about the class of couples: it is indubitable and not difficult to define, whereas the number 2, in any other sense, is a metaphysical entity about which we can never feel sure that it exists or that we have tracked it down. It is therefore more prudent to content ourselves with the class of couples, which we are sure of, than to hunt for a problematical number 2 which must always remain elusive. Accordingly we set up the following definition: –

> The number of a class is the class of all those classes that are similar to it.

Thus the number of a couple will be the class of all couples. In fact, the class of all couples will *be* the number 2, according to our definition. At the expense of a little oddity, this definition secures definiteness and indubitableness; and it is not difficult to prove that numbers so defined have all the properties that we expect numbers to have.

We may now go on to define numbers in general as any one of the bundles into which similarity collects classes. A number will be a set of classes such as that any two are similar to each other, and none outside the set are similar to any inside the set. In other words, a number (in general) is any collection which is the number of one of its members; or, more simply still:

> A number is anything which is the number of some class.

172

Such a definition has a verbal appearance of being circular, but in fact it is not. We define "the number of a given class" without using the notion of number in general; therefore we may define number in general in terms of "the number of a given class" without committing any logical error.

Definitions of this sort are in fact very common. The class of fathers, for example, would have to be defined by first defining what it is to be the father of somebody; then the class of fathers will be all those who are somebody's father. Similarly if we want to define square numbers (say), we must first define what we mean by saying that one number is the square of another, and then define square numbers as those that are the squares of other numbers. This kind of procedure is very common, and it is important to realize that it is legitimate and even often necessary.

We have now given a definition of numbers which will serve for finite collections. It remains to be seen how it will serve for infinite collections. But first we must decide what we mean by "finite" and "infinite," which cannot be done within the limits [here].

III. Mathematics and logic

Mathematics and logic, historically speaking, have been entirely distinct studies. Mathematics has been connected with science, logic with Greek. But both have developed in modern times: logic has become more mathematical and mathematics has become more logical. The consequence is that it has now become wholly impossible to draw a line between the two; in fact, the two are one. They differ as boy and man: logic is the youth of mathematics and mathematics is the manhood of logic. This view is resented by logicians who, having spent their time in the study of classical texts, are incapable of following a piece of symbolic reasoning, and by mathematicians who have learnt a technique without troubling to inquire into its meaning or justification. Both types are now fortunately growing rarer. So much of modern mathematical work is obviously on the borderline of logic, so much of modern logic is symbolic and formal, that the very close relationship of logic and mathematics has become obvious to every instructed student. The proof of their identity is, of course, a matter of detail: starting with premises which would be universally admitted to belong to logic, and arriving by deduction at results which as obviously belong to mathematics, we find that there is no point at which a sharp line can be drawn, with logic to the left and mathematics to the right. If there are still those who do not admit the identity of logic and mathematics, we may challenge them to indicate at what point, in the successive definitions and deductions of *Principia Mathematica,* they consider that logic ends and mathematics begins. It will then be obvious that any answer must be quite arbitrary.

In the earlier chapters of this book, starting from the natural numbers, we have first defined "cardinal number" and shown how to generalise the conception of number, and have then analysed the conceptions involved in the definition, until we found ourselves dealing with the fundamentals of logic. In a synthetic, deductive treatment these fundamentals come first, and the natural numbers are only reached after a long journey. Such treatment, though formally more correct than that which we have adopted, is more difficult for the reader, because the ultimate logical concepts and propositions with which it starts are remote and unfamiliar as compared with the natural numbers. Also they represent the present frontier of knowledge, beyond which is the still unknown; and the dominion of knowledge over them is not as yet very secure.

It used to be said that mathematics is the science of "quantity." "Quantity" is a vague word, but for the sake of argument we may replace it by the word "number." The statement that mathematics is the science of number would be untrue in two different ways. On the one hand, there are recognised branches of mathematics which have nothing to do with number – all geometry that does not use co-ordinates or measurement, for example: projective and descriptive geometry, down to the point at which co-ordinates are introduced, does not have to do with number, or even with quantity in the sense of *greater* or *less*. On the other hand, through the definition of cardinals, through the theory of induction and ancestral relations, through the general theory of series, and through the definitions of the arithmetical operations, it has become possible to generalize much that used to be proved only in connection with numbers. The most elementary properties of numbers are concerned with one-one relations, and similarity between classes. Addition is concerned with the construction of mutually exclusive classes respectively similar to a set of classes which are not known to be mutually exclusive. Multiplication is merged in the theory of "selections," i.e. of a certain kind of one-many relations. Finitude is merged in the general study of ancestral relations, which yields the whole theory of mathematical induction. The ordinal properties of the various kinds of number-series, and the elements of the theory of continuity of functions and the limits of functions, can be generalised so as no longer to involve any essential reference to numbers. It is a principle, in all formal reasoning, to generalize to the utmost, since we thereby secure that a given process of deduction shall have more widely applicable results; we are, therefore, in thus generalizing the reasoning of arithmetic, merely following a precept which is universally admitted in mathematics. And in thus generalizing we have, in effect, created a set of new deductive systems, in which traditional arithmetic is at once dissolved and enlarged; but whether any one of these new deductive systems – for example, the theory of selections – is to be said to belong to logic or to arithmetic is entirely arbitrary, and incapable of being decided rationally.

We are thus brought face to face with the question: What is the subject, which may be called indifferently either mathematics or logic? Is there any way in which we can define it?

Certain characteristics of the subject are clear. To begin with, we do not, in this subject, deal with particular things or particular properties: we deal formally with what can be said about *any* thing or *any* property. We are prepared to say that one and one are two, but not that Socrates and Plato are two, because, in our capacity of logicians or pure mathematicians, we have never heard of Socrates and Plato. A world in which there were no such individuals would still be a world in which one and one are two. It is not open to us, as pure mathematicians or logicians, to mention anything at all, because, if we do so, we introduce something irrelevant and not formal. We may make this clear by applying it to the case of the syllogism. Traditional logic says: "All men are mortal, Socrates is a man, therefore Socrates is mortal." Now it is clear that what we *mean* to assert, to begin with, is only that the premises imply the conclusion, not that premises and conclusion are actually true; even the most traditional logic points out that the actual truth of the premises is irrelevant to logic. Thus the first change to be made in the above traditional syllogism is to state it in the form: "If all men are mortal and Socrates is a man, then Socrates is mortal." We may now observe that it is intended to convey that this argument is valid in virtue of its *form,* not in virtue of the particular terms occurring in it. If we had omitted "Socrates is a man" from our premises, we should have had a non-formal argument, only admissible because Socrates is in fact a man; in that case we could not have generalized the argument. But when, as above, the argument is *formal,* nothing depends upon the terms that occur in it. Thus we may substitute α for *men,* β for *mortals,* and x for Socrates, where α and β are any classes whatever, and x is any individual. We then arrive at the statement: "No matter what possible values x and α and β may have, if all α's are β's and x is an α, then x is a β"; in other words, "the propositional function 'if all α's are β and x is an α, then x is a β' is always true." Here at last we have a proposition of logic – the one which is only *suggested* by the traditional statement about Socrates and men and mortals.

It is clear that, if *formal* reasoning is what we are aiming at, we shall always arrive ultimately at statements like the above, in which no actual things or properties are mentioned; this will happen through the mere desire not to waste our time proving in a particular case what can be proved generally. It would be ridiculous to go through a long argument about Socrates, and then go through precisely the same argument again about Plato. If our argument is one (say) which holds of all men, we shall prove it concerning "x," with the hypothesis "if x is a man." With this hypothesis, the argument will retain its hypothetical validity even when x is not a man. But now we shall find that our argument would still be valid if, instead of supposing x to be a man, we were to suppose him to be a monkey or a goose or a Prime Minister. We shall therefore not waste our time taking as our premiss "x is a man" but shall take "x is an α," where α is any class of individuals, or "ϕx" where ϕ is any propositional function of some assigned type. Thus the absence of all mention of particular things or properties in logic or pure mathematics is a necessary result of the fact that this study is, as we say, "purely formal."

At this point we find ourselves faced with a problem which is easier to state than to solve. The problem is: "What are the constituents of a logical proposition?" I do not know the answer, but I propose to explain how the problem arises.

Take (say) the proposition "Socrates was before Aristotle." Here it seems obvious that we have a relation between two terms, and that the constituents of the proposition (as well as of the corresponding fact) are simply the two terms and the relation, i.e. Socrates, Aristotle, and *before*. (I ignore the fact that Socrates and Aristotle are not simple; also the fact that what appear to be their names are really truncated descriptions. Neither of these facts is relevant to the present issue.) We may represent the general form of such propositions by "xRy," which may be read "x has the relation R to y." This general form may occur in logical propositions, but no particular instance of it can occur. Are we to infer that the general form itself is a constituent of such logical propositions?

Given a proposition, such as "Socrates is before Aristotle," we have certain constituents and also a certain form. But the form is not itself a new constituent; if it were, we should need a new form to embrace both it and the other constituents. We can, in fact, turn *all* the constituents of a proposition into variables, while keeping the form unchanged. This is what we do when we use such a schema as "xRy," which stands for any one of a certain class of propositions, namely, those asserting relations between two terms. We can proceed to general assertions, such as "xRy is sometimes true" – i.e. there are cases where dual relations hold. This assertion will belong to logic (or mathematics) in the sense in which we are using the word. But in this assertion we do not mention any particular things or particular relations; no particular things or relations can ever enter into a proposition of pure logic. We are left with pure *forms* as the only possible constituents of logical propositions.

I do not wish to assert positively that pure forms – e.g. the form "$x R y$" – do actually enter into propositions of the kind we are considering. The question of the analysis of such propositions is a difficult one, with conflicting considerations on the one side and on the other. We cannot embark upon this question now, but we may accept, as a first approximation, the view that *forms* are what enter into logical propositions as their constituents. And we may explain (though not formally define) what we mean by the "form" of a proposition as follows: –

> The "form" of a proposition is that, in it, that remains unchanged when every constituent of the proposition is replaced by another.

Thus "Socrates is earlier than Aristotle" has the same form as "Napoleon is greater than Wellington," though every constituent of the two propositions is different.

We may thus lay down, as a necessary (though not sufficient) characteristic of logical or mathematical propositions, that they are to be such as can be obtained from a proposition containing no variables (i.e. no such words as *all, some, a, the,* etc.) by turning every constituent into a variable and asserting that the result is always true or sometimes true, or that it is always true in respect of some of the variables that the result is sometimes true in respect of the others, or any variant of these forms. And another way of stating the same thing is to say that logic (or mathematics) is concerned only with *forms,* and is concerned with them only in the way of stating that they are always or sometimes true – with all the permutations or "always" and "sometimes" that may occur.

There are in every language some words whose sole function is to indicate form. These words, broadly speaking, are commonest in languages having fewest inflections. Take "Socrates is human." Here "is" is not a constituent of the proposition, but merely indicates the subject-predicate form. Similarly in "Socrates is earlier than Aristotle," "is" and "than" merely indicate form; the proposition is the same as "Socrates precedes Aristotle," in which these words have disappeared and the form is otherwise indicated. Form, as a rule, *can* be indicated otherwise than by specific words: the order of the words can do most of what is wanted. But this principle must not be pressed. For example, it is difficult to see how we could conveniently express molecular forms of propositions (i.e. what we call "truth-functions") without any word at all. We saw ... that one word is enough for this purpose, namely, a word or symbol expressing *incompatibility.* But without even one we should find ourselves in difficulties. This, however, is not the point that is important for our present purpose. What is important for us is to observe that form may be the one concern of a general proposition, even when no word or symbol in that proposition designates the form. If we wish to speak about the form itself, we must have a word for it; but if, as in mathematics, we wish to speak about all propositions that have the form, a word for the form will usually be found not indispensable; probably in theory it is *never* indispensable.

Assuming – as I think we may – that the forms of propositions *can* be represented by the forms of the propositions in which they are expressed without any special word for forms, we should arrive at a language in which everything formal belonged to syntax and not to vocabulary. In such a language we could express *all* the propositions of mathematics even if we did not know one single word of the language. The language of mathematical logic, if it were perfected, would be such a language. We should have symbols for variables, such as "x" and "R" and "y," arranged in various ways; and the way of arrangement would indicate that something was being said to be true of all values or some values of the variables. We should not need to know any words, because they would only be needed for giving values to the variables, which is the business of the applied mathematician, not of the pure mathematician or logician. It is one of the marks of a proposition of logic that, given a suitable language, such a proposition can be asserted in such a language by a person who knows the syntax without knowing a single word of the vocabulary.

But, after all, there are words that express form, such as "is" and "than." And in every symbolism hitherto invented for mathematical logic there are symbols having constant formal meanings. We may take as an example the symbol for incompatibility which is employed in building up truth-functions. Such words or symbols may occur in logic. The question is: How are we to define them?

Such words or symbols express what are called "logical constants." Logical constants may be defined exactly as we defined forms; in fact, they are in essence the same thing. A fundamental logical constant will be that which is in common among a number of propositions, any one of which can result from any other by substitution of terms one for another. For example, "Napoleon is greater than Wellington" results from "Socrates is earlier than Aristotle" by the substitution of "Napoleon" for "Socrates," "Wellington" for "Aristotle," and "greater" for "earlier." Some propositions can be obtained in this way from the prototype "Socrates is earlier than Aristotle" and some cannot; those that can are those that are of the form "xRy," i.e. express dual relations. We cannot obtain from the above prototype by term-for-term substitution such propositions as "Socrates is human" or "the Athenians gave the hemlock to Socrates," because the first is of the subject-predicate form and the second expresses a three-term relation. If we are to have any words in our pure logical language, they must be such as express "logical constants," and "logical constants" will always either be, or be derived from, what is in common among a group of propositions derivable from each other, in the above manner, by term-for-term substitution. And this which is in common is what we call "form."

In this sense all the "constants" that occur in pure mathematics are logical constants. The number 1, for example, is derivative from propositions of the form: "There is a term c such that ϕx is true when, and only when, x is c." This is a function of ϕ, and various different propositions result from giving different values to ϕ. We may (with a little omission of intermediate steps not relevant to our present purpose) take the above function of ϕ as what is meant by "the class determined by ϕ is a unit class" or "the class determined by ϕ is a member of 1" (1 being a class of classes). In this way, propositions in which 1 occurs acquire a meaning which is derived from a certain constant logical form. And the same will be found to be the case with all mathematical constants: all are logical constants, or symbolic abbeviations whose full use in a proper context is defined by means of logical constants.

But although all logical (or mathematical) propositions can be expressed wholly in terms of logical constants together with variables, it is not the case that, conversely, all propositions that can be expressed in this way are logical. We have found so far a necessary but not a sufficient criterion of mathematical propositions. We have sufficiently defined the character of the primitive *ideas* in terms of which all the ideas of mathematics can be *defined,* but not of the primitive *propositions* from which all the propositions of mathematics can be *deduced.* This is a more difficult matter, as to which it is not yet known what the full answer is.

We may take the axiom of infinity as an example of a proposition which, though it can be enunciated in logical terms, cannot be asserted by logic to be true. All the propositions of logic have a characteristic which used to be expressed by saying that they were analytic, or that their contradictories were self-contradictory. This mode of statement, however, is not satisfactory. The law of contradiction is merely one among logical propositions; it has no special pre-eminence; and the proof that the contradictory of some proposition is self-contradictory is likely to require other principles of deduction besides the law of contradiction. Nevertheless, the characteristic of logical propositions that we are in search of is the one which was felt, and intended to be defined, by those who said that it consisted in deducibility from the law of contradiction.

179

This characteristic, which, for the moment, we may call *tautology*, obviously does not belong to the assertion that the number of individuals in the universe is n, whatever number n may be. But for the diversity of types, it would be possible to prove logically that there are classes of n terms, where n is any finite integer; or even that there are classes of $\aleph_0$ terms. But, owing to types, such proofs ... are fallacious. We are left to empirical observation to determine whether there are as many as n individuals in the world. Among "possible" worlds, in the Leibnizian sense, there will be worlds having one, two, three, ... individuals. There does not even seem any logical necessity why there should be even one individual[4] – why, in fact, there should be any world at all. The ontological proof of the existence of God, if it were valid, would establish the logical necessity of at least one individual. But it is generally recognized as invalid, and in fact rests upon a mistaken view of existence – i.e. it fails to realize that existence can only be asserted of something described, not of something named, so that it is meaningless to argue from "this is the so-and-so" and "the so-and-so exists" to "this exists." If we reject the ontological argument, we seem driven to conclude that the existence of a world is an accident – i.e. it is not logically necessary. If that be so, no principle of logic can assert "existence" except under a hypothesis, i.e. none can be of the form "the propositional function so-and-so is sometimes true." Propositions of this form, when they occur in logic, will have to occur as hypotheses or consequences of hypotheses, not as complete asserted propositions. The complete asserted propositions of logic will all be such as affirm that some propositional function is *always* true. For example, it is always true that if p implies q and q implies r then p implies r, or that, if all α's are β's and x is an α then x is a β. Such propositions may occur in logic, and their truth is independent of the existence of the universe. We may lay it down that, if there were no universe, *all* general propositions would be true; for the contradictory of a general proposition ... is a proposition asserting existence, and would therefore always be false if no universe existed.

Logical propositions are such as can be known *a priori*, without study of the actual world. We only know from a study of empirical facts that Socrates is a man, but we know the correctness of the syllogism in its abstract form (i.e. when it is stated in terms of variables) without needing any appeal to experience. This is a characteristic, not of logical propositions in themselves, but of the way in which we know them. It has, however, a bearing upon the question what their nature may be, since there

[4]The primitive propositions in *Principia Mathematica* are such as to allow the inference that at least one individual exists. But I now view this as a defect in logical purity.

are some kinds of propositions which it would be very difficult to suppose we could know without experience.

It is clear that the definition of "logic" or "mathematics" must be sought by trying to give a new definition of the old notion of "analytic" propositions. Although we can no longer be satisfied to define logical propositions as those that follow from the law of contradiction, we can and must still admit that they are a wholly different class of propositions from those that we come to know empirically. They all have the characteristic which, a moment ago, we agreed to call "tautology." This, combined with the fact that they can be expressed wholly in terms of variables and logical constants (a logical constant being something which remains constant in a proposition even when *all* its constituents are changed), will give the definition of logic or pure mathematics. For the moment, I do not know how to define "tautology."[5] It would be easy to offer a definition which might seem satisfactory for a while; but I know of none that I feel to be satisfactory, in spite of feeling thoroughly familiar with the characteristic of which a definition is wanted. At this point, therefore, for the moment, we reach the frontier of knowledge on our backward journey into the logical foundations of mathematics.

We have now come to an end of our somewhat summary introduction to mathematical philosophy. It is impossible to convey adequately the ideas that are concerned in this subject so long as we abstain from the use of logical symbols. Since ordinary language has no words that naturally express exactly what we wish to express, it is necessary, so long as we adhere to ordinary language, to strain words into unusual meanings; and the reader is sure, after a time if not at first, to lapse into attaching the usual meanings to words, thus arriving at wrong notions as to what is intended to be said. Moreover, ordinary grammar and syntax is extraordinarily misleading. This is the case, e.g. as regards numbers; "ten men" is grammatically the same form as "white men," so that 10 might be thought to be an adjective qualifying "men." It is the case, again, wherever propositional functions are involved, and in particular as regards existence and descriptions. Because language is misleading, as well as because it is diffuse and inexact when applied to logic (for which it was never intended), logical symbolism is absolutely necessary to any exact or thorough treatment of our subject. Those readers, therefore, who wish to acquire a mastery of the principles of mathematics, will, it is to be hoped, not shrink from the labour of mastering the symbols –

[5]The importance of "tautology" for a definition of mathematics was pointed out to me by my former pupil Ludwig Wittgenstein, who was working on the problem. I do not know whether he has solved it, or even whether he is alive or dead.

a labour which is, in fact, much less than might be thought. As the above hasty survey must have made evident, there are innumerable unsolved problems in the subject, and much work needs to be done. If any student is led into a serious study of mathematical logic by this little book, it will have served the chief purpose for which it has been written.

On the infinite

DAVID HILBERT

As a result of his penetrating critique, Weierstrass has provided a solid foundation for mathematical analysis. By elucidating many notions, in particular those of minimum, function, and differential quotient, he removed the defects which were still found in the infinitesimal calculus, rid it of all confused notions about the infinitesimal, and thereby completely resolved the difficulties which stem from that concept. If in analysis today there is complete agreement and certitude in employing the deductive methods which are based on the concepts of irrational number and limit, and if in even the most complex questions of the theory of differential and integral equations, notwithstanding the use of the most ingenious and varied combinations of the different kinds of limits, there nevertheless is unanimity with respect to the results obtained, then this happy state of affairs is due primarily to Weierstrass's scientific work.

And yet in spite of the foundation Weierstrass has provided for the infinitesimal calculus, disputes about the foundations of analysis still go on.

These disputes have not terminated because the meaning of the *infinite*, as that concept is used in mathematics, has never been completely clarified. Weierstrass's analysis did indeed eliminate the infinitely large and the infinitely small by reducing statements about them to [statements about] relations between finite magnitudes. Nevertheless the infinite still appears in the infinite numerical series which defines the real numbers and in the concept of the real number system which is thought of as a completed totality existing all at once.

In his foundation for analysis, Weierstrass accepted unreservedly and used repeatedly those forms of logical deduction in which the concept of the infinite comes into play, as when one treats of *all* real numbers with a certain property or when one argues that *there exist* real numbers with a certain property.

Delivered June 4, 1925, before a congress of the Westphalian Mathematical Society in Munster, in honor of Karl Weierstrass. Translated by Erna Putnam and Gerald J. Massey from *Mathematische Annalen* (Berlin) vol. 95 (1926), pp. 161–90. Permission for the translation and inclusion of the article in this volume was kindly granted by the publishers, Springer Verlag.

183

Hence the infinite can reappear in another guise in Weierstrass's theory and thus escape the precision imposed by his critique. It is, therefore, *the problem of the infinite* in the sense just indicated which we need to resolve once and for all. Just as in the limit processes of the infinitesimal calculus, the infinite in the sense of the infinitely large and the infinitely small proved to be merely a figure of speech, so too we must realize that the infinite in the sense of an infinite totality, where we still find it used in deductive methods, is an illusion. Just as operations with the infinitely small were replaced by operations with the finite which yielded exactly the same results and led to exactly the same elegant formal relationships, so in general must deductive methods based on the infinite be replaced by finite procedures which yield exactly the same results; i.e., which make possible the same chains of proofs and the same methods of getting formulas and theorems.

The goal of my theory is to establish once and for all the certitude of mathematical methods. This is a task which was not accomplished even during the critical period of the infinitesimal calculus. This theory should thus complete what Weierstrass hoped to achieve by his foundation for analysis and toward the accomplishment of which he has taken a necessary and important step.

But a still more general perspective is relevant for clarifying the concept of the infinite. A careful reader will find that the literature of mathematics is glutted with inanities and absurdities which have had their source in the infinite. For example, we find writers insisting, as though it were a restrictive condition, that in rigorous mathematics only a *finite* number of deductions are admissible in a proof – as if someone had succeeded in making an infinite number of them.

Also old objections which we supposed long abandoned still reappear in different forms. For example, the following recently appeared: Although it may be possible to introduce a concept without risk, i.e., without getting contradictions, and even though one can prove that its introduction causes no contradictions to arise, still the introduction of the concept is not thereby justified. Is not this exactly the same objection which was once brought against complex-imaginary numbers when it was said: "True, their use doesn't lead to contradictions. Nevertheless their introduction is unwarranted, for imaginary magnitudes do not exist"? If, apart from proving consistency, the question of the justification of a measure is to have any meaning, it can consist only in ascertaining whether the measure is accompanied by commensurate success. Such success is in fact essential, for in mathematics as elsewhere success is the supreme court to whose decisions everyone submits.

On the infinite

As some people see ghosts, another writer seems to see contradictions even where no statements whatsoever have been made, viz., in the concrete world of sensation, the "consistent functioning" of which he takes as special assumption. I myself have always supposed that only statements, and hypotheses insofar as they lead through deductions to statements, could contradict one another. The view that facts and events could themselves be in contradiction seems to me to be a prime example of careless thinking.

The foregoing remarks are intended only to establish the fact that the definitive clarification of *the nature of the infinite,* instead of pertaining just to the sphere of specialized scientific interests, is needed for *the dignity of the human intellect* itself.

From time immemorial, the infinite has stirred men's *emotions* more than any other question. Hardly any other *idea* has stimulated the mind so fruitfully. Yet, no other *concept* needs *clarification* more than it does.

Before turning to the task of clarifying the nature of the infinite, we should first note briefly what meaning is actually given to the infinite. First let us see what we can learn from physics. One's first naïve impression of natural events and of matter is one of permanency, of continuity. When we consider a piece of metal or a volume of liquid, we get the impression that they are unlimitedly divisible, that their smallest parts exhibit the same properties that the whole does. But wherever the methods of investigating the physics of matter have been sufficiently refined, scientists have met divisibility boundaries which do not result from the shortcomings of their efforts but from the very nature of things. Consequently we could even interpret the tendency of modern science as emancipation from the infinitely small. Instead of the old principle *natura non facit saltus,* we might even assert the opposite, viz., "nature makes jumps."

It is common knowledge that all matter is composed of tiny building blocks called "atoms," the combinations and connections of which produce all the variety of macroscopic objects. Still physics did not stop at the atomism of matter. At the end of the last century there appeared the atomism of electricity which seems much more bizarre at first sight. Electricity, which until then had been thought of as a fluid and was considered the model of a continuously active agent, was then shown to be built up of positive and negative *electrons.*

In addition to matter and electricity, there is one other entity in physics for which the law of conservation holds, viz., energy. But it has been established that even energy does not unconditionally admit of infinite divisibility. Planck has discovered *quanta of energy.*

185

Hence, a homogeneous continuum which admits of the sort of divisibility needed to realize the infinitely small is nowhere to be found in reality. The infinite divisibility of a continuum is an operation which exists only in thought. It is merely an idea which is in fact impugned by the results of our observations of nature and of our physical and chemical experiments.

The second place where we encounter the question of whether the infinite is found in nature is in the consideration of the universe as a whole. Here we must consider the expanse of the universe to determine whether it embraces anything infinitely large. But here again modern science, in particular astronomy, has reopened the question and is endeavoring to solve it, not by the defective means of metaphysical speculation, but by reasons which are based on experiment and on the application of the laws of nature. Here, too, serious objections against infinity have been found. *Euclidean* geometry necessarily leads to the postulate that space is infinite. Although euclidean geometry is indeed a consistent conceptual system, it does not thereby follow that euclidean geometry actually holds in reality. Whether or not real space is euclidean can be determined only through observation and experiment. The attempt to prove the infinity of space by pure speculation contains gross errors. From the fact that outside a certain portion of space there is always more space, it follows only that space is unbounded, not that it is infinite. Unboundedness and finiteness are compatible. In so-called *elliptical* geometry, mathematical investigation furnishes the natural model of a finite universe. Today the abandonment of euclidean geometry is no longer merely a mathematical or philosophical speculation but is suggested by considerations which originally had nothing to do with the question of the finiteness of the universe. Einstein has shown that euclidean geometry must be abandoned. On the basis of his gravitational theory, he deals with cosmological questions and shows that a finite universe is possible. Moreover, all the results of astronomy are perfectly compatible with the postulate that the universe is elliptical.

We have established that the universe is finite in two respects, i.e., as regards the infinitely small and the infinitely large. But it may still be the case that the infinite occupies a justified place *in our thinking,* that it plays the role of an indispensable concept. Let us see what the situation is in mathematics. Let us first interrogate that purest and simplest offspring of the human mind, viz., number theory. Consider one formula out of the rich variety of elementary formulas of number theory, e.g., the formula

$$1^2 + 2^2 + 3^2 \cdots + n^2 = \tfrac{1}{6}n(n+1)(2n+1)$$

Since we may substitute any integer whatsoever for n, for example $n = 2$ or $n = 5$, this formula implicitly contains *infinitely many* propositions. This characteristic is essential to a formula. It enables the formula to represent the solution of an arithmetical problem and necessitates a special idea for its proof. On the other hand, the individual numerical equations

$$1^2 + 2^2 = \tfrac{1}{6} \cdot 2 \cdot 3 \cdot 5$$
$$1^2 + 2^2 + 3^2 + 4^2 + 5^2 = \tfrac{1}{6} \cdot 5 \cdot 6 \cdot 11$$

can be verified simply by calculation and hence individually are of no especial interest.

We encounter a completely different and quite unique conception of the notion of infinity in the important and fruitful method of *ideal elements*. The method of ideal elements is used even in elementary plane geometry. The points and straight lines of the plane originally are real, actually existent objects. One of the axioms that hold for them is the axiom of connection: one and only one straight line passes through two points. It follows from this axiom that two straight lines intersect at most at one point. There is no theorem that two straight lines always intersect at some point, however, for the two straight lines might well be parallel. Still we know that by introducing ideal elements, viz., infinitely long lines and points at infinity, we can make the theorem that two straight lines always intersect at one and only one point come out universally true. These ideal "infinite" elements have the advantage of making the system of connection laws as simple and perspicuous as possible. Moreover, because of the symmetry between a point and a straight line, there results the very fruitful principle of duality for geometry.

Another example of the use of ideal elements are the familiar *complex-imaginary* magnitudes of algebra which serve to simplify theorems about the existence and number of the roots of an equation.

Just as infinitely many straight lines, viz., those parallel to each other, are used to define an ideal point in geometry, so certain systems of infinitely many numbers are used to define an *ideal number*. This application of the principle of ideal elements is the most ingenious of all. If we apply this principle systematically throughout an algebra, we obtain exactly the same simple and familiar laws of division which hold for the familiar whole numbers 1, 2, 3, 4, ... We are already in the domain of higher arithmetic.

We now come to the most aesthetic and delicately erected structure of mathematics, viz., analysis. You already know that infinity plays the leading role in analysis. In a certain sense, mathematical analysis is a symphony of the infinite.

The tremendous progress made in the infinitesimal calculus results mainly from operating with mathematical systems of infinitely many elements. But, as it seemed very plausible to identify the infinite with the "very large", there soon arose inconsistencies which were known in part to the ancient sophists, viz., the so-called paradoxes of the infinitesimal calculus. But the recognition that many theorems which hold for the finite (for example, the part is smaller than the whole, the existence of a minimum and a maximum, the interchangeability of the order of the terms of a sum or a product) cannot be immediately and unrestrictedly extended to the infinite, marked fundamental progress. I said at the beginning of this paper that these questions have been completely clarified, notably through Weierstrass's acuity. Today, analysis is not only infallible within its domain but has become a practical instrument for using the infinite.

But analysis alone does not provide us with the deepest insight into the nature of the infinite. This insight is procured for us by a discipline which comes closer to a general philosophical way of thinking and which was designed to cast new light on the whole complex of questions about the infinite. This discipline, created by George Cantor, is set theory. In this paper, we are interested only in that unique and original part of set theory which forms the central core of Cantor's doctrine, viz., the theory of *transfinite* numbers. This theory is, I think, the finest product of mathematical genius and one of the supreme achievements of purely intellectual human activity. What, then, is this theory?

Someone who wished to characterize briefly the new conception of the infinite which Cantor introduced might say that in analysis we deal with the infinitely large and the infinitely small only as limiting concepts, as something becoming, happening, i.e., with the *potential infinite*. But this is not the true infinite. We meet the true infinite when we regard the totality of numbers $1, 2, 3, 4, \ldots$ itself as a completed unity, or when we regard the points of an interval as a totality of things which exists all at once. This kind of infinity is known as *actual infinity*.

Frege and Dedekind, the two mathematicians most celebrated for their work in the foundations of mathematics, independently of each other used the actual infinite to provide a foundation for arithmetic which was independent of both intuition and experience. This foundation was based solely on pure logic and made use only of deductions that were purely logical. Dedekind even went so far as not to take the notion of finite number from intuition but to derive it logically by employing the concept of an infinite set. But is was Cantor who systematically developed the concept of the actual infinite. Consider the two examples of the infinite already mentioned

188

On the infinite

1. $1, 2, 3, 4, \ldots$.
2. The points of the interval 0 to 1 or, what comes to the same thing, the totality of real numbers between 0 and 1.

It is quite natural to treat these examples from the point of view of their size. But such a treatment reveals amazing results with which every mathematician today is familiar. For when we consider the set of all rational numbers, i.e., the fractions $1/2$, $1/3$, $2/3$, $1/4, \ldots, 3/7, \ldots$, we notice that – from the sole standpoint of its size – this set is no larger than the set of integers. Hence we say that the rational numbers can be counted in the usual way; i.e., that they are enumerable. The same holds for the set of all roots of numbers, indeed even for the set of all algebraic numbers. The second example is analogous to the first. Surprisingly enough, the set of all the points of a square or cube is no larger than the set of points of the interval 0 to 1. Similarly for the set of all continuous functions. On learning these facts for the first time, you might think that from the point of view of size there is only one unique infinite. No, indeed! The sets in examples (1) and (2) are not, as we say, "equivalent". Rather, the set (2) cannot be enumerated, for it is larger than the set (1). We meet what is new and characteristic in Cantor's theory at this point. The points of an interval cannot be counted in the usual way, i.e., by counting $1, 2, 3, \ldots$. But, since we admit the actual infinite, we are not obliged to stop here. When we have counted $1, 2, 3, \ldots$, we can regard the objects thus enumerated as an infinite set existing all at once in a particular order. If, following Cantor, we call the type of this order ω, then counting continues naturally with $\omega+1$, $\omega+2, \ldots$ up to $\omega+\omega$ or $\omega \cdot 2$, and then again

$$(\omega \cdot 2) + 1, (\omega \cdot 2) + 2, (\omega \cdot 2) + 3, \ldots (\omega \cdot 2) + \omega \text{ or } \omega \cdot 3,$$

and further

$$\omega \cdot 2, \omega \cdot 3, \omega \cdot 4, \ldots, \omega \cdot \omega \text{ (or } \omega^2), \omega^2 + 1, \ldots,$$

so that we finally get this table:

$$1, 2, 3, \ldots$$
$$\omega, \omega+1, \omega+2, \ldots$$
$$\omega \cdot 2, (\omega \cdot 2)+1, (\omega \cdot 2)+2, \ldots$$
$$\omega \cdot 3, (\omega \cdot 3)+1, (\omega \cdot 3)+2, \ldots$$
$$\vdots$$
$$\omega^2, \omega^2+1, \ldots$$
$$\omega^2+\omega, \omega^2+\omega \cdot 2, \omega^2+\omega \cdot 3, \ldots$$
$$\omega^2 \cdot 2, (\omega^2 \cdot 2)+1, \ldots$$

189

$$(\omega^2 \cdot 2) + \omega, \ (\omega^2 \cdot 2) + (\omega \cdot 2), \ldots$$
$$\omega^3, \ldots$$
$$\omega^4, \ldots$$
$$\vdots$$
$$\omega^\omega, \omega^{\omega^\omega}, \omega^{\omega^{\omega^\omega}}, \ldots$$

These are Cantor's first transfinite numbers, or, as he called them, the numbers of the second number class. We arrive at them simply by extending counting beyond the ordinarily enumerably infinite, i.e., by a natural and uniquely determined consistent continuation of ordinary finite counting. As until now we counted only the first, second, third, ... member of a set, we not count also the ωth, $(\omega + 1)$st, ..., ω^ωth member.

Given these developments one naturally wonders whether or not, by using these transfinite numbers, one can really count those sets which cannot be counted in the ordinary way.

On the basis of these concepts, Cantor developed the theory of transfinite numbers quite successfully and invented a full calculus for them. Thus, thanks to the Herculean collaboration of Frege, Dedekind, and Cantor, the infinite was made king and enjoyed a reign of great triumph. In daring flight, the infinite had reached a dizzy pinnacle of success.

But reaction was not lacking. It took in fact a very dramatic form. It set in perfectly analogously to the way reaction had set in against the development of the infinitesimal calculus. In the joy of discovering new and important results, mathematicians paid too little attention to the validity of their deductive methods. For, simply as a result of employing definitions and deductive methods which had become customary, contradictions began gradually to appear. These contradictions, the so-called paradoxes of set theory, though at first scattered, became progressively more acute and more serious. In particular, a contradiction discovered by Zermelo and Russell had a downright catastrophic effect when it became known throughout the world of mathematics. Confronted by these paradoxes, Dedekind and Frege completely abandoned their point of view and retreated. Dedekind hesitated a long time before permitting a new edition of his epoch-making treatise *Was sind und was sollen die Zahlen* to be published. In an epilogue, Frege too had to acknowledge that the direction of his book *Grundgesetze der Arithmetik* was wrong. Cantor's doctrine, too, was attacked on all sides. So violent was this reaction that even the most ordinary and fruitful concepts and the simplest and most important deductive methods of mathematics were threatened and their employment was on the verge of being declared illicit. The old order had its defenders, of course. Their defensive tactics, however, were too fainthearted and they never formed a united front at the vital spots. Too many different remedies for the paradoxes were offered, and the methods proposed to clarify them were too variegated.

Admittedly, the present state of affairs where we run up against the paradoxes is intolerable. Just think, the definitions and deductive methods which everyone learns, teaches, and uses in mathematics, the paragon of truth and certitude, lead to absurdities! If mathematical thinking is defective, where are we to find truth and certitude?

There is, however, a completely satisfactory way of avoiding the paradoxes without betraying our science. The desires and attitudes which help us find this way and show us what direction to take are these:

1. Wherever there is any hope of salvage, we will carefully investigate fruitful definitions and deductive methods. We will nurse them, strengthen them, and make them useful. No one shall drive us out of the paradise which Cantor has created for us.

2. We must establish throughout mathematics the same certitude for our deductions as exists in ordinary elementary number theory, which no one doubts and where contradictions and paradoxes arise only through our own carelessness.

Obviously these goals can be attained only after we have fully elucidated *the nature of the infinite*.

We have already seen that the infinite is nowhere to be found in reality, no matter what experiences, observations, and knowledge are appealed to. Can thought about things be so much different from things? Can thinking processes be so unlike the actual processes of things? In short, can thought be so far removed from reality? Rather is it not clear that, when we think that we have encountered the infinite in some real sense, we have merely been seduced into thinking so by the fact that we often encounter extremely large and extremely small dimensions in reality?

Does material logical deduction somehow deceive us or leave us in the lurch when we apply it to real things or events?[1] No! Material logical deduction is indispensable. It deceives us only when we form arbitrary abstract definitions, especially those which involve infinitely many objects. In such cases we have illegitimately used material logical deduction; i.e., we have not paid sufficient attention to the preconditions necessary for its valid use. In recognizing that there are such preconditions that must be taken into account, we find ourselves in agreement with the philoso-

---

[1][Throughout this paper the German word 'inhaltlich' has been translated by the words 'material' or 'materially' which are reserved for that purpose and which are used to refer to matter in the sense of the traditional distinction between matter or content and logical form. – Tr.]

phers, notably with Kant. Kant taught – and it is an integral part of his doctrine – that mathematics treats a subject matter which is given independently of logic. Mathematics, therefore, can never be grounded solely on logic. Consequently, Frege's and Dedekind's attempts to so ground it were doomed to failure.

As a further precondition for using logical deduction and carrying out logical operations, something must be given in conception, viz., certain extralogical concrete objects which are intuited as directly experienced prior to all thinking. For logical deduction to be certain, we must be able to see every aspect of these objects, and their properties, differences, sequences, and contiguities must be given, together with the objects themselves, as something which cannot be reduced to something else and which requires no reduction. This is the basic philosophy which I find necessary, not just for mathematics, but for all scientific thinking, understanding, and communicating. The subject matter of mathematics is, in accordance with this theory, the concrete symbols themselves whose structure is immediately clear and recognizable.

Consider the nature and methods of ordinary finitary number theory. It can certainly be constructed from numerical structures through intuitive material considerations. But mathematics surely does not consist solely of numerical equations and surely cannot be reduced to them alone. Still one could argue that mathematics is an apparatus which, when applied to integers, always yields correct numerical equations. But in that event we still need to investigate the structure of this apparatus thoroughly enough to make sure that it in fact always yields correct equations. To carry out such an investigation, we have available only the same concrete material finitary methods as were used to derive numerical equations in the construction of number theory. This scientific requirement can in fact be met, i.e., it is possible to obtain in a purely intuitive and finitary way – the way we attain the truths of number theory – the insights which guarantee the validity of the mathematical apparatus.

Let us consider number theory more closely. In number theory we have the numerical symbols

$$1, 11, 111, 11111$$

where each numerical symbol is intuitively recognizable by the fact it contains only 1's. These numerical symbols which are themselves our subject matter have no significance in themselves. But we require in addition to these symbols, even in elementary number theory, other symbols which have meaning and which serve to facilitate communication; for example the symbol 2 is used as an abbreviation for the numerical symbol 11, and the numerical symbol 3 as an abbreviation for the numerical symbol 111. Moreover, we use symbols like $+$, $=$, and $>$ to communicate statements. $2+3=3+2$ is intended to communicate the fact that $2+3$ and $3+2$, when abbreviations are taken into account, are the self-same numerical symbol, viz., the numerical symbol 11111. Similarly $3>2$ serves to communicate the fact that the symbol 3, i.e., 111, is longer than the symbol 2, i.e., 11; or, in other words, that the latter symbol is a proper part of the former.

We also use the letters $\mathfrak{a}$, $\mathfrak{b}$, $\mathfrak{c}$ for communication. Thus $\mathfrak{b}>\mathfrak{a}$ communicates the fact that the numerical symbol $\mathfrak{b}$ is longer than the numerical symbol $\mathfrak{a}$. From this point of view, $\mathfrak{a}+\mathfrak{b}=\mathfrak{b}+\mathfrak{a}$ communicates only the fact that the numerical symbol $\mathfrak{a}+\mathfrak{b}$ is the same as $\mathfrak{b}+\mathfrak{a}$. The content of this communication can also be proved through material deduction. Indeed, this kind of intuitive material treatment can take us quite far.

But let me give you an example where this intuitive method is outstripped. The largest known prime number is (39 digits)

$$\mathfrak{p}=170\,141\,183\,460\,469\,231\,731\,687\,303\,715\,884\,105\,727$$

By a well-known method due to Euclid we can give a proof, one which remains entirely within our finitary framework, of the statement that between $\mathfrak{p}+1$ and $\mathfrak{p}!+1$ there exists at least one new prime number. The statement itself conforms perfectly to our finitary approach, for the expression 'there exists' serves only to abbreviate the expression: it is certain that $\mathfrak{p}+1$ or $\mathfrak{p}+2$ or $\mathfrak{p}+3\ldots$ or $\mathfrak{p}!+1$ is a prime number. Furthermore, since it obviously comes down to the same thing to say: there exists a prime number which is

1. $>\mathfrak{p}$, and at the same time is
2. $\leqslant\mathfrak{p}!+1$,

we are led to formulate a theorem which expresses only a part of what the euclidean theorem expresses; viz., the theorem that there exists a prime number $>\mathfrak{p}$. Although this theorem is a much weaker statement in terms of content – it asserts only part of what the euclidean theorem asserts – and although the passage from the euclidean theorem to this one seem quite harmless, that passage nonetheless involves a leap into the transfinite when the partial statement is taken out of context and regarded as an independent statement.

How can this be? Because we have an existential statement, 'there exists'! True, we had a similar expression in the euclidean theorem, but there the 'there exists' was, as I already mentioned, an abbreviation for: either $\mathfrak{p}+1$ or $\mathfrak{p}+2$ or $\mathfrak{p}+3\ldots$ or $\mathfrak{p}!+1$ is a prime number – just as when, instead of saying 'either this piece of chalk or this piece or this piece ... or this piece is red' we say briefly 'there exists a red piece of chalk among these pieces'. A statement such as 'there exists' an object with a certain property in a finite totality conforms perfectly to our finitary approach. But a statement like 'either $p+1$ or $p+2$ or $p+3$... or (ad infinitum)... has a certain property' is itself an infinite logical product. Such an extension into the infinite is, unless further explanation and precautions are forthcoming, no more permissible than the extension from finite to infinite products in calculus. Such extensions, accordingly, usually lapse into meaninglessness.

From our finitary point of view, an existential statement of the form 'there exists a number with a certain property' has in general only the significance of a partial statement; i.e., it is regarded as part of a more determinate statement. The more precise formulation may, however, be unnecessary for many purposes.

In analyzing an existential statement whose content cannot be expressed by a finite disjunction, we encounter the infinite. Similarly, by negating a general statement, i.e., one which refers to arbitrary numerical symbols, we obtain a transfinite statement. For example, the statement that if a is a numerical symbol, then $a+1=1+a$ is universally true, is from our finitary perspective *incapable of negation*. We will see this better if we consider that this statement cannot be interpreted as a conjunction of infinitely many numerical equations by means of 'and' but only as a hypothetical judgment which asserts something for the case when a numerical symbol is given.

From our finitary viewpoint, therefore, we cannot argue that an equation like the one just given, where an arbitrary numerical symbol occurs, either holds for every symbol or is disproved by a counter example. Such an argument, being an application of the law of excluded middle, rests on the presupposition that the statement of the universal validity of such an equation is capable of negation.

At any rate, we note the following: if we remain within the domain of finitary statements, as indeed we must, we have as a rule very complicated logical laws. Their complexity becomes unmanageable when the expressions 'all' and 'there exists' are combined and when they occur in expressions nested within other expressions. In short, the logical laws which Aristotle taught and which men have used ever since they began to think do not hold. We could, of course, develop logical laws which do hold for the domain of finitary statements. But it would do us no good to develop such a logic, for we do not want to give up the use of the simple laws of Aristotelian logic. Furthermore, no one, though he speak with the tongues of angels, could keep people from negating general statements, or from forming partial judgments, or from using *tertium non datur*. What, then, are we to do?

On the infinite

Let us remember that *we are mathematicians* and that as mathematicians we have often been in precarious situations from which we have been rescued by the ingenious method of ideal elements. I showed you some illustrious examples of the use of this method at the beginning of this paper. Just as $i = \sqrt{-1}$ was introduced to preserve in simplest form the laws of algebra (for example, the laws about the existence and number of roots of an equation); just as ideal factors were introduced to preserve the simple laws of divisibility for algebraic whole numbers (for example, a common ideal divisor for the numbers 2 and $1 + \sqrt{-5}$ was introduced, though no such divisor really exists); similarly, to preserve the simple formal rules of ordinary Aristotelian logic, we must *supplement the finitary statements with ideal statements*. It is quite ironic that the deductive methods which Kronecker so vehemently attacked are the exact counterpart of what Kronecker himself admired so enthusiastically in Kummer's work on number theory which Kronecker extolled as the highest achievement of mathematics.

How do we obtain *ideal statements?* It is remarkable as well as a favorable and promising fact that to obtain ideal statements, we need only continue in a natural and obvious fashion the development which the theory of the foundations of mathematics has already undergone. Indeed, we should realize that even elementary mathematics goes beyond the standpoint of intuitive number theory. Intuitive, material number theory, as we have been construing it, does not include the method of algebraic computation with letters. Formulas were always used exclusively for communication in intuitive number theory. The letters stood for numerical symbols and an equation communicated the fact that the two symbols coincided. In algebra, on the other hand, we regard expressions containing letters as independent structures which formalize the material theorems of number theory. In place of statements about numerical symbols, we have formulas which are themselves the concrete objects of intuitive study. In place of number-theoretic material proof, we have the derivation of a formula from another formula according to determinate rules.

Hence, as we see even in algebra, a proliferation of finitary objects takes place. Up to now the only objects were numerical symbols like $1, 11, \ldots, 11111$. These alone were the objects of material treatment. But mathematical practice goes further, even in algebra. Indeed, even when from our finitary viewpoint a formula is valid with respect to what it signifies as, for example, the theorem that always

$$\mathfrak{a} + \mathfrak{b} = \mathfrak{b} + \mathfrak{a},$$

where $\mathfrak{a}$ and $\mathfrak{b}$ stand for particular numerical symbols, nevertheless we

195

prefer not to use this form of communication but to replace it instead by the formula

$$a + b = b + a.$$

This latter formula is in no wise an immediate communication of something signified but is rather a certain formal structure whose relation to the old finitary statements,

$$2 + 3 = 3 + 2,$$
$$5 + 7 = 7 + 5,$$

consists in the fact that, when a and b are replaced in the formula by the numerical symbols 2, 3, 5, 7, the individual finitary statements are thereby obtained, i.e., by a proof procedure, albeit a very simple one. We therefore conclude that a, b, $=$, $+$, as well as the whole formula $a + b = b + a$ mean nothing in themselves, no more than the numerical symbols meant anything. Still we can derive from that formula other formulas to which we do ascribe meaning, viz., by interpreting them as communications of finitary statements. Generalizing this conclusion, we conceive mathematics to be a stock of two kinds of formulas: first, those to which the meaningful communications of finitary statements correspond; and, secondly, other formulas which signify nothing and which are the *ideal structures of our theory*.

Now what was our goal? In mathematics, on the one hand, we found finitary statements which contained only numerical symbols, for example,

$$3 > 2, \ 2 + 3 = 3 + 2, \ 2 = 3, \ 1 \neq 1$$

which from our finitary standpoint are immediately intuitable and understandable without recourse to anything else. These statements can be negated, truly or falsely. One can apply Aristotelian logic unrestrictedly to them without taking special precautions. The principle of non-contradiction holds for them; i.e., the negation of one of these statements and the statement itself cannot both be true. *Tertium non datur* holds for them; i.e., either a statement or its negation is true. To say that a statement is false is equivalent to saying that its negation is true. On the other hand, in addition to these elementary statements which present no problems, we also found more problematic finitary statements; e.g., we found finitary statements that could not be split up into partial statements. Finally, we introduced ideal statements in order that the ordinary laws of logic would hold universally. But since these ideal statements, viz., the formulas, do not mean anything insofar as they do not express finitary statements, logical operations cannot be materially applied to them as they can be to finitary statements. It is, therefore, necessary to formalize the logical operations and the mathematical proofs themselves. This formalization necessitates translating logical relations into formulas. Hence, in addition to mathematical symbols, we must also introduce logical symbols such as

$$\& \, , \quad \vee \, , \quad \rightarrow \, , \quad \sim^{2}$$
$$\text{(and)} \quad \text{(or)} \quad \text{(implies)} \quad \text{(not)}$$

and in addition to the mathematical variables $a, b, c, \ldots$ we must also employ logical variables, viz., the propositional variables $A, B, C, \ldots$.

How can this be done? Fortunately that same preestablished harmony which we have so often observed operative in the history of the development of science, the same preestablished harmony which aided Einstein by giving him the general invariant calculus already fully developed for his gravitational theory, comes also to our aid: we find the logical calculus already worked out in advance. To be sure, the logical calculus was originally developed from an altogether different point of view. The symbols of the logical calculus originally were introduced only in order to communicate. Still it is consistent with our finitary viewpoint to deny any meaning to logical symbols, just as we denied meaning to mathematical symbols, and to declare that the formulas of the logical calculus are ideal statements which mean nothing in themselves. We possess in the logical calculus a symbolic language which can transform mathematical statements into formulas and express logical deduction by means of formal procedures. In exact analogy to the transition from material number theory to formal algebra, we now treat the signs and operation symbols of the logical calculus in abstraction from their meaning. Thus we finally obtain, instead of material mathematical knowledge which is communicated in ordinary language, just a set of formulas containing mathematical and logical symbols which are generated successively, according to determinate rules. Certain of the formulas correspond to mathematical axioms. The rules whereby the formulas are derived from one another correspond to material deduction. Material deduction is thus replaced by a formal procedure governed by rules. The rigorous transition from a naïve to a formal treatment is effected, therefore, both for the axioms (which, though originally viewed naïvely as basic truths, have been long treated in modern axiomatics as mere relations between concepts) and for the logical calculus (which originally was supposed to be merely a different language).

We will now explain briefly how *mathematical proofs* are formalized.

[2][Although Hilbert's original paper used '$-$' as the sign for negation, we have substituted '$\sim$' for greater conformity with the notation used in other papers in this collection. – Eds.]

I have already said that certain formulas which serve as building blocks for the formal structure of mathematics are called "axioms." A mathematical proof is a figure which as such must be accessible to our intuition. It consists of deductions made according to the deduction schema

$$\frac{\begin{array}{c}\mathfrak{S}\\ \mathfrak{S} \to \mathfrak{T}\end{array}}{\mathfrak{T}}$$

where each premise, i.e., the formulas $\mathfrak{S}$ and $\mathfrak{S} \to \mathfrak{T}$, either is an axiom, or results from an axiom by substitution, or is the last formula of a previous deduction, or results from such a formula by substitution. A formula is said to be provable if it is the last formula of a proof.

Our program itself guides *the choice of axioms for our theory of proof.* Notwithstanding a certain amount of arbitrariness in the choice of axioms, as in geometry certain groups of axioms are qualitatively distinguishable. Here are some examples taken from each of these groups:

I. Axioms for implication
    (i) $A \to (B \to A)$
        (addition of a hypothesis)
    (ii) $(B \to C) \to \{(A \to B) \to (A \to C)\}$
        (elimination of a statement)
II. Axioms for negation
    (i) $\{A \to (B \,\&\, {\sim}B)\} \to {\sim}A$
        (law of contradiction)
    (ii) ${\sim}{\sim}A \to A$
        (law of double negation)

The axioms in groups I and II are simply the axioms of the propositional calculus.

III. Transfinite axioms
    (i) $(a)A(a) \to A(b)$
        (inference from the universal to the particular; Aristotelian axiom);
    (ii) ${\sim}(a)A(a) \to (\exists a){\sim}A(a)$
        (if a predicate does not apply universally, then there is a counterexample);
    (iii) ${\sim}(\exists a)A(a) \to (a){\sim}A(a)$
        (if there are no instances of a proposition, then the proposition is false for all a).

At this point we discover the very remarkable fact that these transfinite axioms can be derived from a single axiom which contains the gist of the so-called axiom of choice, the most disputed axiom in the literature of mathematics:

$$(\text{i}') \quad A(a) \rightarrow A(\epsilon A)$$

where ϵ is the transfinite, logical choice-function.

Then the following specifically mathematical axioms are added to those just given:

IV. Axioms for identity
  (i) $a = a$
  (ii) $a = b \rightarrow \{A(a) \rightarrow A(b)\}$,

and finally

V. Axioms for number
  (i) $a + 1 \neq 0$
  (ii) The axiom of complete induction.

Thus we are now in a position to carry out our theory of proof and to construct the system of provable formulas, i.e., mathematics. But in our general joy over this achievement and in our particular joy over finding that indispensable tool, the logical calculus, already developed without any effort on our part, we must not forget the essential condition of our work. There is just one condition, albeit an absolutely necessary one, connected with the method of ideal elements. That condition is a *proof of consistency*, for the extension of a domain by the addition of ideal elements is legitimate only if the extension does not cause contradictions to appear in the old, narrower domain, or, in other words, only if the relations that obtain among the old structures when the ideal structures are deleted are always valid in the old domain.

The problem of consistency is easily handled in the present circumstances. It reduces obviously to proving that from our axioms and according to the rules we laid down we cannot get '$1 \neq 1$' as the last formula of a proof, or, in other words, that '$1 \neq 1$' is not a provable formula. This task belongs just as much to the domain of intuitive treatment as does, for example, the task of finding a proof of the irrationality of $\sqrt{2}$ in materially constructed number theory – i.e., a proof that it is impossible to find two numerical symbols $\mathfrak{a}$ and $\mathfrak{b}$ which stand in the relation $\mathfrak{a}^2 = 2\mathfrak{b}^2$, or in other words, that one cannot produce two numerical symbols with a certain property. Similarly, it is incumbent on us to show that one cannot produce a certain kind of proof. A formalized proof, like a numerical symbol, is a concrete and visible object. We can describe it completely. Further, the requisite property of the last formula; viz., that it read '$1 \neq 1$', is a concretely ascertainable property of the proof. And since we can, as a matter of fact, prove that it is impossible to get a proof which has that formula as its last formula, we thereby justify our introduction of ideal statement.

It is also a pleasant surprise to discover that, at the very same time, we have resolved a problem which has plagued mathematicians for a long time, viz., the problem of proving the consistency of the axioms of arithmetic. For, wherever the axiomatic method is used, the problem of proving consistency arises. Surely in choosing, understanding, and using rules and axioms we do not want to rely solely on blind faith. In geometry and physical theory, proof of consistency is effected by reducing their consistency to that of the axioms of arithmetic. But obviously we cannot use this method to prove the consistency of arithmetic itself. Since our theory of proof, based on the method of ideal elements, enables us to take this last important step, it forms the necessary keystone of the doctrinal arch of axiomatics. What we have twice experienced, once with the paradoxes of the infinitesimal calculus and once with the paradoxes of set theory, will not be experienced a third time, nor ever again.

The theory of proof which we have here sketched not only is capable of providing a solid basis for the foundations of mathematics but also, I believe, supplies a general method for treatment fundamental mathematical questions which mathematicians heretofore have been unable to handle.

In a sense, mathematics has become a court of arbitration, a supreme tribunal to decide fundamental questions – on a concrete basis on which everyone can agree and where every statement can be controlled.

The assertions of the new so-called "intuitionism" – modest though they may be – must in my opinion first receive their certificate of validity from this tribunal.

An example of the kind of fundamental questions which can be so handled is the thesis that every mathematical problem is solvable. We are all convinced that it really is so. In fact one of the principal attractions of tackling a mathematical problem is that we always hear this cry within us: There is the problem, find the answer; you can find it just by thinking, for there is no *ignorabimus* in mathematics. Now my theory of proof cannot supply a general method for solving every mathematical problem – there just is no such method. Still the proof (that the assumption that every mathematical problem is solvable is a consistent assumption) falls completely within the scope of our theory.

I will now play my last trump. The acid test of a new theory is its ability to solve problems which, though known for a long time, the theory was not expressly designed to solve. The maxim "By their fruits ye shall know them" applies also to theories. When Cantor discovered his first transfinite numbers, the so-called numbers of the second number class, the question immediately arose, as I already mentioned, whether this transfinite method of counting enables one to count sets known from elsewhere which are not countable in the ordinary sense. The points of an interval figured prominently as such a set. This question – whether the points of an interval, i.e., the real numbers, can be counted by means of the numbers of the table given previously – is the famous continuum problem which Cantor posed but failed to solve. Though some mathematicians have thought that they could dispose of this problem by denying its existence, the following remarks show how wrong they were: The continuum problem is set off from other problems by its uniqueness and inner beauty. Further, it offers the advantage over other problems of combining these two qualities: on the one hand, new methods are required for its solution since the old methods fail to solve it; on the other hand, its solution itself is of the greatest importance because of the results to be obtained.

The theory which I have developed provides a solution of the continuum problem. The proof that every mathematical problem is solvable constitutes the first and most important step toward its solution....[3]

In summary, let us return to our main theme and draw some conclusions from all our thinking about the infinite. Our principal result is that the infinite is nowhere to be found in reality. It neither exists in nature nor provides a legitimate basis for rational thought – a remarkable harmony between being and thought. In contrast to the earlier efforts of Frege and Dedekind, we are convinced that certain intuitive concepts and insights are necessary conditions of scientific knowledge, and logic alone is not sufficient. Operating with the infinite can be made certain only by the finitary.

The role that remains for the infinite to play is solely that of an idea – if one means by an idea, in Kant's terminology, a concept of reason which transcends all experience and which completes the concrete as a totality – that of an idea which we may unhesitatingly trust within the framework erected by our theory.

Lastly, I wish to thank P. Bernays for his intelligent collaboration and valuable help, both technical and editorial, especially with the proof of the continuum theorem.

[3][At this point, Hilbert sketched an attempted solution of the continuum problem. The attempt was, although not devoid of interest, never carried out. We omit it here. – Eds.]

Remarks on the definition and nature
of mathematics

HASKELL B. CURRY

This paper is a discussion, written as a result of a request of Professor Gonseth, of certain points concerning the philosophy of mathematics. It is a revision of my previous discourse, on this subject, which I now regard as inadequate. The argument is based directly on my contact with mathematics without benefit of any technical acquaintance with philosophy. I have not attempted to confine myself with what is novel; but the paper is intended to be self-contained.

The principal thesis is that mathematics may be conceived as an objective science which is independent of any except the most rudimentary philosophical assumptions. It is a body of propositions dealing with a certain subject matter; and these propositions are true insofar as they correspond with the facts. The position taken is a species of formalism, which may be called empirical formalism.

The problem of mathematical truth

There are three principal types of opinion as to the subject matter of mathematics, viz. realism, idealism, and formalism. We shall consider here the realist and intuitionist views, leaving formalism for the next section.

According to realism, mathematical propositions express the most general properties of our physical environment. Although this is the primitive view of mathematics, yet, on account of the essential role played by infinity in mathematics, it is untenable to-day.

On the idealistic view mathematics deals with the properties of mental objects of some sort. There are various varieties of this view according to the nature of these mental objects. The extremes are Platonism, which ascribes a reality to all the infinistic constructions of classical mathematics, and intuitionism, which depends on an *a priori* intuition of temporal sucession. All forms of idealism are subject to the same fundamental criticism: in the first place they are vague, and, in the second

Reprinted with the kind permission of the author and editor from *Dialectica,* 8 (1954), 228–33. The author has indicated to us that, although this paper appeared in 1954, it was written in 1939 and represents his views as of that time.

place they depend on metaphysical assumptions from which mathematics, if it is to have the pre-philosophical character above mentioned, must be free.

It is important to see that this criticism, so obvious in the case of Platonism, applies also to intuitionism. As to the vagueness, Heyting, in his Ergebnisse report, explicitly denies the possibility of an exact description of this mathematical intuition. As to the metaphysical character, it is clear from the intuitionist writings that their "ur-intuition" has the following properties: (1) it is essentially a thinking activity; (2) it is *a priori;* (3) it is independent of language; and (4) it is objective in the sense that it is the same in all thinking beings. The existence of an intuition – temporal or not – satisfying these four conditions is an outright assumption.

The formalist definition of mathematics

According to formalism the central concept in mathematics is that of a formal system. Such a system is defined by a set of conventions, which I shall call its *primitive frame,* specifying the following: first, what the objects of the theory, which I shall call *terms,* shall be; second, how certain propositions, which I shall call *elementary propositions,* may be stated concerning these terms, i.e. what *predicates* (classes, relations, etc.) we shall take as fundamental; and third, which of these elementary propositions are true. The first and third of these sets of conventions are essentially recursive definitions; we do not specify the ultimate nature of the terms, but give simply a list of primitive terms, or tokens, together with operators and rules of formation by means of which all further terms are constructed; likewise we start with a list of elementary propositions, called *axioms,* which are true by definition, and then give *rules of procedure* by means of which further elementary theorems are derived. The proof of an elementary proposition then consists simply in showing that it satisfies the recursive definition of elementary theorem.

It should be noted that in such a formal system it is immaterial what we take for the tokens (and operators) – we may take these as discrete objects, symbols, abstract concepts, variables, or what not. Any such way of understanding a formal system we may call a representation of it. The primitive frame specifies, independently of the representation, which elementary propositions are true, and therefore determines the meaning of the fundamental predicates. In this sense the primitive frame defines the system.

One representative of particular importance is when the tokens are taken as symbols. In this representation, which is insisted on by Frege and his followers (Hilbert, Carnap, the Poles), a formal system becomes essentially equivalent to the formalized syntax of an object language. This representation has certain advantages of definiteness and concreteness. But it also has certain disadvantages. For it is necessary, as Carnap has shown, to distinguish between the symbols used as names of the terms and the specimens of those terms; usually this means that the familiar symbols are used for the latter purpose and more or less outlandish ones for the former. Since we never use the symbols of the object languages in the theory but only in the introduction, this leads to unintelligibility. Why not abolish the object language altogether and understand that the tokens are objects which we can take as symbols if we want to? Again the consideration of other sorts of representations may suggest simplifications which a syntactical representation would not. Thus the syntactical viewpoint gives rise to an extreme nominalism – as shown by the inclusion of parentheses, commas, etc., among the symbols and of expressions which are not "well formed" – which is avoidable and contrary to the spirit of mathematics.

In the study of formal systems we do not confine ourselves to the derivation of elementary propositions step by step. Rather we take the system, defined by its primitive frame, as datum, and then study it by any means at our command. In so doing we may formulate further propositions, which we call metatheoretic propositions. Like the elementary propositions these state, essentially, properties of the system; but they may also involve extraneous considerations. Insomuch as they deal with what is, in view of the primitive frame, a well defined subject matter, the question of their truth involves no difficulties beyond those inherent in compound propositions in general.

The formalist definition of mathematics is then this: mathematics is the science of formal systems. The propositions of mathematics are the propositions, elementary or metatheoretic, of some formal system or set of systems. For each such proposition which does not involve extraneous considerations, we have an objective criterion of truth in the sense that an alleged proof can be checked objectively; but a proposition may be indefinite in the sense that we have no resolution process (*Entscheidungsverfahren*). Intuition is, of course, involved in this; viz. the intuition of recursive definitions, mathematical induction and the like; but the metaphysical nature of this intuition is irrelevant. (If extraneous considerations are involved, then we have to do with applied, not pure, mathematics – the boundary line between mathematics and other sciences is not sharp, and should not be.) It should be noted that we have not confined mathematics to a single formal system; moreover, metatheoretic propositions are included in mathematics. This answers the objections which might be raised on the ground of the incompleteness theorems of Skolem, Gödel, et al.

Truth and acceptability

We now turn to the relation of mathematics to its application. For this purpose we introduce another kind of quasi-truth concept which applies, not to single propositions, but to systems as a whole. I shall call this *acceptability*. By acceptability, then, I mean the considerations which lead us to be interested in one formal system rather than in another.

Acceptability is usually a matter of interpreting the theory in relation to some subject matter. Such an interpretation is to be distinguished from a representation: in a representation the predicates are defined by the primitive frame; in an interpretation we associate them with certain intuitive notions, so that the question arises as to the agreement between the truth of the propositions of the formal system and that of the associated intuitive ones. Acceptability is thus relative to a purpose; and a discussion of acceptability is pointless unless the purpose is stated.

As an illustration of acceptability questions let us consider the acceptability of classical mathematics for the purpose of application in physics.

Among the criteria of acceptability we may mention the following: (1) the intuitive evidence of the premises; (2) consistency (an internal criterion); (3) the usefulness of the theory as a whole.

The intuitionists have made much of the first criterion. They point out that certain propositions of classical mathematics lack intuitive evidence; and they have constructed systems which – we can admit this without swallowing the intuitionistic metaphysics – have greater intuitive evidence than the classical. But these systems are so complicated as to be useless, and are inacceptable by criterion (3). Moreover this is the decisive consideration; for physics is an empirical science, and therefore the question of intuitive evidence is secondary. The acceptability of classical mathematics is an empirical fact, and the proper retort to the intuitionist gibe, that classical mathematics has only a heuristic value, is that so far as physics is concerned, that is all the value that an intuitionist mathematics has either.

The criterion of consistency has been stressed by Hilbert. Presumably the reason for this is that he, like the intuitionists, seeks an *a priori* justification. But aside from the fact that for physics the question of an *a priori* justification is irrelevant, I maintain that a proof of consistency is neither a necessary nor a sufficient condition for acceptability. It is obviously not sufficient. As to necessity, so long as no inconsistency is known,

205

a consistency proof, although it adds to our knowledge about the system, does not alter its usefulness. Even if an inconsistency is discovered this does not mean complete abandonment of the theory, but its modification and refinement. As a matter of fact, essentially this has happened in the past; we now know, for example, that the mathematics of the eighteenth century was inconsistent, but we have not abandoned the results of the eighteenth-century mathematicians. The peculiar position of Hilbert in regard to consistency is thus no part of the formalist conception of mathematics, and it is therefore unfortunate that many persons identify formalism with what should be called Hilbertism.

Let us now cut short this discussion and summarize as follows: Acceptability is relative to a purpose, and a system acceptable for one purpose may not be for another. For example, I agree with Weyl and Gentzen that there are purposes for which intuitionistic systems are acceptable, although they are not acceptable, on empirical grounds, for application to physics. Again, acceptability is a different question from truth; in fact a formalist definition of mathematical truth is compatible with almost any position in regard to acceptability. In this sense formalist mathematics is compatible with various philosophical views; it is an objective science which can form part of the data of philosophy.

Mathematics and logic

In current popular discussions it is said that intuitionism, formalism, and logicism are the three main views in regard to the nature of mathematics; the last is supposed to be the view that mathematics is logic. But we do not have here a third view of mathematics parallel with the other two; for to say that mathematics is logic is merely to replace one undefined term by another. When we go back of the word "logic" to its definition in the logistic systems, we find that they run the gamut from extreme Platonism to pure formalism. The question of the relation of mathematics to logic is thus a different question from the definition of mathematics; on account of the lack of space we cannot go into that question here.

Hilbert's programme

GEORG KREISEL

1. If one may judge from his publications, Hilbert's conception of the problem of foundations underwent marked developments.

In [1932–5, 3: 145–56; orig. 1918] he still concentrated on the "sound" and rather colourless *Independence Problem* which may be formulated as follows: given a branch of knowledge which is so well-developed as to be axiomatized, the problem is to get a clear view of the logical relationships (dependence and independence or, derivability and non-derivability) of statements of the axiomatic theory. Hilbert emphasized the consistency problem which is so to speak the weakest non-derivability result, since it is the problem of showing that there exists at least one statement which is not derivable.

But in later writings (though also in 1905, [1899b, 7th ed.: 247–61] the *Consistency Problem* was associated with the problem of understanding the concept of infinity. He sought such an understanding in understanding the *use of transfinite machinery* from a finitist point of view. And this he saw in the elimination of transfinite (ϵ—) symbols from proofs of formulae not containing such symbols. He was convinced from the start that such an elimination was possible, and expressed it by saying that the problems of foundations were to be *removed* or that doubts were to be eliminated instead of saying that they were to be investigated.

We note at once that there is no evidence in Hilbert's writings of the kind of formalist view suggested by Brouwer when he called Hilbert's approach "formalism." In particular, we could say that Hilbert wanted to eliminate the use of transfinite concepts from proofs of finitist assertions instead of referring to symbols and formulae as above. The symbols were a means of representation. The real opposition between Brouwer's and Hilbert's approach was not at all between formalism and intuitive mathematics, but between (i) the conception of what constitutes

Revised by the author from an earlier version which appeared in *Dialectica* 12 (1958), 346–72. Printed here by kind permission of the author and the editor of *Dialectica*. For this second edition Professor Kreisel has added a Postscript, as well as notes to a number of the sections of the original article. These are collected at the end of the essay. Except as noted otherwise, these additions were made in the autumn of 1978 and represent his views as of that time.

a foundation[1] and (ii) between two informal ways of reasoning, namely finitist and intuitionist. In fact, Bernays repeatedly emphasized the latter point, the lack of evidence in the basic intuitionistic conception of constructive proof [Hilbert 1932–5, 3: 212, or Bernays 1941: 147]: in short, it is not the restrictions imposed *by* intuitionism, but those *on* intuitionism which seem to constitute the most significant differences. Hilbert's own remarks on this opposition seem quite inept.[2]

The view above on the significant differences between Brouwer's and Hilbert's approach does not deny, of course, the popular attraction of a syntactic formulation of foundational problems. Derivability in formal systems (which codify the manipulation of symbolic representations of transfinite concepts) is "down to earth," it refers to palpable acts, to what we "actually do," while the transfinite concepts themselves are "up in the air," they are abstract, and therefore supposed to be inaccessible to exact study. Even though Hilbert was not a strict positivist like Comte [Hilbert 1932–5, 3: 387], his presentation was certainly congenial to a positivist era: at least, his problems were positivistically meaningful for a more liberal conception of positivism, while the transfinite concepts themselves are senseless even for the latter. It might have been expected that deficiencies in our understanding of these concepts would reappear in our inability to solve the syntactic problems, in particular to survey those parts of the corresponding formal system which are devoid of positivist interest, namely all those formulae which contain symbols for transfinite concepts. But equally it was to be expected that specific partial (syntactic) problems could be solved despite a lack of deeper understanding of the abstract concepts. In the following sections we shall try to "reconstruct" Hilbert's programme in the light of these observations.

Hilbert was certainly not a fanatic of crude formalism: thus while his problems concerned syntactic properties of formal systems, the solutions were to be given by intuitively correct reasoning, and he explicitly considered any formalization of *this* reasoning as unnecessary. (For qualifications, cf. Section 18.)

2. The fabric of Hilbert's conception. He asserted that there was a certain type of evident reasoning which was presupposed in all scientific thinking [Hilbert 1932–5, 3: 162–63], and finitist operations were typical

[1]Brouwer ignores non-constructive mathematics altogether and therefore does not have an analogous problem of foundations to Hilbert's.

[2]E.g., [Hilbert 1899b, 7th ed.: 307]: Considering that the intended meaning of the intuitionistic disjunction is different from that of classical disjunction, the rejection of *tertium non datur* is much more like depriving non-commutative algebra of the rule $ab = ba$ than a boxer of the use of his fists.

of this. He believed that there were no essentially different truths in mathematics [Hilbert 1932–5, 3: 157].

These views lead to the hope of a final solution of the problem of foundations [Hilbert 1931: 489, 494] by a reduction of all mathematical reasoning to finitist reasoning: for if the minimum that has to be presupposed suffices for this reduction then we have a complete solution. Conversely, something of this kind is required for a "complete" solution: if there is a plurality of essentially different mathematical truths there is hardly a hope of an enumeration of such truths which convinces us as being complete; furthermore there would be the problem of their interrelations, and so there would be no unique outstanding problem of foundations.

Next, if the finitist truths are the only[3] absolute ones [Hilbert 1932–5, 3: 180], it is at least natural to regard mathematical expressions containing transfinite symbols as "ideal" elements whose sole purpose is the streamlining of the symbolism [Hilbert 1899b, 7th ed.: 280, or 1932–5, 3: 187]. And in this case consistency is all that matters because, in the usual systems, if consistency is established by (finitist) methods, and a formula without transfinite symbols is derived in the system considered, then this formula can be proved by the same methods [Hilbert 1899b, 7th ed.: 304, and Bernays 1941: 154]; in other words, we have the required elimination. Conversely, if consistency is all that we demand then there is no assurance that formulae containing transfinite symbols in an essential way have the intended interpretation, and so they must be regarded as ideal. This is established by Gödel's construction of consistent, but ω-inconsistent, systems,[4] because purely universal formulae are deductively equivalent to formulae without transfinite symbols, and purely existential formulae which are provable in a consistent system need not be true.

We note in passing an interesting aspect of Hilbert's idea of a *paradise:* a characteristic of Cantor's set theory (Hilbert's paradise, 1899b, 7th ed.: 274) is the abundance of transfinite machinery which Hilbert regarded in the same paper as "ideal" elements to be used as gadgets to make life smoother.

---

[3]Though, I think, Hilbert does not deny this elsewhere, his emphasizing that the geometric continuum [1934–5, 3: 159] is a concept in its own right and independent of number seems to weaken the doctrine that all absolute truths are finitist.

[4]The possibility of ω-inconsistent systems was evidently clear to Hilbert, but as late as 1930 [1899b, 7th ed.: 320] he wanted to show that every consistent statement of arithmetic was provable, i.e., that the usual system of arithmetic was complete. Actually he requires more, namely that consistency "looks after the rest", since a system might be complete and yet some of its theorems false.

It is plain that such concepts as "finitist," "essentially the same truth," "reduction" are not at all precise. But if one really believes in the success of the finitist reduction it was not necessary to clarify them in advance. For when the work is done one can examine what methods are actually needed, in what sense we have a reduction, and at this stage one can then decide if it is satisfactory.

3. Critique of detail. There is one point in the above picture which is not convincing even if the basic assumptions are granted. Hilbert regarded complex numbers [1899b, 7th ed.: 269] as a typical example of ideal elements. Yet the reduction to pairs of real numbers does not only ensure consistency, but also gives each formula containing symbols for "ideal" elements a *meaning* in terms of real numbers. It would not seem unreasonable to demand the same for formulae with transfinite symbols, at least in any specific context. At any rate one would thereby extract more "absolute truths" from the formal machinery.

4. Critique of basic assumptions. Of course, the first basic assumption is that a reduction to finitist methods is possible at least in the sense of a finitist consistency proof. If this assumption is false, the problems at the end of para. 2 reappear and more besides.

First, instead of having a single kind of elementary reasoning whereby we understand the use of transfinite symbols, there will now be methods of reasoning involving a hierarchy of conceptions such as, e.g. more and more abstract conceptions of a "construction," and we have a hierarchy of Hilbert programmes of *discovering the appropriate complex of such methods which is needed for understanding the use of transfinite symbols in given systems* (modified Hilbert programme).

Second, it will be necessary to ascertain that the consistency of a particular system cannot be established by finitist means, or whatever other complex of methods is being considered. For such an impossibility result it will be necessary to define the notion of *finitist proof* and even to make precise how *consistency* is to be formulated. The latter point is illustrated by the need for derivability conditions on the arithmetized proof predicate in Gödel's second undecidability theorem.

Third, when we are not dealing with an elimination of the nonfinitist methods, but with a *separation* between them, it is necessary to determine the significance of the distinction between finitist and non-finitist. An analogous problem will arise for each subclass of the constructive methods used in the analysis of the transfinite machinery.

Hilbert's own writings contain little information about the solution of these new problems. For the first problem Gentzen's use of ordinals $< \epsilon_0$

is a good illustration of the kind of subclass of constructive methods which is particularly appropriate for the analysis of a given class of transfinite methods, in this case number theory.

It is difficult to separate the second and the third problem in practice because one can only decide whether e.g. a *definition* of finitist proof is correct if its *significance* (importance) is clear. Hilbert himself is quite unconvincing about the inherent virtues of finitist reasoning. At one time [1932–5, 3: 160, 162] the main purpose of the finitist reduction, in fact the whole need for foundations, consisted for him in clearing the fair name of mathematics which had been sullied by the paradoxes. Now on p. 158 of the same paper he said that the paradoxes simply have nothing to do with the theory of sets of numbers: it is hard to see why this remark, if true, has not cleared the fair name at least of analysis unless one believes that stained reputations can only be cleared with a great deal of ceremony. Von Neumann's line [1927] on the subject is that finitist consistency proofs would reduce strict intuitionism *ad absurdum* (but only if one means by strict intuitionism the view that classical analysis is formally inconsistent and not merely in contradiction with intuitionistic theorems): this is at least less pious than the talk about reputations, but doing down the intuitionists is hardly a grand scientific programme. – Hilbert did emphasize the increase of information contained in proofs of a formula of the form $(Ex)(y)A(x,y)$ [1932–5, 3: 154, 155] from a pure existence proof [intuitionistically $\sim (x) \sim (y)A(x,y)$] to $(Ex)[x \leqslant 12 \ \& \ (y)A(x,y)]$ to $\mu_x(y)A(x,y)=10$. But this gives no clue to the nature of the improvement involved in replacing a non-finitist proof of a universal formula $(x)A(x)$ by a finitist one, and this is the critical case (cf. Kreisel 1958b: 177).

Hilbert sometimes speaks of the reliability (*Sicherheit*) of finitist reasoning. As Bernays has pointed out [Hilbert 1932–5, 3: 210], realistically speaking, almost the opposite is true, the chance of an oversight in long finitist arguments of metamathematics being particularly great. At any rate it seems improbable that a satisfactory characterization of "finitist proof" would be based on this notion of reliability. – We shall take up the subject in the text below.

5. *Critique of basic assumptions* (continued). In the previous paragraph we considered some problems which arise when one attempts to follow Hilbert's aim of understanding the concept of infinity by eliminating the use of transfinite machinery from proofs of finitist or, more generally, constructive assertions. But this view, that understanding a concept consists simply in the technique of reasoning about it in some well-defined context, does not seem quite adequate.

Thus e.g. the first-order theory of the addition of natural numbers has a complete formalization and hence a decision method. So, from the syntactical point of view it leaves nothing to be desired. But by the compactness theorem for the predicate calculus this theory has a model containing "non-standard" integers. Thus we do not get the degree of understanding that we should have with a system which is satisfied only by a finite set. – We may regard this as a limitation on the syntactic approach for an understanding of the concept of infinity or, at least, as illustrating the use of the notion of a *model* in this connection.

Usually Gödel's incompleteness theorems are taken as showing a limitation on the syntactic approach to an understanding of the concept of infinity. We note the superficially paradoxical fact, brought out above, that the completeness of the predicate calculus, so to speak the "adequacy" of the syntactical approach for predicate logic, leads to a limitation too. This is a point of view emphasized by Skolem.

6. Conclusion. In my opinion, Hilbert's programme, including the modified version in section 4 above and the independence problem of section 1, is a rich line of research in foundations. The general problems of section 4 seem important and somewhat more specific problems will be arrived at below in a brief analysis of the work done by Hilbert's school so far. Also, the (modified) Hilbert programme gives scope to a great variety of methods of mathematical logic including those of the topological approach, intuitionism, recursion theory, as will be seen below.

My own attitude towards the original Hilbert programme is this.

As far as piecemeal understanding is concerned, its importance consists in having led to the fruitful study of the constructive aspects of axiomatic systems. But even if it is compared only with other parts of mathematical logic and not with other mathematical disciplines, its role is not unique; cf. the studies of the distinction between first-order and higher-order reasoning, or of the set theoretical aspects of informal mathematics. My own interest in the modified Hilbert programme does not go one way only, i.e. the elimination of non-constructive methods, but I find that greater facility with the non-constructive methods comes from a study of their constructive aspects.[5]

As far as an over-all philosophical understanding is concerned, the origi-

[5]It is remarkable that mathematicians have not yet learned to exploit the non-constructive methods effectively in the following sense: in the famous non-constructive proofs of constructive results (for references, see Kreisel 1958b) the elimination of the non-constructive methods used is finitist, and does not require more sophisticated notions of constructivity. We know that a more essential use of non-constructive methods must be possible, and, I believe, closer study of their constructive aspects may give one a better "feeling" for them.

nal Hilbert programme has failed, and, as is usual with great schemes it gives no hint of what might take its place. When asked: "What is mathematics about?", Hilbert could still have said: about the arithmetico-combinatorial facts of finitist mathematics; even though the latter may raise problems of their own, such a "reduction" could have been satisfying. Hilbert's answer is simply not true even for the very weak sense of "equivalence of content" expressed in statements of formal deducibility and nondeducibility. Furthermore, we have no idea by what sort of investigation we could even hope to find a satisfactory answer to such a question. Hilbert thought it would be supplied by pure mathematics itself [1899b, 7th ed.: 316]. But it seems clear, as Bernays has expressed it, that the totality of pure mathematics (mathematical structures) is not itself a mathematical structure; this is not only a stumbling block to a mathematical treatment of the conception of the whole of mathematics, but even to an exhaustive treatment of the concept of natural or real number because the characterization by means of a least[6] (or largest) class refers explicitly to the totality from which these classes are to be taken. If one may use for orientation the formulation of *finitist proof* sketched below, one would say this: just as it is necessary to use non-finitist concepts to study the totality of finitist proofs, so it is necessary to use non-mathematical concepts, i.e. concepts lacking the precision which permit mathematical manipulation, for a significant approach to foundations. It is not at all a question of concepts which are more "reliable" than those of current mathematics, but of concepts which provide a frame of reference for discussing the status of mathematics (cf. Bernays 1957: 245). I do not believe that at the present time we have any concepts of this kind which we can take seriously.

7. For certain parts of non-constructive mathematics the Hilbert programme has been carried out in the originally intended sense, e.g. arithmetic without induction based on the classical predicate calculus. For this purpose a detailed syntactic analysis of the latter is used. Below we shall describe three methods of syntactic analysis, their mathematical

---

[6] It is of course possible (and natural) to consider the relevant extremal causes as primitive notions in their own right and not as part of set theory, e.g., the "and only those required by the foregoing rules" in the usual definitions of natural numbers or recursive ordinals. This is used, e.g., in Hilbert's reduction [1899b, 7th ed.: 302] of the principle of induction to the reversibility of the formation of numerals. The latter follows from the extremal clause, and in a similar way other principles of proof can be obtained from such clauses (cf. Lorenzen's induction and inversion principles). But in the application of induction the troublesome totalities reappear in the choice of properties to which induction is applied, e.g., by applying induction to certain non-elementary properties such as truth-definitions formulae of classical arithmetic Z can be proved which cannot be proved in Z itself.

significance, including applications to the independence problem, and their bearing on the modified Hilbert programme. We conclude with two observations on finitist proofs and the completeness of predicate logic.

I. Syntactic analysis

8. We begin by considering what seems to be the main novelty of Hilbert's work in mathematical logic, without which his conception of the Hilbert programme would have been wholly unconvincing, namely his proof theory. He wanted *proof* itself to be made the object of mathematical study [1932–5, 3: 165]. Though this is needed for a syntactic (combinatorial) *formulation* of independence results, by itself it is not the crucial point for proof theory. For, clearly, the traditional independence proofs by means of models, as in the case of the parallel axiom, or even the impossibility proofs for certain constructions by means of ruler and compass, are applicable to formalized systems. Thus the consistency of the rules of set theory is proved as follows: when read in the intended manner, the axioms of, e.g., Zermelo's set theory are true of the concept of set and the rules of proof are such that true statements are transformed into true ones. Hence the formal system is consistent.

No, from the point of view of technique the crucial point is that from an early stage Hilbert had in mind a *new type of analysis* in which the detailed structure of the proof is considered. In particular, in the consideration of the so-called transfinite symbols $\epsilon_x A(x)$, one does not define "models" for them once and for all, but different numbers are substituted for a given symbol depending on the particular proof of the system which is analyzed. Briefly: instead of constructing a model for a system as a whole he gives a method for constructing a model for each particular proof of the system.

In short, Hilbert's conception of a mathematical theory in which *proof* itself is an object of study does not restrict the means of independence proofs but on the contrary enlarges it.[7] A restriction comes about only when one restricts the methods to be used in this new theory to so-called finitist or arithmetico-combinatorial methods. We observe in passing the superficially paradoxical character of the fact that up to now the significant independence proofs for classical (and intuitionistic) arithmetic have been obtained by the restricted and not by the potentially more powerful methods. We shall return to this point below.

---

[7]Within the framework of classical constructive mathematics (i.e., non-constructive methods of proof, but only recursive functions and predicates) we can give this the following precise sense: by means of (quantifier-free) double induction syntactic independence proofs for primitive recursive arithmetic can be obtained, but not by means of constructive models [e.g., Kreisel 1958b or Mostowski 1957a].

214

9. Decision problem. A very satisfactory syntactic analysis is got from a practical decision method for the theory considered. In particular it solves the independence problem of para. 1 for finite sets of statements of the theory considered. It is not the most satisfactory solution because, given the decision method, one can now ask whether given statements are independent of a certain *infinite* set of statements of the theory, and this question need not be decidable; in fact, it seems inconceivable that there is an optimal solution of the problem. Similarly, the impossibility of a decision method for a given theory is not a "catastrophe";[8] it sets a limit on how[9] clear a view one may reasonably expect to get of the structure of the theory. – This stoic view of the matter is probably universal now, perhaps largely due to the thoughtful writings of Bernays.

It is clear that if a system is consistent a (finitist) proof of decidability automatically yields a (finitist) consistency proof. As early as 1927, von Neumann doubted the decidability of the predicate calculus, though Hilbert does not seem to have been quite so definite. But in any case his actual investigations aimed at much less than decidability. We shall now describe them.

10. The ϵ-substitution method. The main tool in his work on predicate logic was a reformulation of the predicate calculus by means of the ϵ-symbol.

This reformulation is not specially elegant in practice, e.g. the formula $(Ey)(z)A(y,z)$ in the usual notation is written as

$$A[\epsilon_y A\{y, \epsilon_z \neg A(y,z)\}, \epsilon_z \neg A\{\epsilon_y A[y, \epsilon_u \neg A(y,u)], z\}],$$

but it makes *evident* a fact of logical reasoning which was basic for Hilbert's programme, namely that in logical reasoning one never makes full use of the intended meaning of the transfinite symbols, in the following sense. $\epsilon_z A(y,z)$ is intended as a choice function (some z_1 which satisfies $A(y,z)$ if such a z_1 exists, and an arbitrary value otherwise), and on this meaning the schema

(*) $A(y,b) \rightarrow A[y, \epsilon_z A(y,z)]$

is valid. Also, in an obvious way, the universal and existential quantifier can be defined by the use of the ϵ-symbol, and the schema (*) is enough to derive the usual schemata for the quantifiers. Now the gain is this:

---

[8]The Twenties constantly saw potential catastrophes in the doings of logicians.

[9]It is usual to measure this by the degree of undecidability of the theory. There is quite a different conception of a "partially clear view," namely that obtained by methods enumerating (subsets of the) unprovable formulae, as afforded by incomplete interpretations (cf. para. 15).

we see immediately that in any given proof, since (*) is applied only to finitely many y, one never needs the full extension of the choice function $\epsilon_z A(y, z)$ for all y, and, moreover, in the course of the proof one does not need its "real" values, but, e.g. if (*) were the only application of the schema in the given proof, we could simply take b for $\epsilon_z A(y, z)$ and still have a proof. Such considerations make the elimination of the ϵ-symbols (from a proof of a formula without ϵ-symbols) at least plausible. This idea was developed by Ackermann, presented in detail in [Hilbert and Bernays 1934-9: vol. 2], for predicate logic, and for number theory by Ackerman [1940]. – However, for practical applications it is best to apply the simple idea directly [Kreisel 1958b: 171].

There is a formulation of the substitution problem which does not use the ϵ-symbol at all. If distinct ϵ-matrices are replaced by distinct function symbols f, the ϵ-formulae reduce to the form $\Phi(f_1, \ldots, f_n) = 0$ where Φ is an elementary functional. The problem is to prove $(E f_1) \ldots (E f_n)$ $[\Phi(f_1, \ldots, f_n) = 0]$; it is evident if this is true at all there are functions f^* which are zero except for a finite number of arguments, and can therefore be found by trial and error. The existence of functions f seems assured by the interpretation of ϵ-matrices as choice functions.

It seems understandable that Hilbert assumed that the *proof* of such an elementary matter as the existence of f^* must be a relatively minor task.[10] For, if one regards metamathematical results as the absolute truths of mathematics [Hilbert 1932-5, 3: 180] and the "transfinite" formulae as ideal elements without significance outside the framework of a formal system, it is natural to regard the metamathematical results as significant independently of their proof: though this, of course, does not mean that the proof is easy or even of an elementary character, one may be tempted to think so, e.g. because of the double meaning of "significant" (meaningful, but also: not trivial, not easy).

Open problem. Notwithstanding the interest of alternative analyses of predicate logic and its extensions, to be described below, an examination

---

[10]In [1899b, 7th ed.: 317], Hilbert expresses this by saying that in the case of analysis "only" the proof of the purely arithmetical statement that the method terminates is needed. (He had of course the feeling [1932-5, 3: 187] that a purely arithmetical truth must have a purely arithmetical proof, which is refuted by Gödel's theorem if, e.g., "arithmetical" is interpreted as: expressible in classical number theory.) – Hilbert's "only" is misplaced: for after all, the statement of consistency is also purely arithmetical, so from the very start, "only" the proof of the main contention of his life was needed. However, from a mathematician's point of view, Hilbert's excitement at his reformulation of the consistency problem in terms of the convergence of the substitution method is very natural: the latter problem, and particularly the problem of finding bounds, has the general look of a mathematical problem, while the consistency problem does not, even if it is formulated as a combinatorial problem (cf. end of para. 10).

of the substitution method applied to analysis seems very promising at the present time. Somewhere there is a combinatorial lemma lurking in the proofs which show that the substitution method terminates and which gives information about the solution of the functional equations $\Phi(f_1,\ldots,f_m)=0$ mentioned above.[11]

11. Cut free formalizations. Other syntactic analyses of predicate logic were developed by Herbrand and perfected by Gentzen. They resulted in reformulations of predicate logic specially adapted for proofs of completeness.

Herbrand's *lemme fondamental* becomes much simpler for prenex formulae[12] in contrast to Gentzen's *Hauptsatz*. Be that as it may, both of them gave explicit (primitive recursive) instructions for converting a proof with cuts into one without. But the simplest[13] exposition of, e.g. Herbrand's reformulation of the predicate calculus goes by way of the completeness theorem: if a (prenex) formula is not provable by Herbrand's rules (Hilbert and Bernays 1934–9, 2: 158, *b*) then it is not valid; so if a formula is provable in the ordinary way it must be provable by Herbrand's rules. But though further constructive analysis of the completeness proof is possible (cf. Kreisel 1958b: 168), without it the rule for getting an Herbrand proof is only general recursive.

Given a prenex formula, say $(x)(Ey)(z)A(x,y,z)$, we ask: how could it be false? i.e. $(Ex)(y)(Ez)\neg A(x,y,z)$. There would have to be an element α such that $(y)(Ez)\neg A(\alpha,y,z)$ and without loss of generality we may as well call it 0, i.e. $(y)(Ez)\neg A(0,y,z)$. For $y=0$ we must have $(Ez)\neg A(0,0,z)$; we may as well take $z=1$; for either $z\neq0$, then calling $z=1$ is permissible, if $z=0$, then we must simply regard 1 as another name for the individual called 0; we must not take $z=0$ in general. This explains the disparateness conditions of [Hilbert and Bernays 1934–9, 2: 173]. Now in order that $\neg A(0,0,1)$, there is a certain finite set of truth distributions on the prime formulae of $\neg A(0,0,1)$ which may be recorded in the form of a finitary tree. Let R_i be the conjunction of the

[11]This "lurking lemma" has since been formulated and proved by Tait [1965a] for the substitution method as applied to the formalism of arithmetic. It shows in particular in a natural way how the first ϵ-number enters into the problem: so it represents one of three independent analyses of the role of ϵ_0, the other two being the computational analysis of Gödel's functionals of para. 12 below, given in detail in [Tait 1965b], and the analysis of infinite cut free proof trees by means of ordinals (cf. footnote 15).

[12]The existence of prenex normal forms in classical logic makes this distinction more important in the analogous treatment of intuitionistic logic.

[13]I have not studied Herbrand's own publication. The exposition is suggested by Beth's semantic tableaux [1957] which are a generalization of the criteria of refutability [Hilbert and Bernays 1934–9: vol. 2]. – The proof of Herbrand's theorem [Hilbert and Bernays 1934–9: vol. 2] gives primitive recursive instructions.

ith set of prime formulae and negations of prime formulae with arguments 0, 1 which make $\neg A(0,0,1)$ true. Now, for each i we consider all extensions $R_{i1},\ldots,R_{im}$ of R_i by prime formulae with arguments 0, 1, 2 which make $\neg A(0,1,2)$ true, $\neg A(0,0,1)$ being made true automatically since R_{ij} are extensions of R_i. We record this information in a tree

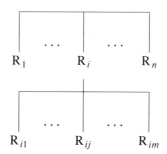

(If there is no extension of R_i, we stop the construction of R_{ij} and consider R_{i+1}, $i < n$.)

If this tree is unbounded, by the *Unendlichkeitslemma,* there is an infinite branch. If the prime formulae of $(x)(Ey)(z)A(x,y,z)$ are given the truth values which they have on such a branch and if the variables range over the natural numbers then $\neg (x)(Ey)(z)A(x,y,z)$. (Completeness.)

If the tree is bounded, the formula $\neg (x)(Ey)(z)A(x,y,z)$ is not satisfiable at all. Now the whole tree can be converted into a proof of $(x)(Ey)(z)A(x,y,z)$. For, regard the numerals as variables. All we need are rules which allow us to infer $(x)(Ey)(z)A(x,y,z)$ from $A(0,0,1) \vee A(0,1,2) \vee \ldots \vee A(0,n,n+1)$ for each n. The rules required are just Herbrand's rules. Thus we have not only a new formalization of, but also a cogent motivation for, the choice of the rules. – Beth [1957] has used the same idea for treating arbitrary formulae and arriving at a variant of Gentzen's rules.

Here one conceives of a counter example to $(x)(Ey)(z)A(x,y,z)$ in terms of satisfying $\neg A[0,n,\varphi(n)]$ for some given disparate function φ and arbitrary numerals n. The impossibility of obtaining one, i.e. the breakdown of the construction above, yields a classical proof of $(x)(Ey)(z)A(x,y,z)$. Another conception of a counter example is to satisfy $\neg (x)(Ey)(z)A(x,y,z)$ in the prenex form $(Ex)(y)(Ez)\neg A(x,y,z)$ by a constant α and a function f, i.e. $(y)\neg A[\alpha,y,f(y)]$. The impossibility of obtaining one is expressed by $(Ey)A[\alpha,y,f(y)]$ and this can be expressed explicitly by means of functionals $\varphi_i(f,\alpha)$ (terms containing α and f) such that $\ldots \vee A[\alpha,\varphi_i(f,\alpha),f\{\varphi_i(f,\alpha)\}] \vee \ldots$[14] — This is only

[14]This is used in [Hilbert and Bernays 1934–9: vol. 2] as an auxiliary in the proof of Herbrand's theorem. Below and in other publications we emphasize the independent sig-

suitable for prenex formulae (unless one uses functionals of higher type) but more suitable for arithmetic than the alternative described above.

12. Gödel's intervention. There is a totally different analysis of classical proofs due to Gödel. First classical proofs are replaced by intuitionistic ones of (classically) equivalent theorems [Gödel 1931–2b], then these proofs are analyzed by means of certain simple functionals of finite type, and these in turn are shown to be well defined by means of transfinite induction ($< \epsilon_0$ for arithmetic with induction, $< \omega^\omega$ for arithmetic without). In this way the syntactic analysis is effected in several manageable steps, each of them of interest in itself.[15] For further details see Gödel's article in *Dialectica* 12 (1958), 280–7.

II. Significance of syntactic analysis

In accordance with para. 3 and following Herbrand's lead we do not formulate the results of syntactic analysis as consistency theorems, but as interpretations in the sense of [Kreisel 1958b].

13. Mathematical significance. To avoid repetition, we refer here to a recent discussion of this matter [Kreisel 1958b], in particular the application of an interpretation to the independence problem and other mathematical questions. To keep matters up to date we note that an interpretation in the sense of [Kreisel 1958b] of classical analysis in a quantifier-free *classical* theory of continuous functionals of finite type has been given whose constants are *recursive* continuous functionals [Kreisel 1959].

As a result we have a new

Open Problem. To give explicit characterizations (schemata) of the particular recursive functionals actually needed for the interpretation of one of the current systems of analysis.[16]

For independence proofs it is desirable to give alternative schemata.

Digression. A totally different type of syntactic study of classical analysis is suggested by so-called predicative enterprises, e.g. [Lorenzen 1955; Spector 1957; and Wang 1954]. It is evident that we do not get a model of classical analysis C simply by letting the variables of higher

nificance of this step specially for the Hilbert programme for arithmetic.

[15] I regard the use of infinite induction by Schütte and Lorenzen as an intermediate step corresponding to the use of intuitionistic arithmetic in Gödel's work: it is to be supplemented by an analysis by means of ordinals.

[16] An important contribution to this problem is made in Spector's paper [1962, section 10].

type[17] in a formula of C range over the sets of Wang's Σ_α for a fixed α[17]; for if α is not a limit number we conflict with the theorem of the least upper bound, and if α is a limit number we conflict with the existence of a non-denumerable set.[18] However, it seems promising to try this: given a proof in classical set theory to index the variables actually occurring in the proof by means of ordinals α so that the proof goes into a set of true statements in Spector's sense [1957]. This would not give an interpretation in the sense of [Kreisel 1958b] because the statement $(Ef)(g)A(f,g)$ is replaced by $(Ef_\alpha)(g_\alpha)A(f,g)$, with special α, from which $(Ef)(g)A(f,g)$ cannot in general be inferred, but nevertheless such an indexing would yield independence proofs if for certain formulae of classical set theory there is no indexing which yields true statements (if indices α less than an appropriate bound are used).

The situation seems not unlike Euclidean geometry. For, just as in analysis, the most natural conception of a point ignores the matter of naming the point, i.e. how the real number is represented or by what constructions the point is reached from given points. But if one wants to assert the impossibility of, say, ruler and compass constructions, one introduces coordinates of a suitably restricted kind. Actually, the analogy is not complete because the field of real square root extensions of the rationals is a model of the Euclidean geometry considered, i.e. without the general continuity axiom, whereas the class of Σ_α sets is not a model of classical analysis if α is recursive.[18]

14. Remark on the methods used in syntactic analysis. Most of the mathematical applications of syntactic analysis, whether achieved or projected, depend little on restrictions of the metamathematical methods, except perhaps for primitive recursive bounds as in [Kreisel 1958b: 165]. In [Łoś, Mostowski, and Rasiowa 1956] there is an attractive exposition of a non-finitist approach to the complex of Herbrand-type theorems[19] for predicate calculi. – It is perhaps fair to say that the most significant

---

[17]*Type* in the sense of the simple theory of types; α is called a *level*.

[18]More precisely: (*a*) For countable α, Σ_α does not satisfy the axioms of classical analysis A if, as in [Wang 1954], Σ_α contains an enumeration of the union of $\{\Sigma_\beta : \beta < \alpha\}$, possibly in addition to the usual ramified hierarchy (Σ_α^-). (*b*) For $\alpha < \omega_1$, i.e. recursive α, such an enumeration is explicitly definable in Σ_α^-. (*c*) Significant parts of A are realised in $\Sigma_{\alpha+\omega}^-$ (for all α) and in $\Sigma_{\omega_1}^-$, but e.g. the Cantor–Bendixson theorem is not realised in any Σ_β^-, $\beta \leqslant \omega_1$; for a survey [cf. Kreisel 1960b]. (*d*) There are denumerable α, α' ($\alpha' > \alpha$) such that, for $\beta > \alpha$, Σ_β^- satisfies A; for $\beta > \alpha'$, Σ_β^- satisfies classical set theory, and, in both cases, strong axioms of constructibility [Cohen 1963b]. Thus the ramified theory is useful for the syntactic study of some subsystems of A by (*c*), for the reduction of extensions of A *to* A by (*d*), but not for the study *of* A itself.

[19]It should be observed that, in contrast to mere consistency problems, Herbrand-type theorems remain significant even if no restriction is imposed on the metamathematical methods in contrast to the consistency "proof" of para. 8.

results in this area were first discovered through a good understanding of finitist or intuitionistic conceptions simply because certain results are evident from a constructive meaning of the formulae involved. But, at least once the results are known, one would expect to prove them more simply by full use of non-constructive methods. – The situation is different at present in the study of systems of arithmetic such as Z (cf. end of para. 8).[20] There is a temptation to regard these results as artificial because the systems studied are fragments of arithmetic (by incompleteness) and may be freaks to which freakish methods are well adapted. – I do not believe that this is a fruitful account of the position; on the contrary, I believe we are here faced with a very good problem of foundations. But I do not have a satisfactory answer: we have come back to the third problem of para. 4 (cf. para. 18).

15. Modified Hilbert programme. We take for granted Gödel's two incompleteness theorems and that finitist proof in its original sense does not go essentially beyond classical arithmetic [Hilbert 1932-5, 3: 212]; the latter would follow if the formulation of finitist proof sketched below is accepted. Thus it is necessary to use a hierarchy of constructive methods for the study (interpretation) of axiomatic systems. As we said above, we regard Herbrand's theorem as the paradigm for such an interpretation.[21]

The question is of course which aspects of Herbrand's theorem are essential for the present purpose, namely the understanding of transfinite symbols, and which are details, perhaps of importance elsewhere.

As in para. 11, according to Herbrand, $\mathfrak{A}$, say $(x)(\mathrm{E}y)(z)\mathrm{A}(x,y,z)$, is provable in the predicate calculus, if and only if one of a sequence of quantifier-free A_n is provable, A_n being of the form

$$\mathrm{A}\{\alpha, \varphi_1(f), f[\varphi_1(f)]\} \vee \ldots \vee \mathrm{A}\{\alpha, \varphi_{p(n)}(f), f[\varphi_{p(n)}(f)]\}$$

Note in passing that quite trivially the sequence could be replaced by a single formula with a constructive metamathematical quantifier $(\exists n)$, namely $(\exists n)\mathrm{A}(n)$, where $\mathrm{A}(n)$ denotes A_n. (In the case of number theory and analysis we do not even need a metamathematical quantifier, but a

---

[20]Cf. [Kreisel 1958b] and footnote 7.

[21]In the sense of [Kreisel 1958b]. The mathematical use of interpretations for independence proofs, etc., is discussed elsewhere and is here taken for granted. Mostowski [1957b] and Tarski have protested against my use of the word "interpretation." There are two issues: (i) It clashes with Tarski's use of the word in [Tarski, Mostowski, and Robinson 1953]. (ii) Does it express at least one important meaning of the word in common use? – As to (i) there is a perfectly good word current for Tarski's meaning, namely model, and in case of doubt, one could use "formal" or "syntactic" model; also that book appeared much later than my first use of the term in [Kreisel 1951-2]. But (ii) raises a serious question if one is interested in the notion of a "reduction"; the discussion below was stimulated by Mostowski's criticism *lc.*

constructive existential quantifier $(\exists\varphi)$, φ ranging over a suitable class of recursive functionals.)

$\mathfrak{A}$ and A_n are in the close relation that $\mathfrak{A}$ can be proved from A_n by means of Herbrand's rules which have the property that it is *decidable*[22] for given $\mathfrak{A}$ and A_n, whether $\mathfrak{A}$ be so provable from A_n. What seems to me significant about this is that the logical relation between $\mathfrak{A}$ and A_n considered is essentially more elementary than the logical relations discussed, namely those of the (undecidable) predicate calculus. In this sense A_n expresses the full content of $\mathfrak{A}$.

Furthermore, as in Bernays' consistency theorem, if $\mathfrak{B}$ is provable from $\mathfrak{A}$ by means of transfinite symbols and $\mathfrak{A}$, interpreted as one of the A_n, is finitistically true, then so is some B_m.[23] – It is natural that a good understanding of the use of transfinite symbols should include the consideration of implications $\mathfrak{A} \vdash \mathfrak{B}$ since their purpose is not only to produce tautologies, but also, as we have said, to step from extra-logical axioms to theorems.

This analysis led very obviously to the notion of interpretation, in particular, in finitist (constructive) systems. The conditions given are intended to express what one would reasonably expect of an understanding of transfinite symbols by, or of a reduction to, finitist (constructive) means. Whether they do so, seems a proper subject for discussion by philosophers. I myself would apply the word *reduction* only to an interpretation of a system in one of its subsystems where the primitive notions and theorems of the latter are a subset of the former. – In particular, I do not call the set theoretical definitions of natural numbers or finiteness (cf. footnote 6) a reduction because, though we now have only the one primitive notion of a set, its content is not clearly comparable with that of the others. – Further I should require the reduction to be established by (sound) methods which can be formalized in the subsystem itself.

For me the "reduction" of primitive notions is not a matter of principle. A reduction does not eliminate them since merely to see that a proposed reduction is correct one has to start with the primitive notion considered anyway. But in practice such reductions can be extraordinarily fertile if for no other reason than that they permit the formulation of new questions: of $(x)(\mathrm{E}y)A(x,y)$ we ask only if it is valid or not, of $A[\alpha, t_1(\alpha)] \vee \ldots \vee A[\alpha, t_n(\alpha)]$ we ask what the complexity of the terms $t_i(\alpha)$ is. This is typical of the increased information contained in Herbrand's theorem. Another example is the arithmetization of the com-

---

[22]Since in any application of Herbrand's rules the number of disjunction symbols or of free variables is reduced.

[23]This is really clear for an applied predicate calculus only, cf. in note on the pure predicate calculus in the remark below.

pleteness theorem in [Hilbert and Bernays 1934–9: vol. 2] compared with
the mere assertion of completeness. What kind of additional information
is regarded as providing a satisfactory answer is a similar kind of question to the one about the significance of finitist proof; see also end of
para. 18.

Open problem. For a consistently constructive interpretation of analysis
(para. 13) it is necessary to prove constructively the existence of the continuous functionals of whatever schema is produced in the solution of the
open problem of para. 13. This problem is analogous to the proof of
existence of Gödel's functionals in para. 12.

16. Remark. It seems desirable to discuss the bearing of an *interpretation*
on Brouwer's objection that consistency leaves open the possibility that
some provable theorems are intuitively false. Remembering the difference between the intuitionistic and truth functional meaning of the
logical constants one must use an appropriate translation of classical
theorems before speaking of their being intuitively false. Gödel elucidated Brouwer's point most elegantly: in as much as consistency does not
ensure ω-consistency, Brouwer was evidently right (even if one translates
classical $(Ex)A(x)$ into intuitive $\sim(x)\sim A(x)$); on the other hand for
classical number theory in particular Brouwer's objection did not arise
because of the translation into Heyting's number theory. In the case of
Herbrand's own interpretation the following points should be noted.
Using Hilbert's terminology it is clear that quantifier-free formulae, i.e.
those of the elementary calculus with free variables [Hilbert and Bernays
1934–9: vol. 2], are regarded as the "real" elements, others as ideal and
in need of an interpretation. Now, Brouwer would certainly not accept
this because on his interpretation of propositional formulae, the theorems
of the classical propositional calculus are not valid; in particular, he
interprets $A \lor B$ as: A is provable or B is provable; also he wishes to substitute for propositional letters incompletely defined propositions, e.g.
propositions containing a parameter ranging over free choice sequences;
under this interpretation $A \lor \sim A$ is evidently not valid. On the other
hand, if a finitist had made Brouwer's objection, Gödel's translation
into Heyting's arithmetic would not have been sufficient, and to answer
it something like the extensions of Herbrand's theorem *lc* would have
been necessary. However, now the formulae of the classical propositional
calculus could be regarded as "real" elements because for finitist propositions the truth-functional interpretation of the logical connectives is
applicable.

We note in passing that just as there was a Hilbert programme for understanding classical transfinite machinery from a finitist point of view so there is an analogous programme for understanding intuitionistic machinery. Certainly as far as independence results are concerned, the latter is more rewarding if for no other reason than that it is less familiar.

Finitist proof

17. Work is in progress[24] on a characterization of finitist proofs in the usual sense: a formal system is described such that (i) each formal proof of the system is recognized as a finitist proof, and (ii) each formula in the notation of the system is asserted to be provable in the system if it is provable by finitist methods at all. The variables of the system are of two types (natural numbers and free function variables from the natural numbers to the natural numbers though the latter can be avoided), the constants are particular numerals, certain constant functionals and (finitist) proof predicates, the last two being introduced only after certain existential statements have already been established. The idea corresponds closely to what Hilbert imagined the whole mathematics to be like, namely an interplay between formal proofs and metamathematics [Hilbert 1932–5, 3: 174–5]. Finitist proofs constitute then the least class of proofs closed under the following condition (and containing a certain obvious minimum): (i) if a proof predicate has been shown by finitist methods to satisfy the relevant existential conditions then it is finitist too, and (ii) if $\mathrm{Prov}(n, m)$ is a finitist proof predicate already introduced, and if, with free variable n, $(\mathrm{E}p)\,\mathrm{Prov}[p, \ulcorner A(0^{(n)})\urcorner]$ is established by a finitist proof, then so is $A(n)$; – by Gödel's theorem this closure condition can be achieved only if no predicate of the system itself is both extensionally equivalent to the proof predicate of the whole system and also satisfies the conditions on a proof predicate imposed in Gödel's second undecidability theorem.

Evidently, there is no reason why the class of theorems should not be recursively enumerable: in fact, it is. A finitist could even conjecture that a particular enumeration gives precisely the class of finitistically provable theorems, e.g. in the notation of primitive recursive arithmetic, but this would be, for him, an empirical conjecture incapable of (finitist) proof.

Just as with the class of recursive functions, the only completeness properties of our class of proofs are certain closure properties, i.e. it is the least class with these closure properties.

The main open "problem" is to discover intuitively really convincing

---

[24]This has since been presented in the sketch [Kreisel 1960b], where the characterization of other informal notions of proof is also considered. – My interest in a definition of "finitist proofs" was reawakened by conversations with K. Gödel, 1955–7.

completeness properties, aided of course by more detailed information about the class itself, cf. para. 4 above.

As pointed out in para. 4, other classes of constructive proofs should be studied in addition to finitist proofs. In particular, as Bernays mentioned [1941: 151], the use of the first ϵ-number is intermediate between finitist and full intuitionist mathematics. It would be interesting to motivate the choice of some subclass of intuitionist proofs which includes the use, made e.g. in Ackermann [1940], of the first ϵ-number. (I have not done this to my own satisfaction even for the notion of finitist proof.)

18. General remarks. Since our understanding of the notion of "constructive proof" and of its special case "finitist proof" is not too detailed, the very meaning of Hilbert's programme is not too precise. However, one can summarize what parts of Hilbert's programme are settled on the basis of our partial understanding of these notions (cf. end of section 1). If one prefers a more formal approach, one would begin with a (partial) axiomatization of the notions of "constructive proof" or "finitist proof" which contains only trivial properties of these notions, and then investigate from what additional axioms for these notions the assertions below can be formally derived.

Hilbert's programme in the wide sense wanted to establish the "adequacy" of deductive formalisms for the representation of intuitive branches of mathematics. Their inadequacy (in the case of arithmetic) is established by Gödel's first incompleteness theorem on the basis of our partial understanding of the notions involved, namely the recursive enumerability of the set of theorems of any deductive formalism in the original sense; then there is an A such that [($\ulcorner A \urcorner$ is a formal theorem) $\rightarrow$ A] is not formally derivable, and so the formalism cannot formally be proved to be sound if it is sound (for this A); further, it is inadequate, where by "adequacy" one means: if A then ($\ulcorner A \urcorner$ is a theorem). It is interesting to observe that this inadequacy of the usual formalisms is connected with their surprising adequacy in another sense, namely the possibility of representing all recursive functions, at numerical arguments, by terms of the usual formal systems. As far as I know, before Gödel's work it was not even realized that all multiply recursive predicates could be defined in first order arithmetic, even when symbols for primitive recursive functions are added; although multiply recursive predicates were accepted as finitist. In fact, the schemata for recursions of higher type which Hilbert considered [1899b, 7th ed.: 295 and elsewhere] could, at least at first sight, be expected to "transcend" even analysis. On the other hand, as we know now, the proofs in present-day number theory can indeed be rather easily formalized in Hilbert's own formal system Z of arithmetic by means of the devices introduced by Gödel. So, while in Hilbert's days the empirical evidence (from "ordinary" number theory) for the adequacy of Z was slight, at the present time it would be overwhelming! It seems clear therefore that Hilbert's grounds for the feasibility of his programme must have rested on general philosophical considerations, perhaps the following: All that we "do" in mathematics (or: in thinking generally) is to operate with symbols, and that is all we "really" communicate to one another. (Hilbert sets great store by the "finiteness of our thoughts," both in principle [1932–5, 3: 187] and in proof theory, cf. para. 10 above.) As far as theorems of a transfinite character are concerned, we have proofs: their "truth" is metaphysical or poetic, and any reference to their truth must be "ultimately" reducible to assertions of formal derivability. The notion of mathematical truth can have no place in mathematics as we know it. – While it would be generally granted that this reflection itself is metaphysical or poetic it is convincing because in "ordinary" mathematics there is not the slightest hint of any practical use of the distinction between "truth" and "formal derivability" (even when rules of derivability are made explicit). The first thing to do, if one takes this distinction at all seriously, is to consider the assertion: ($\ulcorner A \urcorner$ is a formal theorem) $\rightarrow A$. Expressing this assertion by means of an arithmetic formula is an essential step towards the incompleteness results. Also, the truth of each such formula is evident from the interpretation of the formal rules of proof and not from the combinatorial use of these rules. In other words, each such formula is obtained from the intended meaning of the formal systems considered, i.e. accepted on the basis of this meaning, without being formally implied by the given rules (unless A itself is so implied). Each such formula would be accepted as an axiom for arithmetic. Thus we have here an illustration of how one chooses axioms for formal systems from the intended meaning of the formal systems to be considered. The part of mathematical activity concerned with a good choice of axioms had no place in Hilbert's "official" conception of mathematics. If there is any real justification for calling Hilbert's approach "formalist" it is certainly this deficiency of Hilbert's official conception of mathematics and not his use of syntactic formulations in the foundations of mathematics (cf. para. 1).

Next consider Hilbert's programme in the narrow sense, namely to prove the consistency of the usual formalizations of mathematics by finitist means. Strictly speaking, he had a number of intermediate conjectures, too, such as the decidability of arithmetic or the somewhat peculiar assertion (*): it is consistent to assume that every problem of arithmetic is solvable. But a detailed discussion seems of little more than biographical interest, since these conjectures are either settled or genuinely ill formulated: if (*) means that one may consistenty add to formal arithmetic the statement:

> *For every closed formula* A *either* A *is a formal theorem or* ∼A *is a formal theorem,*

then the conjecture follows trivially from Gödel's result that one may even add consistently the assertion that the formal system is inconsistent. It is unlikely that Hilbert was satisfied by this manner of establishing his conjecture. Naturally, for the narrower programme an impossibility proof requires a more detailed analysis of the notions involved than for Hilbert's general programme. In particular, Gödel's second incompleteness theorem is not conclusive until reasons are adduced which show that all finitist theorems are included among the formal theorems of the system considered. As regards a characterization of finitist proof the sketch [Kreisel 1960b] leaves open the problem of upper bounds for the set of finitist theorems, and one may have to be content to obtain bounds in stages. It is, for instance, quite likely that more convincing arguments can be given, i.e. fewer assumptions on the notion of finitist proof are needed, for showing that all finitist functions are ordinal recursive of order $< \epsilon_0$ than, e.g. for showing that the consistency of first order arithmetic cannot be proved by finitist means. (In other words, the impossibility of a finitist proof for the ω-consistency of arithmetic may be genuinely easier to establish.) We may note in passing that current mathematics illustrates the use of different properties of intuitive intensional concepts in different theorems. Thus Gödel's first incompleteness theorem, which is concerned with Hilbert's general programme, uses substantially fewer properties of the intuitive notion of "representability" of (the syntactic) properties in a formal system than the second, which is concerned with the narrower programme. The first theorem requires that the set of theorems be representable, i.e. the formal system considered may be "identified" with its set of theorems (i.e. all we need know about it is its set of theorems), while the second requires certain (internal) properties of the proof relation to be formally derivable. This will be the case if one "identifies" the formal system with its set of production rules but not if one identifies it with the set of theorems since this set can be generated by different rules. (One could state this explicitly in the following formulation: *If a formula can be proved in* S *to express the consistency of* S, *and* S *is consistent, then this formula cannot be proved in* S; certain minimal conditions on the notion "expressing the consistency of S" are formulated in [Hilbert and Bernays 1934–9: vol. 2].)

It is to be emphasized that the remarks above do no more than assert a conviction that the notions of "constructive" and "finitist" are ripe for a systematic study. The reformulations of known theorems given above indicate that our conception of these notions is coherent because these reformulations are not forced.

We conclude with some pragmatic remarks on the choice of a significant class of constructive methods which goes beyond finitist mathematics. Here one has to recognize the following difference between the untutored notion of constructivity of the "ordinary" mathematician (interested in constructivity) and the one that seems to be forced on one if one tries to be coherent. Naïvely, "constructive" is applied to procedures or instructions for manipulating finite configurations (as in the case of finitist mathematics), but not to operations on objects of higher type or logical operations (e.g. on proofs with meaning, as in the case of intuitionistic mathematics). The naïve conception does not get into difficulties in familiar parts of mathematics, because, once constructive definitions are given, the proofs are in general quite unproblematic, cf. footnote 5. On the basis of this experience the mathematician is ill-prepared to judge the constructivity of, e.g. definitions by means of recursion of the form $f(n) = g\{n, f[\tau(n)]\}$ if $\tau(n) < \cdot n$ and $f(n) = g(n, 0)$ if $\tau(n) \not< \cdot n$, where g, τ are constructive functions, $< \cdot$ a well-founded constructive ordering (but may not have been proved to be well founded by "constructive means"). This is precisely the situation in several consistency proofs for arithmetic [e.g., Ackermann 1940]. If one thinks in terms of (idealized) machines, the above definition is a proper instruction, provided $g, \tau, < \cdot$ are constructive, but the *proof* of well-ordering is irrelevant. It is likely that, on this naïve conception, recursion on (the usual orderings whose ordinal is) the first ϵ-number would be more evidently "constructive" than Gödel's simple functionals quoted in para. 12 above, which involve variables of higher type.

On the other hand, both philosophically and pragmatically, there is strong evidence that a more coherent theory of constructivity is possible if this notion is applied also to logic (inference), in particular to the interpretation of logical constants. For instance, philosophically, it is clear that there is a gap in accepting as constructive the definitions by transfinite recursion of the preceding paragraph if no condition is imposed on the proof of well-ordering: one certainly could not claim that the definition has been justified on constructive principles. Pragmatically, i.e. for the purpose of obtaining, in a systematic manner, well-rounded mathematical theories, the use of the primitive notion of a constructive proof has definitely been fruitful. One example is Gödel's use of intuitionistic logic as an intermediary in the consistency proof of arithmetic (cf. para. 12), another is the use of generalized inductive definitions based on intuitionistic logic, which explains the remarkable fact that many functions

228

defined on inductive sets (e.g. Church–Kleene ordinal notations) are recursive: this is mysterious from the non-constructive approach to recursion theory because the inductive sets are highly non-recursive, and one sees no reason why definitions by transfinite induction on such sets should lead to recursive objects. But, also the following more isolated example is instructive.

Suppose we ask for a modification of classical arithmetic which leaves the constructive theorems (of the form $(x)(Ey)A(x,y)$ with quantifier-free A) unchanged, but makes every prenex theorem recursively satisfiable. This would express more or less what the naïve constructive mathematician described above might be after. In such a modification the class of quantifier-free theorems is the same as in classical arithmetic, while the class of theorems containing quantifiers is changed. Who would have thought of modifying only the rules for the propositional connectives, and not for quantifiers, which is precisely what is done in intuitionistic mathematics. However, if one has the concept of constructive proof and Heyting's interpretation of the logical concepts, this step is inevitable. Moreover, while the general problem (of finding rules of proof valid for the constructive interpretation of the logical constants) has a high degree of determinateness,[25] the naïve constructivist's question certainly does not have a unique answer. Thus $[\sim p \to (q \lor r)] \to [(\sim p \to q) \lor (\sim p \to r)]$ could be added to the propositional axiom schemata of Heyting's arithmetic, or, again, for primitive recursive $A(x)$, $\sim (x)A(x) \to (\exists x)\sim A(x)$; both these additions are nonconservative, since the former schema is not a theorem of Heyting's propositional calculus and the latter is deducible in his arithmetic only if $(\exists x)\sim A(x) \lor (x)A(x)$ is too.

More generally, even if one is primarily interested in developing systematically a smooth-running formalism, it is necessary to develop a coherent philosophical notion of constructivity. The conception itself (which applies to functions, proofs, etc.) has a degree of generality which transcends any mathematical treatment, but it is the business of pure mathematics to develop it within suitable specific contexts. In this respect it is like the notions of structure, truth, or proof. It may be remarked that I myself have come to recognize the need for, and usefulness of, a (non-finitist) notion of constructive proof only very reluctantly.

Completeness of the predicate calculus

19. I wish to emphasize two points of interest of the completeness problem; namely it illustrates (i) the mathematical treatment of a concept of

---

[25]Even though the possibility of a (provably) complete formalization is doubtful.

great generality in a specific context, as mentioned in the preceding paragraph, (ii) the limitations of a purely finitist metamathematics.

If $A(P_1, \ldots, P_n)$ is a (closed) formula of the predicate calculus whose predicate symbols are $P_1, \ldots, P_n$, then it is valid if and only if for every domain D of individuals and predicates P_i^o, $1 \leqslant i \leqslant n$, defined in D, $A(P_i^o, \ldots, P_n^o)$ is true on the usual interpretation of the logical symbols. The calculus is complete if every valid formula is provable in it. The reference to arbitrary domains here is a typical case of philosophical generality, reminiscent of the paradoxes. It would be disturbing if the paradoxes led to the formalization of logic and the justification of the formalization led back to the very notions involved in the paradoxes! Now, if one merely wishes to show that every theorem of the calculus is valid, one considers any particular set D, presumed to be well defined, and follows out the rules of the calculus with quantifiers ranging over this set; there is no generation of new sets, which is the typical step in the paradoxes. But, if one shows that every valid formula is provable, the premise itself involves quantification over all sets. As long as one has a precisely limited notation, one may expect to sharpen the result by restricting the sets for which the formula has to be valid. Gödel shows that for each $A(P_1, \ldots, P_n)$ there is a *single* set of predicates P_i^A defined over the domain D_o of natural numbers such that instead of the validity of A we merely need the truth of $A(P_1^A, \ldots, P_n^A)$ in D_o.

From a finitist point of view even this is not enough since in general the P_i^A cannot be chosen to be constructive at all. There are two finitist versions of the result, namely (a) if $A(P_1^A, \ldots, P_n^A)$ is provable in Z then $A(P_1, \ldots, P_n)$ is provable in the predicate calculus, or the stronger form (b) $A(P_1^A, \ldots, P_n^A) \rightarrow$ "$A(P_1, \ldots, P_n)$ is provable in the predicate calculus," the implication being provable in Z for each A where the statement in inverted commas denotes some natural arithmetization of itself. – It should be remembered that these versions are not what Hilbert originally required [1899b, 7th ed.: 322] namely that if $A(P_1, \ldots, P_n)$ is not provable in the predicate calculus, $\neg A(P_1^A, \ldots, P_n^A)$ should actually be provable in Z. There is no recursively enumerable extension of the predicate calculus with this property. (Hilbert evidently assumed Z to be complete and therefore did not distinguish between these versions which are equivalent on his assumption.)

Of course, on occasions the extra information contained in the finitist versions is needed. But the finitist point of view is simply not appropriate for discussing the general problem of completeness.

Postscript (Autumn 1978) to "Hilbert's Programme"

After twenty years it is, fortunately, necessary to add bibliographical references to bring the technical part of the article up to date. This is done in short notes that refer to the relevant chapters of the article (without giving the verse). In addition I wish to stress two general changes in my views over the last twenty, and especially the last ten, years.

First of all, the article seems far too conciliatory on the *finitist philosophy of mathematics*. Obviously this philosophy gets its sting from its negative side, the rejection of non-finitist methods and particularly non-finitist formulations, and much less from its positive side, the mere use of finitist methods. This is so because, quite often, a finitist proof tells us something we want to know that its competition, for example, a slicker non-finitist proof, does not. Thus no grand "philosophy" is behind the almost universal preference for a proof of convergence of a^n $(0 \leqslant a \leqslant 1)$ that computes bounds for the rate of convergence over its competitor; the latter notes monotonicity and boundedness of the sequence a^n, and then appeals to the principle of the least upper bound (cf. also Hilbert's example in the last paragraph but one of §4). As I see things now, it is little short of an evasion to call the switch in §3 from Hilbert's *consistency problem* (which concerns only proofs of purely universal statements) to an *interpretation* (of logically compound formulae) a "critique of detail": this logically fruitful and quite decisive switch draws attention away from the consequences of rejecting non-finitist proofs, for example, the perfectly sensible consistency proof mentioned at the beginning of §8. Behind such proofs there is very real progress, provided, of course, that there is a clear description of the structures for which the axioms considered are true. (For example, the clear description of segments of the cumulative hierarchy of sets has done more for the removal of genuine doubts than all finitist consistency proofs put together.) Presumably, mathematical practice has not been affected much by finitist doctrine: For one thing, there are plenty of interesting problems within those parts of mathematics that are simply *intended* to be finitist. But foundational research has unquestionably been affected negatively, particularly – and ironically – proof theory itself. ("Ironically" because this branch of mathematical logic was created by Hilbert for the sake of a finitist philosophy.) The mechanism at work is quite simple. Consciously or unconsciously, the bulk of publications in proof theory stress those results that, like the consistency of currently used principles, are easy or trivial unless one adopts (the negative side of) the finitist doctrine. For this very reason members of the silent majority, not only of mathematicians but of logicians, are put off by (the dullness of these results in) proof theory.

231

As is usual in such circumstances, the doctrinaire finitists have come to feel isolated and to think of the silent majority as lacking that Higher Sensibility that is needed to appreciate finitist doubts, though, as somebody said, those doubts are more dubious than what is being doubted.

The second major defect of the article is, so to speak, accidental, and related to the following circumstance. During the years 1956–8, I managed to do something of interest both with the notion of finitist proof itself (§17), by use of so-called autonomous progessions, and the more general notion of constructive proof (§18), for example, in connection with the completeness of Heyting's systems. As I see matters now, I was overimpressed by the fact that *anything* precise could be done with these notions, and even more by the fact that the work was most *satisfaisant pour l'esprit*. (The euphoria lasted for some time because closely related considerations worked for the notion of predicative proof and led to refinements of then-current notions concerning ω-models and generalizations of recursion theory, and finally to the proper choice of languages with infinitely long formulae). What I failed to do was to consider in detail any genuine alternative, so to speak competing, categories (of proofs) that might serve the same general purpose as the traditional categories (finitist, constructive, predicative). In particular, I failed to pursue the issue, stated emphatically in §18, whether one should extend to proofs, that is, to logical inferences, the more naïve categories of operations and definitions. In this connection recursion theory had already provided a convincing analysis of (hereditarily) *finite* operations. Sure, from a finitist point of view it is obviously coherent to restrict also the proofs that establish that a proposed rule for an operation is well-defined. But as I see things now – and contrary to §18 – there was little evidence that this finitist aim would be rewarding (compare it to the aims of alchemy, which are only occasionally rewarding for chemistry); all the more since, by §4, the restriction in question to finitist proofs was not expected to increase reliability.

Looking back, I remember a number of such alternative categories that had struck me before 1958 and that have since turned out to lend themselves to logical study. (It is of little interest – even to me personally! – whether I "should" have studied them rather than finitist or intuitionist principles of proof.) For example, there was the matter of *structural complexity* of proofs. This cuts across the traditional categories. It is critical for reliability, since if there is no doubt about the validity of principles of proof, the decisive factor is the probability of error in applying those (correct) principles: Here the complexity of individual proofs is relevant. Evidently, there is not only one (simple) measure of complexity, no more than in the case of physical objects: Their complexity, in the sense of difficulty of grasping them, involves length, volume, weight, shape, even chemical composition, electric charge, and so on. The most one can hope is that a *few* measures are significant for, that is determine, *many* issues that arise in the course of nature. As it happens, there are a couple of abstracts on *relative consistency proofs* (Kreisel 1958c: 109–10 and Kreisel 1976a: 285–6) that illustrate the development strikingly. The first, stressing principles of *proof* only, provides a mere counter-example; the second, stressing *operations* (relating hypothetical proofs of inconsistency in the two systems), gives a satifactory solution.

As another example that certainly fits in with finitist preoccupations and with the thoughts on incompleteness at the end of §14: By and large I neglected subdivisions *within* finitist mathematics or first order arithmetic, and barely considered the mathematical significance of such subdivisions in the sense of §13. Very soon after those preoccupations ended I noticed that van der Waerden's theorem on arithmetic progressions (which I have known very well since my student days) presented an excellent candidate for using *primitive* recursion: All the known proofs of the theorem use double recursion (footnote 7) to bound N in terms of k and l:

If $\{1,\ldots,N\}$ are divided into k classes, at least one class contains an arithmetic progession of length l.

In contrast, derivability of van der Waerden's theorem in first order arithmetic is obvious and of no interest. In short, we have a new criterion for the choice of formal systems which are rewarding to study metamathematically, simply by concentrating on *particular* theorems. This observation was, as a matter of empirical fact, overlooked in §14 and the first paragraph of §15, where the superiority of more traditional logical criteria is tacitly assumed.

In fairness it should be added that, even with present experience, we can find very few corners in the mathematics of the 1950s that lend themselves to the type of logical analysis envisaged (and set out in Kreisel 1958b), particularly, the unwinding of proofs of Π_2^0 theorems in terms of the rate of growth of bounds, let alone of Π_1^0 theorems. This has changed significantly only in the seventies.

The general point of view of this Postscript is presented in a very condensed style in "What Have We Learned from Hilbert's Second Problem?" (Kreisel 1976b). [Added in autumn of 1982:] A leisurely presentation of the general point of view, in a broad context, is to be found in my obituary of Kurt Gödel (Kreisel 1980).

Notes

§4. Not only here, in the third paragraph, but throughout the article I make far too heavy weather of "derivability conditions" on the predicates and functions that are used to code arithmetically syntactic notions and operations concerning *given* formal rules $\mathfrak{F}$. In particular, I associated these conditions, in the sense of Hilbert and Bernays (1934–9, 2: 286) with certain differences between Gödel's and Henkin's self-referential sentences (I am not, resp. I am provable) which I had noticed in "On a Problem of Henkin's" (Kreisel 1953b: 405–6). But starting with (Kreisel 1962b: 243–6) I stressed the existence of *canonical representations* of syntactic notions, provided of course one has made up one's mind on the data used to determine $\mathfrak{F}$ (for example, rules in the style of Post), a perfect parallel to, say, the familiar algebraic representation of geometric notions. In particular, the representation is demonstrably unique up to isomorphism (or: equivalence in the case of predicates) once the notions to be represented are sufficiently analyzed axiomatically. Naturally, certain "brutal" questions may be quite insensitive to the choice of data; for example, recursive decidability for which the "abstract" set of theorems of $\mathfrak{F}$ is adequate – and equally evidently this is not so for proofs of the consistency of $\mathfrak{F}$. The point to remember is that, for example, in Gödel's second theorem, it is not vagaries of representations of a given $\mathfrak{F}$ that are at issue, but so to speak vagaries among formal rules. Specifically, unorthodox "representations" of the kind used by Rosser to improve Gödel's first theorem, for $\mathfrak{F}$, are canonical for different rules $\mathfrak{F}_R$ (where $\mathfrak{F}$ and $\mathfrak{F}_R$ have the same set of theorems if $\mathfrak{F}$ happens to be consistent). And one of the "derivability" conditions (no. (iii), demonstrable completeness for Σ_1^0 sentences) is *not* satisfied by that *canonical* representation. The matter is put straight by use of an example in a leisurely discussion in the second edition of Hilbert and Bernays (1968–70, 2: 298–301) and discussed fully in my joint paper with G. Takeuti, "Formally Self-referential Propositions for Cut-Free Classical Analysis and Related Systems" (Kreisel and Takeuti 1974: App. 3).

§5. The use of the completeness theorem (for predicate logic) to establish incompleteness, in the last sentence, refers to "Note on Arithmetic Models..." (Kreisel 1950: 265–85). In the meantime, I noticed how to extend the argument to get a "model-theoretic" proof of Gödel's second incompleteness theorem, naturally, for classical systems (cf. end of footnote 43 on p. 383 of Kreisel 1968; for a detailed exposition, see Smorynski 1977).

234

§6, especially footnote 5. In the last few years, non-constructive proofs have been given of (constructive) Π_2^0 theorems. But it can hardly be claimed that they were helped by "closer study of ... constructive aspects of non-constructive methods." Logicians are particularly attracted by a variant RT_A, of Ramsey's theorem RT on partitions, which is *known* not to be provable in first order arithmetic (and thus not finitistically provable in the sense of Kreisel 1960a; cf. Paris and Harrington 1977 on "mathematical incompleteness"). But the variant was discovered through a combination of experience in model theory and the partition calculus for "large" cardinals. Actually, less than ten years ago, I was led – by the traditional preoccupation with provability, rather than with particular proofs – to misinterpret a fact that had very much struck me at the time; specifically, as shown by Jockusch (1972: 268–80), the infinite version RT_∞ of RT is not arithmetic and, certainly, by far the easiest proof of (any variant of) RT which combines RT_∞, and a compactness argument is not arithmetic either. It just so happens that RT also has a proof in a fragment of primitive recursive arithmetic. But because of the latter, I dropped closer study of the easy proof of RT via RT_∞. In any case, in the last couple of years, non-constructive proofs of Π_2^0 theorems have been discovered that use principles that go far beyond first order arithmetic, specifically, proofs using so-called generalized inductive definitions familiar from the theorem of Cantor–Bendixson in descriptive set theory. Also – and in contrast to RT_A – these new proofs are very much in the "mainstream" of mathematics (see, for example, Furstenberg 1977: 204–56). [Added Autumn 1982:] More on the unwinding of these proofs is to be found in Kreisel (1982: 50).

§8. A more detailed description of the "new type of analysis" can be found in my review (Kreisel 1962c: 250–5) of *Beweistheorie* (Schütte 1960a). But while it is true that, in principle, this new analysis does not restrict the means of independence proofs, in practice so far it has not enlarged them either! In particular, we now have perfectly good model theoretic proofs of the independence results first discovered by the new (proof theoretic) analysis. *Warning:* Footnote 7 is very much restricted to *constructive* mathematics! Specifically, there are very manageable non-recursive (non-standard) models defined by suitable ultraproducts.

§9. The blithe reference to "a practical decision method" is almost empty, at least when applied to the – full first order – theories that were usually considered in the fifties. None of those theories has a practical decision method, and the problem is to select classes of formulae that do:

By now, the standard example comes from the theory of diophantine equations

$$(\exists x_1 \epsilon \omega) \ldots (\exists x_n \epsilon \omega)[p(x_1,\ldots,x_n)=0]$$

where the polynomials p are selected according to geometric properties of varieties: $p=0$, defined in finite fields! (Cf. also Meyer 1975: 132–54.)

§10, especially footnote 10. Hilbert's ϵ-calculus has been neglected during the last twenty years except for a few asides. At one extreme, it is not known (even without restriction on the metamathematical methods) whether the particular substitution method for analysis proposed in Hilbert and Bernays (1934–9, vol. 2) converges (cf. Kreisel 1965: 168, 3.351). At another extreme, for the set theoretic formalism with primitives: membership, union, pair, empty set, there is a primitive recursive method for deciding whether any formula in the ϵ-formalism can be realized (by means of hereditarily finite sets; cf. Ville 1971: 513–16, extended by Gogol 1978: 289–90). And though no doubt any finite set of "critical" ϵ-formulae and axioms of current set theory can be realized, this fact cannot be proved in set theory.

§11. Knowledge of cut-free systems has improved immensely during the last twenty years. (a) The model-theoretical proof of cut elimination, mentioned in the second paragraph (apparently for the first time), has been refined so as to be formalized in a fragment of arithmetic. Constructive analysis of this fragment yields an α-recursive rule ($\alpha < \omega^{\omega^\omega}$). (b) The notion of *semi-valuation* introduced for type theory by K. Schütte (1960: 305–26) extends naturally to predicate logic. Though, by (a), there is no (extensional) difference between validity in *all* total, resp. semivaluations, there is a difference if valuations of suitably *restricted complexity* $\mathcal{C}$ are considered; equivalently, if the category of proof trees without any infinite paths is extended to those without any path $\epsilon\mathcal{C}$ (Kreisel, Mints, and Simpson 1975: 38–131). (c) Ibid. The 'cogent motivation' mentioned in para. 5 of §11 is made explicit: the (model-theoretic) motive is to construct a tree T_A of formulae that codes *all* countable valuations in which A is false; and the rules (of inference) reverse the rules needed for constructing T_A.

§11, footnote 13. Herbrand's own publication has been found defective (cf. B. Dreben, P. Andrews, S. Aanderaa 1963: 699–706).

§12. The use of infinite induction, that is, of the ω-rule has considerably more significance than is suggested by footnote 15 (cf. Kreisel 1976c: 177–223).

Concerning footnote 14: the (quite essential) use of function variables presents an essential departure from Herbrand's *intentions* whose conception of finitist proof did not allow such variables, and led to his complicated reformulation.

§*13*. Despite the heading, there is an occasional bias toward logical rather than mathematical applications! Specifically, an important relation between the rate of growth of bounds and formal derivability of Π_2^0 formulae from true Π_1^0 sentences is emphasized (Kreisel 1958b: 159, ii). But – in contrast to the Postscript above – I thought of using this relation for establishing the logical property of underivability.

§*14*. The doubts about the use of proof theoretic methods for analyzing the set of theorems prov*able* in fragments of arithmetic were more than justified, as explained in the Postscript and the note concerning §8. The methods have permanent value for analyzing, for example, unwinding *proofs*.

§*15*. The doubts (still unresolved, incidentally) about an adequate (general) analysis of such aims as "understanding by or reduction to limited means" and about a proper choice of limitation do not affect the *alternative* to traditional foundations mentioned at the end of the Postscript, where our starting point is a proof of a particular theorem and we know what more we want to know about it. The *value* of a foundational analysis seems to me to be properly measured by the frequency of particular cases where we don't know this by the light of nature.

§*17, footnote 24*. The exposition of the ideas of §17 given in Kreisel 1965: 171–2 is better than the sketch in Kreisel 1960a.

§*18*. The quite fundamental, and generally neglected, distinction made here, between Hilbert's programme in the wide sense (adequacy of formal systems for representing reasoning) and in the narrow sense (consistency of particular systems), is stated more forcefully at the beginning of "A Survey of Proof Theory" (Kreisel 1968: 321–88). In particular, Gödel's *first* incompleteness theorem is sufficient to refute the programme in the wide sense, especially, if one adds Hilbert's requirement of a "final solution" of foundational problems by mathematical means, mentioned in the last paragraph of §6 of the text. The comments on "representability" of syntactic properties at the end of paragraph 3 of §11, on Hilbert's programme in the narrow sense, are superseded by the note above to §4. In particular, using the type of formal rules called $\mathfrak{F}_R$ (see

p. 234) we get a system, which has (i) exactly the same *derivations,* not only the same theorems as, say, first order arithmetic, say, Z,

> can prove (ii) its own consistency, but for example,
> cannot prove (iii) its completeness for Σ_1^0 sentences

(whereas (iii) demonstrably holds for Z). The difference between $\mathfrak{F}_R$ and Z is that the procedure for checking derivations is different, involving so to speak a general comparison with background knowledge, a look at theorems proved "earlier."

Concerning footnote 25, there is a quite extensive literature beginning with a convincing use of topological "interpretations" to establish deductive completeness of Heyting's system of propositional logic, specifically, for predicates of what are nowadays called lawless sequences (Kreisel 1958d: 369–88). The most recent result in this direction covers the fragment $\{\wedge, \vee, \rightarrow, \forall, \exists\}$ (without negation); cf. Theorem 13 in *Elements of Intuitionism* (Dummett 1977, p. 288). On the negative side, the assumption that all constructive number theoretic functions are recursive (that is, Church's thesis extended to intuitionistic, not only mechanical rules) implies that the set of constructively valid formulae in $\{\neg, \wedge, \vee, \rightarrow, \forall, \exists\}$ is not recursively enumerable (for exposition, cf. A. S. Troelstra 1977b: 39–58). For a connected account of the variants of Heyting's rules mentioned in the last paragraph but one of §18, see the axiomatizations of sentences (demonstrably) valid for various realizability and functional interpretations in the monograph of A. S. Troelstra 1973.

§*19.* The version (a), in the last paragraph but one, is nowadays called "maximality" of predicate logic for (schemata in) Z, since the paper by Dana S. Scott, "Extending the Topological Interpretation to Intuitionistic Analysis II" (1970). An example showing that (b) is 'stronger' is provided by Heyting's predicate calculus, which is maximal for Heyting's arithmetic but, by the note concerning §18, not complete. Put simply, we have maximality because we have forgotten the same valid (logical) rules in pure predicate calculus and in the calculus applied to arithmetic.

A general account of completeness, in terms of so-called informal rigor, which is considered to be readable, is in Kreisel 1967: 138–71.

PART II

The existence of mathematical objects

Empiricism, semantics, and ontology[1]

RUDOLF CARNAP

1. The problem of abstract entities

Empiricists are in general rather suspicious with respect to any kind of abstract entities like properties, classes, relations, numbers, propositions, etc. They usually feel much more in sympathy with nominalists than with realists (in the medieval sense). As far as possible they try to avoid any reference to abstract entities and to restrict themselves to what is sometimes called a nominalistic language, i.e., one not containing such references. However, within certain scientific contexts it seems hardly possible to avoid them. In the case of mathematics, some empiricists try to find a way out by treating the whole of mathematics as a mere calculus, a formal system for which no interpretation is given or can be given. Accordingly, the mathematician is said to speak not about numbers, functions, and infinite classes, but merely about meaningless symbols and formulas manipulated according to given formal rules. In physics it is more difficult to shun the suspected entities, because the language of physics serves for the communication of reports and predictions and hence cannot be taken as a mere calculus. A physicist who is suspicious of abstract entities may perhaps try to declare a certain part of the language of physics as uninterpreted and uninterpretable, that part which refers to real numbers as space-time coordinates or as values of physical magnitudes, to functions, limits, etc. More probably he will just speak about all these things like anybody else but with an uneasy conscience, like a man who in his everyday life does with qualms many things which are not in accord with the high moral principles he professes on Sundays. Recently the problem of abstract entities has arisen again in connection with semantics, the theory of meaning and truth. Some semanticists say that certain expressions designate certain entities, and

Reprinted with the kind permission of the author and publishers from Rudolf Carnap, *Meaning and Necessity,* 2nd ed. (Chicago: The University of Chicago Press, 1956), pp. 205–221, and from *Revue Internationale de Philosophie,* vol. 4 (1950), pp. 20–40. The slightly modified version that was printed in *Meaning and Necessity* appears here.
[1] I have made here some minor changes in the formulations to the effect that the term "framework" is now used only for the system of linguistic expressions, and not for the system of the entities in question.

among these designated entities they include not only concrete material things but also abstract entities, e.g., properties as designated by predicates and propositions as designated by sentences.[2] Others object strongly to this procedure as violating the basic principles of empiricism and leading back to a metaphysical ontology of the Platonic kind.

It is the purpose of this article to clarify this controversial issue. The nature and implications of the acceptance of a language referring to abstract entities will first be discussed in general; it will be shown that using such a language does not imply embracing a Platonic ontology but is perfectly compatible with empiricism and strictly scientific thinking. Then the special question of the role of abstract entities in semantics will be discussed. It is hoped that the clarification of the issue will be useful to those who would like to accept abstract entities in their work in mathematics, physics, semantics, or any other field; it may help them to overcome nominalistic scruples.

2. Linguistic frameworks

Are there properties, classes, numbers, propositions? In order to understand more clearly the nature of these and related problems, it is above all necessary to recognize a fundamental distinction between two kinds of questions concerning the existence or reality of entities. If someone wishes to speak in his language about a new kind of entities, he has to introduce a system of new ways of speaking, subject to new rules; we shall call this procedure the construction of a linguistic *framework* for the new entities in question. And now we must distinguish two kinds of questions of existence: first, questions of the existence of certain entities of the new kind *within the framework;* we call them *internal questions;* and second, questions concerning the existence or reality *of the system of entities as a whole,* called *external questions.* Internal questions and possible answers to them are formulated with the help of the new forms of expressions. The answers may be found either by purely logical methods or by empirical methods, depending upon whether the framework is a logical or a factual one. An external question is of a problematic character which is in need of closer examination.

The world of things. Let us consider as an example the simplest kind of entities dealt with in the everyday language: the spatio-temporally ordered system of observable things and events. Once we have accepted the thing language with its framework for things, we can raise and answer internal questions; e.g., "Is there a white piece of paper on my

---

[2]The terms "sentence" and "statement" are here used synonymously for declarative (indicative, propositional) sentences.

desk?'', ''Did King Arthur actually live?'', ''Are unicorns and centaurs real or merely imaginary?'', and the like. These questions are to be
answered by empirical investigations. Results of observations are eval-
uated according to certain rules as confirming or disconfirming evidence
for possible answers. (This evaluation is usually carried out, of course, as
a matter of habit rather than as a deliberate, rational procedure. But it is
possible, in a rational reconstruction, to lay down explicit rules for the
evaluation. This is one of the main tasks of a pure, as distinguished from
a psychological, epistemology.) The concept of reality occurring in these
internal questions is an empirical, scientific, non-metaphysical concept.
To recognize something as a real thing or event means to succeed in
incorporating it into the system of things at a particular space-time position so that it fits together with the other things recognized as real,
according to the rules of the framework.

From these questions we must distinguish the external question of the
reality of the thing world itself. In contrast to the former questions, this
question is raised neither by the man in the street nor by scientists, but
only by philosophers. Realists give an affirmative answer, subjective
idealists a negative one, and the controversy goes on for centuries without ever being solved. And it cannot be solved because it is framed in a
wrong way. To be real in the scientific sense means to be an element of
the system; hence this concept cannot be meaningfully applied to the
system itself. Those who raise the question of the reality of the thing
world itself have perhaps in mind not a theoretical question as their
formulation seems to suggest, but rather a practical question, a matter of
a practical decision concerning the structure of our language. We have to
make the choice whether or not to accept and use the forms of expression
in the framework in question.

In the case of this particular example, there is usually no deliberate
choice because we all have accepted the thing language early in our lives
as a matter of course. Nevertheless, we may regard it as a matter of decision in this sense: we are free to choose to continue using the thing language or not; in the latter case we could restrict ourselves to a language
of sense-data and other ''phenomenal'' entities, or construct an alternative to the customary thing language with another structure, or, finally,
we could refrain from speaking. If someone decides to accept the thing
language, there is no objection against saying that he has accepted the
world of things. But this must not be interpreted as if it meant his acceptance of a *belief* in the reality of the thing world; there is no such belief
or assertion or assumption, because it is not a theoretical question. To
accept the thing world means nothing more than to accept a certain form
of language, in other words, to accept rules for forming statements and

for testing, accepting, or rejecting them. The acceptance of the thing language leads, on the basis of observations made, also to the acceptance, belief, and assertion of certain statements. But the thesis of the reality of the thing world cannot be among these statements, because it cannot be formulated in the thing language or, it seems, in any other theoretical language.

The decision of accepting the thing language, although itself not of a cognitive nature, will nevertheless usually be influenced by theoretical knowledge, just like any other deliberate decision concerning the acceptance of linguistic or other rules. The purposes for which the language is intended to be used, for instance, the purpose of communicating factual knowledge, will determine which factors are relevant for the decision. The efficiency, fruitfulness, and simplicity of the use of the thing language may be among the decisive factors. And the questions concerning these qualities are indeed of a theoretical nature. But these questions cannot be identified with the question of realism. They are not yes-no questions but questions of degree. The thing language in the customary form works indeed with a high degree of efficiency for most purposes of everyday life. This is a matter of fact, based upon the content of our experiences. However, it would be wrong to describe this situation by saying: "The fact of the efficiency of the thing language is confirming evidence for the reality of the thing world"; we should rather say instead: "This fact makes it advisable to accept the thing language".

The system of numbers. As an example of a system which is of a logical rather than a factual nature let us take the system of natural numbers. The framework for this system is constructed by introducing into the language new expressions with suitable rules: (1) numerals like "five" and sentence forms like "there are five books on the table"; (2) the general term "number" for the new entities, and sentence forms like "five is a number"; (3) expressions for properties of numbers (e.g., "odd", "prime"), relations (e.g., "greater than"), and functions (e.g., "plus"), and sentence forms like "two plus three is five"; (4) numerical variables ("m", "n", etc.) and quantifiers for universal sentences ("for every $n,\ldots$") and existential sentences ("there is an n such that ...") with the customary deductive rules.

Here again there are internal questions, e.g., "Is there a prime number greater than a hundred?" Here, however, the answers are found, not by empirical investigation based on observations, but by logical analysis based on the rules for the new expressions. Therefore the answers are here analytic, i.e., logically true.

What is now the nature of the philosophical question concerning the existence or reality of numbers? To begin with, there is the internal ques-

244

tion which, together with the affirmative answer, can be formulated in the new terms, say by "There are numbers" or, more explicitly, "There is an *n* such that *n* is a number". This statement follows from the analytic statement "five is a number" and is therefore itself analytic. Moreover, it is rather trivial (in contradistinction to a statement like "There is a prime number greater than a million", which is likewise analytic but far from trivial), because it does not say more than that the new system is not empty; but this is immediately seen from the rule which states that words like "five" are substitutable for the new variables. Therefore nobody who meant the question "Are there numbers?" in the internal sense would either assert or even seriously consider a negative answer. This makes it plausible to assume that those philosophers who treat the question of the existence of numbers as a serious philosophical problem and offer lengthy arguments on either side do not have in mind the internal question. And, indeed, if we were to ask them: "Do you mean the question as to whether the framework of numbers, *if* we were to accept it, would be found to be empty or not?", they would probably reply: "Not at all; we mean a question *prior* to the acceptance of the new framework". They might try to explain what they mean by saying that it is a question of the ontological status of numbers; the question whether or not numbers have a certain metaphysical characteristic called reality (but a kind of ideal reality, different from the material reality of the thing world) or subsistence or status of "independent entities". Unfortunately, these philosophers have so far not given a formulation of their question in terms of the common scientific language. Therefore our judgment must be that they have not succeeded in giving to the external question and to the possible answers any cognitive content. Unless and until they supply a clear cognitive interpretation, we are justified in our suspicion that their question is a pseudo-question, that is, one disguised in the form of a theoretical question while in fact it is non-theoretical; in the present case it is the practical problem whether or not to incorporate into the language the new linguistic forms which constitute the framework of numbers.

The system of propositions. New variables, "*p*", "*q*", etc., are introduced with a rule to the effect that any (declarative) sentence may be substituted for a variable of this kind; this includes, in addition to the sentences of the original thing language, also all general sentences with variables of any kind which may have been introduced into the language. Further, the general term "proposition" is introduced. "*p* is a proposition" may be defined by "*p* or not *p*" (or by any other sentence form yielding only analytic sentences). Therefore, every sentence of the form "... is a proposition" (where any sentence may stand in the place of the dots) is analytic. This holds, for example, for the sentence:

(*a*)   "Chicago is large is a proposition".

(We disregard here the fact that the rules of English grammar require not a sentence but a that-clause as the subject of another sentence; accordingly, instead of (*a*) we should have to say "That Chicago is large is a proposition".) Predicates may be admitted whose argument expressions are sentences; these predicates may be either extensional (e.g., the customary truth-functional connectives) or not (e.g., modal predicates like "possible", "necessary", etc.). With the help of the new variables, general sentences may be formed, e.g.,

(*b*)   "For every *p*, either *p* or not-*p*".
(*c*)   "There is a *p* such that *p* is not necessary and not-*p* is not necessary".
(*d*)   "There is a *p* such that *p* is a proposition".

(*c*) and (*d*) are internal assertions of existence. The statement "There are propositions" may be meant in the sense of (*d*); in this case it is analytic (since it follows from (*a*)) and even trivial. If, however, the statement is meant in an external sense, then it is non-cognitive.

It is important to notice that the system of rules for the linguistic expressions of the propositional framework (of which only a few rules have here been briefly indicated) is sufficient for the introduction of the framework. Any further explanations as to the nature of the propositions (i.e., the elements of the system indicated, the values of the variables "*p*", "*q*", etc.) are theoretically unnecessary because, if correct, they follow from the rules. For example, are propositions mental events (as in Russell's theory)? A look at the rules shows us that they are not, because otherwise existential statements would be of the form: "If the mental state of the person in question fulfils such and such conditions, then there is a *p* such that ...". The fact that no references to mental conditions occur in existential statements (like (*c*), (*d*), etc.) shows that propositions are not mental entities. Further, a statement of the existence of linguistic entities (e.g., expressions, classes of expressions, etc.) must contain a reference to a language. The fact that no such reference occurs in the existential statements here shows that propositions are not linguistic entities. The fact that in these statements no reference to a subject (an observer or knower) occurs (nothing like: "There is a *p* which is necessary for Mr. *X*") shows that the propositions (and their properties, like necessity, etc.) are not subjective. Although characterizations of these or

246

similar kinds are, strictly speaking, unnecessary, they may nevertheless be practically useful. If they are given, they should be understood, not as ingredient parts of the system, but merely as marginal notes with the purpose of supplying to the reader helpful hints or convenient pictorial associations which may make his learning of the use of the expressions easier than the bare system of the rules would do. Such a characterization is analogous to an extra-systematic explanation which a physicist sometimes gives to the beginner. He might, for example, tell him to imagine the atoms of a gas as small balls rushing around with great speed, or the electromagnetic field and its oscillations as quasi-elastic tensions and vibrations in an ether. In fact, however, all that can accurately be said about atoms or the field is implicitly contained in the physical laws of the theories in question.[3]

The system of thing properties. The thing language contains words like "red", "hard", "stone", "house", etc., which are used for describing what things are like. Now we may introduce new variables, say "f", "g", etc., for which those words are substitutable and furthermore the general term "property". New rules are laid down which admit sentences like "Red is a property", "Red is a color", "These two pieces of paper have at least one color in common" (i.e., "There is an f such that f is a color, and ..."). The last sentence is an internal assertion. It is of an empirical, factual nature. However, the external statement, the philosophical statement of the reality of properties – a special case of the thesis of the reality of universals – is devoid of cognitive content.

The systems of integers and rational numbers. Into a language con-

---

[3] In my book *Meaning and Necessity* (1947) I have developed a semantic method which takes propositions as entities designated by sentences (more specifically, as intensions of sentences). In order to facilitate the understanding of the systematic development, I added some informal, extra-systematic explanations concerning the nature of propositions. I said that the term "proposition" "is used neither for a linguistic expression nor for a subjective, mental occurrence, but rather for something objective that may or may not be exemplified in nature.... We apply the term 'proposition' to any entities of a certain logical type, namely, those that may be expressed by (declarative) sentences in a language" (p. 27). After some more detailed discussion concerning the relation between propositions and facts, and the nature of false propositions, I added: "It has been the purpose of the preceding remarks to facilitate the understanding of our conception of propositions. If, however, a reader should find these explanations more puzzling than clarifying, or even unacceptable, he may disregard them" (p. 31) (that is, disregard these extra-systematic explanations, not the whole theory of the propositions as intensions of sentences, as one reviewer understood). In spite of this warning, it seems that some of those readers who were puzzled by the explanations did not disregard them but thought that by raising objections against them they could refute the theory. This is analogous to the procedure of some laymen who by (correctly) criticizing the ether picture or other visualizations of physical theories thought they had refuted those theories. Perhaps the discussions in the present paper will help in clarifying the role of the system of linguistic rules for the introduction of a framework for entities on the one hand, and that of extra-systematic explanations concerning the nature of the entities on the other.

taining the framework of natural numbers we may introduce first the (positive and negative) integers as relations among natural numbers and then the rational numbers as relations among integers. This involves introducing new types of variables, expressions substitutable for them, and the general terms "integer" and "rational number".

The system of real numbers. On the basis of the rational numbers, the real numbers may be introduced as classes of a special kind (segments) of rational numbers (according to the method developed by Dedekind and Frege). Here again a new type of variables is introduced, expressions substitutable for them (e.g., "$\sqrt{2}$"), and the general term "real number".

The spatio-temporal coordinate system for physics. The new entities are the space-time points. Each is an ordered quadruple of four real numbers, called its coordinates, consisting of three spatial and one temporal coordinate. The physical state of a spatio-temporal point or region is described either with the help of qualitative predicates (e.g, "hot") or by ascribing numbers as values of a physical magnitude (e.g., mass, temperature, and the like). The step from the system of things (which does not contain space-time points but only extended objects with spatial and temporal relations between them) to the physical coordinate system is again a matter of decision. Our choice of certain features, although itself not theoretical, is suggested by theoretical knowledge, either logical or factual. For example, the choice of real numbers rather than rational numbers or integers as coordinates is not much influenced by the facts of experience but mainly due to considerations of mathematical simplicity. The restriction to rational coordinates would not be in conflict with any experimental knowledge we have, because the result of any measurement is a rational number. However, it would prevent the use of ordinary geometry (which says, e.g., that the diagonal of a square with the side 1 has the irrational value $\sqrt{2}$) and thus lead to great complications. On the other hand, the decision to use three rather than two or four spatial coordinates is strongly suggested, but still not forced upon us, by the result of common observations. If certain events allegedly observed in spiritualistic séances, e.g., a ball moving out of a sealed box, were confirmed beyond any reasonable doubt, it might seem advisable to use four spatial coordinates. Internal questions are here, in general, empirical questions to be answered by empirical investigations. On the other hand, the external questions of the reality of physical space and physical time are pseudo-questions. A question like "Are there (really) space-time points?" is ambiguous. It may be meant as an internal question; then the affirmative answer is, of course, analytic and trivial. Or it may be meant in the external sense: "Shall we introduce such and such forms into our language?"; in this case it is not a theoretical but a practical question, a matter of decision rather than assertion, and hence the proposed formulation would be misleading. Or finally, it may be meant in the following sense: "Are our experiences such that the use of the linguistic forms in question will be expedient and fruitful?" This is a theoretical question of a factual, empirical nature. But it concerns a matter of degree; therefore a formulation in the form "real or not?" would be inadequate.

3. What does acceptance of a kind of entities mean?

Let us now summarize the essential characteristics of situations involving the introduction of a new kind of entities, characteristics which are common to the various examples outlined above.

The acceptance of a new kind of entities is represented in the language by the introduction of a framework of new forms of expressions to be used according to a new set of rules. There may be new names for particular entities of the kind in question; but some such names may already occur in the language before the introduction of the new framework. (Thus, for example, the thing language contains certainly words of the type of "blue" and "house" before the framework of properties is introduced; and it may contain words like "ten" in sentences of the form "I have ten fingers" before the framework of numbers is introduced.) The latter fact shows that the occurrence of constants of the type in question – regarded as names of entities of the new kind after the new framework is introduced – is not a sure sign of the acceptance of the new kind of entities. Therefore the introduction of such constants is not to be regarded as an essential step in the introduction of the framework. The two essential steps are rather the following. First, the introduction of a general term, a predicate of higher level, for the new kind of entities, permitting us to say of any particular entity that it belongs to this kind (e.g., "Red is a *property*", "Five is a *number*"). Second, the introduction of variables of the new type. The new entities are values of these variables; the constants (and the closed compound expressions, if any) are substitutable for the variables.[4] With the help of the variables, general sentences concerning the new entities can be formulated.

After the new forms are introduced into the language, it is possible to formulate with their help internal questions and possible answers to them. A question of this kind may be either empirical or logical; accordingly a true answer is either factually true or analytic.

---

[4]W. V. Quine was the first to recognize the importance of the introduction of variables as indicating the acceptance of entities. "The ontology to which one's use of language commits him comprises simply the objects that he treats as falling ... within the range of values of his variables" (1943: 118; compare also 1939 and 1947).

From the internal questions we must clearly distinguish external questions, i.e., philosophical questions concerning the existence or reality of the total system of the new entities. Many philosophers regard a question of this kind as an ontological question which must be raised and answered *before* the introduction of the new language forms. The latter introduction, they believe, is legitimate only if it can be justified by an ontological insight supplying an affirmative answer to the question of reality. In contrast to this view, we take the position that the introduction of the new ways of speaking does not need any theoretical justification because it does not imply any assertion of reality. We may still speak (and have done so) of "the acceptance of the new entities" since this form of speech is customary; but one must keep in mind that this phrase does not mean for us anything more than acceptance of the new framework, i.e., of the new linguistic forms. Above all, it must not be interpreted as referring to an assumption, belief, or assertion of "the reality of the entities". There is no such assertion. An alleged statement of the reality of the system of entities is a pseudo-statement without cognitive content. To be sure, we have to face at this point an important question; but it is a practical, not a theoretical question; it is the question of whether or not to accept the new linguistic forms. The acceptance cannot be judged as being either true or false because it is not an assertion. It can only be judged as being more or less expedient, fruitful, conducive to the aim for which the language is intended. Judgments of this kind supply the motivation for the decision of accepting or rejecting the kind of entities.[5]

Thus it is clear that the acceptance of a linguistic framework must not be regarded as implying a metaphysical doctrine concerning the reality of the entities in question. It seems to me due to a neglect of this important distinction that some contemporary nominalists label the admission of variables of abstract types as "Platonism".[6] This is, to say the least, an

[5]For a closely related point of view on these questions see the detailed discussions in Feigl 1950: 35–62.

[6]Bernays 1935: 52–69 (reprinted in this volume). W. V. Quine, see previous footnote and a recent paper (1948). Quine does not acknowledge the distinction which I emphasize above, because according to his general conception there are no sharp boundary lines between logical and factual truth, between questions of meaning and questions of fact, between the acceptance of a language structure and the acceptance of an assertion formulated in the language. This conception, which seems to deviate considerably from customary ways of thinking, is explained in his article 1951c. When Quine classifies my logistic conception of mathematics (derived from Frege and Russell) and "platonic realism" (1948: 33), this is meant (according to a personal communication from him) not as ascribing to me agreement with Plato's metaphysical doctrine of universals, but merely as referring to the fact that I accept a language of mathematics containing variables of higher levels. With respect to the basic attitude to take in choosing a language form (an "ontology" in Quine's terminology, which seems to me misleading), there appears now to be agreement between us: "the obvious counsel is tolerance and an experimental spirit" (1948: 38).

extremely misleading terminology. It leads to the absurd consequence that the position of everybody who accepts the language of physics with its real number variables (as a language of communication, not merely as a calculus) would be called Platonistic, even if he is a strict empiricist who rejects Platonic metaphysics.

A brief historical remark may here be inserted. The non-cognitive character of the questions which we have called here external questions was recognized and emphasized already by the Vienna Circle under the leadership of Moritz Schlick, the group from which the movement of logical empiricism originated. Influenced by ideas of Ludwig Wittgenstein, the Circle rejected both the thesis of the reality of the external world and the thesis of its irreality as pseudo-statements;[7] the same was the case for both the thesis of the reality of universals (abstract entities, in our present terminology) and the nominalistic thesis that they are not real and that their alleged names are not names of anything but merely *flatus vocis*. (It is obvious that the apparent negation of a pseudo-statement must also be a pseudo-statement.) It is therefore not correct to classify the members of the Vienna Circle as nominalists, as is sometimes done. However, if we look at the basic anti-metaphysical and pro-scientific attitude of most nominalists (and the same holds for many materialists and realists in the modern sense), disregarding their occasional pseudo-theoretical formulations, then it is, of course, true to say that the Vienna Circle was much closer to those philosophers than to their opponents.

4. Abstract entities in semantics

The problem of the legitimacy and the status of abstract entities has recently again led to controversial discussions in connection with semantics. In a semantical meaning analysis certain expressions in a language are often said to designate (or name or denote or signify or refer to) certain extra-linguistic entities.[8] As long as physical things or events (e.g., Chicago or Caesar's death) are taken as designata (entities designated), no serious doubts arise. But strong objections have been raised, especially by some empiricists, against abstract entities as designata, e.g., against semantical statements of the following kind:

[7]See Carnap 1928b and Schlick 1932.
[8]See Carnap 1942, 1947. The distinction I have drawn in the latter book between the method of the name-relation and the method of intension and extension is not essential for our present discussion. The term "designation" is used in the present article in a neutral way; it may be understood as referring to the name-relation or to the intension-relation or to the extension-relation or to any similar relations used in other semantical methods.

(1) "The word 'red' designates a property of things";
(2) "The word 'color' designates a property of properties of things";
(3) "The word 'five' designates a number";
(4) "The word 'odd' designates a property of numbers";
(5) "The sentence 'Chicago is large' designates a proposition".

Those who criticize these statements do not, of course, reject the use of the expressions in question, like "red" or "five"; nor would they deny that these expressions are meaningful. But to be meaningful, they would say, is not the same as having a meaning in the sense of an entity designated. They reject the belief, which they regard as implicitly presupposed by those semantical statements, that to each expression of the types in question (adjectives like "red", numerals like "five", etc.) there is a particular real entity to which the expression stands in the relation of designation. This belief is rejected as incompatible with the basic principles of empiricism or of scientific thinking. Derogatory labels like "Platonic realism", "hypostatization", or "'Fido'-Fido principle" are attached to it. The latter is the name given by Gilbert Ryle [Meaning] to the criticized belief, which, in his view, arises by a naïve inference of analogy: just as there is an entity well known to me, viz. my dog Fido, which is designated by the name "Fido", thus there must be for every meaningful expression a particular entity to which it stands in the relation of designation or naming, i.e., the relation exemplified by "Fido"-Fido. The belief criticized is thus a case of hypostatization, i.e., of treating as names expressions which are not names. While "Fido" is a name, expressions like "red", "five", etc., are said not to be names, not to designate anything.

Our previous discussion concerning the acceptance of frameworks enables us now to clarify the situation with respect to abstract entities as designata. Let us take as an example the statement:

(a) "'Five' designates a number".

The formulation of this statement presupposes that our language L contains the forms of expressions which we have called the framework of numbers, in particular, numerical variables and the general term "number". If L contains these forms, the following is an analytic statement in L:

(b) "Five is a number".

Further, to make the statement (a) possible, L must contain an expression like "designates" or "is a name of" for the semantical relation of designation. If suitable rules for this term are laid down, the following is likewise analytic:

(*c*) " 'Five' designates five".

(Generally speaking, any expression of the form " '...' designates..."
is an analytic statement provided the term "..." is a constant in an accepted framework. If the latter condition is not fulfilled, the expression
is not a statement.) Since (*a*) follows from (*c*) and (*b*), (*a*) is likewise
analytic.

Thus it is clear that *if* someone accepts the framework of numbers,
then he must acknowledge (*c*) and (*b*) and hence (*a*) as true statements.
Generally speaking, if someone accepts a framework for a certain kind
of entities, then he is bound to admit the entities as possible designata.
Thus the question of the admissibility of entities of a certain type or of
abstract entities in general as designata is reduced to the question of the
acceptability of the linguistic framework for those entities. Both the
nominalistic critics, who refuse the status of designators or names to
expressions like "red", "five", etc., because they deny the existence of
abstract entities, and the skeptics, who express doubts concerning the
existence and demand evidence for it, treat the question of existence as a
theoretical question. They do, of course, not mean the internal question;
the affirmative answer to *this* question is analytic and trivial and too
obvious for doubt or denial, as we have seen. Their doubts refer rather to
the system of entities itself; hence they mean the external question. They
believe that only after making sure that there really is a system of entities
of the kind in question are we justified in accepting the framework by
incorporating the linguistic forms into our language. However, we have
seen that the external question is not a theoretical question but rather the
practical question whether or not to accept those linguistic forms. This
acceptance is not in need of a theoretical justification (except with
respect to expediency and fruitfulness), because it does not imply a belief
or assertion. Ryle says that the "Fido"-Fido principle is "a grotesque
theory". Grotesque or not, Ryle is wrong in calling it a theory. It is
rather the practical decision to accept certain frameworks. Maybe Ryle is
historically right with respect to those whom he mentions as previous
representatives of the principle, viz. John Stuart Mill, Frege, and
Russell. If these philosophers regarded the acceptance of a system of
entities as a theory, an assertion, they were victims of the same old, metaphysical confusion. But it is certainly wrong to regard *my* semantical
method as involving a belief in the reality of abstract entities, since I
reject a thesis of this kind as a metaphysical pseudo-statement.

The critics of the use of abstract entities in semantics overlook the
fundamental difference between the acceptance of a system of entities
and an internal assertion, e.g., an assertion that there are elephants or

253

electrons or prime numbers greater than a million. Whoever makes an internal assertion is certainly obliged to justify it by providing evidence, empirical evidence in the case of electrons, logical proof in the case of the prime numbers. The demand for a theoretical justification, correct in the case of internal assertions, is sometimes wrongly applied to the acceptance of a system of entities. Thus, for example, Ernest Nagel (1948) asks for "evidence relevant for affirming with warrant that there are such entities as infinitesimals or propositions". He characterizes the evidence required in these cases – in distinction to the empirical evidence in the case of electrons – as "in the broad sense logical and dialectical". Beyond this no hint is given as to what might be regarded as relevant evidence. Some nominalists regard the acceptance of abstract entities as a kind of superstition or myth, populating the world with fictitious or at least dubious entities, analogous to the belief in centaurs or demons. This shows again the confusion mentioned, because a superstition or myth is a false (or dubious) internal statement.

Let us take as example the natural numbers as cardinal numbers, i.e., in contexts like "Here are three books". The linguistic forms of the framework of numbers, including variables and the general term "number", are generally used in our common language of communication; and it is easy to formulate explicit rules for their use. Thus the logical characteristics of this framework are sufficiently clear (while many internal questions, i.e., arithmetical questions, are, of course, still open). In spite of this, the controversy concerning the external question of the ontological reality of the system of numbers continues. Suppose that one philosopher says: "I believe that there are numbers as real entities. This gives me the right to use the linguistic forms of the numerical framework and to make semantical statements about numbers as designata of numerals". His nominalistic opponent replies: "You are wrong; there are no numbers. The numerals may still be used as meaningful expressions. But they are not names, there are no entities designated by them. Therefore the word "number" and numerical variables must not be used (unless a way were found to introduce them as merely abbreviating devices, a way of translating them into the nominalistic thing language)." I cannot think of any possible evidence that would be regarded as relevant by both philosophers, and therefore, if actually found, would decide the controversy or at least make one of the opposite theses more probable than the other. (To construe the numbers as classes or properties of the second level, according to the Frege-Russell method, does, of course, not solve the controversy, because the first philosopher would affirm and the second deny the existence of the system of classes or properties of the second level.) Therefore I feel compelled to regard the external question as a pseudo-question, until both parties to the controversy offer a common interpretation of the question as a cognitive question; this would involve an indication of possible evidence regarded as relevant by both sides.

There is a particular kind of misinterpretation of the acceptance of abstract entities in various fields of science and in semantics that needs to be cleared up. Certain early British empiricists (e.g., Berkeley and Hume) denied the existence of abstract entities on the ground that immediate experience presents us only with particulars, not with universals, e.g., with this red patch, but not with Redness or Color-in-General; with this scalene triangle, but not with Scalene Triangularity or Triangularity-in-General. Only entities belonging to a type of which examples were to be found within immediate experience could be accepted as ultimate constituents of reality. Thus, according to this way of thinking, the existence of abstract entities could be asserted only if one could show either that some abstract entities fall within the given, or that abstract entities can be defined in terms of the types of entity which are given. Since these empiricists found no abstract entities within the realm of sense-data, they either denied their existence, or else made a futile attempt to define universals in terms of particulars. Some contemporary philosophers, especially English philosophers following Bertrand Russell, think in basically similar terms. They emphasize a distinction between the data (that which is immediately given in consciousness, e.g., sense-data, immediately past experiences, etc.) and the constructs based on the data. Existence or reality is ascribed only to the data; the constructs are not real entities; the corresponding linguistic expressions are merely ways of speech not actually designating anything (reminiscent of the nominalists' *flatus vocis*). We shall not criticize here this general conception. (As far as it is a principle of accepting certain entities and not accepting others, leaving aside any ontological, phenomenalistic and nominalistic pseudo-statements, there cannot be any theoretical objection to it.) But if this conception leads to the view that other philosophers or scientists who accept abstract entities thereby assert or imply their occurrence as immediate data, then such a view must be rejected as a misinterpretation. References to space-time points, the electromagnetic field, or electrons in physics, to real or complex numbers and their functions in mathematics, to the excitatory potential or unconscious complexes in psychology, to an inflationary trend in economics, and the like, do not imply the assertion that entities of these kinds occur as immediate data. And the same holds for references to abstract entities as designata in semantics. Some of the criticisms by English philosophers against such references give the impression that, probably due to the misinterpretation just indicated, they accuse the

255

semanticist not so much of bad metaphysics (as some nominalists would do) but of bad psychology. The fact that they regard a semantical method involving abstract entities not merely as doubtful and perhaps wrong, but as manifestly absurd, preposterous and grotesque, and that they show a deep horror and indignation against this method, is perhaps to be explained by a misinterpretation of the kind described. In fact, of course, the semanticist does not in the least assert or imply that the abstract entities to which he refers can be experienced as immediately given either by sensation or by a kind of rational intuition. An assertion of this kind would indeed be very dubious psychology. The psychological question as to which kinds of entities do and which do not occur as immediate data is entirely irrelevant for semantics, just as it is for physics, mathematics, economics, etc., with respect to the examples mentioned above.[9]

5. Conclusion

For those who want to develop or use semantical methods, the decisive question is not the alleged ontological question of the existence of abstract entities but rather the question whether the use of abstract linguistic forms or, in technical terms, the use of variables beyond those for things (or phenomenal data) is expedient and fruitful for the purposes for which semantical analyses are made, viz. the analysis, interpretation, clarification, or construction of languages of communication, especially languages of science. This question is here neither decided nor even discussed. It is not a question simply of yes or no, but a matter of degree. Among those philosophers who have carried out semantical analyses and thought about suitable tools for this work, beginning with Plato and Aristotle and, in a more technical way on the basis of modern logic, with C. S. Peirce and Frege, a great majority accepted abstract entities. This does, of course, not prove the case. After all, semantics in the technical sense is still in the initial phases of its development, and we must be prepared for possible fundamental changes in methods. Let us therefore admit that the nominalistic critics may possibly be right. But if so, they will have to offer better arguments than they have so far. Appeal to ontological insight will not carry much weight. The critics will have to show that it is possible to construct a semantical method which avoids all references to abstract entities and achieves by simpler means essentially the same results as the other methods.

The acceptance or rejection of abstract linguistic forms, just as the

[9]Sellars (1949: 496–504; see pp. 502f.) analyzes clearly the roots of the mistake "of taking the designation relation of semantic theory to be a reconstruction of *being present to an experience*".

acceptance or rejection of any other linguistic forms in any branch of science, will finally be decided by their efficiency as instruments, the ratio of the results achieved to the amount and complexity of the efforts required. To decree dogmatic prohibitions of certain linguistic forms instead of testing them by their success or failure in practical use is worse than futile; it is positively harmful because it may obstruct scientific progress. The history of science shows examples of such prohibitions based on prejudices deriving from religious, mythological, metaphysical, or other irrational sources, which slowed up the developments for shorter or longer periods of time. Let us learn from the lessons of history. Let us grant to those who work in any special field of investigation the freedom to use any form of expression which seems useful to them; the work in the field will sooner or later lead to the elimination of those forms which have no useful function. *Let us be cautious in making assertions and critical in examining them, but tolerant in permitting linguistic forms.*

On platonism in mathematics

PAUL BERNAYS

With your permission, I shall now address you on the subject of the present situation in research in the foundations of mathematics.

Since there remain open questions in this field, I am not in a position to paint a definitive picture of it for you. But it must be pointed out that the situation is not so critical as one could think from listening to those who speak of a foundational crisis. From certain points of view, this expression can be justified; but it could give rise to the opinion that mathematical science is shaken at its roots.

The truth is that the mathematical sciences are growing in complete security and harmony. The ideas of Dedekind, Poincaré, and Hilbert have been systematically developed with great success, without any conflict in the results.

It is only from the philosophical point of view that objections have been raised. They bear on certain ways of reasoning peculiar to analysis and set theory. These modes of reasoning were first systematically applied in giving a rigorous form to the methods of the calculus. [According to them,] the objects of a theory are viewed as elements of a totality such that one can reason as follows: For each property expressible using the notions of the theory, it is [an] objectively determinate [fact] whether there is or there is not an element of the totality which possesses this property. Similarly, it follows from this point of view that either all the elements of a set possess a given property, or there is at least one element which does not possess it.

An example of this way of setting up a theory can be found in Hilbert's axiomatization of geometry. If we compare Hilbert's axiom system to Euclid's, ignoring the fact that the Greek geometer fails to include certain [necessary] postulates, we notice that Euclid speaks of figures to be

Lecture delivered June 18, 1934, in the cycle of *Conférences internationales des Sciences mathématiques* organized by the University of Geneva, in the series on Mathematical Logic.

Translated from the French by C. D. Parsons from *L'enseignement mathématique*, 1st ser. vol. 34 (1935), pp. 52–69. Permission for the translation and inclusion of this paper in this book was kindly granted by the author and the editor of *L'enseignement mathématique*.

constructed,[1] whereas, for Hilbert, system of points, straight lines, and planes exist from the outset. Euclid postulates: One can join two points by a straight line; Hilbert states the axiom: Given any two points, there exists a straight line on which both are situated. "Exists" refers here to existence in the system of straight lines.

This example shows already that the tendency of which we are speaking consists in viewing the objects as cut off from all links with the reflecting subject.

Since this tendency asserted itself especially in the philosophy of Plato, allow me to call it "platonism."

The value of platonistically inspired mathematical conceptions is that they furnish models of abstract imagination. These stand out by their simplicity and logical strength. They form representations which extrapolate from certain regions of experience and intuition.

Nonetheless, we know that we can arithmetize the theoretical systems of geometry and physics. For this reason, we shall direct our attention to platonism in arithmetic. But I am referring to arithmetic in a very broad sense, which includes analysis and set theory.

The weakest of the "platonistic" assumptions introduced by arithmetic is that of the totality of integers. The *tertium non datur* for integers follows from it; viz.: if P is a predicate of integers, either P is true of each number, or there is at least one exception.

By the assumption mentioned, this disjunction is an immediate consequence of the logical principle of the excluded middle; in analysis it is almost continually applied.

For example, it is by means of it that one concludes that for two real numbers a and b, given by convergent series, either $a=b$ or $a<b$ or $b<a$; and likewise: a sequence of positive rational numbers either comes as close as you please to zero or there is a positive rational number less than all the members of the sequence.

At first sight, such disjunctions seem trivial, and we must be attentive in order to notice that an assumption slips in.

But analysis is not content with this modest variety of platonism; it reflects it to a stronger degree with respect to the following notions: set of numbers, sequence of numbers, and function. It abstracts from the possibility of giving definitions of sets, sequences, and functions. These notions are used in a "quasi-combinatorial" sense, by which I mean: in the sense of an analogy of the infinite to the finite.

Consider, for example, the different functions which assign to each

[1][Translator's italics.]

member of the finite series $1, 2, \ldots, n$ a number of the same series. There are n^n functions of this sort, and each of them is obtained by n independent determinations. Passing to the infinite case, we imagine functions engendered by an infinity of independent determinations which assign to each integer an integer, and we reason about the totality of these functions.

In the same way, one views a set of integers as the result of infinitely many independent acts deciding for each number whether it should be included or excluded. We add to this the idea of the totality of these sets. Sequences of real numbers and sets of real numbers are envisaged in an analogous manner. From this point of view, constructive definitions of specific functions, sequences, and sets are only ways to pick out an object which exists independently of, and prior to, the construction.

The axiom of choice is an immediate application of the quasi-combinatorial concepts in question. It is generally employed in the theory of real numbers in the following special form. Let

$$M_1, M_2, \ldots$$

be a sequence of non-empty sets of real numbers, then there is a sequence

$$a_1, a_2, \ldots$$

such that for every index n, a_n is an element of M_n.

The principle becomes subject to objections if the effective construction of the sequence of numbers is demanded.

A similar case is that of Poincaré's impredicative definitions. An impredicative definition of a real number appeals to the hypothesis that all real numbers have a certain property P, or the hypothesis that there exists a real number with the property T.

This kind of definition depends on the assumption of [the existence of] the totality of sequences of integers, because a real number is represented by a decimal fraction, that is to say, by a special kind of sequence of integers.

It is used in particular to prove the fundamental theorem that a bounded set of real numbers always has a least upper bound.

In Cantor's theories, platonistic conceptions extend far beyond those of the theory of real numbers. This is done by iterating the use of the quasi-combinatorial concept of a function and adding methods of collection. This is the well-known method of set theory.

The platonistic conceptions of analysis and set theory have also been applied in modern theories of algebra and topology, where they have proved very fertile.

This brief summary will suffice to characterize platonism and its appli-

260

cation to mathematics. This application is so widespread that it is not an exaggeration to say that platonism reigns today in mathematics.

But on the other hand, we see that this tendency has been criticized in principle since its first appearance and has given rise to many discussions. This criticism was reinforced by the paradoxes discovered in set theory, even though these antinomies refute only extreme platonism.

We have set forth only a restricted platonism which does not claim to be more than, so to speak, an ideal projection of a domain of thought. But the matter has not rested there. Several mathematicians and philosophers interpret the methods of platonism in the sense of conceptual realism, postulating the existence of a world of ideal objects containing all the objects and relations of mathematics. It is this absolute platonism which has been shown untenable by the antinomies, particularly by those surrounding the Russell–Zermelo paradox.

If one hears them for the first time, these paradoxes in their purely logical form can seem to be plays on words without serious significance. Nonetheless one must consider that these abbreviated forms of the paradoxes are obtained by following out the consequences of the various requirements of absolute platonism.

The essential importance of these antinomies is to bring out the impossibility of combining the following two things: the idea of the totality of all mathematical objects and the general concepts of set and function; for the totality itself would form a domain of elements for sets, and arguments and values for functions.

We must therefore give up absolute platonism. But it must be observed that this is almost the only injunction which follows from the paradoxes. Some will think that this is regrettable, since the paradoxes are appealed to on every side. But avoiding the paradoxes does not constitute a univocal program. In particular, restricted platonism is not touched at all by the antinomies.

Still, the critique of the foundations of analysis receives new impetus from this source, and among the different possible ways of escaping from the paradoxes, eliminating platonism offered itself as the most radical.

Let us look and see how this elimination can be brought about. It is done in two steps, corresponding to the two essential assumptions introduced by platonism. The first step is to replace by constructive concepts the concepts of a set, a sequence, or a function, which I have called quasi-combinatorial. The idea of an infinity of independent determinations is rejected. One emphasizes that an infinite sequence or a decimal fraction can be given only by an arithmetical law, and one regards the continuum as a set of elements defined by such laws.

This procedure is adapted to the tendency toward a complete arithmetization of analysis. Indeed, it must be conceded that the arithmetization of analysis is not carried through to the end by the usual method. The conceptions which are applied there are not completely reducible, as we have seen, to the notion of integer and logical concepts.

Nonetheless, if we pursue the thought that each real number is defined by an arithmetical law, the idea of the totality of real numbers is no longer indispensable, and the axiom of choice is not at all evident. Also, unless we introduce auxiliary assumptions – as Russell and Whitehead do – we must do without various usual conclusions. Weyl has made these consequences very clear in his book *Das Kontinuum*.

Let us proceed to the second step of the elimination. It consists in renouncing the idea of the totality of integers. This point of view was first defended by Kronecker and then developed systematically by Brouwer.

Although several of you heard in March [1934] an authentic exposition of this method by Professor Brouwer himself, I shall allow myself a few words of explanation.

A misunderstanding about Kronecker must first be dissipated, which could arise from his often-cited aphorism that the integers were created by God, whereas everything else in mathematics is the work of man. If that were really Kronecker's opinion, he ought to admit the concept of the totality of integers.

In fact, Kronecker's method, as well as that of Brouwer, is characterized by the fact that it avoids the supposition that there exists a series of natural numbers forming a determinate ideal object.

According to Kronecker and Brouwer, one can speak of the series of numbers only in the sense of a process that is never finished, surpassing each limit which it reaches.

This point of departure carries with it the other divergences, in particular those concerning the application and interpretation of logical forms: Neither a general judgment about integers nor a judgment of existence can be interpreted as expressing a property of the series of numbers. A general theorem about numbers is to be regarded as a sort of prediction that a property will present itself for each construction of a number; and the affirmation of the existence of a number with a certain property is interpreted as an incomplete communication of a more precise proposition indicating a [particular] number having the property in question or a method for obtaining such a number; Hilbert calls it a "partial judgment."

For the same reasons the negation of a general or existential proposition about integers does not have precise sense. One must strengthen the negation to arrive at a mathematical proposition. For example, it is to give a strengthened negation of a proposition affirming the existence of a number with a property P to say that a number with the property P cannot be given, or further, that the assumption of a number with this property leads to a contradiction. But for such strengthened negations the law of the excluded middle is no longer applicable.

The characteristic complications to be met with in Brouwer's "intuitionistic" method come from this.

For example, one may not generally make use of disjunctions like these: a series of positive terms is either convergent or divergent; two convergent sums represent either the same real number or different ones.

In the theory of integers and of algebraic numbers, we can avoid these difficulties and manage to preserve all the essential theorems and arguments.

In fact, Kronecker has already shown that the core of the theory of algebraic fields can be developed from his methodological point of view without appeal to the totality of integers.[2]

As for analysis, you know that Brouwer has developed it in accord with the requirements of intuitionism. But here one must abandon a number of the usual theorems, for example, the fundamental theorem that every continuous function has a maximum in a closed interval. Very few things in set theory remain valid in intuitionist mathematics.

We would say, roughly, that intuitionism is adapted to the theory of numbers; the semiplatonistic method, which makes use of the idea of the totality of integers but avoids quasi-combinatorial concepts, is adapted to the arithmetic theory of functions, and the usual platonism is adequate for the geometric theory of the continuum.

There is nothing astonishing about this situation, for it is a familiar procedure of the contemporary mathematician to restrict his assumptions in each domain of the science to those which are essential. By this restriction, a theory gains methodological clarity, and it is in this direction that intuitionism proves fruitful.

But as you know, intuitionism is not at all content with such a role; it opposes the usual mathematics and claims to represent the only true mathematics.

---

[2]To this end, Kronecker set forth in his lectures a manner of introducing the notion of algebraic number which has been almost totally forgotten, although it is the most elementary way of defining this notion. This method consists in representing algebraic numbers by the changes of sign of irreducible polynomials in one variable with rational integers as coefficients; starting from that definition, one introduces the elementary operations and relations of magnitude for algebraic numbers and proves that the ordinary laws of calculation hold; finally one shows that a polynomial with algebraic coefficients having values with different signs for two algebraic arguments a and b has a zero between a and b.

On the other hand, mathematicians generally are not at all ready to exchange the well-tested and elegant methods of analysis for more complicated methods unless there is an overriding necessity for it.

We must discuss the question more deeply. Let us try to portray more distinctly the assumptions and philosophic character of the intuitionistic method.

What Brouwer appeals to is evidence. He claims that the basic ideas of intuitionism are given to us in an evident manner by pure intuition. In relying on this, he reveals his partial agreement with Kant. But whereas for Kant there exists a pure intuition with respect to space and time, Brouwer acknowledges only the intuition of time, from which, like Kant, he derives the intuition of number.

As for this philosophic position, it seems to me that one must concede to Brouwer two essential points: first, that the concept of integer is of intuitive origin. In this respect nothing is changed by the investigations of the logicists,[3] to which I shall return later. Second, one ought not to make arithmetic and geometry correspond in the manner in which Kant did. The concept of number is more elementary than the concepts of geometry.

Still it seems a bit hasty to deny completely the existence of a geometrical intuition. But let us leave that question aside here; there are other, more urgent ones. Is it really certain that the evidence given by arithmetical intuition extends exactly as far as the boundaries of intuitionist arithmetic would require? And finally: Is it possible to draw an exact boundary between what is evident and what is only plausible?

I believe that one must answer these two questions negatively. To begin with, you know that men and even scholars do not agree about evidence in general. Also, the same man sometimes rejects suppositions which he previously regarded as evident.

An example of a much-discussed question of evidence, about which there has been controversy up to the present, is that of the axiom of parallels. I think that the criticism which has been directed against that axiom is partly explained by the special place which it has in Euclid's system. Various other axioms had been omitted, so that the parallels axiom stood out from the others by its complexity.

In this matter I shall be content to point out the following: One can have doubts concerning the evidence of geometry, holding that it extends only to topological facts or to the facts expressed by the projective axioms. One can, on the other hand, claim that geometric intuition is not exact. These opinions are self-consistent, and all have arguments in their

---

[3][I have rendered '*logiciens*' throughout as 'logicists'. – Trans.]

favor. But to claim that metric geometry has an evidence restricted to the laws common to Euclidean and Bolyai-Lobachevskian geometry, an exact metrical evidence which yet would not guarantee the existence of a perfect square, seems to me rather artificial. And yet it was the point of view of a number of mathematicians.

Our concern here has been to underline the difficulties to be encountered in trying to describe the limits of evidence.

Nevertheless, these difficulties do not make it impossible that there should be anything evident beyond question, and certainly intuitionism offers some such. But does it confine itself completely within the region of this elementary evidence? This is not completely indubitable, for the following reason: Intuitionism makes no allowance for the possibility that, for very large numbers, the operations required by the recursive method of constructing numbers can cease to have a concrete meaning. From two integers k, l one passes immediately to k^l; this process leads in a few steps to numbers which are far larger than any occurring in experience, e.g., $67^{(257^{729})}$.

Intuitionism, like ordinary mathematics, claims that this number can be represented by an Arabic numeral. Could not one press further the criticism which intuitionism makes of existential assertions and raise the question: What does it mean to claim the existence of an Arabic numeral for the foregoing number, since in practice we are not in a position to obtain it?

Brouwer appeals to intuition, but one can doubt that the evidence for it really is intuitive. Isn't this rather an application of the general method of analogy, consisting in extending to inaccessible numbers the relations which we can concretely verify for accessible numbers? As a matter of fact, the reason for applying this analogy is strengthened by the fact that there is no precise boundary between the numbers which are accessible and those which are not. One could introduce the notion of a "practicable" procedure, and implicitly restrict the import of recursive definitions to practicable operations. To avoid contradictions, it would suffice to abstain from applying the principle of the excluded middle to the notion of practicability. But such abstention goes without saying for intuitionism.

I hope I shall not be misunderstood: I am far from recommending that arithmetic be done with this restriction. I am concerned only to show that intuitionism takes as its basis propositions which one can doubt and in principle do without, although the resulting theory would be rather meager.

It is therefore not absolutely indubitable that the domain of complete evidence extends to all of intuitionism. On the other hand, several mathematicians recognize the complete evidence of intuitionistic arithmetic and moreover maintain that the concept of the series of numbers is evident in the following sense: The affirmation of the existence of a number does not require that one must, directly or recursively, give a bound for this number. Besides, we have just seen how far beyond a really concrete presentation such a limitation would be.

In short, the point of view of intuitive evidence does not decide uniquely in favor of intuitionism.

In addition, one must observe that the evidence which intuitionism uses in its arguments is not always of an immediate character. Abstract reflections are also included. In fact, intuitionists often use statements, containing a general hypothesis, of the form 'If every number n has the property A(n), then B holds.'

Such a statement is interpreted intuitionistically in the following manner: 'If it is proved that every number n possesses the property A(n), then B.' Here we have a hypothesis of an abstract kind, because since the methods of demonstration are not fixed in intuitionism, the condition that something is proved is not intuitively determined.

It is true that one can also interpret the given statement by viewing it as a partial judgment, i.e., as the claim that there exists a proof of B from the given hypothesis, a proof which would be effectively given.[4] (This is approximately the sense of Kolmogorov's interpretation of intuitionism.) In any case, the argument must start from the general hypothesis, which cannot be intuitively fixed. It is therefore an abstract reflection.

In the example just considered, the abstract part is rather limited. The abstract character becomes more pronounced if one superposes hypotheses; i.e., when one formulates propositions like the following: 'If from the hypothesis that A(n) is valid for every n, one can infer B, then C holds,' or 'If from the hypothesis that A leads to a contradiction, a contradiction follows, then B,' or briefly 'If the absurdity of A is absurd, then B.' This abstractness of statements can be still further increased.

It is by the systematic application of these forms of abstract reasoning that Brouwer has gone beyond Kronecker's methods and succeeded in establishing a general intuitionistic logic, which has been systematized by Heyting.

If we consider this intuitionistic logic, in which the notions of consequence are applied without reservation, and we compare the method used here with the usual one, we notice that the characteristic general

---

[4]["... c'est à dire comme une indication d'un raisonnement conduisant de la dite hypothèse à la conclusion B, raisonnement qu'on présente effectivement."]

feature of intuitionism is not that of being founded on pure intuition, but rather [that of being founded] on the relation of the reflecting and acting subject to the whole development of science.

This is an extreme methodological position. It is contrary to the customary manner of doing mathematics, which consists in establishing theories detached as much as possible from the thinking subject.

This realization leads us to doubt that intuitionism is the sole legitimate method of mathematical reasoning. For even if we admit that the tendency away from the [thinking] subject has been pressed too far under the reign of platonism, this does not lead us to believe that the truth lies in the opposite extreme. Keeping both possibilities in mind, we shall rather aim to bring about in each branch of science an adaptation of method to the character of the object investigated.

For example, for number theory the use of the intuitive concept of a number is the most natural. In fact, one can thus establish the theory of numbers without introducing an axiom, such as that of complete induction, or axioms of infinity like those of Dedekind and Russell.

Moreover, in order to avoid the intuitive concept of number, one is led to introduce a more general concept, like that of a proposition, a function, or an arbitrary correspondence, concepts which are in general not objectively defined. It is true that such a concept can be made more definite by the axiomatic method, as in axiomatic set theory, but then the system of axioms is quite complicated.

You know that Frege tried to deduce arithmetic from pure logic by viewing the latter as the general theory of the universe of mathematical objects. Although the foundation of this absolutely platonistic enterprise was undermined by the Russell–Zermelo paradox, the school of logicists has not given up the idea of incorporating arithmetic in a system of logic. In place of absolute platonism, they have introduced some initial assumptions. But because of these, the system loses the character of pure logic.

In the system of *Principia Mathematica,* it is not only the axioms of infinity and reducibility which go beyond pure logic, but also the initial conception of a universal domain of individuals and of a domain of predicates. It is really an *ad hoc* assumption to suppose that we have before us the universe of things divided into subjects and predicates, ready-made for theoretical treatment.

But even with such auxiliary assumptions, one cannot successfully incorporate the whole of arithmetic into the system of logic. For, since this system is developed according to fixed rules, one would have to be able to obtain by means of a fixed series of rules all the theorems of arithmetic. But this is not the case; as Gödel has shown, arithmetic goes beyond each given formalism. (In fact, the same is true of axiomatic set theory.)

Besides, the desire to deduce arithmetic from logic derives from the traditional opinion that the relation of logic to arithmetic is that of general to particular. The truth, it seems to me, is that mathematical abstraction does not have a lesser degree than logical abstraction, but rather another direction.

These considerations do not detract at all from the intrinsic value of that research of logicists which aims at developing logic systematically and formalizing mathematical proofs. We were concerned here only with defending the thesis that for the theory of numbers, the intuitive method is more suitable.

On the other hand, for the theory of the continuum, given by analysis, the intuitionist method seems rather artificial. The idea of the continuum is a geometrical idea which analysis expresses in terms of arithmetic.

Is the intuitionist method of representing the continuum better adapted to the idea of the continuum than the usual one?

Weyl would have us believe this. He reproaches ordinary analysis for decomposing the continuum into single points. But isn't this reproach better addressed to semiplatonism, which views the continuum as a set of arithmetical laws? The fact is that for the usual method there is a completely satisfying analogy between the manner in which a particular point stands out from the continuum and the manner in which a real number defined by an arithmetical law stands out from the set of all real numbers, whose elements are in general only implicitly involved, by virtue of the quasi-combinatorial concept of a sequence.

This analogy seems to me to agree better with the nature of the continuum than that which intuitionism establishes between the fuzzy character of the continuum and the uncertainties arising from unsolved arithmetical problems.

It is true that in the usual analysis the notion of a continuous function, and also that of a differentiable function, have a generality going far beyond our intuitive representation of a curve. Nevertheless, in this analysis we can establish the theorem of the maximum of a continuous function and Rolle's theorem, thus rejoining the intuitive conception.

Intuitionist analysis, even though it begins with a much more restricted notion of a function, does not arrive at such simple theorems; they must instead be replaced by more complex ones. This stems from the fact that on the intuitionistic conception, the continuum does not have the character of a totality, which undeniably belongs to the geometric idea of the continuum. And it is this characteristic of the continuum which would resist perfect arithmetization.

These considerations lead us to notice that the duality of arithmetic and geometry is not unrelated to the opposition between intuitionism and platonism. The concept of number appears in arithmetic. It is of intuitive origin, but then the idea of the totality of numbers is superimposed. On the other hand, in geometry the platonistic idea of space is primordial, and it is against this background that the intuitionist procedures of constructing figures take place.

This suffices to show that the two tendencies, intuitionist and platonist, are both necessary; they complement each other, and it would be doing oneself violence to renounce one or the other.

But the duality of these two tendencies, like that of arithmetic and geometry, is not a perfect symmetry. As we have noted, it is not proper to make arithmetic and geometry correspond completely: the idea of number is more immediate to the mind than the idea of space. Likewise, we must recognize that the assumptions of platonism have a transcendent character which is not found in intuitionism.

It is also this transcendent character which requires us to take certain precautions in regard to each platonistic assumption. For even when such a supposition is not at all arbitrary and presents itself naturally to the mind, it can still be that the principle from which it proceeds permits only a restricted application, outside of which one would fall into contradiction.

We must be all the more careful in the face of this possibility, since the drive for simplicity leads us to make our principles as broad as possible. And the need for a restriction is often not noticed.

This was the case, as we have seen, for the principle of totality, which was pressed too far by absolute platonism. Here it was only the discovery of the Russell–Zermelo paradox which showed that a restriction was necessary.

Thus it is desirable to find a method to make sure that the platonistic assumptions on which mathematics is based do not go beyond permissible limits. The assumptions in question reduce to various forms of the principle of totality and of the principle of analogy or of the permanence of laws. And the condition restricting the application of these principles is none other than that of the consistency of the consequences which are deduced from the fundamental assumptions.

As you know, Hilbert is trying to find ways of giving us such assurances of consistency, and his proof theory has this as its goal.

This theory relies in part on the results of the logicists. They have shown that the arguments applied in arithmetic, analysis, and set theory can be formalized. That is, they can be expressed in symbols and as symbolic processes which unfold according to fixed rules. To primitive propositions correspond initial formulae, and to each logical deduction corresponds a sequence of formulae derivable from one another according to given rules. In this formalism, a platonistic assumption is represented by an initial formula or by a rule establishing a way of passing from formulae already obtained to others. In this way, the investigation of the possibilities of proof reduces to problems like those which are found in elementary number theory. In particular, the consistency of the theory will be proved if one succeeds in proving that it is impossible to deduce two mutually contradictory formulae A and $\bar{A}$ (with the bar representing negation). This statement which is to be proved is of the same structure as that, for example, asserting the impossibility of satisfying the equation $a^2 = 2b^2$ by two integers a and b.

Thus by symbolic reduction, the question of the consistency of a theory reduces to a problem of an elementary arithmetical character.

Starting from this fundamental idea, Hilbert has sketched a detailed program of a theory of proof, indicating the leading ideas of the arguments (for the main consistency proofs). His intention was to confine himself to intuitive and combinatorial considerations; his "finitary point of view" was restricted to these methods.

In this framework, the theory was developed up to a certain point. Several mathematicians have contributed to it: Ackermann, von Neumann, Skolem, Herbrand, Gödel, Gentzen.

Nonetheless, these investigations have remained within a relatively restricted domain. In fact, they did not even reach a proof of the consistency of the axiomatic theory of integers. It is known that the symbolic representation of this theory is obtained by adding to the ordinary logical calculus formalizations of Peano's axioms and the recursive definitions of sum $(a+b)$ and product $(a \cdot b)$.

Light was shed on this situation by a general theorem of Gödel, according to which a proof of the consistency of a formalized theory cannot be represented by means of the formalism considered. From this theorem, the following more special proposition follows: It is impossible to prove by elementary combinatorial methods the consistency of a formalized theory which can express every elementary combinatorial proof of an arithmetical proposition.

Now it seems that this proposition applies to the formalism of the axiomatic theory of numbers. At least, no attempt made up to now has given us any example of an elementary combinatorial proof which cannot be expressed in this formalism, and the methods by which one can, in the cases considered, translate a proof into the aforementioned formalism seem to suffice in general.

Assuming that this is so,[5] we arrive at the conclusion that means more powerful than elementary combinatorial methods are necessary to prove the consistency of the axiomatic theory of numbers. A new discovery of Gödel and Gentzen leads us to such a more powerful method. They have shown (independently of one another) that the consistency of intuitionist arithmetic implies the consistency of the axiomatic theory of numbers. This result was obtained by using Heyting's formalization of intuitionist arithmetic and logic. The argument is conducted by elementary methods, in a rather simple manner. In order to conclude from this result that the axiomatic theory of numbers is consistent, it suffices to assume the consistency of intuitionist arithmetic.

This proof of the consistency of axiomatic number theory shows us, among other things, that intuitionism, by its abstract arguments, goes essentially beyond elementary combinatorial methods.

The question which now arises is whether the strengthening of the method of proof theory obtained by admitting the abstract arguments of intuitionism would put us into a position to prove the consistency of analysis. The answer would be very important and even decisive for proof theory, and even, it seems to me, for the role which is to be attributed to intuitionistic methods.

Research in the foundations of mathematics is still developing. Several basic questions are open, and we do not know what we shall discover in this domain. but these investigations excite our curiosity by their changing perspectives, and that is a sentiment which is not aroused to the same degree by the more classical parts of science, which have attained greater perfection.

I wish to thank Professor Wavre, who was kind enough to help me improve the text of this lecture for publication. I also thank M. Rueff, who was good enough to look over the first draft to improve the French.

[5]In trying to demonstrate the possibility of translating each elementary combinatorial proof of an arithmetical proposition into the formalism of the axiomatic theory of numbers, we are confronted with the difficulty of delimiting precisely the domain of elementary combinatorial methods.

What numbers could not be

PAUL BENACERRAF

The attention of the mathematician focuses primarily upon mathematical structure, and his intellectual delight arises (in part) from seeing that a given theory exhibits such and such a structure, from seeing how one structure is "modelled" in another, or in exhibiting some new structure and showing how it relates to previously studied ones.... But...the mathematician is satisfied so long as he has some "entities" or "objects" (or "sets" or "numbers" or "functions" or "spaces" or "points") to work with, and he does not inquire into their inner character or ontological status.

The philosophical logician, on the other hand, is more sensitive to matters of ontology and will be especially interested in the kind or kinds of entities there are actually.... He will not be satisfied with being told merely that such and such entities exhibit such and such a mathematical structure. He will wish to inquire more deeply into what these entities are, how they relate to other entities.... Also he will wish to ask whether the entity dealt with is *sui generis* or whether it is in some sense *reducible* to (or *constructible* in terms of) other, perhaps more fundamental entities.

– R. M. MARTIN, *Intension and Decision*

We can...by using...[our]...definitions say what is meant by
"the number 1 + 1 belongs to the concept F"
and then, using this, give the sense of the expression
"the number 1 + 1 + 1 belongs to the concept F"
and so on; but we can never...decide by means of our definitions whether any concept has the number Julius Caesar belonging to it, or whether that same familiar conqueror of Gaul is a number or not.

– G. FREGE, *The Foundations of Arithmetic*

I. The education

Imagine Ernie and Johnny, sons of two militant logicists – children who have not been taught in the vulgar (old-fashioned) way but for whom the

Much of the work on this paper was done while the author held a Procter and Gamble Faculty Fellowship at Princeton University. This is gratefully acknowledged. I am indebted pedagogical order of things has been the epistemological order. They did not learn straight off how to count. Instead of beginning their mathematical training with arithmetic as ordinary men know it, they first learned logic – in their case, actually set theory. Then they were told about the numbers. But to tell people in their position about the numbers was an easy task – very much like the one which faced Monsieur Jourdain's tutor (who, oddly enough, was a philosopher). The parents of our imagined children needed only to point out what aspect or part of what the children already knew, under other names, was what ordinary people called "numbers." Learning the numbers merely involved learning new names for familiar sets. Old (set-theoretic) truths took on new (number-theoretic) clothing.

The way in which this was done will, however, bear some scrutiny and re-examination. To facilitate the exposition, I will concentrate on Ernie and follow his arithmetical education to its completion. I will then return to Johnny.

It might have gone as follows. Ernie was told that there was a set whose members were what ordinary people referred to as the (natural) numbers, and that these were what he had known all along as the elements of the (infinite) set $\mathfrak{N}$. He was further told that there was a relation defined on these "numbers" (henceforth I shall usually omit the shudder quotes), the *less-than* relation, under which the numbers were well ordered. This relation, he learned, was really the one, defined on $\mathfrak{N}$, for which he had always used the letter "R." And indeed, speaking intuitively now, Ernie could verify that every nonempty subset of $\mathfrak{N}$ contained a "least" element – that is, one that bore R to every other member of the subset. He could also show that nothing bore R to itself, and that R was transitive, antisymmetric, irreflexive, and connected in $\mathfrak{N}$. In short, the elements of $\mathfrak{N}$ formed a progression, or series, under R.

And then there was 1, the smallest number (for reasons of future convenience we are ignoring 0). Ernie learned that what people had been referring to as 1 was really the element a of $\mathfrak{N}$, the first, or least, element of $\mathfrak{N}$ under R. Talk about "successors" (each number is said to have one) was easily translated in terms of the concept of the "next" member of $\mathfrak{N}$ (under R). At this point, it was no trick to show that the assumptions made by ordinary mortals about numbers were in fact theorems for Ernie. For on the basis of his theory, he could establish Peano's axioms – an advantage which he enjoyed over ordinary mortals, who must more or

to Paul Ziff for his helpful comments on an earlier draft of this paper.

Reprinted with the kind permission of the editors from the *Philosophical Review* 74(1965): 47–73.

less take them as given, or self-evident, or meaningless-but-useful, or what have you.[1]

There are two more things that Ernie had to learn before he could truly be said to be able to speak with the vulgar. It had to be pointed out to him which operations on the members of $\mathfrak{N}$ were the ones referred to as "addition," "multiplication," "exponentiation," and so forth. And here again he was in a position of epistemological superiority. For whereas ordinary folk had to introduce such operations by recursive definition, a euphemism for postulation, he was in a position to show that these operations could be *explicity* defined. So the additional postulates assumed by the number people were also shown to be derivable in his theory, once it was seen which set-theoretic operations addition, multiplication, and so forth really are.

The last element needed to complete Ernie's education was the explanation of the *applications* of these devices: counting and measurement. For they employ concepts beyond those as yet introduced. But fortunately, Ernie was in a position to see what it was that he was doing that corresponded to these activities (we will concentrate on counting, assuming that measurement can be explained either similarly or in terms of counting).

There are two kinds of counting, corresponding to transitive and intransitive uses of the verb "to count." In one, "counting" admits of a direct object, as in "counting the marbles"; in the other it does not. The case I have in mind is that of the preoperative patient being prepared for the operating room. The ether mask is placed over his face and he is told to count, as far as he can. He has not been instructed to count anything at all. He has merely been told to count. A likely story is that we normally learn the first few numbers in connection with sets having that number of members – that is, in terms of *transitive* counting (thereby learning the use of numbers) and then learn how to generate "the rest" of the numbers. Actually, "the rest" always remains a relatively vague matter. Most of us simply learn that we will never run out, that our notation will extend as far as we will ever need to count. Learning these words, and how to repeat them in the right order, is learning *intransitive* counting. Learning their use as measures of sets is learning *transitive* counting. Whether we learn one kind of counting before the other is immaterial so far as the initial numbers are concerned. What is certain, and not immaterial, is that we will have to learn some recursive procedure for generating the *notation* in the proper order before we have learned to count transitively, for to do the latter is, either directly or

---

[1]The details of the proofs need not detain us.

indirectly, to correlate the elements of the number series with the members of the set we are counting. It seems, therefore, that it is possible for someone to learn to count intransitively without learning to count transitively. But not vice versa. This is, I think, a mildly significant point. But what *is* transitive counting, exactly?

To count the members of a set is to determine the cardinality of the set. It is to establish that a particular relation C obtains between the set and one of the numbers – that is, one of the elements of $\mathfrak{N}$ (we will restrict ourselves to counting finite sets here). Practically speaking, and in simple cases, one determines that a set has k elements by taking (sometimes metaphorically) its elements one by one as we say the numbers one by one (starting with 1 and in order of magnitude, the last number we say being k). To count the elements of some k-membered set b is to establish a one-to-one correspondence between the elements of b and the elements of $\mathfrak{N}$ less than or equal to k. The relation "pointing-to-each-member-of-b-in-turn-while-saying-the-numbers-up-to-and-including-k" establishes such a correspondence.

Since Ernie has at his disposal the machinery necessary to show of any two equivalent finite sets that such a correspondence exists between them, it will be a theorem of his system that any set has k members if and only if it can be put into one-to-one correspondence with the set of numbers less than or equal to k.[2]

Before Ernie's education (and the analysis of number) can be said to have been completed, one last condition on R should be mentioned: that R must be at least recursive, and possibly even primitive recursive. I have never seen this condition included in the analysis of number, but it seems to me so obviously required that its inclusion is hardly debatable. We

[2]It is not universally agreed that these last two parts of our account (defining the operations and defining cardinality) are indeed required for an adequate explication of number. W. V. Quine, for one, explicitly denies that anything need be done other than provide a progression to serve as the numbers. He states: "The condition upon all acceptable explications of number ... can be put ...: *any progression* – i.e., any infinite series each of whose members has only finitely many precursors – will do nicely. Russell once held that a further condition had to be met, to the effect that there be a way of applying one's would-be numbers to the measurement of multiplicity: a way of saying that (1) There are n objects x such that Fx. This, however, was a mistake, for any progression can be fitted to that further condition. For (1) can be paraphrased as saying that the numbers less than n [Quine uses 0 as well] admit of correlation with the objects x such that Fx. This requires that our apparatus include enough of the elementary theory of relations for talk of correlation, or one–one relation; but it requires nothing special about numbers except that they form a progression" (Quine 1960: 262–3). I would disagree. The explanation of cardinality – i.e., of the use of numbers for "transitive counting," as I have called it – is part and parcel of the explication of number. Indeed, if it may be excluded on the grounds Quine offers, we might as well say that there are *no* necessary conditions, since the only one he cites is hardly necessary, provided "that our apparatus contain enough of the theory of sets to contain a progression." But I will return to this point.

have already seen that Quine denies (by implication) that this constitutes an additional requirement: "The condition upon all acceptable explications of number...can be put...: any *progression* – i.e., any infinite series each of whose members has only finitely many precursors – will do nicely" (see note 3). But suppose, for example, that one chose the progression $A = a_1, a_2, a_3, \ldots a_n, \ldots$ obtained as follows. Divide the positive integers into two sequences B and C, within each sequence letting the elements come in order of magnitude. Let B (that is, $b_1, b_2, \ldots$) be the sequence of Gödel numbers of valid formulas of quantification theory, under some suitable numbering, and let C (that is, $c_1, c_2 \ldots$) be the sequence of positive integers which are not numbers of valid formulas of quantification theory under that numbering (in order of magnitude in each case). Now in the sequence A, for each n let $a_{2n-1} = b_n$ and $a_{2n} = c_n$. Clearly A, though a progression, is not recursive, much less primitive recursive. Just as clearly, this progression would be unusable as the numbers – and the reason is that we expect that if we know which numbers two expressions designate, we are able to calculate in a finite number of steps which is the "greater" (in this case, which one comes later in A).[3] More dramatically, if told that set b has n members, and that c has m, it should be possible to determine in a finite number of steps which has more members. Yet it is precisely that which is not possible here. This ability (to tell in a finite number of steps which of two numbers is the greater) is connected with (both transitive and intransitive) counting, since its possibility is equivalent to the possibility of generating ("saying") the numbers in order of magnitude (that is, in their order in A). You could not know that you were saying them in order of magnitude since, no recursive rule existing for generating its members, you could not know what their order of magnitude should be. This is, of course, a very strong claim. There are two questions here, both of which are interesting and neither of which could conceivably receive discussion in this paper. (1) Could a human being be a decision procedure for non-

---

[3]There is, of course, a difficulty with the notion of "knowing which numbers two expressions designate." It is the old one illustrated by the following example. Abraham thinks of a number, and so does Isaac. Call Abraham's number a and Isaac's i. Is a greater than i? I know which number a refers to: Abraham's. And similarly with i. But that brings me no closer to deciding which is the greater. This can be avoided, however, by requiring that numbers be given in canonical notation, as follows. Let the usual (recursive) definition of the numbers serve to define the *set* of "numbers," but not to establish their order. Then take the above definition of A as defining the *less-than* relation among the members of that set, thus defining the *progression*. (The fact that the nonrecursive progression that I use is a progression of *numbers* is clearly inessential to the point at issue. I use it here merely to avoid the elaborate circumlocutions that would result from doing everything settheoretically. One could get the same effect by letting the "numbers" be formulas of quantification theory, instead of their Gödel numbers, and using the formulas autonymously.)

recursive sets, or is the human organism at best a Turing machine (in the relevant respect)? If the latter, then there could not exist a human being who could generate the sequence A, much less *know* that this is what he was doing. Even if the answer to (1), however, is that a human being *could* be (act or be used as) such a decision procedure, the following question would still arise and need an answer: (2) could he *know* all truths of the form $i<j$ (in A)? And it seems that what constitutes knowledge might preclude such a possibility.

But I have digressed enough on this issue. The main point is that the "$<$" relation over the numbers must be recursive. Obviously I cannot give a rigorous proof that this is a requirement, because I cannot prove that man is at best a Turing machine. That the requirement is met by the usual "$<$" relation among numbers – the paradigm of a primitive recursive relation – and has also been met in every detailed analysis ever proposed constitutes good evidence for its correctness.[4] I am just making explicit what almost everyone takes for granted. Later in this paper, we will see that one plausible account of why this is taken for granted connects very closely with one of the views I will be urging.

So it was thus that Ernie learned that he had really been doing number theory all his life (I guess in much the same way that *our* children will learn this surprising fact about themselves if the *nouvelle vague* of mathematics teachers manages to drown them all).

It should be clear that Ernie's education is now complete. He has learned to speak with the vulgar, and it should be obvious to all that my earlier description was correct. He had at his disposal all that was needed for the concept of number. One might even say that he already possessed the concepts of number, cardinal, ordinal, and the usual operations on them, and needed only to learn a different vocabulary. It is my claim that there is nothing having to do with the task of "reducing" the concept of number to logic (or set theory) that has not been done above, or that could not be done along the lines already marked out.

To recapitulate: It was necessary (1) to give definitions of "1," "number," and "successor," and "$+$," "$\times$," and so forth, on the basis of which the laws of arithmetic could be derived; and (2) to explain the "extramathematical" uses of numbers, the principal one being counting – thereby introducing the concept of *cardinality* and cardinal number.

I trust that both were done satisfactorily, that the preceding contains all the elements of a correct account, albeit somewhat incompletely.

[4]Needless to say, it is trivially met in any analysis that provides an effective correlation between the names of the "numbers" of the analysis and the more common names under which we know those numbers.

None of the above was essentially new; I apologize for the tedium of expounding these details yet another time, but it will be crucial to my point that the sufficiency of the above account be clearly seen. For if it is sufficient, presumably Ernie *now* knows which sets the numbers are.

II. The dilemma

The story told in the previous section could have been told about Ernie's friend Johnny as well. For his education also satisfied the conditions just mentioned. Delighted with what they had learned, they started proving theorems about numbers. Comparing notes, they soon became aware that something was wrong, for a dispute immediately ensued about whether or not 3 belonged to 17. Ernie said that it did, Johnny that it did not. Attempts to settle this by asking ordinary folk (who had been dealing with numbers *as* numbers for a long time) understandably brought only blank stares. In support of his view, Ernie pointed to his theorem that for any two numbers, x and y, x is less than y if and only if x belongs to y and x is a proper subset of y. Since by common admission 3 is less than 17 it followed that 3 belonged to 17. Johnny, on the other hand, countered that Ernie's "theorem" was mistaken, for given two numbers, x and y, x belongs to y if and only if y is the successor of x. These were clearly incompatible "theorems." Excluding the possibility of the inconsistency of their common set theory, the incompatibility must reside in the definitions. First "less-than." But both held that x is less than y if and only if x bears R to y. A little probing, however, revealed the source of the trouble. For Ernie, the successor under R of a number x was the set consisting of x and all the members of x, while for Johnny the successor of x was simply $[x]$, the unit set of x – the set whose only member is x. Since for each of them 1 was the unit set of the null set, their respective progressions were

(i) $[\emptyset], [\emptyset,[\emptyset]], [\emptyset,[\emptyset],[\emptyset,[\emptyset]]], \ldots$ for Ernie

and

(ii) $[\emptyset], [[\emptyset]], [[[\emptyset]]], \ldots$ for Johnny.

There were further disagreements. As you will recall, Ernie had been able to prove that a set had n members if and only if it could be put into one-to-one correspondence with the set of numbers less than or equal to n. Johnny concurred. But they disagreed when Ernie claimed further that a set had n members if and only if it could be put into one-to-one correspondence with the number n itself. For Johnny, every number is single-membered. In short, their cardinality relations were different. For Ernie, 17 had 17 members, while for Johnny it had only one.[5] And so it went.

Under the circumstances, it became perfectly obvious why these disagreements arose. But what did not become perfectly obvious was how they were to be resolved. For the problem was this:

If the conclusions of the previous section are correct, then both boys have been given correct accounts of the numbers. Each was told by his father which set the set of numbers really was. Each was taught which object – whose independent existence he was able to prove – was the number 3. Each was given an account of the meaning (and reference) of number words to which no exception could be taken and on the basis of which all that we know about or do with numbers could be explained. Each was taught that some particular set of objects contained what people who used number words were really referring to. But the sets were different in each case. And so were the relations defined on these sets – including crucial ones, like cardinality and the like. But if, as I think we agreed, the account of the previous section was correct – not only as far as it went but correct in that it contained conditions which were both necessary and *sufficient* for any correct account of the phenomena under discussion, then the fact that they disagree on which particular sets the numbers are is fatal to the view that each number is some particular set. For if the number 3 is in fact some particular set b, it cannot be that two correct accounts of the meaning of "3" – and therefore also of its reference – assign two different sets to 3. For if it is true that for some set b, $3 = b$, then it cannot be true that for some set c, different from b, $3 = c$. But if Ernie's account is adequate in virtue of satisfying the conditions spelled out in Section I, so is Johnny's, for it too satisfies those conditions. We are left in a quandary. We have two (infinitely many, really) accounts of the meaning of certain words ("number," "one," "seventeen," and so forth) each of which satisfies what appear to be necessary and sufficient conditions for a correct account. Although there are differences between the two accounts, it appears that both are correct in virtue of satisfying common conditions. If so, the differences are incidental and do not affect correctness. Furthermore, in Fregean terminology, each account fixes the *sense* of the words whose analysis it provides. Each account must also, therefore, fix the *reference* of these

[5]Some of their type-theoretical cousins had even more peculiar views – for to be of cardinality 5 a set had to *belong* to one of the numbers 5. I say "some of" because others did not use that definition of cardinality, or of numbers, but sided either with Ernie or with Johnny.

expressions. Yet, as we have seen, one way in which these accounts differ is in the referents assigned to the terms under analysis. This leaves us with the following alternatives:

(A)   Both are right in their contentions: each account contained conditions each of which was necessary and which were jointly sufficient. Therefore $3 = [[[\emptyset]]]$, and $3 = [\emptyset, [\emptyset], [\emptyset]]]$.

(B)   It is not the case that both accounts were correct; that is at least one contained conditions which were not necessary and possibly failed to contain further conditions which, taken together with those remaining, would make a set of sufficient conditions.

(A) is, of course, absurd. So we must explore (B).

The two accounts agree in over-all structure. They disagree when it comes to fixing the referents for the terms in question. Given the identification of the numbers as some particular set of sets, the two accounts generally agree on the relations defined on that set; under both, we have what is demonstrably a recursive progression and a successor function which follows the order of that progression. Furthermore, the notions of cardinality are defined in terms of the progression, insuring that it becomes a theorem for each n that a set has n members if and only if it can be put into one-to-one correspondence with the set of numbers less than or equal to n. Finally, the ordinary arithmetical operations are defined for these "numbers." They do differ in the way in which cardinality is defined, for in Ernie's account the fact that the number n had n members was exploited to define the notion of having n members. In all other respects, however, they agree.

Therefore, if it is not the case that both $3 = [[[\emptyset]]]$ and $3 = [\emptyset, [\emptyset, [\emptyset]]]$, which it surely is not, then at least one of the corresponding accounts is incorrect as a result of containing a condition that is not necessary. It may be incorrect in other respects as well, but at least that much is clear. I can distinguish two possibilities again: either all the conditions just listed, which both of these accounts share, are necessary for a correct and complete account, or some are not. Let us assume that the former is the case, although I reserve the right to discard this assumption if it becomes necessary to question it.

If all the conditions they share are necessary, then the superfluous conditions are to be found among those that are not shared. Again there are two possibilities: either at least one of the accounts satisfying the conditions we are assuming to be necessary, but which assigns a definite set to each number, is correct, or none are. Clearly no two different ones can be, since they are not even extensionally equivalent, much less intensionally. Therefore exactly one is correct or none is. But then the correct one must be the one that picks out which set of sets is *in fact* the numbers. We are now faced with a crucial problem: if there exists such a "correct" account, do there also exist arguments which will show it to be the correct one? Or does there exist a particular set of sets *b*, which is *really* the numbers, but such that there exists no argument one can give to establish that it, and not, say, Ernie's set $\mathfrak{N}$, is really the numbers? It seems altogether too obvious that this latter possibility borders on the absurd. If the numbers constitute one particular set of sets, and not another, then there must be arguments to indicate which. In urging this I am not committing myself to the decidability by proof of every mathematical question – for I consider this neither a mathematical question nor one amenable to proof. The answer to the question I am raising will follow from an analysis of questions of the form "Is $n = \ldots$?" It will suffice for now to point to the difference between our question and

Is there a greatest prime p such that $p + 2$ is also prime?

or even

Does there exist an infinite set of real numbers equivalent with neither the set of integers nor with the set of all real numbers?

In awaiting enlightenment on the true identity of 3 we are not awaiting a proof of some deep theorem. Having gotten as far as we have without settling the identity of 3, we can go no further. We do not know what a proof of that *could* look like. The notion of "correct account" is breaking loose from its moorings if we admit of the possible existence of unjustifiable but correct answers to questions such as this. To take seriously the question "Is $3 = [[[\emptyset]]]$?" *tout court* (and not elliptically for "in Ernie's account?"), in the absence of any way of settling it, is to lose one's bearings completely. No, if such a question has an answer, there are arguments supporting it, and if there are no such arguments, then there *is* no "correct" account that discriminates among all the accounts satisfying the conditions of which we reminded ourselves a couple of pages back.

How then might one distinguish *the* correct account from all the possible ones? Is there a set of sets that has a greater claim to be the numbers than any other? Are there reasons one can offer to single out that set? Frege chose as the number 3 the extension of the concept "equivalent with some 3-membered set"; that is, for Frege a number was an equivalence class – the class of all classes equivalent with a given class. Although an appealing notion, there seems little to recommend it over, say, Ernie's. It has been argued that this is a more fitting account because number words are really class predicates, and that this account reveals that fact. The view is that in saying that there are n F's you are predicating n-hood of F, just as in saying that red is a color you are predicating colorhood of red. I do not think this is true. And neither did Frege (1950: sec. 57). It is certainly true that to say

(1) There are seventeen lions in the zoo

is not to predicate seventeen-hood of each individual lion. I suppose that it is also true that if there are seventeen lions in the zoo and also seventeen tigers in the zoo, the classes of lions-in-the-zoo and tigers-in-the-zoo are in a class together, though we shall return to that. It does not follow from this that (1) predicates seventeen-hood of one of those classes. First of all, the grammatical evidence for this is scanty indeed. The best one can conjure up by way of an example of the occurrence of a number word in predicative position is a rather artificial one like

(2) The lions in the zoo are seventeen.

If we do not interpret this as a statement about the ages of the beasts, we see that such statements do not predicate anything of any individual lion. One might then succumb to the temptation of analyzing (2) as the noun phrase "The lions in the zoo" followed by the verb phrase "are seventeen," where the analysis is parallel to that of

(3) The Cherokees are vanishing

where the noun phrase refers to the class and the verb phrase predicates something of that class. But the parallel is short-lived. For we soon notice that (2) probably comes into the language by deletion from

(4) The lions in the zoo are seventeen in number,

which in turn probably derives from something like

(5) Seventeen lions are in the zoo.

This is no place to explore in detail the grammar of number words. Suffice it to point out that they differ in many important respects from words we do not hesitate to call predicates. Probably the closest thing to a genuine class predicate involving number words is something on the model of "seventeen-membered" or "has seventeen members." But the step from there to "seventeen" being itself a predicate of classes is a long one indeed. In fact, I should think that pointing to the above two predicates gives away the show – for what is to be the analysis of "seventeen" as it occurs in those phrases?

Not only is there scanty grammatical evidence for this view, there seems to be considerable evidence against it, as any scrutiny of the similarity of function among the number words and "many," "few," "all," "some," "any," and so forth will immediately reveal. The proper study of these matters will have to await another context, but the nonpredicative nature of number words can be further seen by noting how different they are from, say, ordinary adjectives, which do function as predicates. We have already seen that there are really no occurrences of number words in typical predicative position (that is, in "is (are) ..."), the only putative cases being along the lines of (2) above, and therefore rather implausible. The other anomaly is that number words normally outrank *all* adjectives (or all other adjectives, if one wants to class them as such) in having to appear at the head of an adjective string, and not inside. This is such a strong ranking that deviation virtually inevitably results in ungrammaticalness:

(6)   The five lovely little square blue tiles

is fine, but any modification of the position of "five" yields an ungrammatical string; the farther to the right, the worse.[6]

Further reason for denying the predicative nature of number words comes from the traditional first-order analysis of sentences such as (1), with which we started. For that is usually analyzed as:

$$(7) \qquad (\exists x_1)\ldots(\exists x_{17})(Lx_1 \cdot Lx_2 \cdot \ldots \cdot Lx_{17} \cdot x_1 \neq x_2 \cdot x_1 \neq x_3 \cdot \ldots$$
$$\cdot x_{16} \neq x_{17} \cdot (y)(Ly \supset . \, y = x_1 \vee y = x_2 \vee \ldots \vee y = x_{17})).$$

The only predicate in (1) which remains is "lion in the zoo," "seventeen" giving way to numerous quantifiers, truth functions, variables, and occurrences of " $=$," unless, of course, one wishes to consider these also to be predicates of classes. But there are slim grounds indeed for the view that (1) or (7) predicates seventeen-hood of the class of lions in the zoo. Number words function so much like operators such as "all," "some," and so forth, that a readiness to make class names of them should be accompanied by a readiness to make the corresponding move with respect to quantifiers, thereby proving (in traditional philosophic

[6]It might be thought that constructions such as

(i)   The hungry five went home

constitute counterexamples to the thesis that number words must come first in an adjective string. But they do not. For in (i) and similar cases, the number word occurs as a noun, and not as an adjective, probably deriving from

(ii)   The five hungry NP_{pl} went home

by the obvious transformation, and should be understood as such. There are certain genuine counterexamples, but the matter is too complicated for discussion here.

fashion) the existence of the one, the many, the few, the all, the some, the any, the every, the several, and the each.[7]

But then, what support *does* this view have? Well, this much: if two classes each have seventeen members, there probably exists a class which contains them both in virtue of that fact. I say "probably" because this varies from set theory to set theory. For example, this is not the case with type theory, since the two classes have both to be of the same type. But in no consistent theory is there a class of all classes with seventeen members, at least not alongside the other standard set-theoretical apparatus. The existence of the paradoxes is itself a good reason to deny to "seventeen" this univocal role of designating the class of all classes with seventeen members.

I think, therefore, that we may conclude that "seventeen" *need* not be considered a predicate of classes, and there is similarly no necessity to view 3 as the set of all triplets. This is not to deny that "is a class having three members" is a predicate of classes; but that is a different matter indeed. For that follows from all of the accounts under consideration.[8] Our present problem is to see if there is one account which can be established to the exclusion of all others, thereby settling the issue of which sets the numbers really are. And it should be clear by now that there is not. Any purpose we may have in giving an account of the notion of number and of the individual numbers, other than the question-begging one of proving of the right set of sets that *it* is the set of numbers, will be equally well (or badly) served by any one of the infinitely many accounts satisfying the conditions we set out so tediously. There is little need to examine all the possibilities in detail, once the traditionally favored one of Frege and Russell has been seen not to be uniquely suitable.

Where does that leave us? I have argued that at most one of the infinitely many different accounts satisfying our conditions can be correct, on the grounds that they are not even extensionally equivalent, and therefore at least all but one, and possibly all, contain conditions that are not necessary and that lead to the identification of the numbers with some particular set of sets. If numbers are sets, then they must be *particular sets,* for each set is some particular set. But if the number 3 is really one set rather than another, it must be possible to give some cogent reason for thinking so; for the position that this is an unknowable truth is hardly tenable. But there seems to be little to choose among the accounts. Relative to our purposes in giving an account of these matters, one will do as

---

[7]And indeed why not "I am the one who gave his all in fighting for the few against the many"?

[8]Within the bounds imposed by consistency.

well as another, stylistic preferences aside. There is no way connected with the reference of number words that will allow us to choose among them, *for the accounts differ at places where there is no connection whatever between features of the accounts and our uses of the words in question.* If all the above is cogent, then there is little to conclude except that any feature of an account that identifies 3 with a set is a superfluous one – and that therefore 3, and its fellow numbers, could not be sets at all.

III. Way out

In this third and final section, I shall examine and urge some considerations that I hope will lend plausibility to the conclusion of the previous section, if only by contrast. The issues involved are evidently so numerous and complex, and cover such a broad spectrum of philosophic problems, that in this paper I can do no more than indicate what I think they are and how, in general, I think they may be resolved. I hope nevertheless that a more positive account will emerge from these considerations.

A. Identity. Throughout the first two sections, I have treated expressions of the form

(8) $$n = s,$$

where n is a number expression and s a set expression as if I thought that they made perfectly good sense, and that it was our job to sort the true from the false.[9] And it might appear that I had concluded that all such statements were false. I did this to dramatize the kind of answer that a Fregean might give to the request for an analysis of number – to point up the kind of question Frege took it to be. For he clearly wanted the analysis to determine a truth value for each such identity. In fact, he wanted to determine a sense for the result of replacing s with any name or description whatsoever (while an expression ordinarily believed to name a number occupied the position of n). Given the symmetry and transitivity of identity, there were three kinds of identities satisfying these conditions, corresponding to the three kinds of expressions that can appear on the right:

(a) with some arithmetical expression on the right as well as on the left (for example, "$2^{17} = 4,892$," and so forth);

---

[9] I was pleased to find that several of the points in my discussion of Frege have been made quite independently by Charles Parsons (1965). I am indebted to his discussion for a number of improvements.

(b)   with an expression designating a number, but not in a standard arithmetical way, as "the number of apples in the pot," or "the number of F's" (for example, 7 = the number of the dwarfs);

(c)   with a referring expression on the right which is of neither of the above sorts, such as "Julius Caesar," "[[∅]]" (for example, 17 = [[[∅]]]).

The requirement that the usual laws of arithmetic follow from the account takes care of all identities of the first sort. Adding an explication of the concept of cardinality will then suffice for those of kind (b). But to include those of kind (c), Frege felt it necessary to find some "objects" for number words to name and with which numbers could be identical. It was at this point that questions about which set of objects the numbers *really* were began to appear to need answering for, evidently, the simple answer "numbers" would not do. To speak from Frege's standpoint, there is a world of objects – that is, the designata or referents of names, descriptions, and so forth – in which the identity relation had free reign. It made sense for Frege to ask of *any* two names (or descriptions) whether they named the same object or different ones. Hence the complaint at one point in his argument that, thus far, one could not tell from his definitions whether Julius Caesar was a number.

I rather doubt that in order to explicate the use and meaning of number words one will have to decide whether Julius Caesar was (is?) or was not the number 43. Frege's insistence that this needed to be done stemmed, I think, from his (demonstrably) inconsistent logic (interpreted sufficiently broadly to encompass set theory). All items (names) in the universe were on a par, and the question whether two names had the same referent always presumably had an answer – yes or no. The inconsistency of the logic from which this stems is of course *some* reason to regard the view with suspicion. But it is hardly a refutation, since one might grant the meaningfulness of all identity statements, the existence of a universal set as the range of the relation, and still have principles of set existence sufficiently restrictive to avoid inconsistency. But such a view, divorced from the naïve set theory from which it stems, loses much of its appeal. I suggest, tentatively, that we look at the matter differently.

I propose to deny that all identities are meaningful, in particular to discard all questions of the form of (c) above as senseless or "unsemantical" (they are not totally senseless, for we grasp enough of their sense to explain why they are senseless). Identity statements make sense only in contexts where there exist possible individuating conditions. If an expression of the form "$x = y$" is to have a sense, it can be only in contexts where it is clear that both x and y are of some kind or category C, and

286

that it is the conditions which individuate things *as the same C* which are operative and determine its truth value. An example might help clarify the point. If we know x and y to be lampposts (possibly the same, but nothing in the way they are designated decides the issue) we can ask if they are *the same lamppost*. It will be their color, history, mass, position, and so forth which will determine if they are indeed the same lamppost. Similarly, if we know z and w to be numbers, then we can ask if they are *the same number*. And it will be whether they are prime, greater than 17, and so forth which will decide if they are indeed the same number. But just as we cannot individuate a lamppost in terms of these latter predicates, neither can we individuate a number in terms of its mass, color, or similar considerations. What determines that something is a *particular lamppost* could not individuate it as a *particular number*. I am arguing that questions of the identity of a particular "entity" do not make sense. "Entity" is too broad. For such questions to make sense, there must be a well-entrenched predicate C, in terms of which one then asks about the identity of a *particular C*, and the conditions associated with identifying C's as *the same C* will be the deciding ones. Therefore, if for two predicates F and G there is no third predicate C which subsumes both and which has associated with it some uniform conditions for identifying two putative elements as the same (or different) C's, the identity statements crossing the F and G boundary will not make sense.[10] For example, it will make sense to ask of something x (which is in fact a chair) if it is the same ... as y (which in fact is a table). For we can fill the blank with a predicate, "piece of furniture," and we know what it is for a and b to be the same or different pieces of furniture. To put the point differently, questions of identity contain the presupposition that the "entities" inquired about both belong to some general category. This presupposition is normally carried by the context or theory (that is, a more systematic context). To say that they are both "entities" is to make no presuppositions at all – for everything purports to be at least that. "Entity," "thing," "object" are words having a role in the language; they are place fillers whose function is analogous to that of pronouns (and, in more formalized contexts, to variables of quantification).

Identity *is* id-entity, but only within narrowly restricted contexts. Alternatively, what constitutes an entity is category or theory dependent. There are really two correlative ways of looking at the problem. One

---

[10]To give a precise account, it will be necessary to explain "uniform conditions" in such a way as to rule out the obvious counterexamples generated by constructed *ad hoc* disjunctive conditions. But to discuss the way to do this would take us too far afield. I do not pretend to know the answer in any detail.

might conclude that identity is systematically ambiguous, or else one might agree with Frege, that identity is unambiguous, always meaning sameness of object, but that (contra-Frege now) the notion of an *object* varies from theory to theory, category to category – and therefore that his mistake lay in failure to realize this fact. This last is what I am urging, for it has the virtue of preserving identity as a general logical relation whose application in any given well-defined context (that is, one within which the notion of object is univocal) remains unproblematic. Logic can then still be seen as the most general of disciplines, applicable in the same way to and within any given theory. It remains the tool applicable to all disciplines and theories, the difference being only that it is left to the discipline or theory to determine what shall count as an "object" or "individual."

That this is not an implausible view is also suggested by the language. Contexts of the form "the same *G*" abound, and indeed it is in terms of them that identity should be explained, for what will be counted as the same *G* will depend heavily on *G*. The same *man* will have to be an individual man; "the same *act*" is a description that can be satisfied by many individual acts, or by only one, for the individuating conditions for acts make them sometimes types, sometimes tokens. Very rare in the language are contexts open to (satisfiable by) any kind of "thing" whatsoever. There are some – for example, "Sam referred to ...," "Helen thought of ..." – and it seems perfectly all right to ask if what Sam referred to on some occasion was what Helen thought of. But these contexts are very few, and they all seem to be intensional, which casts a referentially opaque shadow over the role that identity plays in them.

Some will want to argue that identities of type (c) are not senseless or unsemantical, but simply false – on the grounds that the distinction of categories is one that cannot be drawn. I have only the following argument to counter such a view. It will be just as hard to explain how one *knows* that they are false as it would be to explain how one knows that they are senseless, for normally we know the falsity of some identity "$x=y$" only if we know of x (or y) that it has some characteristic that we know y (or x) *not* to have. I know that $2 \neq 3$ because I know, for example, that 3 is odd and 2 is not, yet it seems clearly wrong to argue that we know that $3 \neq [[[\emptyset]]]$ because, say, we know that 3 has no (or seventeen, or infinitely many) members while $[[[\emptyset]]]$ has exactly one. We know no such thing. We do not know that it does. But that does not constitute knowing that it does not. What is enticing about the view that these are all false is, of course, that they hardly seem to be open questions to which we may find the answer any day. Clearly, all the evidence is in; if no decision is possible on the basis of it, none will ever be pos-

288

sible. But for the purposes at hand the difference between these two views is not a very serious one. I should certainly be happy with the conclusion that all identities of type (c) are either senseless or false.

B. Explication and reduction. I would like now to approach the question from a slightly different angle. Throughout this paper, I have been discussing what was substantially Frege's view, in an effort to cast some light on the meaning of number words by exposing the difficulties involved in trying to determine which objects the numbers really are. The analyses we have considered all contain the condition that numbers are sets, and that therefore each individual number is some individual set. We concluded at the end of Section II that numbers could not be sets at all – on the grounds that there are no good reasons to say that any particular number is some particular set. To bolster our argument, it might be instructive to look briefly at two activities closely related to that of stating that numbers *are* sets – those of explication and reduction.

In putting forth an explication of number, a philosopher may have as part of his explication the statement that $3 = [[[\emptyset]]]$. Does it follow that he is making the kind of mistake of which I accused Frege? I think not. For there is a difference between *asserting* that 3 *is* the set of all triplets and *identifying* 3 with that set, which last is what might be done in the context of some explication. I certainly do not wish what I am arguing in this paper to militate against identifying 3 with anything you like. The difference lies in that, normally, one who identifies 3 with some particular set does so for the purpose of presenting some theory and does not claim that he has *discovered* which object 3 really is. We might want to know whether some set (and relations and so forth) would do as number surrogates. In investigating this it would be entirely legitimate to state that making such an identification we can do with that set (and those relations) what we now do with the numbers. Hence we find Quine saying:

> Frege dealt with the question "What is a number?" by showing how the work for which the objects in question might be wanted could be done by objects whose nature was presumed to be less in question. (Quine 1960: 262)

Ignoring whether this is a correct interpretation of Frege, surely someone who says this would not claim that, since the answer turned out to be "Yes," it is now clear that numbers had really been sets all along. In such a context, the adequacy of some system of objects to the task is a very real question and one which can be settled. Under our analysis, *any* system of objects, sets or not, that forms a recursive progression must be adequate. Thus, discovering that a different system will do the job very nicely cannot be to discover which objects the numbers are .... Explication in the above reductionistic sense, is therefore neutral with respect to the sort of problem we have been discussing, but it does cast some sobering light on what it is to be an individual number.

There is another reason to deny that it would be legitimate to use the reducibility of arithmetic to set theory as a reason to assert that numbers are really sets after all. Gaisi Takeuti has shown that the Gödel-von Neumann-Bernays set theory is in a strong sense *reducible to* the theory of ordinal numbers less than the least inaccessible number (1954). No wonder numbers are sets; sets are really (ordinal) numbers, after all. *But now, which is really which?*

These brief comments on reduction, explication, and what they might be said to achieve in mathematics lead us back to the quotation from Richard Martin which heads this paper. Martin correctly points out that the mathematician's interest stops at the level of structure. If one theory can be modeled in another (that is, reduced to another) then further questions about whether the individuals of one theory are really those of the second just do not arise. In the same passage, Martin goes on to point out (approvingly, I take it) that the philosopher is not satisfied with this limited view of things. He wants to know more and does ask the questions in which the mathematician professes no interest. I agree. He does. And mistakenly so. It will be the burden of the rest of this paper to argue that such questions miss the point of what arithmetic, at least, is all about.

C. Conclusion: numbers and objects. It was pointed out above that any system of objects, whether sets or not, that forms a recursive progression must be adequate. But this is odd, for any recursive set can be arranged in a recursive progression. So what matters, really, is not any condition on the *objects* (that is, on the set) but rather a condition on the relation under which they form a progression. To put the point differently – and this is the crux of the matter – that any recursive sequence whatever would do suggests that what is important is not the individuality of each element but the structure which they jointly exhibit. This is an extremely striking feature. One would be led to expect from this fact alone that the question of whether a particular "object" – for example, [[[Ø]]] – would do as a replacement for the number 3 would be pointless in the extreme, as indeed it is. "Objects" do not do the job of numbers singly; the whole system performs the job or nothing does. I therefore argue, extending the argument that led to the conclusion that numbers could not be sets, that numbers could not be objects at all; for there is no more reason to identify any individual number with any one particular object than with any other (not already known to be a number).

The pointlessness of trying to determine which objects the numbers are thus derives directly from the pointlessness of asking the question of any individual number. For arithmetical purposes the properties of numbers which do not stem from the relations they bear to one another in virtue of being arranged in a progression are of no consequence whatsoever. But it would be only these properties that would single out a number as this object or that.

Therefore, numbers are not objects at all, because in giving the properties (that is, necessary and sufficient) of numbers you merely characterize an *abstract structure* – and the distinction lies in the fact that the "elements" of the structure have no properties other than those relating them to other "elements" of the same structure. If we identify an abstract structure with a system of relations (in intension, of course, or else with the set of all relations in extension isomorphic to a given system of relations), we get arithmetic elaborating the properties of the "less-than" relation, or of all systems of objects (that is, *concrete* structures) exhibiting that abstract structure. That a system of objects exhibits the structure of the integers implies that the elements of that system have some properties not dependent on structure. It must be possible to individuate those objects independently of the role they play in that structure. But this is precisely what cannot be done with the numbers. To *be* the number 3 is no more and no less than to be preceded by 2, 1, and possibly 0, and to be followed by 4, 5, and so forth. And to *be* the number 4 is no more and no less than to be preceded by 3, 2, 1, and possibly 0, and to be followed by .... *Any* object can *play the role of* 3; that is, any object can be the third element in some progression. What is peculiar to 3 is that it defines that role – not by being a paradigm of any object which plays it, but by representing the relation that any third member of a progression bears to the rest of the progression.

Arithmetic is therefore the science that elaborates the abstract structure that all progressions have in common merely in virtue of being progressions. It is not a science concerned with particular objects – the numbers. The search for which independently identifiable particular objects the numbers really are (sets? Julius Caesars?) is a misguided one.

On this view many things that puzzled us in this paper seem to fall into place. Why so many interpretations of number theory are possible without any being uniquely singled out becomes obvious: there is no unique set of objects that are the numbers. Number theory is the elaboration of the properties of *all* structures of the order type of the numbers. The number words do not have single referents. Furthermore, the reason identification of numbers with objects works wholesale but fails utterly object by object is the fact that the theory is elaborating an abstract structure and not the properties of independent individuals, any one of which could be characterized without reference to its relations to the rest. Only when we are considering a particular sequence as being, not the numbers, but *of the structure of the numbers* does the question of which element is, or rather *corresponds to,* 3 begin to make any sense.

Slogans like "Arithmetic is about numbers," "Number words refer to numbers," when properly urged, may be interpreted as pointing out two quite distinct things: (1) that number words are not names of special non-numerical entities, like sets, tomatoes, or Gila monsters; and (2) that a purely formalistic view that fails to assign any meaning whatsoever to the statements of number theory is also wrong. They need not be incompatible with what I am urging here.

This last formalism is too extreme. But there is a modified form of it, also denying that number words are names, which constitutes a plausible and tempting extension of the view I have been arguing. Let me suggest it here. On this view the sequence of number words is just that – a sequence of words or expressions with certain properties. There are not two kinds of things, numbers and number words, but just one, the words themselves. Most languages contain such a sequence, and any such sequence (of words or terms) will serve the purposes for which we have ours, provided it is recursive in the relevant respect. In counting, we do not correlate sets with initial segments of the numbers as extralinguistic entities, but correlate sets with initial segments of the sequence of number *words.* The central idea is that this recursive sequence is a sort of yardstick which we use to measure sets. Questions of the identification of the referents of number words should be dismissed as misguided in just the way that a question about the referents of the parts of a ruler would be seen as misguided. Although any sequence of expressions with the proper structure would do the job for which we employ our present number words, there is still some reason for having one, relatively uniform, notation: ordinary communication. Too many sequences in common use would make it necessary for us to learn too many different equivalences. The usual objection to such an account – that there is a distinction between numbers and number words which it fails to make will, I think, not do. It is made on the grounds that "two," "*zwei*," "*deux*," "2" are all supposed to "stand for" the same number but yet are *different* words (one of them not a word at all). One can mark the differences among the expressions in question, and the similarities as well, without conjuring up some extralinguistic objects for them to name. One need only point to the similarity of function: within any numbering system, what will be important will be what place in the system any particular expression is used to mark. All the above expressions share this feature with one another – and with the binary use of "10," but not with its decimal employment. The "ambiguity" of "10" is thus easily explained. Here again we see the series-related character of individual numbers, except that it is now mapped a little closer to home. One cannot tell what number a particular expression represents without being given the sequence of which it forms a part. It will then be from its place in that sequence – that is, from its relation to other members of the sequence, *and from the rules governing the use of the sequence in counting* – that it will derive its individuality. It is for this last reason that I urged, contra Quine, that the account of cardinality must explicitly be included in the account of number (see note 3).

Furthermore, other things fall into place as well. The requirement, discussed in Section I, that the "less-than" relation be recursive is most easily explained in terms of a recursive notation. After all, the whole theory of recursive functions makes most sense when viewed in close connection with notations rather than with extralinguistic objects. This makes itself most obvious in three places: the development of the theory by Post systems, by Turing machines, and in the theory of constructive ordinals, where the concern is frankly with recursive notations for ordinals. I do not see why this should not be true of the finite ordinals as well. For a set of *numbers* is recursive if and only if a machine of a particular sort could be programmed to generate them in order of magnitude – that is, to generate the standard or canonical notations for those numbers following the (reverse) order of the "less-than" relation. If that relation over the notation were not recursive, the above theorem would not hold.

It also becomes obvious why every analysis of number ever presented has had a recursive "less-than" relation. If what we are generating is a notation, the most natural way for generating it is by giving recursive rules for getting the next element from any element you may have – and you would have to go far out of your way (and be slightly mad) to generate the notation and then define "less than" as I did on page 276, above, in discussing the requirement of recursiveness.

Furthermore, on this view, we learn the elementary arithmetical operations as the cardinal operations on small sets, and extend them by the usual algorithms. Arithmetic then becomes cardinal arithmetic at the earlier levels in the obvious way, and the more advanced statements become easily interpretable as *projections* via truth functions, quantifiers, and the recursive rules governing the operations. One can therefore be this sort of formalist without denying that there is such a thing as

293

arithmetical truth other than derivability within some given system. One can even explain what the ordinary formalist apparently cannot – why these axioms were chosen and which of two possible consistent extensions we should adopt in any given case.

But I must stop here. I cannot defend this view in detail without writing a book. To return in closing to our poor abandoned children, I think we must conclude that their education was badly mismanaged – not from the mathematical point of view, since we have concluded that there is no mathematically significant difference between what they were taught and what ordinary mortals know, but from the philosophical point of view. They think that numbers are really sets of sets while, if the truth be known, there are no such things as numbers; which is not to say that there are not at least two prime numbers between 15 and 20.

Mathematics without foundations

HILARY PUTNAM

Philosophers and logicians have been so busy trying to provide mathematics with a 'foundation' in the past half-century that only rarely have a few timid voices dared to voice the suggestion that it does not need one. I wish here to urge with some seriousness the view of the timid voices. I don't think mathematics is unclear; I don't think mathematics has a crisis in its foundations; indeed, I do not believe mathematics either has or needs 'foundations'. The much touted problems in the philosophy of mathematics seem to me, without exception, to be problems internal to the thought of various system builders. The systems are doubtless interesting as intellectual exercises; debate between the systems and research within the systems doubtless will and should continue; but I would like to convince you (of course I won't, but one can always hope) that the various systems of mathematical philosophy, without exception, need not be taken seriously.

By way of comparison, it may be salutory to consider the various 'crises' that philosophy has pretended to discover in the past. It is impressive to remember that at the turn of the century there was a large measure of agreement among philosophers – far more than there is now – on certain fundamentals. Virtually all philosophers were idealists of one sort or another. But even the nonidealists were in a large measure of agreement with the idealists. It was generally agreed any property of material objects – say, *redness* or *length* – could be ascribed to the object, if at all, only as a power to produce certain sorts of sensory experiences. When the man on the street thinks of a material object, according to this traditional view, he really thinks of a subjective object, not a real 'external' object. If there are external objects, we cannot really imagine what they are like; we know and can conceive only their powers. Either there are no external objects at all (Berkeley) – i.e. no objects 'external' to minds and their ideas – or there are, but they are *Dinge an sich*. In sum, then, philosophy flattered itself to have discovered not just a crisis, but a fundamental mistake, not in some special science, but in our most common-sense convictions about material objects. To put it

Reprinted with the kind permission of the editors from the *Journal of Philosophy* 64 (1967): 5–22.

crudely, philosophy thought itself to have shown that no one has ever really perceived a material object and that, if material objects exist at all (which was thought to be highly problematical), then no one *could* perceive, or even imagine, one.

Anyone maintaining at the turn of the century that the notions 'red' and 'hard' (or, more abstractly 'material object') were reasonably clear notions; that redness and hardness are *non*dispositional properties of material objects; that we see red things and see *that* they are red; and that *of course* we can imagine red objects, know what a red object is, etc., would have seemed unutterably foolish. After all, the most brilliant philosophers in the world all found difficulties with these notions. Clearly, the man is just too stupid to see the difficulties. Yet today this 'stupid' view is the view of many sophisticated philosophers, and the increasingly prevalent opinion is that it was the arguments purporting to show a contradiction in the view, and not the view itself, that were profoundly wrong. Moral: not everything that passes – in philosophy anyway – as a difficulty with a concept is one. And second moral: the fact that philosophers all agree that a notion is 'unclear' doesn't mean that it *is* unclear.

More recently there was a large measure of agreement among philosophers of science – far more than there is now – that, in some sense, talk about theoretical entities and physical magnitudes is 'highly derived talk' which, in the last analysis, reduces to talk about observables. Just a few years ago, we were being told that 'electron' is a 'partially interpreted' term, whereas 'red' is 'completely interpreted'. Today it is becoming increasingly clear that 'electron' is a term that has complete 'meaning' in every sense in which 'red' has 'meaning'; that the 'purpose' of talk about electrons is not simply to make successful predictions in observation language any more than the 'purpose' of talk about red things is to make true deductions about electrons; and that the whole question about how we 'introduce' theoretical terms was a mare's nest. I refrain from drawing another moral.

Today there is a large measure of agreement among philosophers of mathematics that the concept of a 'set' is unclear. I hope the above short review of some history of philosophy will indicate why I am less than overawed by this agreement. When philosophy discovers something wrong with science, sometimes science has to be changed – Russell's paradox comes to mind, as does Berkeley's attack on the actual infinitesimal – but more often it is philosophy that has to be changed. I do not think that the difficulties that philosophy finds with classical mathematics today are genuine difficulties; and I think that the philosophical interpretations of mathematics that we are being offered on every hand are wrong, and that 'philosophical interpretation' is just what mathematics doesn't need. And I include my own past efforts in this direction.

I do not, however, mean to disparage the value of philosophical inquiry. If philosophy got itself into difficulties with the concept of a material object, it also got itself out; and the result is some modest but significant increase in our clarity about perception and knowledge. It is this sort of clarity about mathematical truth, mathematical 'objects', and mathematical necessity that I should like to see us attain; but I do not think the famous 'isms' in the philosophy of mathematics represent the road to that clarity. Let us therefore make a fresh start.

A sketch of my view

I think that the least mystifying way for me to discuss this topic is as follows: first to give a very cursory and superficial sketch of my own views, so that you will at least be able to guess at the positive position that underlies my criticism of others, and then to survey the alleged difficulties in set theory. Of course, any philosopher hates ever to say briefly, let alone superficially, what his own view on any topic is (although he is delighted to give such a statement to the view of any philosopher with whom he disagrees), because a superficial statement may make his view seem naive or even downright stupid. But such a statement is a great help to others, at least in getting an initial orientation, and for that reason I shall accept the risk involved.

In my view the chief characteristic of mathematical propositions is the very wide variety of equivalent formulations that they possess. I don't mean this in the trivial sense of cardinality: of course, every proposition possesses infinitely many equivalent formulations; what I mean is rather that in mathematics the number of ways of expressing what is in some sense the same fact (if the proposition is true) while apparently not talking about the same objects is especially striking.

The same situation does sometimes arise in empirical science, that is, the situation that what is in some sense the same fact can be expressed in two strikingly different ways, the most famous example being wave-particle duality in quantum mechanics. Reichenbach coined the happy expression 'equivalent descriptions' for this situation. The description of the world as a system of particles, not in the classical sense but in the peculiar quantum-mechanical sense, may be associated with a different picture than the description of the world as a system of waves, again not in the classical sense but in the quantum-mechanical sense; but the two theories are thoroughly intertranslatable, and should be viewed as having the same physical content. The same fact can be expressed either by saying that the electron is a wave with a definite wavelength λ or by saying that the electron is a particle with a sharp momentum *p* and an indeterminate position. What 'same fact' comes to here is, I admit, obscure. Obviously what is *not* being claimed is *synonymy* of *sentences*. It would be absurd to claim that the *sentence* 'there is an electron-wave with the wavelength λ' is *synonymous* with the *sentence* 'there is a particle electron with the momentum *h*/λ and a totally indeterminate position'. What is rather being claimed is this: that the two theories are compatible, not incompatible, given the way in which the theoretical primitives of each theory are now being understood; that indeed, they are not merely compatible but equivalent: the primitive terms of each admit of definition by means of the primitive terms of the other theory, and then each theory is a deductive consequence of the other. Moreover, there is no particular advantage to taking one of the two theories as fundamental and regarding the other one as *derived*. The two theories are, so to speak, on the same explanatory level. Any fact that can be explained by means of one can equally well be explained by means of the other. And in view of the systematic equivalence of statements in the one theory with statements in the other theory, there is no longer any point to regarding the formulation of a given fact in terms of the notions of one theory as more fundamental than (or even as *significantly* different from) the formulation of the fact in terms of the notions of the other theory. In short, what has happened is that the systematic equivalences between the sentences of the two theories have become so well known that they *function* virtually as synonymies in the actual practice of science.

Of course, the fact that two theories can be related in this way is not by itself either surprising or important. It would not be worth remarking that two theories are related in this way if the pictures associated with the two theories were not apparently incompatible or at least very different. In mathematics, the different equivalent formulations of a given mathematical proposition do not call to mind apparently *incompatible* pictures as do the different equivalent formulations of the quantum theory, but they do sometimes call to mind radically different pictures, and I think that the way in which a given philosopher of mathematics proceeds is often determined by which of these pictures he has in mind, and this in turn is often determined by which of the equivalent formulations of the mathematical propositions with which he deals he takes as primary.

Of the many possible 'equivalent descriptions' of the realm of mathematical facts, there are two which seem to me to have especial importance. I shall refer to these, somewhat misleadingly, I admit, by the titles 'Mathematics as Modal Logic' and 'Mathematics as Set Theory', The second, I take it, needs no explanation. Everyone is today familiar with the conception of mathematics as the description of a 'universe' of 'mathematical objects' – and, in particular, with the conception of mathematics as describing relations among *sets*. However, the picture would not be significantly different if one said 'sets and numbers' – that numbers can themselves be 'identified' with sets seems today a matter of minor importance; the important thing about the picture is that mathematics describes 'objects'. The other conception is less familiar, and I shall say a few words about it.

Consider the assertion that there is a counterexample to Fermat's 'last theorem'; i.e. that there is an nth power which is the sum of two nth powers, $2 < n$, all three numbers positive. Abbreviate the standard formula that expresses this statement in first-order arithmetic as '$\sim Fermat$'. If $\sim Fermat$ is provable, then, in fact, $\sim Fermat$ is provable already from a certain easily specified finite subset of the theorems of first-order arithmetic. (N.B., this is owing to the fact that it takes only one counterexample to refute a generalization. So the portion of first-order arithmetic in which we can prove all true statements of the form $x^n + y^n \neq z^n$, x, y, z, n *constant* integers, is certainly strong enough to *disprove* Fermat's last theorem if the last theorem be false, notwithstanding the fact that *all* of first-order arithmetic may be too weak to *prove* Fermat's last theorem if the last theorem be true. And the portion of first-order arithmetic just alluded to is known to be finitely axiomatizable.) Let 'AX' abbreviate the conjunction of the axioms of the finitely axiomatizable subtheory of first-order arithmetic just alluded to. Then Fermat's last theorem is *false* just in case '$AX \supset \sim Fermat$' is valid, i.e. just in case

(1) $$\Box (AX \supset \sim Fermat)$$

Since the truth of (1), in case (1) *is* true, does not depend upon the meaning of the arithmetical primitives, let us suppose these to be replaced by 'dummy letters' (predicate letters). To fix our ideas imagine that the primitives in terms of which AX and $\sim Fermat$ are written are the two three-term relations 'x is the sum of y and z' and 'x is the product of y and z' (exponentiation is known to be first-order-definable from these, and so, of course, are *zero* and *successor*). Let $\text{AX}(S, T)$ and $\sim \text{FERMAT}(S, T)$ be like AX and $\sim Fermat$ except for containing the 'dummy' triadic predicate letters S, T, where AX and $\sim Fermat$ contain the constant predicates 'x is the sum of y and z' and 'x is the product of y and z'. Then (1) is essentially a truth of pure modal logic (if it is true), since the constant predicates occur 'inessentially'; and this can be brought out by replacing (1) by the abstract schema:

299

(2) $\Box [\text{AX}(S, T) \supset \sim \text{FERMAT}(S, T)]$

– and this is a schema of pure first-order modal logic.

Now then, the mathematical content of the assertion (2) is certainly the same as that of the assertion that *there exist numbers* x, y, z, n ($2 < n$, $x, y, z \neq 0$) such that $x^n + y^n = z^n$. Even if the expressions involved are not synonymous, the mathematical equivalence is so obvious that they might as well be synonymous, as far as the mathematician is concerned. Yet the pictures in the mind called up by these two ways of formulating what one might as well consider to be the same mathematical assertion can be quite different. When one speaks of the 'existence of numbers' one gets the picture of mathematics as describing eternal objects; while (2) simply says that $\text{AX}(S, T)$ entails $\text{FERMAT}(S, T)$, no matter how one may interpret the predicate letters 'S' and 'T', and this scarcely seems to be about 'objects' at all. Of course, one can strain after objects if one wants. One can, for example, interpret the dummy letters 'S' and 'T' as quantifiers over 'the totality of all properties', if one wishes. But this is hardly necessary, since one can find a particular substitution instance of (2), even in a nominalistic language (apart from the ' $\Box$ ') which is equivalent to (2) (just choose predicates S^* and T^* to put for S and T such that it is not mathematically impossible that the objects in their field should form an ω-sequence, and such that, if the objects in their field did form an ω-sequence, S^* would be isomorphic to addition of integers, and T^* to multiplication, in the obvious sense). Or one can interpret ' $\Box$ ' as a predicate of statements, rather than as a statement connective, in which case what (2) asserts is that a certain object, namely the statement '$\text{AX}(S, T) \supset \sim \text{FERMAT}(S, T)$' has a certain property ('being necessary'). But still, the only 'object' this commits us to is the statement '$\text{AX}(S, T) \supset \sim \text{FERMAT}(S, T)$', and one has to be pretty compulsive about one's nominalistic cleanliness to scruple about *this*. In short, if one fastens on the first picture (the 'object' picture), then mathematics is wholly extensional, but presupposes a vast totality of eternal objects; while if one fastens on the second picture (the 'modal' picture), then mathematics has *no* special objects of its own, but simply tells us what follows from what. If 'Platonism' has appeared to be *the* issue in the philosophy of mathematics of recent years, I suggest that it is because we have been too much in the grip of the first picture.

So far I have only indicated how one very special mathematical proposition can be treated as a statement involving modalities, but not special objects. I believe that, by making a more complex and iterated use of modal notions, one can analyze the notion of *a standard model for set theory,* and thus extend the objects – modalities duality that I am discussing to the whole of classical mathematics. I shall not show this now; but, needless to say, I would not deal at such length with this one special example if I did not believe it to represent, in some sense, the general situation. For the moment, I shall ask you to accept it on faith that this extension to the general case can be carried out.

What follows, I believe, is that each of these two ways of looking at mathematics can be used to clarify the other. If one is puzzled by the modalities (and I am concerned here with necessity in Quine's narrower sense of logical validity, excluding necessities that depend on alleged synonymy relations in natural languages), then one can be helped by the set-theoretic notion of a *model* (necessity = truth in all models; possibility = truth in some model). On the other hand, if one is puzzled by the question recently raised by Benacerraf (1965; reprinted in this volume): how numbers can be 'objects' if they have *no* properties except order in a particular ω-sequence, then, I believe, one can be helped by the answer: call them 'objects' if you like (they *are* objects in the sense of being things one can quantify over); but remember that these objects have the special property that each fact about them is, in an equivalent formulation, simply a fact about *any* ω-sequence. 'Numbers exist'; but all this comes to, for mathematics anyway, is that (1) ω-sequences are *possible* (mathematically speaking); and (2) there are *necessary* truths of the form 'if α is an ω-sequence, then...' (whether any *concrete* example of an ω-sequence exists or not). Similarly, there is not, from a mathematical point of view, any significant difference between the assertion that *there exists a set of integers* satisfying an arithmetical condition and the assertion that *it is possible to select* integers so as to satisfy the condition. Sets, if you will forgive me for parodying John Stuart Mill, are permanent possibilities of selection.

The question of decidability

The sense that there is a 'crisis in the foundations' of mathematics has many sources. Morris Kline cites the development of non-Euclidean geometry (which shook the idea that the axioms of a mathematical discipline must be *truths*), the lack of a consistency proof for mathematics, and the lack of a universally acceptable solution to the antinomies. In addition to these, one might mention Gödel's theorem (Kline does mention it, in fact, in connection with the consistency problem). For Gödel's theorem suggests that the truth or falsity of some mathematical statements might be impossible in principle to ascertain, and this has led some to wonder if we even know what we mean by 'truth' and 'falsity' in such a context.

301

Now, the example of non-Euclidean geometry does show, I believe, that our notions of what is 'self-evident' have to be subject to revision, not just in the light of new observations, but in the light of new *theories*. The intuitive evidence for the proposition that two lines cannot be a constant distance apart for half their length (i.e. in one half-plane) and then start to approach each other (as geodesics can in General Relativity, e.g. light rays which come in from infinity parallel and then approach each other as they pass on opposite sides of the sun) is as great as the intuitive evidence for the axioms of number theory. I believe that under certain circumstances revisions in the axioms of arithmetic, or even of propositional calculus (e.g. the adoption of a modular logic as a way out of the difficulties in quantum mechanics), is fully conceivable. The philosophical ploy which consists in saying 'then terms would have changed meaning' is uninteresting – except as a question in the philosophy of linguistics, of course – unless one can show that in their 'old meaning' the sentences of the theory in question can still (after the transition to non-Euclidean geometry, or non-Archimedean arithmetic, or modular logic) be admitted to have formerly expressed propositions that are clear and true. If in some sense there are 'Euclidean straight lines' in our space, then the transition to, say, Riemannian geometry *could* (not necessarily *should*) be regarded as a mere 'change of meaning'. But (1) there are *no* curves in space (if the world is Riemannian) that satisfy Euclid's theorems about straight lines; and (2) even if the world is Lobatchevskian, there are no *unique* such curves – to choose any particular remetricization which leads to Euclidean geometry and say '*this* is what "distance", "straight line", etc., *used* to mean' would be arbitrary. In short, the price one pays for the adoption of non-Euclidean geometry is to deny that there are *any* propositions which might *plausibly* have been in the minds of the people who believed in Euclidean geometry and which are simultaneously clear and true. Similarly, if one accepts the interpretation of quantum mechanics that is based on modular logic, then one has to deny that there has been a change in the meaning of the relevant sentences, or else deny that there are any unique propositions which might have been in the minds of those who formerly used those sentences and which were both clear and true. You can't have your conceptual revolution and minimize it too!

Yet all this does not, I think, mean that there is a crisis in the foundations of mathematics. It does not even mean that mathematics becomes an empirical science in the ordinary sense of that term. For the chief characteristic of empirical science is that for each theory there are usually alternatives in the field, or at least alternatives struggling to be born. As long as the major parts of classical logic and number theory and analysis have no alternatives in the field – alternatives which require a change in the axioms and which effect the simplicity of total science, including empirical science, so that a choice has to be made – the situation will be what it has always been. We will be justified in accepting classical propositional calculus or Peano number theory not because the relevant statements are 'unrevisable in principle' but because a great deal of science presupposes these statements and because no real alternative is in the field. Mathematics, on this view, does become 'empirical' in the sense that one is allowed to try to *put* alternatives into the field. Mathematics can be wrong, and not just in the sense that the proofs might be fallacious or that the axioms might not (if we reflected more deeply) be really self-evident. Mathematics (or rather, some mathematical theory) might be wrong in the sense that the 'self-evident' axioms might be false, and the axioms that are true might not be 'evident' at all. But this does not make the pursuit of truth impossible in mathematics any more than it has in empirical science, nor does it mean that we should not trust our intuitions when we have nothing better to go on. After all, a mathematical theory that has become the basis of a successful and powerful scientific system, including many important empirical applications, is not being accepted *merely* because it is 'intuitive', and if someone objects to it we have the right to say 'propose something better!' What this does do, rather, is make the 'foundational' view of mathematical knowledge as suspect as the 'foundational' view of empirical knowledge (if one cares to retain the 'mathematical–empirical' distinction at all).

Again, I cannot weep bitter tears about the lack of a consistency proof for classical mathematics. Even if such a proof were possible, it would only be a development within mathematics and not a foundation for mathematics. Not only would it be possible to raise philosophical questions about the branch of mathematics that was used for the consistency proof; but, in any case, science demands much more of a mathematical theory than that it should merely be *consistent,* as the example of the various alternative systems of geometry already dramatizes.

The question of the significance of the antinomies, and of what to do about the existence of several different approaches to overcoming them, is far more difficult. I propose to defer this question for a moment and to consider first the significance of Gödel's theorem and, more generally, of the existence of mathematically undecidable propositions.

Strictly speaking, all Gödel's theorem shows is that, in any particular consistent axiomatizable extension of certain finitely axiomatizable subtheories of Peano arithmetic, there are propositions of number theory that can neither be proved nor disproved. (I think it is fair to call this 'Gödel's theorem', even though this statement of it incorporates strengthenings due to Rosser and Tarski, Mostowski, Robinson.) It does not follow that any proposition of number theory is, in some sense, absolutely undecidable. However, it may well be the case that some proposition of elementary number theory is neither provable nor refutable in any system whose axioms rational beings will ever have any good reason to accept. This has caused some to doubt whether every mathematical proposition, or even every proposition of the elementary theory of numbers, can be thought of as having a truth value.

A similar consideration is raised by Paul Cohen's recent work in set theory, when that work is taken together with Gödel's classical relative consistency proof of the axiom $V = L$ (which implies the axiom of choice and the generalized continuum hypothesis). Together these results of Gödel and Cohen establish the full independence of the continuum hypothesis (for example) from the other axioms of set theory, assuming those other axioms to be consistent. A striking feature of both proofs is their invariance under small (or even moderately large) perturbations of the axioms. It appears quite possible today that no decisive consideration will ever appear (such as a set-theoretic axiom we have 'overlooked') which will reveal that a system in which the continuum hypothesis is provable is the correct one, and that no consideration will ever appear which will reveal that a system in which the continuum hypothesis is refutable is the correct one. In short, the truth value of the continuum hypothesis – assuming it has a truth value – may be undiscoverable by rational beings, or at least by the 'rational beings' that actually do exist, or ever will exist. Then, what reason is there to think that it has a truth value?

This 'argument' is sometimes taken to show that the notion of a set is unclear. For, since the argument 'shows' (sic!) that the continuum hypothesis has no truth value and the continuum hypothesis involves the concept of a set, the only plausible explanation of the truth-value failure is some unclarity in the notion of a set. (It would be an interesting exercise to find *all* the faults in this particular bit of reasoning. It is horrible, isn't it?)

The first point to notice is that the existence of propositions whose truth value we have no way of discovering is not at all peculiar to mathematics. Consider the assertion that there are infinitely many binary stars (considering the entire space–time universe, i.e. counting binary stars past, present, and future). It is not at all clear that we can discover the truth value of this assertion. Sometimes it is argued that such an assertion is 'verifiable (or at least confirmable) in principle', because it may *follow from a theory*. It is true that in one case we can discover the truth value of this proposition. Namely, if either it or its negation is derivable from laws of nature that we can confirm, then its truth value can be discovered. But it could just happen that there are infinitely many binary stars, without this being required by any law. Moreover, the distribution might be quite irregular, so that ordinary statistical inference could not discover it. Indeed, at some point I cease to understand the question 'Is it always possible *in principle* to discover the truth value of this proposition?' – for the methods of inquiry permitted ('inductive' methods) are just too ill defined a set. But I suspect that, given any *formalizable* inductive logic, one could describe a logically possible world in which (1) there were infinitely many binary stars; and (2) one could never discover this fact using that inductive logic. (Of course, the argument that the proposition is 'confirmable in principle' because it could follow from a theory does not even purport to show that in every possible world the truth or falsity of this statement could be induced from a finite amount of observational material using some inductive method; rather it shows that in *some* possible world the truth of this statement (or its falsity) could be induced from a finite amount of observational material.) Yet I, for one, see no reason – not even a prima facie one – to suspect that this proposition does not have a truth value. Why *should* all truths, even all empirical truths, be discoverable by probabilistic automata (which is what I suspect we are) using a finite amount of observational material? Why does the fact that the truth value of a proposition may be undiscoverable by us suggest to some philosophers – indeed, why does it count as a *proof* for some philosophers – that the proposition in question doesn't *have* a truth value? Surely, some kind of idealistic metaphysics must be lurking in the underbrush!

What is even more startling is that philosophers who would agree with me with respect to propositions about material objects should feel differently about propositions of mathematics. (Perhaps this is due to the pernicious tendency to think of mathematics solely in terms of the mathematical–objects picture. If one doesn't understand the nature of these objects – i.e. that they don't have a 'nature', that talk about them is equivalent to talk about what is impossible – then talk about them may seem like a form of theology, and if one is anti-theological, that may be a reason for rejecting mathematics as a make-believe.) Surely, the *mere* fact that we may never know whether the continuum hypothesis is true or false is by itself just *no* reason to think that it doesn't have a truth value!

'But what does it *mean* to say that the continuum hypothesis is true?' someone will ask. It means that if S is a set of real numbers, and S is not finite and not denumerably infinite, then S can be put in one-to-one

305

correspondence with the unit interval. Or, equivalently, it means that the sentence I have just written holds in any standard model for fourth-order number theory (actually, it can be expressed in third-order number theory). 'But what is a *standard* model?' It is one with the properties that (1) the 'integers' of the model form an ω-sequence under the $<$ of the model – i.e. it is not *possible* to select positive 'integers' $a_1, a_2, a_3, \ldots$ from the model so that, for all i, $a_{i+1} < a_i$ – and (2) the model is maximal with this property – i.e. it is not *possible* to add more 'sets' of 'integers' or 'sets of sets' of 'integers' or 'sets of sets of sets' of 'integers' to the model. (This last explanation contains the germ of the idea which is used in expressing the notion of a 'standard model' in modal-logical, as opposed to set-theoretic, language.)

I think that one can see what is going on more clearly if we imagine, for a moment, that physics has discovered that the physical universe is finite in both space and time and that all physical magnitudes are discrete (finiteness 'in the small'). That this is a possibility we must take into account was already emphasized by Hilbert in his famous article on the infinite (1926; reprinted in this volume) – it may well be, Hilbert pointed out, that we cannot argue for the consistency of any theory whose models are all infinite by arguing that physical space, or physical time, or anything else physical, provides a model for the theory, since physics is increasingly tending to replace infinities and continuities by finites and discretes.

If the whole physical universe is thoroughly finite, both in the large and in the small, then the statement '$10^{100} + 1$ is a prime number' may be one whose truth value we can never know. For, if the statement is true (and even intuitionist mathematicians regard this decidable statement as possessing a truth value), then to verify that it is true by using any sieve method might well be physically impossible. And, if the shortest proof from axioms that rational beings will ever have any reason to accept is too long to be physically written out, then it might be physically impossible for beings to whom only those things are 'evident' that are in fact 'evident' (or ever will be 'evident' or that we will ever in fact have good reason to believe) to know that the statement is true.

Now, although many people doubt that the continuum hypothesis has a truth value, everyone believes that the statement '$10^{100} + 1$ is a prime number' has a truth value. Why? 'Because the statement is decidable.' But what does that mean, 'the statement is decidable'? It means that it is *possible* to try out all the pairs of possible factors and see if any of them 'work'. It means that it is *possible* to decide the statement. Thus, the man who asserts that this statement is decidable, is simply making an assertion of mathematical possibility. Moreover, he believes that just one of the two statements:

> If all pairs n, m $(n, m < 10^{100} + 1)$ were 'tried' by actually computing the product nm, then in some case the product would be found to equal $10^{100} + 1$. (3)
>
> If all pairs $n, m, \ldots$ [same as in (3)], then in no case would the product be found to equal $10^{100} + 1$. (4)

expresses a *necessary* truth, although it may be *physically* impossible to discover which one. Yet this same mathematician or philosopher, who is quite happy in this context with the notion of mathematical possibility (and who does not ask for any nominalistic reduction) and who treats mathematical necessity as well defined in this case, for a reason which is essentially circular, regards it as 'platonistic' to suppose that the continuum hypothesis has a truth value.[1] I realize that this is an ad hominem argument, but still – if there is such an intellectual sin as 'platonism' (and it is remarkably unclear what this supposed sin consists of), why is it not already to commit it, if one supposes that '$10^{100} + 1$ is a prime number' has a truth value, even if no nominalistic reduction of this statement can be offered? (When one is defending a commonsense position, very often the only argument is ad hominem – for one has to keep throwing the burden of the argument back to the other side, by asking to be told *precisely* what is 'unclear' about the notions being attacked, or why a 'reduction' of the kind being demanded is necessary, or why a 'foundation' for the science in question is needed.)

In passing, I should like to remark that the following two principles, which many people seem to accept, can be shown to be inconsistent, by applying the Gödel theorem:

(I)   That, even if some arithmetical (or set-theoretical) statements have no truth value, still, to say of any arithmetical (or set-theoretical) statement that it has (or lacks) a truth value is itself always either true or false (i.e. the statement either has a truth value or it doesn't).

(II) All and only the decidable statements have a truth value.

---

[1] Incidentally, it may also be 'platonism' to treat statements of physical possibility or counterfactual conditionals as well defined. For (1) 'physical possibility' is *compatibility* with the laws of nature. But the relation of compatibility is interdefinable with the modal notions of possibility and necessity, and, of course, the laws of nature themselves require many mathematical notions for their statement. (2) A counterfactual conditional is true just in case the consequent *follows* from the antecedent, together with certain other statements that hold both in the actual and in the hypothetical world under consideration. And, of course, no nominalistic reduction has ever succeeded, either for the notion of physical possibility or for the subjunctive conditional.

For the statement that a mathematical statement S is decidable may itself be undecidable. Then, by (II), it has no truth value to say 'S is decidable'. But, by (I), it has a truth value to say 'S has a truth value' (in fact, *falsity*; since if S has a truth value, then S is decidable, by (II), and, if S is decidable, then 'S is decidable' is also decidable). Since it is false (by the previous parenthetical remark) to say 'S has a truth value' and since we accept the equivalence of 'S has a truth value' and 'S is decidable', then it must also be *false* to say 'S is decidable'. But it has no truth value to say 'S is decidable'. Contradiction.

The significance of the antinomies

The most difficult question in the philosophy of mathematics is, perhaps, the question raised by the antinomies and by the plurality of conflicting set theories. Part of the paradox is this: the antinomies do not at all seem to affect the notion 'set of sets of integers', etc. Yet they *do* seem to affect the notion '*all* sets'. How are we to understand this situation?

One way out might be this: to conclude that we understand the notion 'set' in some contexts (e.g. 'set of integers', 'set of sets of integers'), but to conclude that we do not understand it in the context 'all sets'. But we do seem to understand *some* statements about all sets, e.g. 'for every set x and every set y, there is a set z which is the union of x and y'. So must we really abandon hope of making sense of the locution 'all sets'?

It is at this point that I urge we attend to the objects–modalities duality that I pointed out a few pages ago. The notion of a set has been used by a number of authors to clarify the notions of mathematical possibility and necessity. For example, if we identify the notion of a 'possible world' with the notion of a model (or, more correctly, with the notion of a structure of the appropriate type), then the rationale of the modal system S5 can easily be explained (as, for instance, by Carnap in *Meaning and Necessity*), and this explanation can be extended to the case of quantified modal logic by methods due to Kripke, Hintikka, and others. Here, however, I wish to go in the reverse direction, and assuming that the notions of mathematical possibility and necessity are clear (and there is no paradox associated with the notion of necessity as long as we take the '□' as a statement connective (in the degenerate sense of 'unary connective') and not – in spite of Quine's urging – as a predicate of sentences), I wish to employ these notions to try to give a clear sense to talk about 'all sets'.

My purpose is not to start a *new* school in the foundations of mathematics (say, 'modalism'). Even if in some contexts the modal-logic picture is more helpful than the mathematical-objects picture, in other contexts the reverse is the case. Sometimes we have a clearer notion of what 'possible' means than of what 'set' means; in other cases the reverse is true; and in many, many cases both notions seem as clear as notions ever get in science. Looking at things from the standpoint of many different 'equivalent descriptions', considering what is suggested by *all* the pictures, is both a healthy antidote to foundationalism and of real heuristic value in the study of scientific questions.

Now, the natural way to interpret set-theoretic statements in the model-logical language is to interpret them as statements of what would necessarily be the case if there were standard models for the set theories in question. Since the models for von Neumann–Bernays set theory and its strengthenings (e.g. the system recently proposed by Bernays) are also models for Zermelo set theory, let me concentrate on Zermelo set theory. In order to 'concretize' the notion of a model, let us think of a model as a graph. The 'sets' of the model will then be pencil points (or some higher-dimensional analogue of pencil points, in the case of models of large cardinality), and the relation of membership will be indicated by 'arrows'. (I assume that there is nothing inconceivable about the idea of a physical space of arbitrarily high cardinality; so models of this kind need not necessarily be denumerable, and may even be standard.) Such a model will be called a 'concrete model' (or a 'standard concrete model', if it be standard) for Zermelo set theory. The model will be called standard if (1) there are no infinite-descending 'arrow' paths; and (2) it is not possible to extend the model by adding more 'sets' without adding to the number of 'ranks' in the model. (A 'rank' consists of all the sets of a given – possibly transfinite – type. 'Ranks' are cumulative types; i.e. every set of a given rank is also a set of every higher rank. It is a theorem of set theory that every set belongs to some rank.) A statement that refers only to sets of less than some given rank – say, to sets of rank less than $\omega \times 2$ – will be called a statement of 'bounded rank'. I ask the reader to accept it on faith that the statement that a certain graph G is a *standard* model for Zermelo set theory can be expressed using no 'non-nominalistic' notions except the '$\square$'.

If S is a statement of bounded rank and if we can characterize the 'given rank' in question in some invariant way (invariant with respect to standard models of Zermelo set theory), then the statement S can easily be translated into modal-logical language. The translation is just the statement that if G is any standard model for Zermelo set theory – i.e. any standard concrete model – and G contains the invariantly characterized rank in question, then necessarily S holds in G. (It is trivial to express '*S* holds in *G* ' for any *particular S* without employing the set-theoretic notion of 'holding'.) Our problem, then, is how to translate statements of *un*bounded rank into modal-logical language.

The method is best indicated by means of an example. If the statement has the form $(x)(\exists y)(z)Mxyz$, where *M* is quantifier-free, then the translation is this:

If *G* is any standard concrete model for Zermelo set theory and if *P* is any point in *G*, then it is possible that there is a graph *G* ' that extends *G* (i.e. *G* is a subgraph of *G* ') and a point *y* in *G* ' such that *G* ' is a standard concrete model for Zermelo set theory and such that

(if *G* " is any graph that extends *G* ' and such that *G* " is a standard concrete model for Zermelo set theory and if *z* is any point in *G* ", then *Mxyz* holds in *G* ").

Obviously this method can be extended to an arbitrary set-theoretic statement.

So much for technical matters. I apologize for this brief lapse into technicality, but actually this was only the merest sketch of the technical development, and this much detail is necessary for my discussion. The real question is this: what, if any, is the philosophical significance of such translations?

If there be any philosophical significance to such translations – and I don't claim a great deal – it lies in this: I did not assume that any standard concrete model for Zermelo set theory is maximal. Indeed, I would be inclined to say that no concrete model could be maximal – nor any *non*concrete model either, as far as that goes. Even God could not make a model for Zermelo set theory that it would be *mathematically* impossible to extend, and no matter what 'stuff' He might use. Yet I succeeded in giving a clear sense to statements about 'all sets' (clear relative to the notions I assumed to start with) *without* assuming a maximal model. In metaphysical language, it is not necessary to think of sets as one system of objects in some one possible world in order to follow assertions about all sets.

Furthermore, in construing statements about sets as statements about standard concrete models for set theory, I did not introduce possible concrete models (or even possible worlds) as objects. Introducing the modal connectives ' □ ', ' ◇ ', ' ⇥ ' is not introducing new kinds of objects, but rather extending the kinds of things we can say about ordinary objects and sorts of objects. (Of course, one *can* construe the statement that it is possible that there is a graph *G* satisfying a condition *C* as meaning that *there exists a possible graph G* satisfying the condition *C*; that is one way of smoothing the transition from the modal-logic picture to the mathematical-objects picture.)

The importance of Zermelo set theory and of the other set theories based upon the notion of 'rank' lies in this: we have a strong intuitive conviction that whenever As are possible, so is a structure that we might call 'the family of all sets of As'. Zermelo set theory assumes only this intuition and the intuition that the process of unioning such structures can be extended into the transfinite. Of course, this intuitive conviction *may* be mistaken; it could turn out that Zermelo set theory has no standard models (even if Zermelo set theory is consistent – e.g. the discovery of an ω-inconsistency would show that there are no standard models). But so could the intuitive conviction upon which number theory is based be mistaken. If we wish to be cautious, we can assume only predicative set theory up to some 'low' transfinite type. (It is necessary to extend predicative type theory 'just past' the constructive ordinals if we wish to be able to define *validity* of schemata that contain the quantifiers 'there are infinitely many x such that' and 'there are at most a finite number of x such that', for example.) Such a weak set theory may well give us all the sets we need for physics, and also the basic notions of validity and satisfiability that we need for logic, as well as arithmetic and a weak version of classical analysis. But the fact that we do have an intuitive conviction that standard models of Zermelo set theory, or of other set theories based upon the notion of 'rank' are *mathematically possible structures* is a perfectly good reason for asking what statements necessarily hold in such structures – e.g. for asking whether the continuum hypothesis necessarily holds in such structures.

The real significance of the Russell paradox, from the standpoint of the modal-logic picture, is this: it shows that *no* concrete structure can be a standard model for the naive conception of the totality of all sets; for any concrete structure has a possible extension that contains more 'sets'. (If we identify sets with the points that represent them in the various possible concrete structures, we might say: it is not possible for all *possible* sets to exist in any one world!) Yet set theory does not become impossible. Rather, set theory becomes the study of what must hold in, e.g. any standard model for Zermelo set theory.

PART III

Mathematical truth

The *a priori*

ALFRED JULES AYER

The view of philosophy which we have adopted may, I think, fairly be described as a form of empiricism. For it is characteristic of an empiricist to eschew metaphysics, on the ground that every factual proposition must refer to sense-experience. And even if the conception of philosophizing as an activity of analysis is not to be discovered in the traditional theories of empiricists, we have seen that it is implicit in their practice. At the same time, it must be made clear that, in calling ourselves empiricists, we are not avowing a belief in any of the psychological doctrines which are commonly associated with empiricism. For, even if these doctrines were valid, their validity would be independent of the validity of any philosophical thesis. It could be established only by observation, and not by the purely logical considerations upon which our empiricism rests.

Having admitted that we are empiricists, we must now deal with the objection that is commonly brought against all forms of empiricism; the objection, namely, that it is impossible on empiricist principles to account for our knowledge of necessary truths. For, as Hume conclusively showed, no general proposition whose validity is subject to the test of actual experience can ever be logically certain. No matter how often it is verified in practice, there still remains the possibility that it will be confuted on some future occasion. The fact that a law has been substantiated in $n-1$ cases affords no logical guarantee that it will be substantiated in the nth case also, no matter how large we take n to be. And this means that no general proposition referring to a matter of fact can ever be shown to be necessarily and universally true. It can at best be a probable hypothesis. And this, we shall find, applies not only to general propositions, but to all propositions which have a factual content. They can none of them ever become logically certain. This conclusion, which we shall elaborate later on, is one which must be accepted by every consistent empiricist. It is often thought to involve him in complete scepticism; but this is not the case. For the fact that the validity of a proposition cannot be logically guaranteed in no way entails that it is irrational

Excerpted and reprinted with the kind permission of the author and publishers from Alfred Jules Ayer, *Language, Truth and Logic* (London: Victor Gollancz, Ltd., 1956; New York: Dover Publications, Inc.), pp. 71–87.

for us to believe it. On the contrary, what is irrational is to look for a guarantee where none can be forthcoming; to demand certainty where probability is all that is obtainable. We have already remarked upon this, in referring to the work of Hume. And we shall make the point clearer when we come to treat of probability, in explaining the use which we make of empirical propositions. We shall discover that there is nothing perverse or paradoxical about the view that all the "truths" of science and common sense are hypotheses; and consequently that the fact that it involves this view constitutes no objection to the empiricist thesis.

Where the empiricist does encounter difficulty is in connection with the truths of formal logic and mathematics. For whereas a scientific generalization is readily admitted to be fallible, the truths of mathematics and logic appear to everyone to be necessary and certain. But if empiricism is correct no proposition which has a factual content can be necessary or certain. Accordingly the empiricists must deal with the truths of logic and mathematics in one of the two following ways: he must say either that they are not necessary truths, in which case he must account for the universal conviction that they are; or he must say that they have no factual content, and then he must explain how a proposition which is empty of all factual content can be true and useful and surprising.

If neither of these courses proves satisfactory, we shall be obliged to give way to rationalism. We shall be obliged to admit that there are some truths about the world which we can know independently of experience; that there are some properties which we can ascribe to all objects, even though we cannot conceivably observe that all objects have them. And we shall have to accept it as a mysterious inexplicable fact that our thought has this power to reveal to us authoritatively the nature of objects which we have never observed. Or else we must accept the Kantian explanation which, apart from the epistemological difficulties which we have already touched on, only pushes the mystery a stage further back.

It is clear that any such concession to rationalism would upset the main argument of this book. For the admission that there were some facts about the world which could be known independently of experience would be incompatible with our fundamental contention that a sentence says nothing unless it is empirically verifiable. And thus the whole force of our attack on metaphysics would be destroyed. It is vital, therefore, for us to be able to show that one or other of the empiricist accounts of the propositions of logic and mathematics is correct. If we are successful in this, we shall have destroyed the foundations of rationalism. For the fundamental tenet of rationalism is that thought is an independent source of knowledge, and is moreover a more trustworthy source of knowledge than experience; indeed some rationalists have gone so far as to say that thought is the only source of knowledge. And the ground for this view is simply that the only necessary truths about the world which are known to us are known through thought and not through experience. So that if we can show either that the truths in question are not necessary or that they are not "truths about the world," we shall be taking away the support on which rationalism rests. We shall be making good the empiricist contention that there are no "truths of reason" which refer to matters of fact.

The course of maintaining that the truths of logic and mathematics are not necessary or certain was adopted by Mill. He maintained that these propositions were inductive generalizations based on an extremely large number of instances. The fact that the number of supporting instances was so very large accounted, in his view, for our believing these generalizations to be necessarily and universally true. The evidence in their favor was so strong that it seemed incredible to us that a contrary instance should ever arise. Nevertheless it was in principle possible for such generalizations to be confuted. They were highly probable, but, being inductive generalizations, they were not certain. The difference between them and the hypotheses of natural science was a difference in degree and not in kind. Experience gave us very good reason to suppose that a "truth" of mathematics or logic was true universally; but we were not possessed of a guarantee. For these "truths" were only empirical hypotheses which had worked particularly well in the past; and, like all empirical hypotheses, they were theoretically fallible.

I do not think that this solution of the empiricist's difficulty with regard to the propositions of logic and mathematics is acceptable. In discussing it, it is necessary to make a distinction which is perhaps already enshrined in Kant's famous dictum that, although there can be no doubt that all our knowledge begins with experience, it does not follow that it all arises out of experience (Kant 1881: Introduction, section i). When we say that the truths of logic are known independently of experience, we are not of course saying that they are innate, in the sense that we are born knowing them. It is obvious that mathematics and logic have to be learned in the same way as chemistry and history have to be learned. Nor are we denying that the first person to discover a given logical or mathematical truth was led to it by an inductive procedure. It is very probable, for example, that the principle of the syllogism was formulated not before but after the validity of syllogistic reasoning had been observed in a number of particular cases. What we are discussing, however, when we say that logical and mathematical truths are known independently of experience, is not a historical question concerning the way in which these truths were originally discovered, not a psychological question concerning the way in which each of us comes to learn them, but an epistemological question. The contention of Mill's which we reject is that the propositions of logic and mathematics have the same status as empirical hypotheses; that their validity is determined in the same way. We maintain that they are independent of experience in the sense that they do not owe their validity to empirical verification. We may come to discover them through an inductive process; but once we have apprehended them we see that they are necessarily true, that they hold good for every conceivable instance. And this serves to distinguish them from empirical generalizations. For we know that a proposition whose validity depends upon experience cannot be seen to be necessarily and universally true.

In rejecting Mill's theory, we are obliged to be somewhat dogmatic. We can do no more than state the issue clearly and then trust that his contention will be seen to be discrepant with the relevant logical facts. The following considerations may serve to show that of the two ways of dealing with logic and mathematics which are open to the empiricist, the one which Mill adopted is not the one which is correct.

The best way to substantiate our assertion that the truths of formal logic and pure mathematics are necessarily true is to examine cases in which they might seem to be confuted. It might easily happen, for example, that when I came to count what I had taken to be five pairs of objects, I found that they amounted only to nine. And if I wished to mislead people I might say that on this occasion twice five was not ten. But in that case I should not be using the complex sign "$2 \times 5 = 10$" in the way in which it is ordinarily used. I should be taking it not as the expression of a purely mathematical proposition, but as the expression of an empirical generalization, to the effect that whenever I counted what appeared to be to be five pairs of objects I discovered that they were ten in number. This generalization may very well be false. But if it proved false in a given case, one would not say that the mathematical proposition "$2 \times 5 = 10$" had been confuted. One would say that I was wrong in supposing that there were five pairs of objects to start with, or that one of the objects had been taken away while I was counting, or that two of them had coalesced, or that I had counted wrongly. One would adopt as an explanation whatever empirical hypothesis fitted in best with the accredited facts. The one explanation which would in no circumstances be adopted is that ten is not always the product of two and five.

To take another example: if what appears to be a Euclidean triangle is found by measurement not to have angles totalling 180 degrees, we do not say that we have met with an instance which invalidates the mathematical proposition that the sum of the three angles of a Euclidean triangle is 180 degrees. We say that we have measured wrongly, or, more probably, that the triangle we have been measuring is not Euclidean. And this is our procedure in every case in which a mathematical truth might appear to be confuted. We always preserve its validity by adopting some other explanation of the occurrence.

The same thing applies to the principles of formal logic. We may take an example relating to the so-called law of excluded middle, which states that a proposition must be either true or false, or, in other words, that it is impossible that a proposition and its contradictory should neither of them be true. One might suppose that a proposition of the form "*x* has stopped doing *y*" would in certain cases constitute an exception to this law. For instance, if my friend has never yet written to me, it seems fair to say that it is neither true nor false that he has stopped writing to me. But in fact one would refuse to accept such an instance as an invalidation of the law of excluded middle. One would point out that the proposition "My friend has stopped writing to me" is not a simple proposition, but the conjunction of the two propositions "My friend wrote to me in the past" and "My friend does not write to me now": and, furthermore, that the proposition "My friend has not stopped writing to me" is not, as it appears to be, contradictory to "My friend has stopped writing to me," but only contrary to it. For it means "My friend wrote to me in the past, and he still writes to me." When, therefore, we say that such a proposition as "My friend has stopped writing to me" is sometimes neither true nor false, we are speaking inaccurately. For we seem to be saying that neither it nor its contradictory is true. Whereas what we mean, or anyhow should mean, is that neither it nor its apparent contradictory is true. And its apparent contradictory is really only its contrary. Thus we preserve the law of excluded middle by showing that the negating of a sentence does not always yield the contradictory of the proposition originally expressed.

There is no need to give further examples. Whatever instance we care to take, we shall always find that the situations in which a logical or mathematical principle might appear to be confuted are accounted for in such a way as to leave the principle unassailed. And this indicates that Mill was wrong in supposing that a situation could arise which would overthrow a mathematical truth. The principles of logic and mathematics are true universally simply because we never allow them to be anything else. And the reason for this is that we cannot abandon them without contradicting ourselves, without sinning against the rules which govern the use of language, and so making our utterances self-stultifying. In other words, the truths of logic and mathematics are analytic propositions or tautologies. In saying this we are making what will be held to be an extremely controversial statement, and we must now proceed to make its implications clear.

The most familiar definition of an analytic proposition, or judgment, as he called it, is that given by Kant. He said that an analytic judgment was one in which the predicate B belonged to the subject A as something which was covertly contained in the concept of A (Kant 1881: Introduction, sections iv and v). He contrasted analytic with synthetic judgments, in which the predicate B lay outside the subject A, although it did stand in connection with it. Analytic judgments, he explains, "add nothing through the predicate to the concept of the subject, but merely break it up into those constituent concepts that have all along been thought in it, although confusedly." Synthetic judgments, on the other hand, "add to the concept of the subject a predicate which has not been in any wise thought in it, and which no analysis could possibly extract from it." Kant gives "all bodies are extended" as an example of an analytic judgment, on the ground that the required predicate can be extracted from the concept of "body," "in accordance with the principle of contradiction"; as an example of a synthetic judgment, he gives "all bodies are heavy." He refers also to "$7+5=12$" as a synthetic judgment, on the ground that the concept of twelve is by no means already thought in merely thinking the union of seven and five. And he appears to regard this as tantamount to saying that the judgment does not rest on the principle of contradiction alone. He holds, also, that through analytic judgments our knowledge is not extended as it is through synthetic judgments. For in analytic judgments "the concept which I already have is merely set forth and made intelligible to me."

I think that this is a fair summary of Kant's account of the distinction between analytic and synthetic propositions, but I do not think that it succeeds in making the distinction clear. For even if we pass over the difficulties which arise out of the use of the vague term "concept," and the unwarranted assumption that every judgment, as well as every German or English sentence, can be said to have a subject and a predicate, there remains still this crucial defect. Kant does not give one straightforward criterion for distinguishing between analytic and synthetic propositions; he gives two distinct criteria, which are by no means equivalent. Thus his ground for holding that the proposition "$7+5=12$" is synthetic is, as we have seen, that the subjective intension of "$7+5$" does not comprise the subjective intension of "12"; whereas his ground for holding that "all bodies are extended" is an analytic proposition is that it rests on the principle of contradiction alone. That is, he employs a psychological criterion in the first of these examples, and a logical criterion in the second, and takes their equivalence for granted. But, in fact, a proposition which is synthetic according to the former criterion may very well be analytic according to the latter. For, as we have already pointed out, it is possible for symbols to be synonymous without having the same intentional meaning for anyone: and accordingly from the fact that one can think of the sum of seven and five without necessarily thinking of twelve, it by no means follows that the proposition "7 + 5 = 12" can be denied without self-contradiction. From the rest of his argument, it is clear that it is this logical proposition, and not any psychological proposition, that Kant is really anxious to establish. His use of the psychological criterion leads him to think that he has established it, when he has not.

I think that we can preserve the logical import of Kant's distinction between analytic and synthetic propositions, while avoiding the confusions which mar his actual account of it, if we say that a proposition is analytic when its validity depends solely on the definitions of the symbols it contains, and synthetic when its validity is determined by the facts of experience. Thus, the proposition "There are ants which have established a system of slavery" is a synthetic proposition. For we cannot tell whether it is true or false merely by considering the definitions of the symbols which constitute it. We have to resort to actual observation of the behaviour of ants. On the other hand, the proposition "Either some ants are parasitic or none are" is an analytic proposition. For one need not resort to observation to discover that there either are or are not ants which are parasitic. If one knows what is the function of the words "either," "or," and "not," then one can see that any proposition of the form "Either p is true or p is not true" is valid, independently of experience. Accordingly, all such propositions are analytic.

It is to be noticed that the proposition "Either some ants are parasitic or none are" provides no information whatsoever about the behavior of ants, or, indeed, about any matter of fact. And this applies to all analytic propositions. They none of them provide any information about any matter of fact. In other words, they are entirely devoid of factual content. And it is for this reason that no experience can confute them.

When we say that analytic propositions are devoid of factual content; and consequently that they say nothing, we are not suggesting that they are senseless in the way that metaphysical utterances are senseless. For, although they give us no information about any empirical situation, they do enlighten us by illustrating the way in which we use certain symbols. Thus if I say, "Nothing can be colored in different ways at the same time with respect to the same part of itself," I am not saying anything about the properties of any actual thing; but I am not talking nonsense. I am expressing an analytic proposition, which records our determination to

321

call a color expanse which differs in quality from a neighboring color expanse a different part of a given thing. In other words, I am simply calling attention to the implications of a certain linguistic usage. Similarly, in saying that if all Bretons are Frenchmen, and all Frenchmen Europeans, then all Bretons are Europeans, I am not describing any matter of fact. But I am showing that in the statement that all Bretons are Frenchmen, and all Frenchmen Europeans, the further statement that all Bretons are Europeans is implicitly contained. And I am thereby indicating the convention which governs our usage of the words "if" and "all."

We see, then, that there is a sense in which analytic propositions do give us new knowledge. They call attention to linguistic usages, of which we might otherwise not be conscious, and they reveal unsuspected implications in our assertions and beliefs. But we can see also that there is a sense in which they may be said to add nothing to our knowledge. For they tell us only what we may be said to know already. Thus, if I know that the existence of May Queens is a relic of tree-worship, and I discover that May Queens still exist in England, I can employ the tautology "If p implies q, and p is true, q is true" to show that there still exists a relic of tree-worship in England. But in saying that there are still May Queens in England, and that the existence of May Queens is a relic of tree-worship, I have already asserted the existence in England of a relic of tree-worship. The use of the tautology does, indeed, enable me to make this concealed assertion explicit. But it does not provide me with any new knowledge, in the sense in which empirical evidence that the election of May Queens had been forbidden by law would provide me with new knowledge. If one had to set forth all the information one possessed, with regard to matters of fact, one would not write down any analytic propositions. But one would make use of analytic propositions in compiling one's encyclopaedia, and would thus come to include propositions which one would otherwise have overlooked. And, besides enabling one to make one's list of information complete, the formulation of analytic propositions would enable one to make sure that the synthetic propositions of which the list was composed formed a self-consistent system. By showing which ways of combining propositions resulted in contradictions, they would prevent one from including incompatible propositions and so making the list self-stultifying. But insofar as we had actually used words as "all" and "or" and "not" without falling into self-contradiction, we might be said already to know what was revealed in the formulation of analytic propositions illustrating the rules which govern our usage of these logical particles. So that here again we are justified in saying that analytic propositions do not increase our knowledge.

322

The analytic character of the truths of formal logic was obscured in the traditional logic through its being insufficiently formalized. For in speaking always of judgments, instead of propositions, and introducing irrelevant psychological questions, the traditional logic gave the impression of being concerned in some specially intimate way with the workings of thought. What it was actually concerned with was the formal relationship of classes, as is shown by the fact that all its principles of inference are subsumed in the Boolean class-calculus, which is subsumed in its turn in the propositional calculus of Russell and Whitehead (cf. Menger 1933: 94–6; and Lewis and Langford 1932, chap. 5). Their system, expounded in *Principia Mathematica,* makes it clear that formal logic is not concerned with the properties of men's minds, much less with the properties of material objects, but simply with the possibility of combining propositions by means of logical particles into analytic propositions, and with studying the formal relationship of these analytic propositions, in virtue of which one is deducible from another. Their procedure is to exhibit the propositions of formal logic as a deductive system, based on five primitive propositions, subsequently reduced in number to one. Hereby the distinction between logical truths and principles of inference, which was maintained in the Aristotelian logic, very properly disappears. Every principle of inference is put forward as a logical truth and every logical truth can serve as a principle of inference. The three Aristotelian "laws of thought," the law of identity, the law of excluded middle, and the law of non-contradiction, are incorporated in the system, but they are not considered more important than the other analytic propositions. They are not reckoned among the premises of the system. And the system of Russell and Whitehead itself is probably only one among many possible logics, each of which is composed of tautologies as interesting to the logician as the arbitrarily selected Aristotelian "laws of thought" (cf. Lewis and Langford 1932, chap. 7).

A point which is not sufficiently brought out by Russell, if indeed it is recognized by him at all, is that every logical proposition is valid in its own right. Its validity does not depend on its being incorporated in a system, and deduced from certain propositions which are taken as self-evident. The construction of systems of logic is useful as a means of discovering and certifying analytic propositions, but it is not in principle essential even for this purpose. For it is possible to conceive of a symbolism in which every analytic proposition could be seen to be analytic in virtue of its form alone.

The fact that the validity of an analytic proposition in no way depends on its being deducible from other analytic propositions is our justification for disregarding the question whether the propositions of mathematics are reducible to propositions of formal logic, in the way that Russell supposed (1919, chap. 2) [pp. 167–73 in this volume]. For even if it is the case that the definition of a cardinal number as a class of classes similar to a given class is circular, and it is not possible to reduce mathematical notions to purely logical notions, it will still remain true that the propositions of mathematics are analytic propositions. They will form a special class of analytic propositions, containing special terms, but they will be none the less analytic for that. For the criterion of an analytic proposition is that its validity should follow simply from the definition of the terms contained in it, and this condition is fulfilled by the propositions of pure mathematics.

The mathematical propositions which one might most pardonably suppose to be synthetic are the propositions of geometry. For it is natural for us to think, as Kant thought, that geometry is the study of the properties of physical space, and consequently that its propositions have factual content. And if we believe this, and also recognize that the truths of geometry are necessary and certain, then we may be inclined to accept Kant's hypothesis that space is the form of intuition of our outer sense, a form imposed by us on the matter of sensation, as the only possible explanation of our *a priori* knowledge of these synthetic propositions. But while the view that pure geometry is concerned with physical space was plausible enough in Kant's day, when the geometry of Euclid was the only geometry known, the subsequent invention of non-Euclidean geometries has shown it to be mistaken. We see now that the axioms of a geometry are simply definitions, and that the theorems of a geometry are simply the logical consequences of these definitions. A geometry is not in itself about physical space; in itself it cannot be said to be "about" anything. But we can use a geometry to reason about physical space. That is to say, once we have given the axioms a physical interpretation, we can proceed to apply the theorems to the objects which satisfy the axioms (cf. Poincaré 1903: pt. 2, chap. 3). Whether a geometry can be applied to the actual physical world or not, is an empirical question which falls outside the scope of the geometry itself. There is no sense, therefore, in asking which of the various geometries known to us are false, and which are true. Insofar as they are all free from contradiction, they are all true. What one can ask is which of them is the most useful on any given occasion, which of them can be applied most easily and most fruitfully to an actual empirical situation. But the proposition which states that a certain application of a geometry is possible is not itself a proposition of that geometry. All that the geometry itself tells us is that if anything can be brought under the definitions, it will also satisfy the theorems. It is therefore a purely logical system, and its propositions are purely analytic propositions.

It might be objected that the use made of diagrams in geometrical treatises shows that geometrical reasoning is not purely abstract and logical, but depends on our intuition of the properties of figures. In fact, however, the use of diagrams is not essential to completely rigorous geometry. The diagrams are introduced as an aid to our reason. They provide us with a particular application of the geometry, and so assist us to perceive the more general truth that the axioms of the geometry involve certain consequences. But the fact that most of us need the help of an example to make us aware of those consequences does not show that the relation between them and the axioms is not a purely logical relation. It shows merely that our intellects are unequal to the task of carrying out very abstract processes of reasoning without the assistance of intuition. In other words, it has no bearing on the nature of geometrical propositions, but is simply an empirical fact about ourselves. Moreover, the appeal to intuition, though generally of psychological value, is also a source of danger to the geometer. He is tempted to make assumptions which are accidentally true of the particular figure he is taking as an illustration, but do not follow from his axioms. It has, indeed, been shown that Euclid himself was guilty of this, and consequently that the presence of the figure is essential to some of his proofs (cf. Black 1933: 154). This shows that his system is not, as he presents it, completely rigorous, although of course it can be made so. It does not show that the presence of the figure is essential to a truly rigorous geometrical proof. To suppose that it did would be to take as a necessary feature of all geometries what is really only an incidental defect in one particular geometrical system.

We conclude, then, that the propositions of pure geometry are analytic. And this leads us to reject Kant's hypothesis that geometry deals with the form of intuition of our outer sense. For the ground for this hypothesis was that it alone explained how the propositions of geometry could be both true *a priori* and synthetic: and we have seen that they are not synthetic. Similarly our view that the propositions of arithmetic are not synthetic but analytic leads us to reject the Kantian hypothesis[1] that arithmetic is concerned with our pure intuition of time, the form of our inner sense. And thus we are able to dismiss Kant's transcendental aesthetic without having to bring forward the epistemological difficulties which it is commonly said to involve. For the only argument which can

---

[1]This hypothesis is not mentioned in the *Critique of Pure Reason,* but was maintained by Kant at an earlier date.

be brought in favour of Kant's theory is that it alone explains certain "facts." And now we have found that the "facts" which it purports to explain are not facts at all. For while it is true that we have *a priori* knowledge of necessary propositions, it is not true, as Kant supposed, that any of these necessary propositions are synthetic. They are without exception analytic propositions, or, in other words, tautologies.

We have already explained how it is that these analytic propositions are necessary and certain. We saw that the reason why they cannot be confuted in experience is that they do not make any assertion about the empirical world. They simply record our determination to use words in a certain fashion. We cannot deny them without infringing the conventions which are presupposed by our very denial, and so falling into self-contradiction. And this is the sole ground of their necessity. As Wittgenstein puts it, our justification for holding that the world could not conceivably disobey the laws of logic is simply that we could not say of an unlogical world how it would look (1922: 3.01). And just as the validity of an analytic proposition is independent of the nature of the external world; so is it independent of the nature of our minds. It is perfectly conceivable that we should have employed different linguistic conventions from those which we actually do employ. But whatever these conventions might be, the tautologies in which we recorded them would always be necessary. For any denial of them would be self-stultifying.

We see, then, that there is nothing mysterious about the apodeictic certainty of logic and mathematics. Our knowledge that no observation can ever confute the proposition "$7 + 5 = 12$" depends simply on the fact that the symbolic expression "$7 + 5$" is synonymous with "12," just as our knowledge that every oculist is an eye-doctor depends on the fact that the symbol "eye-doctor" is synonymous with "oculist." And the same explanation holds good for every other *a priori* truth.

What is mysterious at first sight is that these tautologies should on occasion be so surprising, that there should be in mathematics and logic the possibility of invention and discovery. As Poincaré says: "If all the assertions which mathematics puts forward can be derived from one another by formal logic, mathematics cannot amount to anything more than an immense tautology. Logical inference can teach us nothing essentially new, and if everything is to proceed from the principle of identity, everything must be reducible to it. But can we really allow that these theorems which fill so many books serve no other purpose than to say in a round-about fashion 'A = A'?" (Poincaré 1903: pt. 1, chap. 1). Poincaré finds this incredible. His own theory is that the sense of invention and discovery in mathematics belongs to it in virtue of mathematical induction, the principle that what is true for the number 1, and true for

326

$n + 1$ when it is true for n,[2] is true for all numbers. And he claims that this is a synthetic *a priori* principle. It is, in fact, *a priori*, but it is not synthetic. It is a defining principle of the natural numbers, serving to distinguish them from such numbers as the infinite cardinal numbers, to which it cannot be applied (cf. Russell 1919: 27). Moreover, we must remember that discoveries can be made, not only in arithmetic, but also in geometry and formal logic, where no use is made of mathematical induction. So that even if Poincaré were right about mathematical induction, he would not have provided a satisfactory explanation of the paradox that a mere body of tautologies can be so interesting and so surprising.

The true explanation is very simple. The power of logic and mathematics to surprise us depends, like their usefulness, on the limitations of our reason. A being whose intellect was infinitely powerful would take no interest in logic and mathematics (cf. Hahn 1933: 18). For he would be able to see at a glance everything that his definitions implied, and, accordingly, could never learn anything from logical inference which he was not fully conscious of already. But our intellects are not of this order. It is only a minute proportion of the consequences of our definitions that we are able to detect at a glance. Even so simple a tautology as "$91 \times 79 = 7189$" is beyond the scope of our immediate apprehension. To assure ourselves that "7189" is synonymous with "91×79" we have to resort to calculation, which is simply a process of tautological transformation – that is, a process by which we change the form of expressions without altering their significance. The multiplication tables are rules for carrying out this process in arithmetic, just as the laws of logic are rules for the tautological transformation of sentences expressed in logical symbolism or in ordinary language. As the process of calculation is carried out more or less mechanically, it is easy for us to make a slip and so unwittingly contradict ourselves. And this accounts for the existence of logical and mathematical "falsehoods," which otherwise might appear paradoxical. Clearly the risk of error in logical reasoning is proportionate to the length and the complexity of the process of calculation. And in the same way, the more complex an analytic proposition is, the more chance it has of interesting and surprising us.

It is easy to see that the danger of error in logical reasoning can be minimized by the introduction of symbolic devices, which enable us to express highly complex tautologies in a conveniently simple form. And this gives us an opportunity for the exercise of invention in the pursuit of logical enquiries. For a well-chosen definition will call our attention to

---

[2] This was wrongly stated in previous editions as "true for n when it is true for $n + 1$."

analytic truths, which would otherwise have escaped us. And the framing of definitions which are useful and fruitful may well be regarded as a creative act.

Having thus shown that there is no inexplicable paradox involved in the view that the truths of logic and mathematics are all of them analytic, we may safely adopt it as the only satisfactory explanation of their *a priori* necessity. And in adopting it we vindicate the empiricist claim that there can be no *a priori* knowledge of reality. For we show that the truths of pure reason, the proportions which we know to be valid independently of all experience, are so only in virtue of their lack of factual content. To say that a proposition is true *a priori* is to say that it is a tautology. And tautologies, though they may serve to guide us in our empirical search for knowledge, do not in themselves contain any information about any matter of fact.

Truth by convention

W. V. QUINE

The less a science has advanced, the more its terminology tends to rest on an uncritical assumption of mutual understanding. With increase of rigor this basis is replaced piecemeal by the introduction of definitions. The interrelationships recruited for these definitions gain the status of analytic principles; what was once regarded as a theory about the world becomes reconstrued as a convention of language. Thus it is that some flow from the theoretical to the conventional is an adjunct of progress in the logical foundations of any science. The concept of simultaneity at a distance affords a stock example of such development: in supplanting the uncritical use of this phrase by a definition, Einstein so chose the definitive relationship as to verify conventionally the previously paradoxical principle of the absoluteness of the speed of light. But whereas the physical sciences are generally recognized as capable only of incomplete evolution in this direction, and as destined to retain always a nonconventional kernel of doctrine, developments of the past few decades have led to a widespread conviction that logic and mathematics are purely analytic or conventional. It is less the purpose of the present inquiry to question the validity of this contrast than to question its sense.

I

A definition, strictly, is a convention of notational abbreviation (cf. Russell 1903: 429). A *simple* definition introduces some specific expression, e.g., 'kilometer', or 'e', called the *definiendum*, as arbitrary shorthand for some complex expression, e.g., 'a thousand meters' or '$\lim_{n\to\infty}(1+1/n)^n$', called the *definiens*. A *contextual* definition sets up indefinitely many mutually analogous pairs of definienda and definientia according to some general scheme; an example is the definition whereby expressions of the form 'sin———/cos———' are abbreviated as 'tan———'. From a formal standpoint the signs thus introduced are

First published in Otis H. Lee, ed., *Philosophical Essays for A. N. Whitehead* (New York: Longmans, Green & Company, Inc., 1936). Reprinted with the kind permission of the author and David McKay Company. Some minor corrections have been made by the author, particularly on p. 339.

wholly arbitrary; all that is required of a definition is that it be theoretically immaterial, i.e., that the shorthand which it introduces admit in every case of unambiguous elimination in favor of the antecedent longhand.[1]

Functionally a definition is not a premiss to theory, but a license for rewriting theory by putting definiens for definiendum or vice versa. By allowing such replacements a definition transmits truth: it allows true statements to be translated into new statements which are true by the same token. Given the truth of the statement 'The altitude of Kibo exceeds six thousand meters', the definition of 'kilometer' makes for the truth of the statement 'The altitude of Kibo exceeds six kilometers'; given the truth of the statement 'sin π/cos π = sin π/cos π,' of which logic assures us in its earliest pages, the contextual definition cited above makes for the truth of the statement 'tan π = sin j/cos j.' In each case the statement inferred through the definition is true only because it is shorthand for another statement which was true independently of the definition. Considered in isolation from all doctrine, including logic, a definition is incapable of grounding the most trivial statement; even 'tan π = sin π/cos π' is a definitional transformation of an antecedent self-identity, rather than a spontaneous consequence of the definition.

What is loosely called a logical consequence of definitions is therefore more exactly describable as a logical truth definitionally abbreviated: a statement which becomes a truth of logic when definienda are replaced by definientia. In this sense 'tan π = sin π/cos π' is a logical consequence of the contextual definition of the tangent. 'The altitude of Kibo exceeds six kilometers' is not *ipso facto* a logical consequence of the given definition of 'kilometer'; on the other hand it would be a logical consequence of a quite suitable but unlikely definition introducing 'Kibo' as an abbreviation of the phrase 'the totality of such African terrain as exceeds six kilometers in altitude', for under this definition the statement in question is an abbreviation of a truth of logic, viz., 'The altitude of the totality of such African terrain as exceeds six kilometers in altitude exceeds six kilometers.'

Whatever may be agreed upon as the exact scope of logic, we may expect definitional abbreviations of logical truths to be reckoned as logical rather than extralogical truths. This being the case, the preceding

---

[1]From the present point of view a contextual definition may be recursive, but can then count among its definienda only those expressions in which the argument of recursion has a constant value, since otherwise the requirement of eliminability is violated. Such considerations are of little consequence, however, since any recursive definition can be turned into a direct one by purely logical methods. Cf. Carnap 1934b: 23, 79.

conclusion shows logical consequences of definitions to be themselves truths of logic. To claim that mathematical truths are conventional in the sense of following logically from definitions is therefore to claim that mathematics is part of logic. The latter claim does not represent an arbitrary extension of the term 'logic' to include mathematics; agreement as to what belongs to logic and what belongs to mathematics is supposed at the outset, and it is then claimed that definitions of mathematical expressions can so be framed on the basis of logical ones that all mathematical truths become abbreviations of logical ones.

Although signs introduced by definition are formally arbitrary, more than such arbitrary notational convention is involved in questions of definability; otherwise any expression might be said to be definable on the basis of any expressions whatever. When we speak of definability, or of finding a definition for a given sign, we have in mind some traditional usage of the sign antecedent to the definition in question. To be satisfactory in this sense a definition of the sign not only must fulfill the formal requirement of unambiguous eliminability, but must also conform to the traditional usage in question. For such conformity it is necessary and sufficient that every context of the sign which was true and every context which was false under traditional usage be construed by the definition as an abbreviation of some other statement which is correspondingly true or false under the established meanings of its signs. Thus when definitions of mathematical expressions on the basis of logical ones are said to have been framed, what is meant is that definitions have been set up whereby every statement which so involves those mathematical expressions as to be recognized traditionally as true, or as false, is construed as an abbreviation of another correspondingly true or false statement which lacks those mathematical expressions and exhibits only logical expressions in their stead.[2]

An expression will be said to occur *vacuously* in a given statement if its replacement therein by any and every other grammatically admissible expression leaves the truth or falsehood of the statement unchanged. Thus for any statement containing some expressions vacuously there is a class of statements, describable as *vacuous variants* of the given statement, which are like it in point of truth or falsehood, like it also in point of a certain skeleton of symbolic make-up, but diverse in exhibiting all grammatically possible variations upon the vacuous constituents of the

---

[2]Note than an expression is said to be defined, in terms, e.g., of logic, not only when it is a single sign whose elimination from a context in favor of logical expressions is accomplished by a single application of one definition, but also when it is a complex expression whose elimination calls for successive application of many definitions.

given statement. An expression will be said to occur *essentially* in a statement if it occurs in all the vacuous variants of the statement, i.e., if it forms part of the aforementioned skeleton. (Note that though an expression occurs non-vacuously in a statement it may fail of essential occurrence because some of its parts occur vacuously in the statement.)

Now let S be a truth, let the expressions E_i occur vacuously in S, and let the statements S_i be the vacuous variants of S. Thus the S_i will likewise be true. On the sole basis of the expressions belonging to a certain class α, let us frame a definition for one of the expressions F occurring in S outside the E_i. S and the S_i thereby become abbreviations of certain statements S' and S_i' which exhibit only members of α instead of those occurrences of F, but which remain so related that the S_i' are all the results of replacing the E_i in S_i' by any other grammatically admissible expressions. Now since our definition of F is supposed to conform to usage, S' and the S_i' will, like S and the S_i, be uniformly true; hence the S_i' will be vacuous variants of S', and the occurrences of the E_i in S' will be vacuous. The definition thus makes S an abbreviation of a truth S' which, like S, involves the E_i vacuously, but which differs from S in exhibiting only members of α instead of the occurrences of F outside the E_i. Now it is obvious that an expression cannot occur essentially in a statement if it occurs only within expressions which occur vacuously in the statement; consequently F, occurring in S' as it does only within the E_i if at all, does not occur essentially in S'; members of α occur essentially in its stead. Thus if we take F as any non-member of α occurring essentially in S, and repeat the above reasoning for each such expression, we see that, through definitions of all such expressions in terms of members of α, S becomes an abbreviation of a truth S'' involving only members of α essentially.

Thus if in particular we take α as the class of all logical expressions, the above tells us that if logical definitions be framed for all non-logical expressions occurring essentially in the true statement S, S becomes an abbreviation of a truth S'' involving only logical expressions essentially. But if S'' involves only logical expressions essentially, and hence remains true when everything except that skeleton of logical expressions is changed in all grammatically possible ways, then S'' depends for its truth upon those logical constituents alone, and is thus a truth of logic. It is therefore established that if all non-logical expressions occurring essentially in a true statement S be given definitions on the basis solely of logic, then S becomes an abbreviation of a truth S'' of logic. In particular, then, if all mathematical expressions be defined in terms of logic, all truths involving only mathematical and logical expressions essentially become definitional abbreviations of truths of logic.

Now a mathematical truth, e.g., 'Smith's age plus Brown's equals Brown's age plus Smith's,' may contain non-logical, non-mathematical expressions. Still any such mathematical truth, or another whereof it is a definitional abbreviation, will consist of a skeleton of mathematical or logical expressions filled in with non-logical, non-mathematical expressions all of which occur vacuously. Thus every mathematical truth either is a truth in which only mathematical and logical expressions occur essentially, or is a definitional abbreviation of such a truth. Hence, granted definitions of all mathematical expressions in terms of logic, the preceding conclusion shows that all mathematical truths become definitional abbreviations of truths of logic – therefore truths of logic in turn. For the thesis that mathematics is logic it is thus sufficient that all mathematical notation be defined on the basis of logical notation.

If on the other hand some mathematical expressions resist definition on the basis of logical ones, then every mathematical truth containing such recalcitrant expressions must contain them only inessentially, or be a definitional abbreviation of a truth containing such expressions only inessentially, if all mathematics is to be logic: for though a logical truth, e.g., the above one about Africa, may involve non-logical expressions, it or some other logical truth whereof it is an abbreviation must involve only logical expressions essentially. It is of this alternative that those avail themselves who regard mathematical truths, insofar as they depend upon non-logical notions, as elliptical for hypothetical statements containing as tacit hypotheses all the postulates of the branch of mathematics in question (e.g., Russell 1903: 420–30; Behmann 1934: 8–10). Thus, suppose the geometrical terms 'sphere' and 'includes' to be undefined on the basis of logical expressions, and suppose all further geometrical expressions defined on the basis of logical expressions together with 'sphere' and 'includes', as with Huntington (1913: 522–59). Let Huntington's postulates for (Euclidean) geometry, and all the theorems, be expanded by thoroughgoing replacement of definienda by definientia, so that they come to contain only logical expressions and 'sphere' and 'includes', and let the conjunction of the thus expanded postulates be represented as 'Hunt (sphere, includes).' Then, where 'Φ (sphere, includes)' is any of the theorems, similarly expanded into primitive terms, the point of view under consideration is that 'Φ (sphere, includes),' insofar as it is conceived as a mathematical truth, is to be construed as an ellipsis for 'If Hunt (spheres, includes) then Φ (sphere, includes).' Since 'Φ (sphere, includes)' is a logical consequence of Huntington's postulates, the above hypothetical statement is a truth of logic; it involves the expressions 'sphere' and 'includes' inessentially, in fact vacuously, since the logical deducibility of the theorems from the

333

postulates is independent of the meanings of 'sphere' and 'includes' and survives the replacement of those expressions by any other grammatically admissible expressions whatever. Since, granted the fitness of Huntington's postulates, all and only those geometrical statements are truths of geometry which are logical consequences in this fashion of 'Hunt (sphere, includes),' all geometry becomes logic when interpreted in the above manner as a conventional ellipsis for a body of hypothetical statements.

But if, as a truth of mathematics, 'Φ (sphere, includes)' is short for 'If Hunt (sphere, includes) then Φ (sphere, includes),' still there remains, as part of this expanded statement, the original statement 'Φ (sphere, includes)'; this remains as a presumably true statement within some body of doctrine, say for the moment "non-mathematical geometry," even if the title of mathematical truth be restricted to the entire hypothetical statement in question. The body of all such hypothetical statements, describable as the "theory of deduction of non-mathematical geometry," is of course a part of logic; but the same is true of any "theory of deduction of sociology," "theory of deduction of Greek mythology," etc., which we might construct in parallel fashion with the aid of any set of postulates suited to sociology or to Greek mythology. The point of view toward geometry which is under consideration thus reduces merely to an exclusion of geometry from mathematics, a relegation of geometry to the status of sociology or Greek mythology; the labelling of the "theory of deduction of non-mathematical geometry" as "mathematical geometry" is a verbal *tour de force* which is equally applicable in the case of sociology or Greek mythology. To incorporate mathematics into logic by regarding all recalcitrant mathematical truths as elliptical hypothetical statements is thus in effect merely to restrict the term 'mathematics' to exclude those recalcitrant branches. But we are not interested in renaming. Those disciplines, geometry and the rest, which have traditionally been grouped under mathematics are the objects of the present discussion, and it is with the doctrine that mathematics in this sense is logic that we are here concerned.[3]

Discarding this alternative and returning, then, we see that if some mathematical expressions resist definition on the basis of logical ones, mathematics will reduce to logic only if, under a literal reading and without the gratuitous annexation of hypotheses, every mathematical truth contains (or is an abbreviation of one which contains) such recalcitrant expressions only inessentially if at all. But a mathematical expression

[3]Obviously the foregoing discussion has no bearing upon postulate method as such, nor upon Huntington's work.

sufficiently troublesome to have resisted trivial contextual definition in terms of logic can hardly be expected to occur thus idly in all its mathematical contexts. It would thus appear that for the tenability of the thesis that mathematics is logic it is not only sufficient but also necessary that all mathematical expressions be capable of definition on the basis solely of logical ones.

Though in framing logical definitions of mathematical expressions the ultimate objective be to make all mathematical truths logical truths, attention is not to be confined to mathematical and logical truths in testing the conformity of the definitions to usage. Mathematical expressions belong to the general language, and they are to be so defined that all statements containing them, whether mathematical truths, historical truths, or falsehoods under traditional usage, come to be construed as abbreviations of other statements which are correspondingly true or false. The definition introducing 'plus' must be such that the mathematical truth 'Smith's age plus Brown's equals Brown's age plus Smith's' becomes an abbreviation of a logical truth, as observed earlier; but it must also be such that 'Smith's age plus Brown's age equals Jones' age' becomes an abbreviation of a statement which is empirically true or false in conformity with the county records and the traditional usage of 'plus'. A definition which fails in this latter respect is no less Pickwickian than one which fails in the former; in either case nothing is achieved beyond the transient pleasure of a verbal recreation.

But for these considerations, contextual definitions of any mathematical expressions whatever could be framed immediately in purely logical terms, on the basis of any set of postulates adequate to the branch of mathematics in question. Thus, consider again Huntington's systematization of geometry. It was remarked that, granted the fitness of Huntington's postulates, a statement will be a truth of geometry if and only if it is logically deducible from 'Hunt (sphere, includes)' without regard to the meanings of 'sphere' and 'includes'. Thus 'Φ (sphere, includes)' will be a truth of geometry if and only if the following is a truth of logic: 'If α is any class and R any relation such that Hunt (α, R), then Φ (α, R).' For 'sphere' and 'includes' we might then adopt the following contextual definition: Where '———' is any statement containing 'α' or 'R' or both, let the statement 'If α is any class and R any relation such that Hunt (α, R), then ———' be abbreviated as that expression which is got from '———' by putting 'sphere' for 'α' and 'includes' for 'R' throughout. (In the case of a compound statement involving 'sphere' and 'includes', this definition does not specify whether it is the entire statement or each of its constituent statements that is to be

335

accounted as shorthand in the described fashion; but this ambiguity can be eliminated by stipulating that the convention apply only to whole contexts.) 'Sphere' and 'includes' thus receive contextual definition in terms exclusively of logic, for any statement containing one or both of those expressions is construed by the definition as an abbreviation of a statement containing only logical expressions (plus whatever expressions the original statement may have contained other than 'sphere' and 'includes'). The definition satisfies past usage of 'sphere' and 'includes' to the extent of verifying all truths and falsifying all falsehoods of geometry; all those statements of geometry which are true, and only those, become abbreviations of truths of logic.

The same procedure could be followed in any other branch of mathematics, with the help of a satisfactory set of postulates for the branch. Thus nothing further would appear to be wanting for the thesis that mathematics is logic. And the royal road runs beyond that thesis, for the described method of logicizing a mathematical discipline can be applied likewise to any nonmathematical theory. But the whole procedure rests on failure to conform the definitions to usage; what is logicized is not the intended subject-matter. It is readily seen, e.g., that the suggested contextual definition of 'sphere' and 'includes', though transforming purely geometrical truths and falsehoods respectively into logical truths and falsehoods, transforms certain empirical truths into falsehoods and vice versa. Consider, e.g., the true statement 'A baseball is roughly a sphere,' more rigorously 'The whole of a baseball, except for a certain very thin, irregular peripheral layer, constitutes a sphere.' According to the contextual definition, this statement is an abbreviation for the following: 'If α is any class and R any relation such that Hunt (α, R), then the whole of a baseball, except for a thin peripheral layer, constitutes an [a member of] α.' This tells us that the whole of a baseball, except for a thin peripheral layer, belongs to every class α for which a relation R can be found such that Huntington's postulates are true of α and R. Now it happens that 'Hunt $(\alpha, \text{includes})$' is true not only when α is taken as the class of all spheres, but also when α is restricted to the class of spheres a foot or more in diameter (cf. Huntington 1913: 540); yet the whole of a baseball, except for a thin peripheral layer, can hardly be said to constitute a sphere a foot or more in diameter. The statement is therefore false, whereas the preceding statement, supposedly an abbreviation of this one, was true under ordinary usage of words. The thus logicized rendering of any other discipline can be shown in analogous fashion to yield the sort of discrepancy observed just now for geometry, provided only that the postulates of the discipline admit, like those of geometry, of alternative applications; and such multiple applicability is to be expected of any postulate set.[4]

Definition of mathematical notions on the basis of logical ones is thus a more arduous undertaking than would appear from a consideration solely of the truths and falsehoods of pure mathematics. Viewed *in vacuo,* mathematics is trivially reducible to logic through erection of postulate systems into contextual definitions; but "cette science n'a pas uniquement pour objet de contempler éternellement son propre nombril" (Poincaré 1908b: 199). When mathematics is recognized as capable of use, and as forming an integral part of general language, the definition of mathematical notions in terms of logic becomes a task whose completion, if theoretically possible at all, calls for mathematical genius of a high order. It was primarily to this task that Whitehead and Russell addressed themselves in their *Principia Mathematica.* They adopt a meager logical language as primitive, and on its basis alone they undertake to endow mathematical expressions with definitions which conform to usage in the full sense described above: definitions which not only reduce mathematical truths and falsehoods to logical ones, but reduce *all* statements, containing the mathematical expressions in question, to equivalent statements involving logical expressions instead of the mathematical ones. Within *Principia* the program has been advanced to such a point as to suggest that no fundamental difficulties stand in the way of completing the process. The foundations of arithmetic are developed in *Principia,* and therewith those branches of mathematics are accommodated which, like analysis and theory of number, spring from arithmetic. Abstract algebra proceeds readily from the relation theory of *Principia.* Only geometry remains untouched, and this field can be brought into line simply by identifying *n*-dimensional figures with those *n*-adic arithmetical relations ("equations in *n* variables") with which they are correlated through analytic geometry (cf. Study 1914: 86–92). Some question Whitehead and Russell's reduction of mathematics to logic (cf., e.g., Dubislav 1925: 193–208; Hilbert 1928: 12, 21), on grounds for whose exposition and criticism there is not space; the thesis that all mathematics reduces to logic is, however, substantiated by *Principia* to a degree satisfactory to most of us. There is no need here to adopt a final stand in the matter.

If for the moment we grant that all mathematics is thus definitionally constructible from logic, then mathematics becomes true by convention

---

[4]Note that a postulate set is superfluous if it *demonstrably* admits of one and only one application: for it then embodies an adequate defining property for each of its constituent primitive terms. Cf. Tarski 1935a: 85 (proposition 2).

in a relative sense: mathematical truths become conventional transcriptions of logical truths. Perhaps this is all that many of us mean to assert when we assert that mathematics is true by convention; at least, an *analytic* statement is commonly explained merely as one which proceeds from logic and definitions, or as one which, on replacement of definienda by definientia, becomes a truth of logic.[5] But in strictness we cannot regard mathematics as true purely by convention unless all those logical principles to which mathematics is supposed to reduce are likewise true by convention. And the doctrine that mathematics is *analytic* accomplishes a less fundamental simplification for philosophy than would at first appear, if it asserts only that mathematics is a conventional transcription of logic and not that logic is convention in turn: for if in the end we are to countenance any *a priori* principles at all which are independent of convention, we should not scruple to admit a few more, nor attribute crucial importance to conventions which serve only to diminish the number of such principles by reducing some to others.

But if we are to construe logic also as true by convention, we must rest logic ultimately upon some manner of convention other than definition: for it was noted earlier that definitions are available only for transforming truths, not for founding them. The same applies to any truths of mathematics which, contrary to the supposition of a moment ago, may resist definitional reduction to logic; if such truths are to proceed from convention, without merely being reduced to antecedent truths, they must proceed from conventions other than definitions. Such a second sort of convention, generating truths rather than merely transforming them, has long been recognized in the use of postulates.[6] Application of this method to logic will occupy the next section; customary ways of rendering postulates and rules of inference will be departed from, however, in favor of giving the whole scheme the explicit form of linguistic convention.

II

Let us suppose an approximate maximum of definition to have been accomplished for logic, so that we are left with about as meager as possible an array of primitive notational devices. There are indefinitely many ways of framing the definitions, all conforming to the same usage of the expressions in question; apart from the objective of defining much

[5] Cf. Frege 1884: 4; Behmann 1934: 5. Carnap, 1934b, uses the term in essentially the same sense but subject to more subtle and rigorous treatment.

[6] The function of postulates as conventions seems to have been first recognized by Gergonne (1819). His designation of them as 'implicit definitions,'' which has had some following in the literature, is avoided here.

in terms of little, choice among these ways is guided by convenience or chance. Different choices involve different sets of primitives. Let us suppose our procedure to be such as to reckon among the primitive devices the *not*-idiom, the *if*-idiom ('If...then...'), the *every*-idiom ('No matter what x may be, ————x————'), and one or two more as required. On the basis of this much, then, all further logical notation is to be supposed defined; all statements involving any further logical notation become construed as abbreviations of statements whose logical constituents are limited to those primitives.

'Or', as a connective joining statements to form new statements, is amenable to the following contextual definition in terms of the *not*-idiom and the *if*-idiom: A pair of statements with 'or' between is an abbreviation of the statement made up successively of these ingredients: first, 'If'; second, the first statement of the pair, with 'not' inserted to govern the main verb (or, with 'it is false that' prefixed); third, 'then'; fourth, the second statement of the pair. The convention becomes clearer if we use the prefix ' ~ ' as an artificial notation for denial, thus writing ' ~ ice is hot' instead of 'Ice is not hot' or 'It is false that ice is hot.' Where '————' and '———' are any statements, our definition then introduces '———— or ———' as an abbreviation of 'If ~———— then ———.' Again 'and', as a connective joining statements, can be defined contextually by construing '———— and ———' as an abbreviation for ' ~ if ———— then ~———.' Every such idiom is what is known as a *truth-function*, and is characterized by the fact that the truth or falsehood of the complex statement which it generates is uniquely determined by the truth or falsehood of the several statements which it combines. All truth-functions are known to be constructible in terms of the *not*- and *if*-idioms as in the above examples.[7] On the basis of the truth-functions, then, together with our further primitives – the *every*-idiom *et al.* – all further logical devices are supposed defined.

A word may, through historical or other accidents, evoke a train of ideas bearing no relevance to the truth or falsehood of its context; in point of *meaning*, however, as distinct from connotation, a word may be said to be determined to whatever extent the truth or falsehood of its contexts is determined. Such determination of truth or falsehood may be outright, and to that extent the meaning of the word is absolutely determined; or it may be relative to the truth or falsehood of statements containing other words, and to that extent the meaning of the word is determined relatively to those other words. A definition endows a word with

[7]Sheffer (1913: 481–8) has shown ways of constructing these two, in turn, in terms of one; strictly, therefore, such a one should supplant the two in our ostensibly minimal set of logical primitives. Exposition will be facilitated, however, by retaining the redundancy.

complete determinacy of meaning relative to other words. But the alternative is open to us, on introducing a new word, of determining its meaning *absolutely* to whatever extent we like by specifying contexts which are to be true and contexts which are to be false. In fact, we need specify only the former: for falsehood may be regarded as a derivative property depending on the word ' ~ ', in such wise that falsehood of '———' means simply truth of ' ~ ———.' Since all contexts of our new word are meaningless to begin with, neither true nor false, we are free to run through the list of such contexts and pick out as true such ones as we like; those selected become true by fiat, by linguistic convention. For those who would question them we have always the same answer, 'You use the word differently.' The reader may protest that our arbitrary selection of contexts as true is subject to restrictions imposed by the requirement of *consistency* – e.g., that we must not select both '———' and ' ~ ———'; but this consideration, which will receive a clearer status a few pages hence, will be passed over for the moment.

Now suppose in particular that we abstract from existing usage of the locutions 'if-then', 'not' (or ' ~ '), and the rest of our logical primitives, so that for the time being these become meaningless marks, and the erstwhile statements containing them lose their status as statements and become likewise meaningless, neither true nor false; and suppose we run through all those erstwhile statements, or as many of them as we like, segregating various of them arbitrarily as true. To whatever extent we carry this process, we to that extent determine meaning for the initially meaningless marks 'if', 'then', ' ~ ', and the rest. Such contexts as we render true are true by convention.

We saw earlier that if all expressions occurring essentially in a true statement S and not belonging to a class α are given definitions in terms solely of members of α, then S becomes a definitional abbreviation of a truth S'' involving only members of a α essentially. Now let α comprise just our logical primitives, and let S be a statement which, under ordinary usage, is true and involves only logical expressions essentially. Since all logical expressions other than the primitives are defined in terms of the primitives, it then follows that S is an abbreviation of a truth S'' involving only the primitives essentially. But if one statement S is a definitional abbreviation of another S'', the truth of S proceeds wholly from linguistic convention if the truth of S'' does so. Hence if, in the above process of arbitrarily segregating statements as true by way of endowing our logical primitives with meaning, *we assign truth to those statements which, according to ordinary usage, are true and involve only our primitives essentially,* then not only will the latter statements be true by convention, but so will all statements which are true under ordinary usage and involve only logical expressions essentially. Since, as remarked earlier, every logical truth involves (or is an abbreviation of another which involves) only logical expressions essentially, the described scheme of assigning truth makes all logic true by convention.

Not only does such assignment of truth suffice to make all those statements true by convention which are true under ordinary usage and involve only logical expressions essentially, but it serves also to make all those statements false by convention which are false under ordinary usage and involve only logical expressions essentially. This follows from our explanation of the falsehood of '———' as the truth of '~ ———', since '———' will be false under ordinary usage if and only if '~ ———' is true under ordinary usage. The described assignment of truth thus goes far towards fixing all logical expressions in point of meaning, and fixing them in conformity with usage. Still many statements containing logical expressions remain unaffected by the described assignments: all those statements which, from the standpoint of ordinary usage, involve some non-logical expressions essentially. There is hence room for supplementary conventions of one sort or another over and above the described truth-assignments, by way of completely fixing the meanings of our primitives – and fixing them, it is to be hoped, in conformity with ordinary usage. Such supplementation need not concern us now; the described truth-assignments provide partial determinations which, as far as they go, conform to usage and which go far enough to make all logic true by convention.

But we must not be deceived by schematism. It would appear that we sit down to a list of expressions and check off as arbitrarily true all those which, under ordinary usage, are true statements involving only our logical primitives essentially; but this picture wanes when we reflect that the number of such statements is infinite. If the convention whereby those statements are singled out as true is to be formulated in finite terms, we must avail ourselves of conditions finite in length which determine infinite classes of expressions.[8]

Such conditions are ready at hand. One, determining an infinite class of expressions all of which, under ordinary usage, are true statements involving only our primitive *if*-idiom essentially, is the condition of being obtainable from

[8]Such a condition, if effective, constitutes a *formal system*. Usually we assign such meanings to the signs as to construe the expressions of the class as statements, specifically true statements, theorems; but this is neither intrinsic to the system nor necessary in all cases for a useful application of the system.

(1)   'If if p then q then if if q then r then if p then r'

by putting a statement for 'p', a statement for 'q', and a statement for 'r'. In more customary language the form (1) would be expanded, for clarity, in some such fashion as this: 'If it is the case that if p then q, then, if it is the case further that if q then r, then, if p, r.' The form (1) is thus seen to be the principle of the syllogism. Obviously it is true under ordinary usage for all substitutions of statements for 'p', 'q', and 'r'; hence such results of substitution are, under ordinary usage, true statements involving only the *if*-idiom essentially. One infinite part of our program of assigning truth to all expressions which, under ordinary usage, are true statements involving only our logical primitives essentially, is thus accomplished by the following convention:

(I)   Let all results of putting a statement for 'p', a statement for 'q', and a statement for 'r' in (1) be true.

Another infinite part of the program is disposed of by adding this convention:

(II)   Let any expression be true which yields a truth when put for 'q' in the result of putting a truth for 'p' in 'If p then q.'

Given truths '———' and 'If ——— then ———', (II) yields the truth of '———'. That (II) conforms to usage, i.e., that from statements which are true under ordinary usage (II) leads only to statements which are likewise true under ordinary usage, is seen from the fact that under ordinary usage a statement '———' is always true if statements '———' and 'If ——— then ———' are true. Given all the truths yielded by (I), (II) yields another infinity of truths which, like the former, are under ordinary usage truths involving only the *if*-idiom essentially. How this comes about is seen roughly as follows. The truths yielded by (I), being of the form of (1), are complex statements of the form 'If ——— then ———'. The statement '———' here may in particular be of the form (1) in turn, and hence likewise be true according to (I). Then, by (II), '———' becomes true. In general '———' will not be of the form (1), hence would not have been obtainable by (I) alone. Still '———' will in every such case be a statement which, under ordinary usage, is true and involves only the *if*-idiom essentially; this follows from the observed conformity of (I) and (II) to usage, together with the fact that the above derivation of '———' demands nothing of '———' beyond proper structure in terms of 'if-then'. Now our stock of truths embraces not only those yielded by (I) alone, i.e., those having the form (1), but also all those thence derivable by (II) in the manner in which '——' has just now been supposed derived.[9] From this increased stock we can derive yet further ones by (II), and these likewise will, under ordinary usage, be true and involve only the *if*-idiom essentially. The generation proceeds in this fashion *ad infinitum*.

When provided only with (I) as an auxiliary source of truth, (II) thus yields only truths which under ordinary usage are truths involving only the *if*-idiom essentially. When provided with further auxiliary sources of truths, however, e.g., the convention (III) which is to follow, (II) yields truths involving further locutions essentially. Indeed, the effect of (II) is not even confined to statements which, under ordinary usage, involve only logical locutions essentially; (II) also legislates regarding other statements, to the extent of specifying that no two statements '———' and 'If ——— then ——' can both be true unless '——' is true. But this overflow need not disturb us, since it also conforms to ordinary usage. In fact, it was remarked earlier that room remained for supplementary conventions, over and above the described truth-assignments, by way of further determining the meanings of our primitives. This overflow accomplishes just that for the *if*-idiom; it provides, with regard even to a statement 'If ——— then ——' which from the standpoint of ordinary usage involves non-logical expressions essentially, that the statement is not to be true if '———' is true and '——' not.

But present concern is with statements which, under ordinary usage, involve only our logical primitives essentially; by (I) and (II) we have provided for the truth of an infinite number of such statements, but by no means all. The following convention provides for the truth of another infinite set of such statements; these, in contrast to the preceding, involve not only the *if*-idiom but also the *not*-idiom essentially (under ordinary usage).

(III)   Let all results of putting a statement for 'p' and a statement for 'q', in 'If p then if $\sim p$ then q' or 'If if $\sim p$ then p then p,' be true.[10]

Statements generated thus by substitution in 'If p then if $\sim p$ then q' are statements of hypothetical form in which two mutually contradictory statements occur as premises; obviously such statements are trivially true, under ordinary usage, no matter what may figure as conclusion. Statements generated by substitution in 'If [it is the case that] if $\sim p$ then p, then p' are likewise true under ordinary usage, for one reasons as

[9]The latter in fact comprise all and only those statements which have the form 'If if if if q then r then if p then r then s then if if p then q then s'.

[10](1) and the two formulae in (III) are Łukasiewicz's three postulates for the propositional calculus.

follows: Grant the hypothesis, viz., that if $\sim p$ then p; then we must admit the conclusion, viz., that p, since even denying it we admit it. Thus all the results of substitution referred to in (III) are true under ordinary usage no matter what the substituted statements may be; hence such results of substitution are, under ordinary usage, true statements, involving nothing essentially beyond the *if*-idiom and the *not*-idiom ('$\sim$').

From the infinity of truths adopted in (III), together with those already at hand from (I) and (II), infinitely more truths are generated by (II). It happens, curiously enough, that (III) adds even to our stock of statements which involve only the *if*-idiom essentially (under ordinary usage); there are truths of that description which, though lacking the *not*-idiom, are reached by (I)–(III) and not by (I) and (II). This is true, e.g., of any instance of the principle of identity, say

(2) 'If time is money then time is money'.

It will be instructive to derive (2) from (I)–(III), as an illustration of the general manner in which truths are generated by those conventions. (III), to begin with, directs that we adopt these statements as true:

(3) 'If time is money then if time is not money then time is money'.
(4) 'If if time is not money then time is money then time is money'.

(I) directs that we adopt this as true:

(5) 'If if time is money then if time is not money then time is money then if if if time is not money then time is money then time is money then if time is money then time is money'.

(II) tells us that, in view of the truth of (5) and (3), this is true:

(6) 'If if if time is not money then time is money then time is money then if time is money then time is money'.

Finally (II) tells us that, in view of the truth of (6) and (4), (2) is true.

If a statement S is generated by (I)–(III), obviously only the structure of S in terms of 'if-then' and '$\sim$' was relevant to the generation; hence all those variants S_i of S which are obtainable by any grammatically admissible substitutions upon constituents of S not containing 'if', 'then', or '$\sim$', are likewise generated by (I)–(III). Now it has been observed that (I)–(III) conform to usage, i.e., generate only statements which are true under ordinary usage; hence S and all the S_i are uniformly true under ordinary usage, the S_i are therefore vacuous variants of S, and hence only 'if', 'then', and '$\sim$' occur essentially in S. Thus (I)–(III) generate only statements which under ordinary usage are truths involving only the *if*-idiom and the *not*-idiom essentially.

It can be shown also that (I)–(III) generate *all* such statements.[11] Consequently (I)–(III), aided by our definitions of logical locutions in terms of our primitives, are adequate to the generation of all statements which under ordinary usage are truths which involve any of the so-called truth-functions but nothing else essentially: for it has been remarked that all the truth-functions are definable on the basis of the *if*-idiom and the *not*-idiom. All such truths thus become true by convention. They comprise all those statements which are instances of any of the principles of the so-called propositional calculus.

To (I)–(III) we may now add a further convention or two to cover another of our logical primitives – say the *every*-idiom. A little more in this direction, by way of providing for our remaining primitives, and the program is completed; all logic, in some sense, becomes true by convention. The conventions with which (I)–(III) would thus be supplemented would be more complex than (I)–(III). The set of conventions would be an adaptation of one of the various existing systematizations of general logistic, in the same way in which (I)–(III) are an adaptation of a systematization of the propositional calculus.

The systematization chosen must indeed leave some logical statements undecided, by Gödel's theorem, if we set generous bounds to the logical vocabulary. But no matter; logic still becomes true by convention insofar as it gets reckoned as true on any account.

Let us now consider the protest which the reader raised earlier, viz., that our freedom in assigning truth by convention is subject to restrictions imposed by the requirement of consistency (see, e.g., Poincaré 1908b: 162–3, 195–8; Schlick 1918, 2nd ed.: 36, 327). Under the fiction, implicit in an earlier stage of our discussion, that we check off our truths one by one in an exhaustive list of expressions, consistency in the assign-

---

[11]The proof rests essentially upon Łukasiewicz's proof (1929) that his three postulates for the propositional calculus, viz., (1) and the formulae in (III), are *complete*. Adaptation of his result to present purposes depends upon the fact, readily established, that any formula generable by his two rules of inference (the so-called rule of substitution and a rule answering to (II)) can be generated by applying the rules in such order that all applications of the rule of substitution precede all applications of the other rule. This fact is relevant because of the manner in which the rule of substitution has been absorbed, here, into (I) and (III). The adaptation involves also two further steps, which however present no difficulty: we must make connection between Łukasiewicz's *formulae,* containing variables '*p*', '*q*', etc., and the concrete *statements* which constitute the present subject-matter; also between *completeness*, in the sense (Post's) in which Łukasiewicz uses the term, and the generability of all statements which under ordinary usage are truths involving only the *if*-idiom or the *not*-idiom essentially.

ment of truth is nothing more than a special case of conformity to usage. If we make a mark in the margin opposite an expression '———', and another opposite ' ~ ———', we sin only against the established usage of ' ~ ' as a denial sign. Under the latter usage '———' and ' ~ ———' are not both true; in taking them both by convention as true we merely endow the sign ' ~ ', roughly speaking, with a meaning other than denial. Indeed, we might so conduct our assignments of truth as to allow no sign of our language to behave analogously to the denial locution of ordinary usage; perhaps the resulting language would be inconvenient, but conventions are often inconvenient. It is only the objective of ending up with our mother tongue that dissuades us from marking both '———' and ' ~ ———', and this objective would dissuade us also from marking 'It is always cold on Thursday.'

The requirement of consistency still retains the above status when we assign truth wholesale through general conventions such as (I)–(III). Each such convention assigns truth to an infinite sheaf of the entries in our fictive list, and in this function the conventions cannot conflict; by overlapping in their effects they reinforce one another, by not overlapping they remain indifferent to one another. If some of the conventions specified entries to which truth was *not* to be assigned, genuine conflict might be apprehended; such negative conventions, however, have not been suggested. (II) was, indeed, described earlier as specifying that 'If ——— then ——' is not to be true if '———' is true and '——' not; but within the framework of the conventions of truth-assignment this apparent proscription is ineffectual without antecedent proscription of '——'. Thus any inconsistency among the general conventions will be of the sort previously considered, viz., the arbitrary adoption of both '———' and ' ~ ———' as true; and the adoption of these was seen merely to impose some meaning other than denial upon the sign ' ~ '. As theoretical restrictions upon our freedom in the conventional assignment of truth, requirements of consistency thus disappear. Preconceived usage may lead us to stack the cards, but does not enter the rules of the game.

III

Circumscription of our logical primitives in point of meaning, through conventional assignment of truth to various of their contexts, has been seen to render all logic true by convention. Then if we grant the thesis that mathematics is logic, i.e., that all mathematical truths are definitional abbreviations of logical truths, it follows that mathematics is true by convention.

346

If on the other hand, contrary to the thesis that mathematics is logic, some mathematical expressions resist definition in terms of logical ones, we can extend the foregoing method into the domain of these recalcitrant expressions: we can circumscribe the latter through conventional assignment of truth to various of their contexts, and thus render mathematics conventionally true in the same fashion in which logic has been rendered so. Thus, suppose some mathematical expressions to resist logical definition, and suppose them to be reduced to as meager as possible a set of mathematical primitives. In terms of these and our logical primitives, then, all further mathematical devices are supposed defined; all statements containing the latter become abbreviations of statements containing by way of mathematical notation only the primitives. Here, as remarked earlier in the case of logic, there are alternative courses of definition and therewith alternative sets of primitives; but suppose our procedure to be such as to count 'sphere' and 'includes' among the mathematical primitives. So far we have a set of conventions, (I)–(III) and a few more, let us call them (IV)–(VII), which together circumscribe our logical primitives and yield all logic. By way of circumscribing the further primitives 'sphere' and 'includes', let us now add this convention to the set:

(VIII)   Let 'Hunt (sphere, includes)' be true.

Now we saw earlier that where 'Φ (sphere, includes)' is any truth of geometry, supposed expanded into primitive terms, the statement

(7)   'If Hunt (sphere, includes) then Φ (sphere, includes);

is a truth of logic. Hence (7) is one of the expressions to which truth is assigned by the conventions (I)–(VII). Now (II) instructs us, in view of convention (VIII) and the truth of (7), to adopt 'Φ (sphere, includes)' as true. In this way each truth of geometry is seen to be present among the statements to which truth is assigned by the conventions (I)–(VII).

We have considered four ways of construing geometry. One way consisted of straightforward definition of geometrical expressions in terms of logical ones, within the direction of development represented by *Principia Mathematica*; this way, presumably, would depend upon identification of geometry with algebra through the correlations of analytic geometry, and definition of algebraic expressions on the basis of logical ones as in *Principia Mathematica*. By way of concession to those who have fault to find with certain technical points in *Principia*, this possibility was allowed to retain a tentative status. The other three ways all made use of Huntington's postulates, but are sharply to be distinguished from one another. The first was to include geometry in logic by construing geometrical truths as elliptical for hypothetical statements bearing 'Hunt (sphere, includes)' as hypothesis; this was seen to be a mere evasion, tantamount, under its verbal disguise, to the concession that geometry is not logic after all. The next procedure was to define 'sphere' and 'includes' contextually in terms of logical expressions by construing 'Φ (sphere, includes)' in every case as an abbreviation of 'If α is any class and R any relation such that Hunt (α, R), then Φ (α, R)'. This definition was condemned on the grounds that it fails to yield the intended usage of the defined terms. The last procedure finally, just now presented, renders geometry true by convention without making it part of logic. Here 'Hunt (sphere, includes)' is made true by fiat, by way of conventionally delimiting the meanings of "sphere" and "includes." The truths of geometry then emerge not as truths of logic, but in parallel fashion to the truths of logic.

This last method of accommodating geometry is available also for any other branch of mathematics which may resist definitional reduction to logic. In each case we merely set up a conjunction of postulates for that branch as true by fiat, as a conventional circumscription of the meanings of the constituent primitives, and all the theorems of the branch thereby become true by convention: the convention thus newly adopted together with the conventions (I)–(VII). In this way all mathematics becomes conventionally true, not by becoming a definitional transcription of logic, but by proceeding from linguistic convention in the same way as does logic.

But the method can even be carried beyond mathematics, into the so-called empirical sciences. Having framed a maximum of definitions in the latter realm, we can circumscribe as many of our "empirical" primitives as we like by adding further conventions to the set adopted for logic and mathematics; a correspondingly portion of "empirical" science then becomes conventionally true in precisely the manner observed above for geometry.

The impossibility of defining any of the "empirical" expressions in terms exclusively of logical and mathematical ones may be recognized at the outset: for if any proved to be so definable, there can be no question but that it would thenceforward be recognized as belonging to pure mathematics. On the other hand vast numbers of "empirical" expressions are of course definable on the basis of logical and mathematical ones together with other "empirical" ones. Thus 'momentum' is defined as 'mass times velocity'; 'event' may be defined as 'referent of the *later*-relation', i.e., 'whatever is later than something'; 'instant' may be defined as 'maximal class of events no one of which is later than any other event of the class' (Russell 1914: 126); 'time' as 'the class of all instants'; and so on. In these examples 'momentum' is defined on the basis of mathematical expressions together with the further expressions 'mass' and 'velocity'; 'event', 'instant', and 'time' are all defined on the basis ultimately of logical expressions together with the one further expression 'later than'.

Now suppose definition to have been performed to the utmost among such non-logical, non-mathematical expressions, so that the latter are reduced to as few "empirical" primitives as possible.[12] *All* statements then become abbreviations of statements containing nothing beyond the logical and mathematical primitives and these "empirical" ones. Here, as before, there are alternatives of definition and therewith alternative sets of primitives; but supppose our primitives to be such as to include 'later than', and consider the totality of those statements which under ordinary usage are truths involving only 'later than' and mathematical or logical expressions essentially. Examples of such statements are 'Nothing is later than itself'; 'If Pompey died later than Brutus and Brutus died later than Caesar then Pompey died later than Caesar.' All such statements will be either very general principles, like the first example, or else instances of such principles, like the second example. Now it is a simple matter to frame a small set of general statements from which all and only the statements under consideration can be derived by means of logic and mathematics. The conjunction of these few general statements can then be adopted as true by fiat, as 'Hunt (sphere, includes)' was adopted in (VIII); their adoption is a conventional circumscription of the meaning of the primitive 'later than'. Adoption of this convention renders all those statements conventionally true which under ordinary usage are truths essentially involving any logical or mathematical expressions, or 'later than', or any of the expressions which, like 'event', 'instant', and 'time', are defined on the basis of the foregoing, and inessentially involving anything else.

Now we can pick another of our "empirical" primitives, perhaps 'body' or 'mass' or 'energy', and repeat the process. We can continue in this fashion to any desired point, circumscribing one primitive after another by convention, and rendering conventionally true all statements which under ordinary usage are truths essentially involving only the locutions treated up to that point. If in disposing successively of our

---

[12]Carnap has pursued his program with such amazing success as to provide grounds for expecting all the expressions to be definable ultimately in terms of logic and mathematics plus just one "empirical" primitive, representing a certain dyadic relation described as *recollection of resemblance* (1928a). But for the present cursory considerations no such spectacular reducibility need be presupposed.

"empirical" primitives in the above fashion we take them up in an order roughly describable as leading from the general to the special, then as we progress we may expect to have to deal more and more with statements which are true under ordinary usage only with reservations, only with a probability recognized as short of certainty. But such reservations need not deter us from rendering a statement true by convention; so long as under ordinary usage the presumption is rather for than against the statement, our convention conforms to usage in verifying it. In thus elevating the statement from putative to conventional truth, we still retain the right to falsify the statement tomorrow if those events should be observed which would have occasioned its repudiation while it was still putative: for conventions are commonly revised when new observations show the revision to be convenient.

If in describing logic and mathematics as true by convention what is meant is that the primitives *can* be conventionally circumscribed in such fashion as to generate all and only the so-called truths of logic and mathematics, the characterization is empty; our last considerations show that the same might be said of any other body of doctrine as well. If on the other hand it is meant merely that the speaker adopts such conventions for those fields but not for others, the characterization is uninteresting; while if it is meant that it is a general practice to adopt such conventions explicitly for those fields but not for others, the first part of the characterization is false.

Still, there is the apparent contrast between logico-mathematical truths and others that the former are *a priori,* the latter *a posteriori*; the former have "the character of an inward necessity," in Kant's phrase, the latter do not. Viewed behavioristically and without reference to a metaphysical system, this contrast retains reality as a contrast between more and less firmly accepted statements; and it obtains antecedently to any *post facto* fashioning of conventions. There are statements which we choose to surrender last, if at all, in the course of revamping our sciences in the face of new discoveries; and among these there are some which we will not surrender at all, so basic are they to our whole conceptual scheme. Among the latter are to be counted the so-called truths of logic and mathematics, regardless of what further we may have to say of their status in the course of a subsequent sophicated philosophy. Now since these statements are destined to be maintained independently of our observations of the world, we may as well make use here of our technique of conventional truth-assignment and thereby forestall awkward metaphysical questions as to our *a priori* insight into necessary truths. On the other hand this purpose would not motivate extension of the truth-assignment process into the realm of erstwhile contingent statements. On such grounds, then, logic and mathematics may be held to be conventional while other fields are not; it may be held that it is philosophically important to circumscribe the logical and mathematical primitives by conventions of truth-assignment which yield logical and mathematical truths, but that it is idle elaboration to carry the process further. Such a characterization of logic and mathematics is perhaps neither empty nor uninteresting nor false.

In the adoption of the very conventions (I)–(III) etc. whereby logic itself is set up, however, a difficulty remains to be faced. Each of these conventions is general, announcing the truth of every one of an infinity of statements conforming to a certain description; derivation of the truth of any specific statement from the general convention thus requires a logical inference, and this involves us in an infinite regress. E.g., in deriving (6) from (3) and (5) on the authority of (II) we *infer*, from the general announcement (II) and the specific premiss that (3) and (5) are true statements, the conclusion that

(7)   (6) is to be true.

An examination of this inference will reveal the regress. For present purposes it will be simpler to rewrite (II) thus:

(II′)   No matter what *x* may be, no matter what *y* may be, no matter what *z* may be, if *x* and *z* are true [statements] and *z* is the result of putting *x* for '*p*' and *y* for '*q*' in 'If *p* then *q*' then *y* is to be true.

We are to take (II′) as a premiss, then, and in addition the premiss that (3) and (5) are true. We may also grant it as known that (5) is the result of putting (3) for '*p*' and (6) for '*q*' in 'If *p* then *q*.' Our second premiss may thus be rendered compositely as follows:

(8)   (3) and (5) are true and (5) is the result of putting (3) for '*p*' and (6) for '*q*' in 'If *p* then *q*.'

From these two premisses we propose to infer (7). This inference is obviously sound logic; as logic, however, it involves use of (II′) and others of the conventions from which logic is supposed to spring. Let us try to perform the inference on the basis of those conventions. Suppose that our convention (IV), passed over earlier, is such as to enable us to infer specific instances from statements which, like (II′), involve the *every*-idiom; i.e., suppose that (IV) entitles us in general to drop the prefix 'No matter what *x* [or *y*, etc.] may be' and simultaneously to introduce a concrete designation instead of '*x*' [or '*y*', etc.] in the sequel. By invoking (IV) three times, then, we can infer the following from (II′):

(9)   If (3) and (5) are true and (5) is the result of putting (3) for '*p*' and (6) for '*q*' in 'If *p* then *q*' then (6) is to be true.

It remains to infer (7) from (8) and (9). But this is an inference of the kind for which (II') is needed; from the fact that

(10)   (8) and (9) are true and (9) is the result of putting (8) for '*p*' and (7) for '*q*' in 'If *p* then *q*'.

we are to infer (7) with help of (II'). But the task of getting (7) from (10) and (II') is exactly analogous to our original task of getting (6) from (8) and (II'); the regress is thus under way (cf. Dodgson 1895: 278–80). (Incidentally the derivation of (9) from (II') by (IV), granted just now for the sake of argument, would encounter a similar obstacle; so also the various unanalyzed steps in the derivation of (8).)

In a word, the difficulty is that if logic is to proceed *mediately* from conventions, logic is needed for inferring logic from the conventions. Alternatively, the difficulty which appears thus as a self-presupposition of doctrine can be framed as turning upon a self-presupposition of primitives. It is supposed that the *if*-idiom, the *not*-idiom, the *every*-idiom, and so on, mean nothing to us initially, and that we adopt the conventions (I)–(VII) by way of circumscribing their meaning; and the difficulty is that communication of (I)–(VII) themselves depends upon free use of those very idioms which we are attempting to circumscribe, and can succeed only if we are already conversant with the idioms. This becomes clear as soon as (I)–(VII) are rephrased in rudimentary language, after the manner of (II').[13] It is important to note that this difficulty besets only the method of wholesale truth-assignment, not that of definition. It is true, e.g., that the contextual definition of 'or' presented at the beginning of the second section was communicated with the help of logical and other expressions which cannot be expected to have been endowed with meaning at the stage where logical expressions are first being introduced. But a definition has the peculiarity of being theoretically dispensable; it introduces a scheme of abbreviation, and we are free, if we like, to

---

[13] Incidentally the conventions presuppose also some further locutions, e.g., 'true' ('a true statement'), 'the result of putting ... for ... in ...', and various nouns formed by displaying expressions in quotation marks. The linguistic presuppositions can of course be reduced to a minimum by careful rephrasing; (II'), e.g., can be improved to the following extent:

(II')   No matter what *x* may be, no matter what *y* may be, no matter what *z* may be, if *x* is true then if *z* is true then if *z* is the result of putting *x* for '*p*' in the result of putting *y* for '*q*' in 'If *p* then *q*' then *y* is true.

This involves just the *every*-idiom, the *if*-idiom, 'is', and the further locutions mentioned above.

forego the brevity which it affords until enough primitives have been endowed with meaning, through the method of truth-assignment or otherwise, to accommodate full exposition of the definition. On the other hand the conventions of truth-assignment cannot be thus withheld until preparations are complete, because they are needed in the preparations.

If the truth-assignments were made one by one, rather than an infinite number at a time, the above difficulty would disappear; truths of logic such as (2) would simply be asserted severally by fiat, and the problem of inferring them from more general conventions would not arise. This course was seen to be closed to us, however, by the infinitude of the truths of logic.

It may still be held that the conventions (I)–(VIII), etc., are *observed* from the start, and that logic and mathematics thereby become conventional. It may be held that we can adopt conventions through behavior, without first announcing them in words; and that we can return and formulate our conventions verbally afterward, if we choose, when a full language is at our disposal. It may be held that the verbal formulation of conventions is no more a prerequisite of the adoption of the conventions than the writing of a grammar is a prerequisite of speech; that explicit exposition of conventions is merely one of many important uses of a completed language. So conceived, the conventions no longer involve us in vicious regress. Inference from general conventions is no longer demanded initially, but remains to the subsequent sophisticated stage where we frame general statements of the conventions and show how various specific conventional truths, used all along, fit into the general conventions as thus formulated.

It must be conceded that this account accords well with what we actually do. We discourse without first phrasing the conventions; afterwards, in writings such as this, we formulate them to fit our behavior. On the other hand it is not clear wherein an adoption of the conventions, antecedently to their formulation, consists; such behavior is difficult to distinguish from that in which conventions are disregarded. When we first agree to understand 'Cambridge' as referring to Cambridge in England failing a suffix to the contrary, and then discourse accordingly, the rôle of linguistic convention is intelligible; but when a convention is incapable of being communicated until after its adoption, its rôle is not so clear. In dropping the attributes of deliberateness and explicitness from the notion of linguistic convention we risk depriving the latter of any explanatory force and reducing it to an idle label. We may wonder what one adds to the bare statement that the truths of logic and mathematics are *a priori*, or to the still barer behavioristic statement that they are firmly accepted, when he characterizes them as true by convention in such a sense.

The more restricted thesis discussed in the first section, viz., that mathematics is a conventional transcription of logic, is far from trivial; its demonstration is a highly technical undertaking and an important one, irrespectively of what its relevance may be to fundamental principles of philosophy. It is valuable to show the reducibility of any principle to another through definition of erstwhile primitives, for every such achievement reduces the number of our presuppositions and simplifies and integrates the structure of our theories. But as to the larger thesis that mathematics and logic proceed wholly from linguistic conventions, only further clarification can assure us that this asserts anything at all.

Carnap and logical truth

W. V. QUINE

I

Kant's question "How are synthetic judgments a priori possible?" precipitated the *Critique of Pure Reason*. Question and answer notwithstanding, Mill and others persisted in doubting that such judgments were possible at all. At length some of Kant's own clearest purported instances, drawn from arithmetic, were sweepingly disqualified (or so it seemed; but see §II) by Frege's reduction of arithmetic to logic. Attention was thus forced upon the less tendentious and indeed logically prior question, "How is logical certainty possible?" It was largely this latter question that precipitated the form of empiricism which we associate with between-war Vienna – a movement which began with Wittgenstein's *Tractatus* and reached its maturity in the work of Carnap.

Mill's position on the second question had been that logic and mathematics were based on empirical generalizations, despite their superficial appearance to the contrary. This doctrine may well have been felt to do less than justice to the palpable surface differences between the deductive sciences of logic and mathematics, on the one hand, and the empirical sciences ordinarily so-called on the other. Worse, the doctrine derogated from the certainty of logic and mathematics; but Mill may not have been one to be excessively disturbed by such a consequence. Perhaps classical mathematics did lie closer to experience then than now; at any rate the infinitistic reaches of set theory, which are so fraught with speculation and so remote from any possible experience, were unexplored in his day. And it is against just these latter-day mathematical extravagances that empiricists outside the Vienna Circle have since been known to inveigh

Written early in 1954 for P. A. Schilpp, ed., *The Philosophy of Rudolf Carnap* (La Salle, Ill.: Open Court, 1963) at the request of the editor. It appeared in Italian translation in *Rivista di Filosofia,* 1957, and selected portions amounting to somewhat less than half appeared also in Sidney Hook, ed., *American Philosophers at Work* (New York: Criterion, 1956). Its first appearance whole in English was in the Carnap jubilee issue of *Synthèse* (Volume 12, 1960), which was subsequently reissued as a book: B. H. Kazemier and D. Vuysje, eds., *Logic and Language* (Dordrecht, Holland: D. Reidel Publishing Co., 1962).

Reprinted with the kind permission of the author, editor, and publisher from *The Philosophy of Rudolf Carnap,* P. A. Schilpp, ed., Library of Living Philosophers, Open Court 1963.

(Bridgman 1934), in much the spirit in which the empiricists of Vienna and elsewhere have inveighed against metaphysics.

What now of the empiricist who would grant certainty to logic, and to the whole of mathematics, and yet would make a clean sweep of other non-empirical theories under the name of metaphysics? The Viennese solution of this nice problem was predicated on language. Metaphysics was meaningless through misuse of language; logic was certain through tautologous use of language.

As an answer to the question "How is logical certainty possible?" this linguistic doctrine of logical truth has its attractions. For there can be no doubt that sheer verbal usage is in general a major determinant of truth. Even so factual a sentence as 'Brutus killed Caesar' owes its truth not only to the killing but equally to our using the component words as we do. Why then should a logically true sentence on the same topic, e.g., 'Brutus killed Caesar or did not kill Caesar', not be said to owe its truth *purely* to the fact that we use our words (in this case 'or' and 'not') as we do? – for it depends not at all for its truth upon the killing.

The suggestion is not, of course, that the logically true sentence is a contingent truth about verbal usage; but rather that it is a sentence which, given the language, automatically becomes true, whereas 'Brutus killed Caesar', given the language, becomes true only contingently on the alleged killing.

Further plausibility accrues to the linguistic doctrine of logical truth when we reflect on the question of alternative logics. Suppose someone puts forward and uses a consistent logic the principles of which are contrary to our own. We are then clearly free to say that he is merely using the familiar particles 'and', 'all', or whatever, in other than the familiar senses, and hence that no real contrariety is present after all. There may of course still be an important failure of intertranslatability, in that the behavior of certain of our logical particles is incapable of being duplicated by paraphrases in his system or vice versa. If the translation in this sense is possible, from his system into ours, then we are pretty sure to protest that he was wantonly using the familiar particles 'and' and 'all' (say) where we might unmisleadingly have used such-and-such other familiar phrasing. This reflection goes to support the view that the truths of logic have no content over and above the meanings they confer on the logical vocabulary.

Much the same point can be brought out by a caricature of a doctrine of Lévy-Bruhl, according to which there are pre-logical peoples who accept certain simple self-contradictions as true. Oversimplifying, no doubt, let us suppose it claimed that these natives accept as true a certain sentence of the form 'p and not p'. Or – not to oversimplify too much –

356

that they accept as true a certain heathen sentence of the form '*q* ka bu *q*' the English translation of which has the form '*p* and not *p*'. But now just how good a translation is this, and what may the lexicographer's method have been? If any evidence can count against a lexicographer's adoption of 'and' and 'not' as translations of 'ka' and 'bu', certainly the natives' acceptance of '*q* ka bu *q*' as true counts overwhelmingly. We are left with the meaninglessness of the doctrine of there being pre-logical peoples; pre-logicality is a trait injected by bad translators. This is one more illustration of the inseparability of the truths of logic from the meanings of the logical vocabulary.

We thus see that there is something to be said for the naturalness of the linguistic doctrine of logical truth. But before we can get much further we shall have to become more explicit concerning our subject matter.

II

Without thought of any epistemological doctrine, either the linguistic doctrine or another, we may mark out the intended scope of the term 'logical truth', within that of the broader term 'truth', in the following way. First we suppose indicated, by enumeration if not otherwise, what words are to be called logical words; typical ones are 'or', 'not', 'if', 'then', 'and', 'all', 'every', 'only', 'some'. The logical truths, then, are those true sentences which involve only logical words *essentially*. What this means is that any other words, though they may also occur in a logical truth (as witness 'Brutus', 'kill', and 'Caesar' in 'Brutus killed or did not kill Caesar'), can be varied at will without engendering falsity.[1]

Though formulated with reference to language, the above clarification does not of itself hint that logical truths owe their truth to language. What we have thus far is only a delimitation of the class, *per accidens* if you please. Afterward the linguistic doctrine of logical truth, which is an epistemological doctrine, goes on to say that logical truths are true by virtue purely of the intended meanings, or intended usage, of the logical words. Obviously if logical truths *are* true by virtue purely of language, the logical words are the only part of the language that can be concerned in the matter; for these are the only ones that occur essentially.

[1]Substantially this formulation is traced back a century and a quarter, by Bar-Hillel, to Bolzano. But note that the formulation fails of its purpose unless the phrase "can be varied at will," above, is understood to provide for varying the words not only singly but also two or more at a time. E.g., the sentence 'If some men are angels some animals are angels' can be turned into a falsehood by simultaneous substitution for 'men' and 'angels', but not by any substitution for 'angels' alone, nor for 'men', nor for 'animals' (granted the non-existence of angels). For this observation and illustration I am indebted to John R. Myhill, who expresses some indebtedness in turn to Benson Mates. – I added most of this footnote in May 1955, a year after the rest of the essay left my hands.

Elementary logic, as commonly systematized nowadays, comprises truth-function theory, quantification theory, and identity theory. The logical vocabulary for this part, as commonly rendered for technical purposes, consists of truth-function signs (corresponding to 'or', 'and', 'not', etc.), quantifiers and their variables, and ' = '.

The further part of logic is set theory, which requires there to be classes among the values of its variables of quantification. The one sign needed in set theory, beyond those appropriate to elementary logic, is the connective ' ∈ ' of membership. Additional signs, though commonly used for convenience, can be eliminated in well-known ways.

In this dichotomy I leave metatheory, or logical syntax, out of account. For, either it treats of special objects of an extralogical kind, viz., notational expressions, or else, if these are made to give way to numbers by arithmetization, it is reducible via number theory to set theory.

I will not here review the important contrasts between elementary logic and set theory, except for the following one. Every truth of elementary logic is obvious (whatever this really means), or can be made so by some series of individually obvious steps. Set theory, in its present state anyway, is otherwise. I am not alluding here to Gödel's incompleteness principle, but to something right on the surface. Set theory was straining at the leash of intuition ever since Cantor discovered the higher infinites; and with the added impetus of the paradoxes of set theory the leash was snapped. Comparative set theory has now long been the trend; for, so far as is known, no consistent set theory is both adequate to the purposes envisaged for set theory and capable of substantiation by steps of obvious reasoning from obviously true principles. What we do is develop one or another set theory by obvious reasoning, or elementary logic, from unobvious first principles which are set down, whether for good or for the time being, by something very like convention.

Altogether, the contrasts between elementary logic and set theory are so fundamental that one might well limit the word 'logic' to the former (though I shall not), and speak of set theory as mathematics in a sense exclusive of logic. To adopt this course is merely to deprive ' ∈ ' of the status of a logical word. Frege's derivation of arithmetic would then cease to count as a derivation from logic; for he used set theory. At any rate we should be prepared to find that the linguistic doctrine of logical truth holds for elementary logic and fails for set theory, or vice versa. Kant's readiness to see logic as analytic and arithmetic as synthetic, in particular, is not superseded by Frege's work (as Frege supposed; see Frege 1950: sections 87f., 109) if 'logic' be taken as elementary logic. And for Kant logic certainly did not include set theory.

III

Where someone disagrees with us as to the truth of a sentence, it often happens that we can convince him by getting the sentence from other sentences, which he does accept, by a series of steps each of which he accepts. Disagreement which cannot be thus resolved I shall call *deductively irresoluble.* Now if we try to warp the linguistic doctrine of logical truth around into something like an experimental thesis, perhaps a first approximation will run thus: *Deductively irresoluble disagreement as to a logical truth is evidence of deviation in usage* (*or meanings*) *of words.* This is not yet experimentally phrased, since one term of the affirmed relationship, viz., 'usage' (or 'meanings'), is in dire need of an independent criterion. However, the formulation would seem to be fair enough within its limits; so let us go ahead with it, not seeking more subtlety until need arises.

Already the obviousness or potential obviousness of elementary logic can be seen to present an insuperable obstacle to our assigning any experimental meaning to the linguistic doctrine of elementary logical truth. Deductively irresoluble dissent from an elementary logical truth *would* count as evidence of deviation over meanings if anything can, but simply because dissent from a logical truism is as extreme as dissent can get.

The philosopher, like the beginner in algebra, works in danger of finding that his solution-in-progress reduces to '$0=0$'. Such is the threat to the linguistic theory of elementary logical truth. For, that theory now seems to imply nothing that is not already implied by the fact that elementary logic is obvious or can be resolved into obvious steps.

The considerations which were adduced in §I, to show the naturalness of the linguistic doctrine, are likewise seen to be empty when scrutinized in the present spirit. One was the circumstance that alternative logics are inseparable practically from mere change in usage of logical words. Another was that illogical cultures are indistinguishable from ill-translated ones. But both of these circumstances are adequately accounted for by mere obviousness of logical principles, without help of a linguistic doctrine of logical truth. For, there can be no stronger evidence of a change in usage than the repudiation of what had been obvious, and no stronger evidence of bad translation than that it translates earnest affirmations into obvious falsehoods.

Another point in §I was that true sentences generally depend for their truth on the traits of their language in addition to the traits of their subject matter; and that logical truths then fit neatly in as the limiting

359

case where the dependence on traits of the subject matter is nil. Consider, however, the logical truth 'Everything is self-identical', or '$(x)(x=x)$'. We *can* say that it depends for its truth on traits of the language (specifically on the usage of ' $=$ '), and not on traits of its subject matter; but we can also say, alternatively, that it depends on an obvious trait, viz., self-identity, of its subject matter, viz., everything. The tendency of our present reflections is that there is no difference.

I have been using the vaguely psychological word "obvious" nontechnically, assigning it no explanatory value. My suggestion is merely that the linguistic doctrine of elementary logical truth likewise leaves explanation unbegun. I do not suggest that the linguistic doctrine is false and some doctrine of ultimate and inexplicable insight into the obvious traits of reality is true, but only that there is no real difference between these two pseudodoctrines.

Turning away now from elementary logic, let us see how the linguistic doctrine of logical truth fares in application to set theory. As noted in §II, we may think of ' $\in$ ' as the one sign for set theory in addition to those of elementary logic. Accordingly the version of the linguistic doctrine which was italicized at the beginning of the present section becomes, in application to set theory, this: Among persons who are already in agreement on elementary logic, any deductively irresoluble disagreement as to a truth of set theory is evidence of deviation in usage (or meaning) of ' $\in$ '.

This thesis is not trivial in quite the way in which the parallel thesis for elementary logic was seen to be. It is not indeed experimentally significant as it stands, simply because of the lack, noted earlier, of a separate criterion for usage or meaning. But it does seem reasonable, by the following reasoning.

Any acceptable evidence of usage or meaning of words must reside surely either in the observable circumstances under which the words are uttered (in the case of concrete terms referring to observable individuals) or in the affirmation and denial of sentences in which the words occur. Only the second alternative is relevant to ' $\in$ '. Therefore any evidence of deviation in usage or meaning of ' $\in$ ' must reside in disagreement on sentences containing ' $\in$ '. This is not, of course, to say of *every* sentence containing ' $\in$ ' that disagreement over it establishes deviation in usage or meaning of ' $\in$ '. We have to assume in the first place that the speaker under investigation agrees with us on the meanings of words other than ' $\in$ ' in the sentences in question. And it might well be that, even from among the sentences containing only ' $\in$ ' and words on whose meanings he agrees with us, there is only a select species S which is so fundamental that he cannot dissent from them without betraying deviation in his usage or meaning of ' ∈ '. But *S* may be expected surely to include some (if not all) of the sentences which contain *nothing* but ' ∈ ' and the elementary logical particles; for it is these sentences, insofar as true, that constitute (pure, or unapplied) set theory. But it is difficult to conceive of how to be other than democratic toward the truths of set theory. In exposition we may select some of these truths as so-called postulates and deduce others from them, but this is subjective discrimination, variable at will, expository and not set-theoretic. We do not change our meaning of ' ∈ ' between the page where we show that one particular truth is deducible by elementary logic from another and the page where we show the converse. Given this democratic outlook, finally, the law of sufficient reason leads us to look upon *S* as including *all* the sentences which contain only ' ∈ ' and the elementary logical particles. It then follows that anyone in agreement on elementary logic and in irresoluble disagreement on set theory is in deviation with respect to the usage or meaning of ' ∈ '; and this was the thesis.

The effect of our effort to inject content into the linguistic doctrine of logical truth has been, up to now, to suggest that the doctrine says nothing worth saying about elementary logical truth, but that when applied to set-theoretic truth it makes for a reasonable partial condensation of the otherwise vaporous notion of meaning as applied to ' ∈ '.

IV

The linguistic doctrine of logical truth is sometimes expressed by saying that such truths are true by linguistic convention. Now if this be so, certainly the conventions are not in general explicit. Relatively few persons, before the time of Carnap, had ever seen any convention that engendered truths of elementary logic. Nor can this circumstance be ascribed merely to the slipshod ways of our predecessors. For it is impossible in principle, even in an ideal state, to get even the most elementary part of logic exclusively by the explicit application of conventions stated in advance. The difficulty is the vicious regress, familiar from Lewis Carroll, which I have elaborated elsewhere (1936) [reprinted in this volume]. Briefly the point is that the logical truths, being infinite in number, must be given by general conventions rather than singly; and logic is needed then to begin with, in the metatheory, in order to apply the general conventions to individual cases.

"In dropping the attributes of deliberateness and explicitness from the notion of linguistic convention," I went on to complain in the aforementioned paper, "we risk depriving the latter of any explanatory force and reducing it to an idle label." It would seem that to call elementary logic true by convention is to add nothing but a metaphor to the linguistic doctrine of logical truth which, as applied to elementary logic, has itself come to seem rather an empty figure (cf. §III).

The case of set theory, however, is different on both counts. For set theory the linguistic doctrine has seemed less empty (cf. §III); in set theory, moreover, convention in quite the ordinary sense seems to be pretty much what goes on (cf. §II). Conventionalism has a serious claim to attention in the philosophy of mathematics, if only because of set theory. Historically, though, conventionalism was encouraged in the philosophy of mathematics rather by the non-Euclidean geometries and abstract algebras, with little good reason. We can contribute to subsequent purposes by surveying this situation. Further talk of set theory is deferred to §V.

In the beginning there was Euclidean geometry, a compendium of truths about form and void; and its truths were not based on convention (except as a conventionalist might, begging the present question, apply this tag to everything mathematical). Its truths were in practice presented by deduction from so-called postulates (including axioms; I shall not distinguish); and the selection of truths for this role of postulate, out of the totality of truths of Euclidean geometry, was indeed a matter of convention. But this is not *truth* by convention. The truths were there, and what was conventional was merely the separation of them into those to be taken as starting point (for purposes of the exposition at hand) and those to be deduced from them.

The non-Euclidean geometries came of artificial deviations from Euclid's postulates, without thought (to begin with) of true interpretation. These departures were doubly conventional; for Euclid's postulates were a conventional selection from among the truths of geometry, and then the departures were arbitrarily or conventionally devised in turn. But still there was no truth by convention, because there was no truth.

Playing within a non-Euclidean geometry, one might conveniently make believe that his theorems were interpreted and true; but even such conventional make-believe is not truth by convention. For it is not really truth at all; and what is conventionally pretended is that the theorems are true by non-convention.

Non-Euclidean geometries have, in the fullness of time, received serious interpretations. This means that ways have been found of so construing the hitherto unconstrued terms as to identify the at first conventionally chosen set of non-sentences with some genuine truths, and truths presumably not by convention. The status of an interpreted non-Euclidean geometry differs in no basic way from the original status of Euclidean geometry, noted above.

362

Uninterpreted systems became quite the fashion after the advent of non-Euclidean geometries. This fashion helped to cause, and was in turn encouraged by, an increasingly formal approach to mathematics. Methods had to become more formal to make up for the unavailability, in uninterpreted systems, of intuition. Conversely, disinterpretation served as a crude but useful device (until Frege's syntactical approach came to be appreciated) for achieving formal rigor uncorrupted by intuition.

The tendency to look upon non-Euclidean geometries as true by convention applied to uninterpreted systems generally, and then carried over from these to mathematical systems generally. A tendency indeed developed to look upon all mathematical systems as, qua mathematical, uninterpreted. This tendency can be accounted for by the increase of formality, together with the use of disinterpretation as a heuristic aid to formalization. Finally, in an effort to make some sense of mathematics thus drained of all interpretation, recourse was had to the shocking quibble of identifying mathematics merely with the elementary logic which leads from uninterpreted postulates to uninterpreted theorems (see my 1936: section I) [pp. 329–38 in this volume]. What is shocking about this is that it puts arithmetic qua interpreted theory of number, and analysis qua interpreted theory of functions, and geometry qua interpreted theory of space, outside mathematics altogether.

The substantive reduction of mathematics to logic by Frege, Whitehead, and Russell is of course quite another thing. It is a reduction not to elementary logic but to set theory; and it is a reduction of genuine interpreted mathematics, from arithmetic onward.

V

Let us then put aside these confusions and get back to set theory. Set theory is pursued as interpreted mathematics, like arithmetic and analysis; indeed, it is to set theory that those further branches are reducible. In set theory we discourse about certain immaterial entities, real or erroneously alleged, viz., sets, or classes. And it is in the effort to make up our minds about genuine truth and falsity of sentences about these objects that we find ourselves engaged in something very like convention in an ordinary non-metaphorical sense of the word. We find ourselves making deliberate choices and setting them forth unaccompanied by any attempt at justification other than in terms of elegance and convenience. These adoptions, called postulates, and their logical consequences (via elementary logic), are true until further notice.

363

So here is a case where postulation can plausibly be looked on as constituting truth by convention. But in §IV we have seen how the philosophy of mathematics can be corrupted by supposing that postulates always play that role. Insofar as we would epistemologize and not just mathematize, we might divide postulation as follows. Uninterpreted postulates may be put aside, as no longer concerning us; and on the interpreted side we may distinguish between *legislative* and *discursive* postulation. Legislative postulation institutes truth by convention, and seems plausibly illustrated in contemporary set theory. On the other hand discursive postulation is mere selection, from a pre-existing body of truths, of certain ones for use as a basis from which to derive others, initially known or unknown. What discursive postulation fixes is not truth, but only some particular ordering of the truths, for purposes perhaps of pedagogy or perhaps of inquiry into logical relationships (logical in the sense of elementary logic). All postulation is of course conventional, but only legislative postulation properly hints of *truth* by convention.

It is well to recognize, if only for its distinctness, yet a further way in which convention can enter; viz., in the adoption of new notations for old ones, without, as one tends to say, change of theory. Truths containing the new notation are conventional transcriptions of sentences true apart from the convention in question. They depend for their truth partly on language, but then so did 'Brutus killed Caesar' (cf. §I). They come into being through a conventional adoption of a new sign, and they become true through conventional definition of that sign *together with* whatever made the corresponding sentences in the old notation true.

Definition, in a properly narrow sense of the word, is convention in a properly narrow sense of the word. But the phrase 'true by definition' must be taken cautiously; in its strictest usage it refers to a transcription, by the definition, of a truth of elementary logic. Whether such a sentence is true by convention depends on whether the logical truths themselves be reckoned as true by convention. Even an outright equation or biconditional connection of the definiens and the definiendum is a definitional transcription of a prior logical truth of the form '$x=x$' or '$p \equiv p$'.

Definition commonly so-called is not thus narrowly conceived, and must for present purposes be divided, as postulation was divided, into legislative and discursive. Legislative definition introduces a notation hitherto unused, or used only at variance with the practice proposed, or used also at variance, so that a convention is wanted to settle the ambiguity. Discursive definition, on the other hand, sets forth a pre-existing relation of interchangeability or coextensiveness between notations in already familiar usage. A frequent purpose of this activity is to show how some chosen part of language can be made to serve the purposes of a wider part. Another frequent purpose is language instruction.

It is only legislative definition, and not discursive definition or discursive postulation, that makes a conventional contribution to the truth of sentences. Legislative postulation, finally, affords truth by convention unalloyed.

Increasingly the word 'definition' connotes the formulas of definition which appear in connection with formal systems, signaled by some extrasystematic sign such as '$=_{df}$'. Such definitions are best looked upon as correlating two systems, two notations, one of which is prized for its economical lexicon and the other for its brevity or familiarity of expression (see my 1953a: 26f.). Definitions so used can be either legislative or discursive in their inception. But this distinction is in practice left unindicated, and wisely; for it is a distinction only between particular acts of definition, and not germane to the definition as an enduring channel of intertranslation.

The distinction between the legislative and the discursive refers thus to the act, and not to its enduring consequence, in the case of postulation as in the case of definition. This is because we are taking the notion of truth by convention fairly literally and simple-mindedly, for lack of an intelligible alternative. So conceived, conventionality is a passing trait, significant at the moving front of science but useless in classifying the sentences behind the lines. It is a trait of events and not of sentences.

Might we not still project a derivative trait upon the sentences themselves, thus speaking of a sentence as forever true by convention if its first adoption as true was a convention? No; this, if done seriously, involves us in the most unrewarding historical conjecture. Legislative postulation contributes truths which become integral to the corpus of truths; the artificiality of their origin does not linger as a localized quality, but suffuses the corpus. If a subsequent expositor singles out those once legislatively postulated truths again as postulates, this signifies nothing; he is engaged only in discursive postulation. He could as well choose his postulates from elsewhere in the corpus, and will if he thinks this serves his expository ends.

VI

Set theory, currently so caught up in legislative postulation, may some day gain a norm – even a strain of obviousness, perhaps – and lose all trace of the conventions in its history. A day could likewise have been when our elementary logic was itself instituted as a deliberately conventional deviation from something earlier, instead of evolving, as it did, mainly by unplanned shifts of form and emphasis coupled with casual novelties of notation.

Today indeed there are dissident logicians even at the elementary level, propounding deviations from the law of the excluded middle. These deviations, insofar as meant for serious use and not just as uninterpreted systems, are as clear cases of legislative postulation as the ones in set theory. For here we have again, quite as in set theory, the propounding of a deliberate choice unaccompanied (conceivably) by any attempt at justification other than in terms of convenience.

This example from elementary logic controverts no conclusion we have reached. According to §§I and III, the departure from the law of the excluded middle would count as evidence of revised usage of 'or' and 'not'. (This judgment was upheld in §III, though disqualified as evidence for the linguistic doctrine of logical truth.) For the deviating logician the words 'or' and 'not' are unfamiliar, or defamiliarized; and his decisions regarding truth values for their proposed contexts can then be just as genuinely a matter of deliberate convention as the decisions of the creative set theorist regarding contexts of ' ∈ '.

The two cases are indeed much alike. Not only is departure from the classical logic of 'or' and 'not' evidence of revised usage of 'or' and 'not'; likewise, as argued at length in §III, divergences between set theorists may reasonably be reckoned to revised usage of ' ∈ '. Any such revised usage is conspicuously a matter of convention, and can be declared by legislative postulation.

We have been at a loss to give substance to the linguistic doctrine, particularly of elementary logical truth, or to the doctrine that the familiar truths of logic are true by convention. We have found some sense in the notion of truth by convention, but only as attaching to a process of adoption, viz., legislative postulation, and not as a significant lingering trait of the legislatively postulated sentence. Surveying current events, we note legislative postulation in set theory and, at a more elementary level, in connection with the law of the excluded middle.

And do we not find the same continually in the theoretical hypotheses of natural science itself? What seemed to smack of convention in set theory (§V), at any rate, was "deliberate choice, set forth unaccompanied by any attempt at justification other than in terms of elegance and convenience"; and to what theoretical hypothesis of natural science might not this same character be attributed? For surely the justification of any theoretical hypothesis can, at the time of hypothesis, consist in no more than the elegance or convenience which the hypothesis brings to the containing body of laws and data. How then are we to delimit the category of legislative postulation, short of including under it every new act of scientific hypothesis?

The situation may seem to be saved, for ordinary hypotheses in natural science, by there being some indirect but eventual confrontation with empirical data. However, this confrontation can be remote; and, conversely, some such remote confrontation with experience may be claimed even for pure mathematics and elementary logic. The semblance of a difference in this respect is largely due to overemphasis of departmental boundaries. For a self-contained theory which we can check with experience includes, in point of fact, not only its various theoretical hypotheses of so-called natural science but also such portions of logic and mathematics as it makes use of. Hence I do not see how a line is to be drawn between hypotheses which confer truth by convention and hypotheses which do not, short of reckoning *all* hypotheses to the former category save perhaps those actually derivable or refutable by elementary logic from what Carnap used to call protocol sentences. But this version, besides depending to an unwelcome degree on the debatable notion of protocol sentences, is far too inclusive to suit anyone.

Evidently our troubles are waxing. We had been trying to make sense of the role of convention in a priori knowledge. Now the very distinction between a priori and empirical begins to waver and dissolve, at least as a distinction between sentences. (It could of course still hold as a distinction between factors in one's adoption of a sentence, but both factors might be operative everywhere.)

VII

Whatever our difficulties over the relevant distinctions, it must be conceded that logic and mathematics do seem qualitatively different from the rest of science. Logic and mathematics hold conspicuously aloof from any express appeal, certainly, to observation and experiment. Having thus nothing external to look to, logicians and mathematicians look closely to notation and explicit notational operations: to expressions, terms, substitution, transposition, cancellation, clearing of fractions, and the like. This concern of logicians and mathematicians with syntax (as Carnap calls it) is perennial, but in modern times it has become increasingly searching and explicit, and has even prompted, as we see, a linguistic philosophy of logical and mathematical truth.

On the other hand an effect of these same formal developments in modern logic, curiously, has been to show how to divorce mathematics (other than elementary logic) from any peculiarly notational considerations not equally relevant to natural science. By this I mean that mathematics can be handled (insofar as it can be handled at all) by axiomatization, outwardly quite like any system of hypotheses elsewhere in science; and elementary logic can then be left to extract the theorems.

The consequent affinity between mathematics and systematized natural science was recognized by Carnap when he propounded his P-rules alongside his L-rules or meaning postulates. Yet he did not look upon the P-rules as engendering analytic sentences, sentences true purely by language. How to sustain this distinction has been very much our problem in these pages, and one on which we have found little encouragement.

Carnap appreciated this problem, in *Logical Syntax*, as a problem of finding a difference in kind between the P-rules (or the truths thereby specified) and the L-rules (or the L-truths, analytic sentences, thereby specified). Moreover he proposed an ingenious solution (1937; section 50). In effect he characterized the logical (including mathematical) vocabulary as the largest vocabulary such that (1) there are sentences which contain only that vocabulary and (2) all such sentences are determinable as true or false by a purely syntactical condition – i.e., by a condition which speaks only of concatenation of marks. Then he limited the L-truths in effect to those involving just the logical vocabulary essentially.[2]

Truths given by P-rules were supposedly excluded from the category of logical truth under this criterion, because, though the rules specifying them are formally stated, the vocabulary involved can also be recombined to give sentences whose truth values are not determinate under any set of rules formally formulable in advance.

At this point one can object (pending a further expedient of Carnap's, which I shall next explain) that the criterion based on (1) and (2) fails of its purpose. For, consider to begin with the totality of those sentences which are expressed purely within what Carnap (or anyone) would want to count as logical (and mathematical) vocabulary. Suppose, in conformity with (2), that the division of this totality into the true and the false is reproducible in purely syntactical terms. Now surely the adding of one general term of an extra-logical kind, say 'heavier than', is not going to alter the situation. The truths which are expressible in terms of just 'heavier than', together with the logical vocabulary, will be truths of only the most general kind, such as '$(\exists x)(\exists y)(x$ is heavier than $y)$', '$(x) \sim (x$ is heavier than $x)$', and '$(x)(y)(z)(x$ is heavier than $y . y$ is heavier than $z . \supset . x$ is heavier than $z)$'. The division of the truths from

---

[2]Cf. §I above. Also, for certain reservations conveniently postponed at the moment, see §IX on "essential predication."

the falsehoods in this supplementary domain can probably be reproduced in syntactical terms if the division of the original totality could. But then, under the criterion based on (1) and (2), 'heavier than' qualifies for the logical vocabulary. And it is hard to see what whole collection of general terms of natural science might not qualify likewise.

The further expedient, by which Carnap met this difficulty, was his use of Cartesian co-ordinates (1937: sections 3, 15). Under this procedure, each spatio-temporal particular c becomes associated with a class K of quadruples of real numbers, viz., the class of those quadruples which are the co-ordinates of component point events of c. Further let us write $K[t]$ for the class of triples which with t appended belong to K; thus $K[t]$ is that class of triples of real numbers which is associated with the momentary state of object c at time t. Then, in order to say, e.g., that c_1 is heavier than c_2 at time t, we say '$H(K_1[t], K_2[t])$', which might be translated as 'The momentary object associated with $K_1[t]$ is heavier than that associated with $K_2[t]$'. Now $K_1[t]$ and $K_2[t]$ are, in every particular case, purely mathematical objects; viz., classes of triples of real numbers. So let us consider all the true and false sentences of the form '$H(K_1[t], K_2[t])$' where, in place of '$K_1[t]$' and '$K_2[t]$', we have purely logico-mathematical designations of particular classes of triples of real numbers. There is no reason to suppose that all the truths of *this* domain can be exactly segregated in purely syntactical terms. Thus inclusion of 'H' does violate (2), and therefore 'H' fails to qualify as logical vocabulary. By adhering to the method of co-ordinates and thus reconstruing all predicates of natural science in the manner here illustrated by 'H', Carnap overcomes the objection noted in the preceding paragraph.

To sum up very roughly, this theory characterizes logic (and mathematics) as the largest part of science within which the true–false dichotomy *can* be reproduced in syntactical terms. This version may seem rather thinner than the claim that logic and mathematics are somehow true by linguistic convention, but at any rate it is more intelligible, and, if true, perhaps interesting and important. To become sure of its truth, interest, and importance, however, we must look more closely at this term 'syntax'.

As used in the passage: "The terms 'sentence' and 'direct consequence' are the two primitive terms of logical syntax" (Carnap 1935c: 47), the term 'syntax' is of course irrelevant to a thesis. The relevant sense is that rather in which it connotes discourse about marks and their succession. But here still we must distinguish degrees of inclusiveness; two different degrees are exemplified in *Logical Syntax,* according as the object language is Carnap's highly restricted Language I or his more powerful Language II. For the former, Carnap's formulation of logical truth is narrowly syntactical in the manner of familiar formalizations of logical systems by axioms and rules of inference. But Gödel's proof of the incompletability of elementary number theory shows that no such approach can be adequate to mathematics in general, nor in particular to set theory, nor to Language II. For Language II, in consequence, Carnap's formulation of logical truth proceeded along the lines rather of Tarski's technique of truth definition.[3] The result was still a purely syntactical specification of the logical truths, but only in this more liberal sense of 'syntactical': it was couched in a vocabulary consisting (in effect) of (a) names of signs, (b) an operator expressing concatenation of expressions, and (c), by way of auxiliary machinery, the whole logical (and mathematical) vocabulary itself.

So construed, however, the thesis that logico-mathematical truth is syntactically specifiable becomes uninteresting. For, what it says is that logico-mathematical truth is specifiable in a notation consisting solely of (a), (b), *and* the whole logico-mathematical vocabulary itself. But *this* thesis would hold equally if 'logico-mathematical' were broadened (at *both* places in the thesis) to include physics, economics, and anything else under the sun; Tarski's routine of truth definition would still carry through just as well. No special trait of logic and mathematics has been singled out after all.

Strictly speaking, the position is weaker still. The mathematics appealed to in (c) must, as Tarski shows, be a yet more inclusive mathematical theory in certain respects than that for which truth is being defined. It was largely because of his increasing concern over this self-stultifying situation that Carnap relaxed his stress on syntax, in the years following *Logical Syntax,* in favor of semantics.

VIII

Even if logical truth were specifiable in syntactical terms, this would not show that it was grounded in language. Any *finite* class of truths (to take an extreme example) is clearly reproducible by a membership condition couched in as narrowly syntactical terms as you please; yet we certainly cannot say of every finite class of truths that its members are true purely by language. Thus the ill-starred doctrine of syntactical specifiability of logical truth was always something other than the linguistic doctrine of

---

[3]*Logical Syntax,* especially §§34a–i, 60a–d, 71a–d. These sections had been omitted from the German edition, but only for lack of space; cf. p. xi of the English edition. Meanwhile they had appeared as articles: "die Antinomien ..." and "ein Gültigkeitskriterium ..." At that time Carnap had had only partial access to Tarski's ideas (cf. "Gültigkeitskriterium," footnote 3), the full details of which reached the non-Slavic world in 1935 in Tarski's *Wahrheitsbegriff.*''

logical truth, if this be conceived as the doctrine that logical truth is grounded in language. In any event the doctrine of syntactical specifiability, which we found pleasure in being able to make comparatively clear sense of, has unhappily had to go by the board. The linguistic doctrine of logical truth, on the other hand, goes sturdily on.

The notion of logical truth is now counted by Carnap as semantical. This of course does not of itself mean that logical truth is gounded in language; for note that the general notion of truth is also semantical, though truth in general is not grounded purely in language. But the semantical attribute of logical truth, in particular, *is* one which, according to Carnap, is grounded in language: in convention, fiat, meaning. Such support as he hints for this doctrine, aside from ground covered in §§I–VI, seems to depend on an analogy with what goes on in the propounding of artificial languages; and I shall now try to show why I think the analogy mistaken.

I may best schematize the point by considering a case, not directly concerned with logical truth, where one might typically produce an artificial language as a step in an argument. This is the imaginary case of a logical positivist, say Ixmann, who is out to defend scientists against the demands of a metaphysician. The metaphysician argues that science presupposes metaphysical principles, or raises metaphysical problems, and that the scientists should therefore show due concern. Ixmann's answer consists in showing in detail how people (on Mars, say) might speak a language quite adequate to all of our science but, unlike our language, incapable of expressing the alleged metaphysical issues. (I applaud this answer, and think it embodies the most telling component of Carnap's own anti-metaphysical representations; but here I digress.) Now how does our hypothetical Ixmann specify that doubly hypothetical language? By telling us, at least to the extent needed for his argument, what these Martians are to be imagined as uttering and what they are thereby to be understood to mean. Here is Carnap's familiar duality of formation rules and transformation rules (or meaning postulates), as rules of language. But these rules are part only of Ixmann's narrative machinery, not part of what he is portraying. He is not representing his hypothetical Martians themselves as somehow explicit on formation and transformation rules. Nor is he representing there to be any intrinsic difference between those truths which happen to be disclosed to us by his partial specifications (his transformation rules) and those further truths, hypothetically likewise known to the Martians of his parable, which he did not trouble to sketch in.

The threat of fallacy lurks in the fact that Ixmann's rules are indeed arbitrary fiats, as is his whole Martian parable. The fallacy consists in confusing levels, projecting the conventional character of the rules into the story, and so misconstruing Ixmann's parable as attributing truth legislation to his hypothetical Martians.

The case of a non-hypothetical artificial language is in principle the same. Being a new invention, the language has to be explained; and the explanation will proceed by what may certainly be called formation and transformation rules. These rules will hold by arbitrary fiat, the artifex being boss. But all we can reasonably ask of these rules is that they enable us to find corresponding to each of his sentences a sentence of like truth value in familiar ordinary language. There is no (to me) intelligible additional decree that we can demand of him as to the boundary between analytic and synthetic, logic and fact, among his truths. We may well decide to extend our word 'analytic' or 'logically true' to sentences of his language which he in his explanations has paired off fairly directly with English sentences so classified by us; but this is our decree, regarding our word 'analytic' or 'logically true'.

IX

We had in §II to form some rough idea of what logical truth was supposed to take in, before we could get on with the linguistic doctrine of logical truth. This we did, with help of the general notion of truth[4] together with a partial enumeration of the logical vocabulary of a particular language. In §VII we found hope of a less provincial and accidental characterization of logical vocabulary; but it failed. Still, the position is not intolerable. We well know from modern logic how to devise *a* technical notation which is admirably suited to the business of 'or', 'not', 'and', 'all', 'only', and such other particles as we would care to count as logical; and to enumerate the signs and constructions of that technical notation, or a theoretically adequate subset of them, is the work of a moment (cf. §II). Insofar as we are content to think of all science as fitted within that stereotyped logical framework – and there is no hardship in so doing – our notion of logical vocabulary is precise. And so, derivatively, is our notion of logical truth. But only in point of extent. There is no epistemological corollary as to the *ground* of logical truth (cf. §II).

Even this halfway tolerable situation obtains only for logical truth in a relatively narrow sense, omitting truths by "essential predication" (Aristotle) such as 'No bachelor is married'. I tend to reserve the term 'logically true' for the narrower domain, and to use the term 'analytic'

---

[4]In defense of this general notion, in invidious contrast to that of analyticity, see my *From a Logical Point of View,* pp. 137f.

for the more inclusive domain which includes truths by essential predication. Carnap on the contrary has used both terms in the broader sense. But the problems of the two subdivisions of the analytic class differ in such a way that it has been convenient up to now in this essay to treat mainly of logical truth in the narrower sense.

The truths by essential predication are sentences which can be turned into logical truths by supplanting certain simple predicates (e.g., 'bachelor') by complex synonyms (e.g., 'man not married'). This formulation is not inadequate to such further examples as 'if A is part of B and B is part of C then A is part of C'; this case can be managed by using for 'is part of' the synonym 'overlaps nothing save what overlaps' (after Goodman 1951). The relevant notion of synonymy is simply *analytic* co-extensiveness (however circular this might be as a definition).

To count analyticity a genus of logical truth is to grant, it may seem, the linguistic doctrine of logical truth; for the term 'analytic' directly suggests truth by language. But this suggestion can be adjusted, in parallel to what was said of 'true by definition' in §V. 'Analytic' means true by synonymy and logic, hence no doubt true by language and logic, and simply true by language *if* the linguistic doctrine of logical truth is right. Logic itself, throughout these remarks, may be taken as including or excluding set theory (and hence mathematics), depending on further details of one's position.

What has made it so difficult for us to make satisfactory sense of the linguistic doctrine is the obscurity of 'true by language'. Now 'synonymous' lies within that same central obscurity; for, about the best we can say of synonymous predicates is that they are somehow "co-extensive by language." The obscurity extends, of course, to 'analytic'.

One quickly identifies certain seemingly transparent cases of synonymy, such as 'bachelor' and 'man not married', and senses the triviality of associated sentences such as 'No bachelor is married'. Conceivably the mechanism of such recognition, when better understood, might be made the basis of a definition of synonymy and analyticity in terms of linguistic behavior. On the other hand such an approach might make sense only of something like degrees of synonymy and analyticity. I see no reason to expect that the full-width analyticity which Carnap and others make such heavy demands upon can be fitted to such a foundation in even an approximate way. In any event, we at present lack any tenable general suggestion, either rough and practical or remotely theoretical, as to what it is to be an analytic sentence. All we have are purported illustrations, and claims that the truths of elementary logic, with or without the rest of mathematics, should be counted in. Wherever there has been a semblance of a general criterion, to my knowledge, there has been either some drastic failure such as tended to admit all or no sentences as analytic, or there has been a circularity of the kind noted three paragraphs back, or there has been a dependence on terms like 'meaning', 'possible', 'conceivable', and the like, which are at least as mysterious (and in the same way) as what we want to define. I have expatiated on these troubles elsewhere (1951d), as has White.

Logical truth (in my sense, excluding the additional category of essential predication) is, we saw, well enough definable (relatively to a fixed logical notation). *Elementary* logical truth can even be given a narrowly syntactical formulation, such as Carnap once envisaged for logic and mathematics as a whole (cf. §VII); for the deductive system of elementary logic is known to be complete. But when we would supplement the logical truths by the rest of the so-called analytic truths, true by essential predication, then we are no longer able even to say what we are talking about. The distinction itself, and not merely an epistemological question concerning it, is what is then in question.

What of settling the limits of the broad class of analytic truths by fixing on a standard language as we did for logical truth? No, the matter is very different. Once given the logical vocabulary, we have a means of clearly marking off the species logical truth within the genus truth. But the intermediate genus analyticity is not parallel, for it does not consist of the truths which contain just a certain vocabulary essentially (in the sense of §II). To segregate analyticity we should need rather some sort of accounting of synonymies throughout a universal or all-purpose language. No regimented universal language is at hand, however, for consideration; what Carnap has propounded in this direction have of course been only illustrative samples, fragmentary in scope. And even if there were one, it is not clear by what standards we would care to settle questions of synonymy and analyticity within it.

X

Carnap's present position (particularly 1952) is that one has specified a language quite rigorously only when he has fixed, by dint of so-called meaning postulates, what sentences are to count as analytic. The proponent is supposed to distinguish between those of his declarations which count as meaning postulates, and thus engender analyticity, and those which do not. This he does, presumably, by attaching the label 'meaning postulate'.

But the sense of this label is far less clear to me than four causes of its seeming to be clear. Which of these causes has worked on Carnap, if any,

I cannot say; but I have no doubt that all four have worked on his readers. One of these causes is misevaluation of the role of convention in connection with artificial language; thus note the unattributed fallacy described in §VIII. Another is misevaluation of the conventionality of postulates: failure to appreciate that postulates, though they are postulates always by fiat, are not *therefore* true by fiat (cf. §§IV–V). A third is over-estimation of the distinctive nature of postulates, and of definitions, because of conspicuous and peculiar roles which postulates and definitions have played in situations not really relevant to present concerns: postulates in uninterpreted systems (cf. §IV), and definitions in double systems of notation (cf. §V). A fourth is misevaluation of legislative postulation and legislative definition themselves, in two respects: failure to appreciate that this legislative trait is a trait of scientific hypotheses very generally (cf. §VI), and failure to appreciate that it is a trait of the passing event rather than of the truth which is thereby instituted (cf. end of §V).

Suppose a scientist introduces a new term, for a certain substance or force. He introduces it by an act either of legislative definition or of legislative postulation. Progressing, he evolves hypotheses regarding further traits of the named substance or force. Suppose now that some such eventual hypothesis, well attested, identifies this substance or force with one named by a complex term built up of other portions of his scientific vocabulary. We all know that this new identity will figure in the ensuing developments quite on a par with the identity which first came of the act of legislative definition, if any, or on a par with the law which first came of the act of legislative postulation. Revisions, in the course of further progress, can touch any of these affirmations equally. Now I urge that scientists, proceeding thus, are not thereby slurring over any meaningful distinction. Legislative acts occur again and again; on the other hand a dichotomy of the resulting truths themselves into analytic and synthetic, truths by meaning postulate and truths by force of nature, has been given no tolerably clear meaning even as a methodological ideal.

One conspicuous consequence of Carnap's belief in this dichotomy may be seen in his attitude toward philosophical issues as to what there is (Quine 1951a). It is only by assuming the cleavage between analytic and synthetic truths that he is able to declare the problem of universals to be a matter not of theory but of linguistic decision. Now I am as impressed as anyone with the vastness of what language contributes to science and to one's whole view of the world; and in particular I grant that one's hypothesis as to what there is, e.g., as to there being universals, is at bottom just as arbitrary or pragmatic a matter as one's adoption of a new brand of set theory or even a new system of bookkeeping. Carnap in turn recognizes that such decisions, however conventional, "will never-theless usually be influenced by theoretical knowledge" (1950: §2). But what impresses me more than it does Carnap is how well this whole atti-tude is suited also to the theoretical hypotheses of natural science itself, and how little basis there is for a distinction.

The lore of our fathers is a fabric of sentences. In our hands it develops and changes, through more or less arbitrary and deliberate revisions and additions of our own, more or less directly occasioned by the continuing stimulation of our sense organs. It is a pale gray lore, black with fact and white with convention. But I have found no substantial reasons for con-cluding that there are any quite black threads in it, or any white ones.

On the nature of mathematical truth

CARL G. HEMPEL

1. The problem

It is a basic principle of scientific inquiry that no proposition and no theory is to be accepted without adequate grounds. In empirical science, which includes both the natural and the social sciences, the grounds for the acceptance of a theory consist in the agreement of predictions based on the theory with empirical evidence obtained either by experiment or by systematic observation. But what are the grounds which sanction the acceptance of mathematics? That is the question I propose to discuss in the present paper. For reasons which will become clear subsequently, I shall use the term "mathematics" here to refer to arithmetic, algebra, and analysis – to the exclusion, in particular, of geometry.[1]

2. Are the propositions of mathematics self-evident truths?

One of the several answers which have been given to our problem asserts that the truths of mathematics, in contradistinction to the hypotheses of empirical science, require neither factual evidence nor any other justification because they are "self-evident." This view, however, which ultimately relegates decisions as to mathematical truth to a feeling of self-evidence, encounters various difficulties. First of all, many mathematical theorems are so hard to establish that even to the specialist in the particular field they appear as anything but self-evident. Secondly, it is well known that some of the most interesting results of mathematics – especially in such fields as abstract set theory and topology – run counter to deeply ingrained intuitions and the customary kind of feeling of self-evidence. Thirdly, the existence of mathematical conjectures such as those of Goldbach and of Fermat, which are quite elementary in content and yet undecided up to this day, certainly shows that not all mathematical truths can be self-evident. And finally, even if self-evidence were

Reprinted from *The American Mathematical Monthly,* vol. 52 (1945), pp. 543–556, with the kind permission of the author and The Mathematical Association of America.
[1] A discussion of the status of geometry is given in Hempel 1945a: 7–17.

attributed only to the basic postulates of mathematics, from which all other mathematical propositions can be deduced, it would be pertinent to remark that judgments as to what may be considered as self-evident, are subjective; they may vary from person to person and certainly cannot constitute an adequate basis for decisions as to the objective validity of mathematical propositions.

3. Is mathematics the most general empirical science?

According to another view, advocated especially by John Stuart Mill, mathematics is itself an empirical science which differs from the other branches, such as astronomy, physics, chemistry, etc., mainly in two respects: its subject matter is more general than that of any other field of scientific research, and its propositions have been tested and confirmed to a greater extent than those of even the most firmly established sections of astronomy or physics. Indeed, according to this view, the degree to which the laws of mathematics have been borne out by the past experiences of mankind is so overwhelming that – unjustifiably – we have come to think of mathematical theorems as qualitatively different from the well confirmed hypotheses or theories of other branches of science: we consider them as certain, while other theories are thought of as at best "very probable" or very highly confirmed.

But this view, too, is open to serious objections. From a hypothesis which is empirical in character – such as, for example, Newton's law of gravitation – it is possible to derive predictions to the effect that under certain specified conditions certain specified observable phenomena will occur. The actual occurrence of these phenomena constitutes confirming evidence, their non-occurrence disconfirming evidence for the hypothesis. It follows in particular that an empirical hypothesis is theoretically disconfirmable; i.e., it is possible to indicate what kind of evidence, if actually encountered, would disconfirm the hypothesis. In the light of this remark, consider now a simple "hypothesis" from arithmetic: $3+2=5$. If this is actually an empirical generalization of past experiences, then it must be possible to state what kind of evidence would oblige us to concede the hypothesis was not generally true after all. If any disconfirming evidence for the given proposition can be thought of, the following illustration might well be typical of it: We place some microbes on a slide, putting down first three of them and then another two. Afterwards we count all the microbes to test whether in this instance 3 and 2 actually added up to 5. Suppose now that we counted 6 microbes altogether. Would we consider this as an empirical disconfirmation of the given proposition, or at least as a proof that it does not apply to

378

microbes? Clearly not; rather, we would assume we had made a mistake in counting or that one of the microbes had split in two between the first and the second count. But under no circumstances could the phenomenon just described invalidate the arithmetical proposition in question; for the latter asserts nothing whatever about the behavior of microbes; it merely states that any set consisting of $3+2$ objects may also be said to consist of 5 objects. And this is so because the symbols "$3+2$" and "5" denote the same number: they are synonymous by virtue of the fact that the symbols "2," "3," "5," and "+" are *defined* (or tacitly understood) in such a way that the above identity holds as a consequence of the meaning attached to the concepts involved in it.

4. The analytic character of mathematical propositions

The statement that $3+2=5$, then, is true for similar reasons as, say, the assertion that no sexagenarian is 45 years of age. Both are true simply by virtue of definitions or of similar stipulations which determine the meaning of the key terms involved. Statements of this kind share certain important characteristics: Their validation naturally requires no empirical evidence; they can be shown to be true by a mere analysis of the meaning attached to the terms which occur in them. In the language of logic, sentences of this kind are called analytic or true *a priori*, which is to indicate that their truth is logically independent of, or logically prior to, any experiential evidence.[2] And while the statements of empirical science, which are synthetic and can be validated only a posteriori, are constantly subject to revision in the light of new evidence, the truth of an analytic statement can be established definitely, once and for all. However, this characteristic "theoretical certainty" of analytic propositions has to be paid for at a high price: An analytic statement conveys no factual information. Our statement about sexagenarians, for example, asserts nothing that could possibly conflict with any factual evidence: it has no factual implications, no empirical content; and it is precisely for this reason that the statement can be validated without recourse to empirical evidence.

[2]The objection is sometimes raised that without certain types of experience, such as encountering several objects of the same kind, the integers and the arithmetical operations with them would never have been invented, and that therefore the propositions of arithmetic do have an empirical basis. This type of argument, however, involves a confusion of the logical and the psychological meaning of the term "basis." It may very well be the case that certain experiences occasion psychologically the formation of arithmetical ideas and in this sense form an empirical "basis" for them; but this point is entirely irrelevant for the logical questions as to the *grounds* on which the propositions of arithmetic may be accepted as true. The point made above is that no empirical "basis" or evidence whatever is needed to establish the truth of the propositions of arithmetic.

Let us illustrate this view of the nature of mathematical propositions by reference to another, frequently cited, example of a mathematical – or rather logical – truth, namely the proposition that whenever $a = b$ and $b = c$ then $a = c$. On what grounds can this so-called "transitivity of identity" be asserted? Is it of an empirical nature and hence at least theoretically disconfirmable by empirical evidence? Suppose, for example, that a, b, c are certain shades of green, and that as far as we can see, $a = b$ and $b = c$, but clearly $a \neq c$. This phenomenon actually occurs under certain conditions; do we consider it as disconfirming evidence for the proposition under consideration? Undoubtedly not; we would argue that if $a \neq c$, it is impossible that $a = b$ and also $b = c$; between the terms of at least one of these latter pairs, there must obtain a difference, though perhaps only a subliminal one. And we would dismiss the possibility of empirical disconfirmation, and indeed the idea that an empirical test should be relevant here, on the grounds that identity is a transitive relation by virtue of its definition or by virtue of the basic postulates governing it.[3] Hence the principle in question is true *a priori*.

5. Mathematics as an axiomatized deductive system

I have argued so far that the validity of mathematics rests neither on its alleged self-evidential character nor on any empirical basis, but derives from the stipulations which determine the meaning of the mathematical concepts, and that the propositions of mathematics are therefore essentially "true by definition." This latter statement, however, is obviously oversimplified and needs restatement and a more careful justification.

For the rigorous development of a mathematical theory proceeds not simply from a set of definitions but rather from a set of non-definitional propositions which are not proved within the theory; these are the postulates or axioms of the theory.[4] They are formulated in terms of certain basic or primitive concepts for which no definitions are provided within the theory. It is sometimes asserted that the postulates themselves represent "implicit definitions" of the primitive terms. Such a characterization of the postulates, however, is misleading. For while the postulates do limit, in a specific sense, the meanings that can possibly be ascribed to the primitives, any self-consistent postulate system admits, nevertheless, many different interpretations of the primitive terms (this will soon be illustrated), whereas a set of definitions in the strict sense of the word determines the meanings of the definienda in a unique fashion.

[3]A precise account of the definition and the essential characteristics of the identity relation may be found in Tarski 1941: chap. 3.
[4]For a lucid and concise account of the axiomatic method, see Tarski 1941: chap. 6.

Once the primitive terms and the postulates have been laid down, the entire theory is completely determined; it is derivable from its postulational basis in the following sense: Every term of the theory is definable in terms of the primitives, and every proposition of the theory is logically deducible from the postulates. To be entirely precise, it is necessary also to specify the principles of logic which are to be used in the proof of the propositions, i.e. in their deduction from the postulates. These principles can be stated quite explicitly. They fall into two groups: Primitive sentences, or postulates, of logic (such as: If p and q is the case, then p is the case), and rules of deduction or inference (including, for example, the familiar modus ponens rule and the rules of substitution which make it possible to infer, from a general proposition, any one of its substitution instances). A more detailed discussion of the structure and content of logic would, however, lead too far afield in the context of this article.

6. Peano's axiom system as a basis for mathematics

Let us now consider a postulate system from which the entire arithmetic of the natural numbers can be derived. This system was devised by the Italian mathematician and logician G. Peano (1858–1932). The primitives of this system are the terms "0," "number," and "successor." While, of course, no definition of these terms is given within the theory, the symbol "0" is intended to designate the number 0 in its usual meaning, while the term "number" is meant to refer to the natural numbers $0, 1, 2, 3 \cdots$ exclusively. By the successor of a natural number n, which will sometimes briefly be called n', is meant the natural number immediately following n in the natural order. Peano's system contains the following 5 postulates:

P1. 0 is a number.
P2. The successor of any number is a number.
P3. No two numbers have the same successor.
P4. 0 is not the successor of any number.
P5. If P is a property such that (a) 0 has the property P, and (b) whenever a number n has the property P, then the successor of n also has the property P, then every number has the property P.

The last postulate embodies the principle of mathematical induction and illustrates in a very obvious manner the enforcement of a mathematical "truth" by stipulation. The construction of elementary arithmetic on this basis begins with the definition of the various natural numbers. 1 is defined as the successor of 0, or briefly as $0'$; 2 as $1'$, 3 as $2'$, and so on. By virtue of P2, this process can be continued indefinitely; because of P3 (in combination with P5), it never leads back to one of the

381

numbers previously defined, and in view of P4, it does not lead back to 0 either.

As the next step, we can set up a definition of addition which expresses in a precise form the idea that the addition of any natural number to some given number may be considered as a repeated addition of 1; the latter operation is readily expressible by means of the successor relation. This definition of addition runs as follows:

D1.   (a) $n+0=n$;   (b) $n+k'=(n+k)'$.

The two stipulations of this recursive definition completely determine the sum of any two integers. Consider, for example, the sum $3+2$. According to the definitions of the numbers 2 and 1, we have $3+2=3+1'=3+(0')'$; by D1 (b), $3+(0')'=(3+0')'=((3+0)')'$; but by D1 (a), and by the definitions of the numbers 4 and 5, $((3+0)')'=(3')'=4'=5$. This proof also renders more explicit and precise the comments made earlier in this paper on the truth of the proposition that $3+2=5$: Within the Peano system of arithmetic, its truth flows not merely from the definition of the concepts involved, but also from the postulates that govern these various concepts. (In our specific example, the postulates P1 and P2 are presupposed to guarantee that $1,2,3,4,5$ are numbers in Peano's system; the general proof that D1 determines the sum of any two numbers also makes use of P5.) If we call the postulates and definitions of an axiomatized theory the "stipulations" concerning the concepts of that theory, then we may say now that the propositions of the arithmetic of the natural numbers are true by virtue of the stipulations which have been laid down initially for the arithmetical concepts. (Note, incidentally, that our proof of the formula "$3+2=5$" repeatedly made use of the transitivity of identity; the latter is accepted here as one of the rules of logic which may be used in the proof of any arithmetical theorem; it is, therefore, included among Peano's postulates no more than any other principle of logic.)

Now, the multiplication of natural numbers may be defined by means of the following recursive definition, which expresses in a rigorous form the idea that a product nk of two integers may be considered as the sum of k terms each of which equals n.

D2.   (a) $n\cdot0=0$;   (b) $n\cdot k'=n\cdot k+n$.

It now is possible to prove the familiar general laws governing addition and multiplication, such as the commutative, associative, and distributive laws $(n+k=k+n, \ n\cdot k=k\cdot n; \ n+(k+l)=(n+k)+l, \ n\cdot(k\cdot l)=(n\cdot k)\cdot l; \ n\cdot(k+l)=(n\cdot k)+(n\cdot l))$. – In terms of addition and multiplication, the inverse operations of subtraction and division can then be defined. But it turns out that these "cannot always be performed"; i.e., in contradistinction to the sum and the product, the difference and the quotient are not defined for every couple of numbers; for example, $7-10$ and $7 \div 10$ are undefined. This situation suggests an enlargement of the number system by the introduction of negative and of rational numbers.

It is sometimes held that in order to effect this enlargement, we have to "assume" or else to "postulate" the existence of the desired additional kinds of numbers with properties that make them fit to fill the gaps of subtraction and division. This method of simply postulating what we want has its advantages; but, as Bertrand Russell puts it, they are the same as the advantages of theft over honest toil (1919: 71); and it is a remarkable fact that the negative as well as the rational numbers can be obtained from Peano's primitives by the honest toil of constructing explicit definitions for them, without the introduction of any new postulates or assumptions whatsoever. Every positive and negative integer – in contradistinction to a natural number which has no sign – is definable as a certain set of ordered couples of natural numbers; thus, the integer $+2$ is definable as the set of all ordered couples (m, n) of natural numbers where $m = n + 2$; the integer -2 is the set of all ordered couples (m, n) of natural numbers with $n = m + 2$. – Similarly, rational numbers are defined as classes of ordered couples of integers. – The various arithmetical operations can then be defined with reference to these new types of numbers, and the validity of all the arithmetical laws governing these operations can be proved by virtue of nothing more than Peano's postulates and the definitions of the various arithmetical concepts involved.

The much broader system thus obtained is still incomplete in the sense that not every number in it has a square root, and more generally, not every algebraic equation whose coefficients are all numbers of the system has a solution in the system. This suggests further expansions of the number system by the introduction of real and finally of complex numbers. Again, this enormous extension can be effected by mere definition, without the introduction of a single new postulate.[5] On the basis thus obtained, the various arithmetical and algebraic operations can be defined for the numbers of the new system, the concepts of function, of limit, of derivative and integral can be introduced, and the familiar theorems pertaining to these concepts can be proved, so that finally the huge system of mathematics as here delimited rests on the narrow basis of Peano's system: Every concept of mathematics can be defined by

[5]For a more detailed account of the construction of the number system on Peano's basis, cf. Russell 1919: esp. chaps. 1 and 7 [chapters 1, 2, and 18 are reprinted in this volume. Eds.]. – A rigorous and concise presentation of that construction, beginning, however, with the set of all integers rather than that of the natural numbers, may be found in Birkhoff and MacLane 1941: chaps. 1, 2, 3, 5. – For a general survey of the construction of the number system, cf. also Young 1911: esp. lectures 10, 11, 12.

means of Peano's three primitives, and every proposition of mathematics can be deduced from the five postulates enriched by the definitions of the non-primitive terms.[6] These deductions can be carried out, in most cases, by means of nothing more than the principles of formal logic; the proof of some theorems concerning real numbers, however, requires one assumption which is not usually included among the latter. This is the so-called axiom of choice. It asserts that given a class of mutually exclusive classes, none of which is empty, there exists at least one class which has exactly one element in common with each of the given classes. By virtue of this principle and the rules of formal logic, the content of all of mathematics can thus be derived from Peano's modest system – a remarkable achievement in systematizing the content of mathematics and clarifying the foundations of its validity.

7. Interpretations of Peano's primitives

As a consequence of this result, the whole system of mathematics might be said to be true by virtue of mere definitions (namely, of the non-primitive mathematical terms) provided that the five Peano postulates are true. However, strictly speaking, we cannot, at this juncture, refer to the Peano postulates as propositions which are either true or false, for they contain three primitive terms which have not been assigned any specific meaning. All we can assert so far is that any specific interpretation of the primitives which satisfies the five postulates – i.e., turns them into true statements – will also satisfy all the theorems deduced from them. But for Peano's system, there are several – indeed, infinitely many – interpretations which will do this. For example, let us understand by 0 the origin of a half-line, by the successor of a point on that half-line the

[6] As a result of very deep-reaching investigations carried out by K. Gödel it is known that arithmetic, and *a fortiori* mathematics, is an incomplete theory in the following sense: While all those propositions which belong to the classical systems of arithmetic, algebra, and analysis can indeed be derived, in the sense characterized above, from the Peano postulates, there exist nevertheless other propositions which can be expressed in purely arithmetical terms, and which are true, but which cannot be derived from the Peano system. And more generally: For any postulate system of arithmetic (or of mathematics for that matter) which is not self-contradictory, there exist propositions which are true, and which can be stated in purely arithmetical terms, but which cannot be derived from that postulate system. In other words, it is impossible to construct a postulate system which is not self-contradictory, and which contains among its consequences all true propositions which can be formulated within the language of arithmetic.

This fact does not, however, affect the result outlined above, namely, that it is possible to deduce, from the Peano postulates and the additional definitions of non-primitive terms, all those propositions which constitute the classical theory of arithmetic, algebra, and analysis; and it is to these propositions that I refer above and subsequently as the propositions of mathematics.

point 1 cm. behind it, counting from the origin, and by a number any point which is either the origin or can be reached from it by a finite succession of steps each of which leads from one point to its successor. It can then readily be seen that all the Peano postulates as well as the ensuing theorems turn into true propositions, although the interpretation given to the primitives is certainly not the customary one, which was mentioned earlier. More generally, it can be shown that every progression of elements of any kind provides a true interpretation, or a "model," of the Peano system. This example illustrates our earlier observation that a postulate system cannot be regarded as a set of "implicit definitions" for the primitive terms: The Peano system permits of many different interpretations, whereas in everyday as well as in scientific language, we attach one specific meaning to the concepts of arithmetic. Thus, e.g., in scientific and in everyday discourse, the concept 2 is understood in such a way that from the statement "Mr. Brown as well as Mr. Cope, but no one else is in the office, and Mr. Brown is not the same person as Mr. Cope," the conclusion "Exactly two persons are in the office" may be validly inferred. But the stipulations laid down in Peano's system for the natural numbers, and for the number 2 in particular, do not enable us to draw this conclusion; they do not "implicitly determine" the customary meaning of the concept 2 or of the other arithmetical concepts. And the mathematician cannot acquiesce at this deficiency by arguing that he is not concerned with the customary meaning of the mathematical concepts; for in proving, say, that every positive real number has exactly two real square roots, he is himself using the concept 2 in its customary meaning, and his very theorem cannot be proved unless we presuppose more about the number 2 than is stipulated in the Peano system.

If therefore mathematics is to be a correct theory of the mathematical concepts in their intended meaning, it is not sufficient for its validation to have shown that the entire system is derivable from the Peano postulates plus suitable definitions; rather, we have to inquire further whether the Peano postulates are actually true when the primitives are understood in their customary meaning. This question, of course, can be answered only after the customary meaning of the terms "0," "natural number," and "successor" has been clearly defined. To this task we now turn.

8. Definition of the customary meaning of the concepts of arithmetic in purely logical terms

At first blush, it might seem a hopeless undertaking to try to define these basic arithmetical concepts without presupposing other terms of arithmetic, which would involve us in a circular procedure. However, quite rigorous definitions of the desired kind can indeed be formulated, and it can be shown that for the concepts so defined, all Peano postulates turn into true statements. This important result is due to the research of the German logician G. Frege (1848–1925) and to the subsequent systematic and detailed work of the contemporary English logicians and philosophers B. Russell and A. N. Whitehead. Let us consider briefly the basic ideas underlying these definitions.[7]

A natural number – or, in Peano's term, a number – in its customary meaning can be considered as a characteristic of certain *classes* of objects. Thus, e.g., the class of the apostles has the number 12, the class of the Dionne quintuplets the number 5, any couple the number 2, and so on. Let us now express precisely the meaning of the assertion that a certain class C has the number 2, or briefly, that $n(C)=2$. Brief reflection will show that the following definiens is adequate in the sense of the customary meaning of the concept 2: There is some object x and some object y such that (1) $x \in C$ (i.e., x is an element of C) and $y \in C$, (2) $x \neq y$, and (3) if z is any object such that $z \in C$, then either $z=x$ or $z=y$. (Note that on the basis of this definition it becomes indeed possible to infer the statement "The number of persons in the office is 2" from "Mr. Brown as well as Mr. Cope, but no one else is in the office, and Mr. Brown is not identical with Mr. Cope"; C is here the class of persons in the office.) Analogously, the meaning of the statement that $n(C)=1$ can be defined thus: There is some x such that $x \in C$, and any object y such that $y \in C$, is identical with x. Similarly, the customary meaning of the statement that $n(C)=0$ is this: There is no object such that $x \in C$.

The general pattern of these definitions clearly lends itself to the definition of any natural number. Let us note especially that in the definitions thus obtained, the definiens never contains any arithmetical term, but merely expressions taken from the field of formal logic, including the signs of identity and difference. So far, we have defined only the meaning of such phrases as "$n(C)=2$," but we have given no definition for the numbers $0, 1, 2, \ldots$ apart from this context. This desideratum can be met on the basis of the consideration that 2 is that property which is common to all couples, i.e., to all classes C such that $n(C)=2$. This common property may be conceptually represented by the class of all those classes which share this property. Thus we arrive at the definition: 2 is the class

[7]For a more detailed discussion, cf. Russell, 1919: chaps. 2, 3, 4. A complete technical development of the idea can be found in the great standard work in mathematical logic, Whitehead and Russell 1910–13. – For a very precise recent development of the theory, see Quine 1940. – A specific discussion of the Peano system and its interpretations from the viewpoint of semantics is included in Carnap 1939: esp. sections 14, 17, 18.

of all couples, i.e., the class of all classes C for which $n(C) = 2$. – This definition is by no means circular because the concept of couple – in other words, the meaning of "$n(C) = 2$" – has been previously defined without any reference to the number 2. Analogously, 1 is the class of all unit classes, i.e., the class of all classes C for which $n(C) = 1$. Finally, 0 is the class of all null classes, i.e., the class of all classes without elements. And as there is only one such class, 0 is simply the class whose only element is the null class. Clearly, the customary meaning of any given natural number can be defined in this fashion.[8] In order to characterize the intended interpretation of Peano's primitives, we actually need, of all the definitions here referred to, only that of the number 0. It remains to define the terms "successor" and "integer."

The definition of "successor," whose precise formulation involves too many niceties to be stated here, is a careful expression of a simple idea which is illustrated by the following example: Consider the number 5, i.e., the class of all quintuplets. Let us select an arbitrary one of these quintuplets and add to it an object which is not yet one of its members. $5'$, the successor of 5, may then be defined as the number applying to the set thus obtained (which, of course, is a sextuplet). Finally, it is possible to formulate a definition of the customary meaning of the concept of natural number; this definition, which again cannot be given here, expresses, in a rigorous form, the idea that the class of the natural numbers consists of the number 0, its successor, the successor of that successor, and so on.

If the definitions here characterized are carefully written out – this is one of the cases where the techniques of symbolic, or mathematical, logic prove indispensable – it is seen that the definiens of every one of them contains exclusively terms from the field of pure logic. In fact, it is possible to state the customary interpretation of Peano's primitives, and thus also the meaning of every concept definable by means of them – and that includes every concept of mathematics – in terms of the following 7 expressions, in addition to variables such as "x" and "C": *not, and, if – then; for every object x* it is the case that . . .; *there is some object x* such that . . .; *x* is an *element* of class C; *the class of all things x such that* . . . .

[8] The assertion that the definitions given above state the "customary" meaning of the arithmetical terms involved is to be understood in the logical, not the psychological sense of the term "meaning." It would obviously be absurd to claim that the above definitions express "what everybody has in mind" when talking about numbers and the various operations that can be performed with them. What is achieved by those definitions is rather a "logical reconstruction" of the concepts of arithmetic in the sense that if the definitions are accepted, then those statements in science and everyday discourse which involve arithmetical terms can be interpreted coherently and systematically in such a manner that they are capable of objective validation. The statement about the two persons in the office provides a very elementary illustration of what is meant here.

And it is even possible to reduce the number of logical concepts needed to a mere four: the first three of the concepts just mentioned are all definable in terms of *"neither-nor,"* and the fifth is definable by means of the fourth and *"neither-nor,"* thus, all the concepts of mathematics prove definable in terms of four concepts of pure logic. (The definition of one of the more complex concepts of mathematics in terms of the four primitives just mentioned may well fill hundreds or even thousands of pages; but clearly this affects in no way the theoretical importance of the result just obtained; it does, however, show the great convenience and indeed practical indispensability for mathematics of having a large system of highly complex defined concepts available.)

9. The truth of Peano's postulates in their customary interpretation

The definitions characterized in the preceding section may be said to render precise and explicit the customary meaning of the concepts of arithmetic. Moreover – and this is crucial for the question of the validity of mathematics – it can be shown that the Peano postulates all turn into true propositions if the primitives are construed in accordance with the definitions just considered.

Thus, P1 (0 is a number) is true because the class of all numbers – i.e., natural numbers – was defined as consisting of 0 and all its successors. The truth of P2 (The successor of any number is a number) follows from the same definition. This is true also of P5, the principle of mathematical induction. To prove this, however, we would have to resort to the precise definition of "integer" rather than the loose description given of that definition above. P4 (0 is not the successor of any number) is seen to be true as follows: By virtue of the definition of "successor," a number which is a successor of some number can apply only to classes which contain at least one element; but the number 0, by definition, applies to a class if and only if that class is empty. – While the truth of P1, P2, P4, P5 can be inferred from the above definitions simply by means of the principles of logic, the proof of P3 (No two numbers have the same successor) presents a certain difficulty. As was mentioned in the preceding section, the definition of the successor of a number n is based on the process of adding, to a class of n elements, one element not yet contained in that class. Now if there should exist only a finite number of things altogether then this process could not be continued indefinitely, and P3, which (in conjunction with P1 and P2) implies that the integers form an infinite set, would be false. Russell's way of meeting this difficulty was to introduce a special "axiom of infinity," which stipulates, in effect, the existence of infinitely many objects and thus makes P3 demonstrable (cf. Russell 1919: 24 and chap. 13). The axiom of infinity does not belong to the generally recognized laws of logic; but it is capable of expression in purely logical terms and may be treated as an additional postulate of logic.

10. Mathematics as a branch of logic

As was pointed out earlier, all the theorems of arithmetic, algebra, and analysis can be deduced from the Peano postulates and the definitions of those mathematical terms which are not primitives in Peano's system. This deduction requires only the principles of logic plus, in certain cases, the axiom of choice, which asserts that for any set of mutually exclusive non-empty sets $\alpha, \beta, \ldots$, there exists at least one set which contains exactly one element from each of the sets $\alpha, \beta, \ldots$, and which contains no other elements.[9] By combining this result with what has just been said about the Peano system, the following conclusion is obtained, which is also known as *the thesis of logicism concerning the nature of mathematics*:

Mathematics is a branch of logic. It can be derived from logic in the following sense:

a. all the concepts of mathematics, i.e. of arithmetic, algebra, and analysis, can be defined in terms of four concepts of pure logic.

b. All the theorems of mathematics can be deduced from those definitions by means of the principles of logic (including the axioms of infinity and choice).[10]

In this sense it can be said that the propositions of the system of mathematics as here delimited are true by virtue of the definitions of the mathematical concepts involved, or that they make explicit certain characteristics with which we have endowed our mathematical concepts by definition. The propositions of mathematics have, therefore, the same unquestionable certainty which is typical of such propositions as "All

[9]This only apparently self-evident postulate is used in proving certain theorems of set theory and of real and complex analysis; for a discussion of its significance and of its problematic aspects, see Russell 1919: chap. 12 (where it is called the multiplicative axiom), and Fraenkel 1919 (Dover): §16, sections 7 and 8.

[10]The principles of logic developed in Quine's work and in similar modern systems of formal logic embody certain restrictions as compared with those logical rules which had been rather generally accepted as sound until about the turn of the 20th century. At that time, the discovery of the famous paradoxes of logic, especially of Russell's paradox (cf. Russell 1919: chap. 13) revealed the fact that the logical principles implicit in customary mathematical reasoning involved contradictions and therefore had to be curtailed in one manner or another.

bachelors are unmarried,'' but they also share the complete lack of empirical content which is associated with that certainty: The propositions of mathematics are devoid of all factual content; they convey no information whatever on any empirical subject matter.

11. On the applicability of mathematics to empirical subject matter

This result seems to be irreconcilable with the fact that after all mathematics has proved to be eminently applicable to empirical subject matter, and that indeed the greater part of present-day scientific knowledge has been reached only through continual reliance on and application of the propositions of mathematics. – Let us try to clarify this apparent paradox by reference to some examples.

Suppose that we are examining a certain amount of some gas, whose volume v, at a certain fixed temperature, is found to be 9 cubic feet when the pressure p is 4 atmospheres. And let us assume further that the volume of the gas for the same temperature and $p = 6$ at., is predicted by means of Boyle's law. Using elementary arithmetic we reason thus: For corresponding values of v and p, $vp = c$, and $v = 9$ when $p = 4$; hence $c = 36$: Therefore, when $p = 6$, then $v = 6$. Suppose that this prediction is borne out by subsequent test. Does that show that the arithmetic used has a predictive power of its own, that its propositions have factual implications? Certainly not. All the predictive power here deployed, all the empirical content exhibited stems from the initial data and from Boyle's law, which asserts that $vp = c$ for *any* two corresponding values of v and p, hence also for $v = 9$, $p = 4$, and for $p = 6$ and the corresponding value of v.[11] The function of the mathematics here applied is not predictive at all; rather, it is analytic or explicative: it renders explicit certain assumptions or assertions which are included in the content of the premises of the argument (in our case, these consist of Boyle's law plus the additional data); mathematical reasoning reveals that those premises contain – hidden in them, as it were, – an assertion about the case as yet unobserved. In accepting our premises – so arithmetic reveals – we have – knowingly or unknowingly – already accepted the implication that the p-value in question is 6. Mathematical as well as logical reasoning is a conceptual technique of making explicit what is implicitly contained in a set of premises. The conclusions to which this technique leads assert nothing that is *theoretically new* in the sense of not being contained in the content of the premises. But the results obtained may

[11]Note that we may say ''hence'' by virtue of the rule of substitution, which is one of the rules of logical inference.

well be *psychologically new*: we may not have been aware, before using the techniques of logic and mathematics, what we committed ourselves to in accepting a certain set of assumptions or assertions.

A similar analysis is possible in all other cases of applied mathematics, including those involving, say, the calculus. Consider, for example, the hypothesis that a certain object, moving in a specified electric field, will undergo a constant acceleration of 5 feet/sec^2. For the purpose of testing this hypothesis, we might derive from it, by means of two successive integrations, the prediction that if the object is at rest at the beginning of the motion, then the distance covered by it at any time t is $(5/2)t^2$ feet. This conclusion may clearly be psychologically new to a person not acquainted with the subject, but it is not theoretically new; the content of the conclusion is already contained in that of the hypothesis about the constant acceleration. And indeed, here as well as in the case of the compression of a gas, a failure of the prediction to come true would be considered as indicative of the factual incorrectness of at least one of the premises involved (*f.ex.*, of Boyle's law in its application to the particular gas), but never as a sign that the logical and mathematical principles involved might be unsound.

Thus, in the establishment of empirical knowledge, mathematics (as well as logic) has, so to speak, the function of a theoretical juice extractor: the techniques of mathematical and logical theory can produce no more juice of factual information than is contained in the assumptions to which they are applied; but they may produce a great deal more juice of this kind than might have been anticipated upon a first intuitive inspection of those assumptions which form the raw material for the extractor.

At this point, it may be well to consider briefly the status of those mathematical disciplines which are not outgrowths of arithmetic and thus of logic; these include in particular topology, geometry, and the various branches of abstract algebra, such as the theory of groups, lattices, fields, etc. Each of these disciplines can be developed as a purely deductive system on the basis of a suitable set of postulates. If P be the conjunction of the postulates for a given theory, then the proof of a proposition T of that theory consists in deducing T from P by means of the principles of formal logic. What is established by the proof is therefore not the truth of T, but rather the fact that T is true provided that the postulates are. But since both P and T contain certain primitive terms of the theory, to which no specific meaning is assigned, it is not strictly possible to speak of the truth of either P or T; it is therefore more adequate to state the point as follows: If proposition T is logically deduced from P, then every specific interpretation of the primitives which turns all the postulates of P into true statements, will also render T a true statement. – Up to this point, the analysis is exactly analogous to that of arithmetic as based on Peano's set of postulates. In the case of arithmetic, however, it proved possible to go a step further, namely to define the customary meanings of the primitives in terms of purely logical concepts and to show that the postulates – and therefore also the theorems – of arithmetic are unconditionally true by virtue of these definitions. An analogous procedure is not applicable to those disciplines which are not outgrowths of arithmetic: The primitives of the various branches of abstract algebra have no specific "customary meaning"; and if geometry in its customary interpretation is thought of as a theory of the structure of physical space, then its primitives have to be construed as referring to certain types of physical entities, and the question of the truth of a geometrical theory in this interpretation turns into an *empirical* problem.[12] For the purpose of applying any one of these nonarithmetical disciplines to some specific field of mathematics or empirical science, it is therefore necessary first to assign to the primitives some specific meaning and then to ascertain whether in this interpretation the postulates turn into true statements. If this is the case, then we can be sure that all the theorems are true statements too, because they are logically derived from the postulates and thus simply explicate the content of the latter in the given interpretation. – In their application to empirical subject matter, therefore, these mathematical theories no less than those which grow out of arithmetic and ultimately out of pure logic, have the function of an analytic tool, which brings to light the implications of a given set of assumptions but adds nothing to their content.

But while mathematics in no case contributes anything to the content of our knowledge of empirical matters, it is entirely indispensable as an instrument for the validation and even for the linguistic expression of such knowledge: the majority of the more far-reaching theories in empirical science – including those which lend themselves most eminently to prediction or to practical application – are stated with the help of mathematical concepts; the formulation of these theories makes use, in particular, of the number system, and of functional relationships among different metrical variables. Furthermore, the scientific test of these theories, the establishment of predictions by means of them, and finally their practical application, all require the deduction, from the general theory, of certain specific consequences; and such deduction would be entirely impossible without the techniques of mathematics which reveal

---

[12]For a more detailed discussion of this point, cf. Hempel 1945a.

what the given general theory implicitly asserts about a certain special case.

Thus, the analysis outlined on these pages exhibits the system of mathematics as a vast and ingenious conceptual structure without empirical content and yet an indispensable and powerful theoretical instrument for the scientific understanding and mastery of the world of our experience.

On the nature of mathematical reasoning

HENRI POINCARE

I

The very possibility of mathematical science seems an insoluble contradiction. If this science is only deductive in appearance, from whence is derived that perfect rigour which is challenged by none? If, on the contrary, all the propositions which it enunciates may be derived in order by the rules of formal logic, how is it that mathematics is not reduced to a gigantic tautology? The syllogism can teach us nothing essentially new, and if everything must spring from the principle of identity, then everything should be capable of being reduced to that principle. Are we then to admit that the enunciations of all the theorems with which so many volumes are filled are only indirect ways of saying that A is A?

No doubt we may refer back to axioms which are at the source of all these reasonings. If it is felt that they cannot be reduced to the principle of contradiction, if we decline to see in them any more than experimental facts which have no part or lot in mathematical necessity, there is still one resource left to us: we may class them among *a priori* synthetic views. But this is no solution of the difficulty – it is merely giving it a name; and even if the nature of the synthetic views had no longer for us any mystery, the contradiction would not have disappeared; it would have only been shirked. Syllogistic reasoning remains incapable of adding anything to the data that are given it; the data are reduced to axioms, and that is all we should find in the conclusions.

No theorem can be new unless a new axiom intervenes in its demonstration; reasoning can only give us immediately evident truths borrowed from direct intuition; it would only be an intermediary parasite. Should we not therefore have reason for asking if the syllogistic apparatus serves only to disguise what we have borrowed?

The contradiction will strike us the more if we open any book on mathematics; on every page the author announces his intention of generalizing some proposition already known. Does the mathematical

Excerpted and reprinted with the kind permission of the publisher from Henri Poincaré. *Science and Hypothesis* (New York: Dover Publications, Inc., 1952), pp. 1–19.

method proceed from the particular to the general, and, if so, how can it be called deductive?

Finally, if the science of number were merely analytical, or could be analytically derived from a few synthetic intuitions, it seems that a sufficiently powerful mind could with a single glance perceive all its truths; nay, one might even hope that some day a language would be invented simple enough for these truths to be made evident to any person of ordinary intelligence.

Even if these consequences are challenged, it must be granted that mathematical reasoning has of itself a kind of creative virtue, and is therefore to be distinguished from the syllogism. The difference must be profound. We shall not, for instance, find the key to the mystery in the frequent use of the rule by which the same uniform operation applied to two equal numbers will give identical results. All these modes of reasoning, whether or not reducible to the syllogism, properly so called, retain the analytical character, and *ipso facto,* lose their power.

II

The argument is an old one. Let us see how Leibnitz tried to show that two and two make four. I assume the number one to be defined, and also the operation $x+1$ – i.e., the adding of unity to a given number x. These definitions, whatever they may be, do not enter into the subsequent reasoning. I next define the numbers 2, 3, 4 by the equalities: –

(1) $1+1=2$;   (2) $2+1=3$;   (3) $3+1=4$,

and in the same way I define the operation $x+2$ by the relation;

(4) $x+2=(x+1)+1.$

Given this, we have: –

$$2+2=(2+1)+1; \quad \text{(def. 4)}.$$
$$(2+1)+1=3+1 \quad \text{(def. 2)}.$$
$$3+1=4 \quad \text{(def. 3)}.$$

whence

$$2+2=4 \qquad \text{Q.E.D.}$$

It cannot be denied that this reasoning is purely analytical. But if we ask a mathematician, he will reply: "This is not a demonstration properly so called; it is a verification." We have confined ourselves to bringing together one or other of two purely conventional definitions, and we have verified their identity; nothing new has been learned. *Verification* differs from proof precisely because it is analytical, and because it leads to nothing. It leads to nothing because the conclusion is nothing but the premisses translated into another language. A real proof, on the other hand, is fruitful, because the conclusion is in a sense more general than the premisses. The equality $2+2=4$ can be verified because it is particular. Each individual enunciation in mathematics may be always verified in the same way. But if mathematics could be reduced to a series of such verifications it would not be a science. A chess-player, for instance, does not create a science by winning a piece. There is no science but the science of the general. It may even be said that the object of the exact sciences is to dispense with these direct verifications.

III

Let us now see the geometer at work, and try to surprise some of his methods. The task is not without difficulty; it is not enough to open a book at random and to analyse any proof we may come across. First of all, geometry must be excluded, or the question becomes complicated by difficult problems relating to the rôle of the postulates, the nature and the origin of the idea of space. For analogous reasons we cannot avail ourselves of the infinitesimal calculus. We must seek mathematical thought where it has remained pure – i.e., in Arithmetic. But we still have to choose; in the higher parts of the theory of numbers the primitive mathematical ideas have already undergone so profound an elaboration that it becomes difficult to analyse them.

It is therefore at the beginning of Arithmetic that we must expect to find the explanation we seek; but it happens that it is precisely in the proofs of the most elementary theorems that the authors of classic treatises have displayed the least precision and rigour. We may not impute this to them as a crime; they have obeyed a necessity. Beginners are not prepared for real mathematical rigour; they would see in it nothing but empty, tedious subtleties. It would be waste of time to try to make them more exacting; they have to pass rapidly and without stopping over the road which was trodden slowly by the founders of the science.

Why is so long a preparation necessary to habituate oneself to this perfect rigour, which it would seem should naturally be imposed on all minds? This is a logical and psychological problem which is well worthy of study. But we shall not dwell on it; it is foreign to our subject. All I wish to insist on is, that we shall fail in our purpose unless we reconstruct the proofs of the elementary theorems, and give them, not the rough

On the nature of mathematical reasoning

form in which they are left so as not to weary the beginner, but the form which will satisfy the skilled geometer.

Definition of addition

I assume that the operation $x+1$ has been defined; it consists in adding the number 1 to a given number x. Whatever may be said of this definition, it does not enter into the subsequent reasoning.

We now have to define the operation $x+a$, which consists in adding the number a to any given number x. Suppose that we have defined the operation $x+(a-1)$; the operation $x+a$ will be defined by the equality: (1) $x+a=[x+(a-1)]+1$. We shall know what $x+a$ is when we know what $x+(a-1)$ is, and as I have assumed that to start with we know what $x+1$ is, we can define successively and "by recurrence" the operations $x+2$, $x+3$, etc. This definition deserves a moment's attention; it is of a particular nature which distinguishes it even at this stage from the purely logical definition; the equality (1), in fact, contains an infinite number of distinct definitions, each having only one meaning when we know the meaning of its predecessor.

Properties of addition

Associative. – I say that $a+(b+c)=(a+b)+c$; in fact, the theorem is true for $c=1$. It may then be written $a+(b+1)=(a+b)+1$; which, remembering the difference of notation, is nothing but the equality (1) by which I have just defined addition. Assume the theorem true for $c=\gamma$, I say that it will be true for $c=\gamma+1$. Let $(a+b)+\gamma=a+(b+\gamma)$, it follows that $[(a+b)+\gamma]+1=[a+(b+\gamma)]+1$; or by def. (1) – $(a+b)+(\gamma+1)=a+(b+\gamma+1)=a+[b+(\gamma+1)]$, which shows by a series of purely analytical deductions that the theorem is true for $\gamma+1$. Being true for $c=1$, we see that it is successively true for $c=2$, $c=3$, etc.

Commutative. – (1) I say that $a+1=1+a$. The theorem is evidently true for $a=1$; we can *verify* by purely analytical reasoning that if it is true for $a=\gamma$ it will be true for $a=\gamma+1$.[1] Now, it is true for $a=1$, and therefore is true for $a=2$, $a=3$, and so on. This is what is meant by saying that the proof is demonstrated "by recurrence."

(2) I say that $a+b=b+a$. The theorem has just been shown to hold good for $b=1$, and it may be verified analytically that if it is true for

[1] For $(\gamma+1)+1=(1+\gamma)+1=1+(\gamma+1)$. – [Tr.]

397

$b=\beta$, it will be true for $b=\beta+1$. The proposition is thus established by recurrence.

Definition of multiplication

We shall define multiplication by the equalities: (1) $a \times 1 = a$. (2) $a \times b = [a \times (b-1)] + a$. Both of these include an infinite number of definitions; having defined $a \times 1$, it enables us to define in succession $a \times 2$, $a \times 3$, and so on.

Properties of multiplication

Distributive. – I say that $(a+b) \times c = (a \times c) + (b \times c)$. We can verify analytically that the theorem is true for $c=1$; then if it is true for $c=\gamma$, it will be true for $c=\gamma+1$. The proposition is then proved by recurrence.

Commutative. – (1) I say that $a \times 1 = 1 \times a$. the theorem is obvious for $a=1$. We can verify analytically that if it is true for $a=a$, it will be true for $a=\alpha+1$.

(2) I say that $a \times b = b \times a$. The theorem has just been proved for $b=1$. We can verify analytically that if it be true for $b=\beta$ it will be true for $b=\beta+1$.

IV

This monotonous series of reasonings may now be laid aside; but their very monotony brings vividly to light the process, which is uniform, and is met again at every step. The process is proof by recurrence. We first show that a theorem is true for $n=1$; we then show that if it is true for $n-1$ it is true for n, and we conclude that it is true for all integers. We have now seen how it may be used for the proof of the rules of addition and multiplication – that is to say, for the rules of the algebraical calculus. This calculus is an instrument of transformation which lends itself to many more different combinations than the simple syllogism; but it is still a purely analytical instrument, and is incapable of teaching us anything new. If mathematics had no other instrument, it would immediately be arrested in its development; but it has recourse anew to the same process – i.e., to reasoning by recurrence, and it can continue its forward march. Then if we look carefully, we find this mode of reasoning at every step, either under the simple form which we have just given to it, or under a more or less modified form. It is therefore mathematical reasoning *par excellence*, and we must examine it closer.

V

The essential characteristic of reasoning by recurrence is that it contains, condensed, so to speak, in a single formula, an infinite number of syllogisms. We shall see this more clearly if we enunciate the syllogisms one after another. They follow one another, if one may use the expression, in a cascade. The following are the hypothetical syllogisms: – The theorem is true of the number 1. Now, if it is true of 1, it is true of 2; therefore it is true of 2. Now, if it is true of 2, it is true of 3; hence it is true of 3, and so on. We see that the conclusion of each syllogism serves as the minor of its successor. Further, the majors of all our syllogisms may be reduced to a single form. If the theorem is true of $n-1$, it is true of n.

We see, then, that in reasoning by recurrence we confine ourselves to the enunciation of the minor of the first syllogism, and the general formula which contains as particular cases all the majors. This unending series of syllogisms is thus reduced to a phrase of a few lines.

It is now easy to understand why every particular consequence of a theorem may, as I have above explained, be verified by purely analytical processes. If, instead of proving that our theorem is true for all numbers, we only wish to show that it is true for the number 6 for instance, it will be enough to establish the first five syllogisms in our cascade. We shall require 9 if we wish to prove it for the number 10; for a greater number we shall require more still; but however great the number may be we shall always reach it, and the analytical verification will always be possible. But however far we went we should never reach the general theorem applicable to all numbers, which alone is the object of science. To reach it we should require an infinite number of syllogisms, and we should have to cross an abyss which the patience of the analyst, restricted to the resources of formal logic, will never succeed in crossing.

I asked at the outset why we cannot conceive of a mind powerful enough to see at a glance the whole body of mathematical truth. The answer is now easy. A chess-player can combine for four or five mores ahead; but, however extraordinary a player he may be, he cannot prepare for more than a finite number of moves. If he applies his faculties to Arithmetic, he cannot conceive its general truths by direct intuition alone; to prove even the smallest theorem he must use reasoning by recurrence, for that is the only instrument which enable us to pass from the finite to the infinite. This instrument is always useful, for it enables us to leap over as many stages as we wish; it frees us from the necessity of long, tedious, and monotonous verifications which would rapidly become impracticable. Then when we take in hand the general theorem it becomes indispensable, for otherwise we should ever by approaching the analytical verification without every actually reaching it. In this domain of Arithmetic we may think ourselves very far from the infinitesimal analysis, but the idea of mathematical infinity is already playing a pre-ponderating part, and without it there would be no science at all, because there would be nothing general.

VI

The views upon which reasoning by recurrence is based may be exhibited in other forms; we may say, for instance, that in any finite collection of different integers there is always one which is smaller than any other. We may readily pass from one enunciation to another, and thus give our-selves the illusion of having proved that reasoning by recurrence is legiti-mate. But we shall always be brought to a full stop – we shall always come to an indemonstrable axiom, which will at bottom be but the prop-osition we had to prove translated into another language. We cannot therefore escape the conclusion that the rule of reasoning by recurrence is irreducible to the principle of contradiction. Nor can the rule come to us from experiment. Experiment may teach us that the rule is true for the first ten or the first hundred numbers, for instance; it will not bring us to the indefinite series of numbers, but only to a more or less long, but always limited, portion of the series.

Now, if that were all that is in question, the principle of contradiction would be sufficient, it would always enable us to develop as many syllo-gisms as we wished. It is only when it is a question of a single formula to embrace an infinite number of syllogisms that this principle breaks down, and there, too, experiment is powerless to aid. This rule, inac-cessible to analytical proof and to experiment, is the exact type of the *a priori* synthetic intuition. On the other hand, we cannot see in it a con-vention as in the case of the postulates of geometry.

Why then is this view imposed upon us with such an irresistible weight of evidence? It is because it is only the affirmation of the power of the mind which knows it can conceive of the indefinite repetition of the same act, when the act is once possible. The mind has a direct intuition of this power, and experiment can only be for it an opportunity of using it, and thereby of becoming conscious of it.

But it will be said, if the legitimacy of reasoning by recurrence cannot be established by experiment alone, is it so with experiment aided by induction? We see successively that a theorem is true of the number 1, of the number 2, of the number 3, and so on – the law is manifest, we say, and it is so on the same ground that every physical law is true which is based on a very large but limited number of observations.

It cannot escape our notice that here is a striking analogy with the usual processes of induction. But an essential difference exists. Induction applied to the physical sciences is always uncertain, because it is based on the belief in a general order of the universe, an order which is external to us. Mathematical induction – i.e., proof by recurrence – is, on the contrary, necessarily imposed on us, because it is only the affirmation of a property of the mind itself.

VII

Mathematicians, as I have said before, always endeavour to generalize the propositions they have obtained. To seek no further example, we have just shown the equality, $a+1=1+a$, and we then used it to establish the equality, $a+b=b+a$, which is obviously more general. Mathematics may, therefore, like the other sciences, proceed from the particular to the general. This is a fact which might otherwise have appeared incomprehensible to us at the beginning of this study, but which has no longer anything mysterious about it, since we have ascertained the analogies between proof by recurrence and ordinary induction.

No doubt mathematical recurrent reasoning and physical inductive reasoning are based on different foundations, but they move in parallel lines and in the same direction – namely, from the particular to the general.

Let us examine the case a little more closely. To prove the equality $a+2=2+a\ldots(1)$, we need only apply the rule $a+1=1+a$, twice, and write $a+2=a+1+1=1+a+1=1+1+a=2+a\ldots(2)$.

The equality thus deduced by purely analytical means is not, however, a simple particular case. It is something quite different. We may not therefore even say in the really analytical and deductive part of mathematical reasoning that we proceed from the general to the particular in the ordinary sense of the words. The two sides of the equality (2) are merely more complicated combinations than the two sides of the equality (1), and analysis only serves to separate the elements which enter into these combinations and to study their relations.

Mathematicians therefore proceed "by construction," they "construct" more complicated combinations. When they analyze these combinations, these aggregates, so to speak, into their primitive elements, they see the relations of the elements and deduce the relations of the aggregates themselves. The process is purely analytical, but it is not a passing from the general to the particular, for the aggregates obviously cannot be regarded as more particular than their elements.

Great importance has been rightly attached to this process of "construction," and some claim to see in it the necessary and sufficient condition of the progress of the exact sciences. Necessary, no doubt, but not sufficient! For a construction to be useful and not mere waste of mental effort, for it to serve as a stepping-stone to higher things, it must first of all possess a kind of unity enabling us to see something more than the juxtaposition of its elements. Or more accurately, there must be some advantage in considering the construction rather than the elements themselves. What can this advantage be? Why reason on a polygon, for instance, which is always decomposable into triangles, and not on elementary triangles? It is because there are properties of polygons of any number of sides, and they can be immediately applied to any particular kind of polygon. In most cases it is only after long efforts that those properties can be discovered, by directly studying the relations of elementary triangles. If the quadrilateral is anything more than the juxtaposition of two triangles, it is because it is of the polygon type.

A construction only becomes interesting when it can be placed side by side with other analogous constructions for forming species of the same genus. To do this we must necessarily go back from the particular to the general, ascending one or more steps. The analytical process "by construction" does not compel us to descend, but it leaves us at the same level. We can only ascend by mathematical induction, for from it alone we can learn something new. Without the aid of this induction, which in certain respects differs from, but is as fruitful as, physical induction, construction would be powerless to create science.

Let me observe, in conclusion, that this induction is only possible if the same operation can be repeated indefinitely. That is why the theory of chess can never become a science, for the different moves of the same piece are limited and do not resemble each other.

Mathematical truth

PAUL BENACERRAF

Although this symposium is entitled "Mathematical Truth," I will also discuss issues which are somewhat broader but which nevertheless have the notion of mathematical truth at their core, which themselves depend on how truth in mathematics is properly explained. The most important of these is mathematical knowledge. It is my contention that two quite distinct kinds of concerns have separately motivated accounts of the nature of mathematical truth: (1) the concern for having a homogeneous semantical theory in which semantics for the propositions of mathematics parallel the semantics for the rest of the language,[1] and (2) the concern that the account of mathematical truth mesh with a reasonable epistemology. It will be my general thesis that almost all accounts of the concept of mathematical truth can be identified with serving one or another of these masters *at the expense of the other.* Since I believe further that both concerns must be met by any adequate account, I find myself deeply dissatisfied with any package of semantics and epistemology that purports to account for truth and knowledge both within and outside of mathematics. For, as I will suggest, accounts of truth that treat mathematical and nonmathematical discourse in relevantly similar ways do so at the cost of leaving it unintelligible how we can have any mathematical knowledge whatsoever; whereas those which attribute to

Presented at a symposium on Mathematical Truth, sponsored jointly by the American Philosophical Association, Eastern Division, and the Association for Symbolic Logic, December 27, 1973. Various segments of an early (1967) version of this paper have been read at Berkeley, Harvard, Chicago Circle, Johns Hopkins, New York University, Princeton, and Yale. I am grateful for the help I received on these occasions, as well as for many comments from my colleagues at Princeton, both students and faculty. I am particularly indebted to Richard Grandy, Hartry Field, Adam Morton, and Mark Steiner. That these have not resulted in more significant improvements is due entirely to my own stubbornness. The present version is an attempt to summarize the essentials of the longer paper while making minor improvements along the way. The original version was written during 1967/68 with the generous support of the John Simon Guggenheim Foundation and Princeton University. This is gratefully acknowledged.

Reprinted with the kind permission of the editors from the *Journal of Philosophy* 70 (1973): 661–80.

[1]I am indulging here in the fiction that we *have* semantics for "the rest of language," or, more precisely, that the proponents of the views that take their impetus from this concern often think of themselves as having such semantics, at least for philosophically important segments of the language.

mathematical propositions the kinds of truth conditions we can clearly know to obtain, do so at the expense of failing to connect these conditions with any analysis of the sentences which shows how the assigned conditions are conditions of their *truth*. What this means must ultimately be spelled out in some detail if I am to make out my case, and I cannot hope to do that within this limited context. But I will try to make it sufficiently clear to permit you to judge whether or not there is likely to be anything in the claim.

I take it to be obvious that any philosophically satisfactory account of truth, reference, meaning, and knowledge must embrace them all and must be adequate for all the propositions to which these concepts apply.[2] An account of knowledge that *seems* to work for certain empirical propositions about medium-sized physical objects but which fails to account for more theoretical knowledge is unsatisfactory – not only because it is incomplete, but because it may be incorrect as well, even as an account of the things it seems to cover quite adequately. To think otherwise would be, among other things, to ignore the interdependence of our knowledge in different areas. And similarly for accounts of truth and reference. A theory of truth for the language we speak, argue in, theorize in, mathematize in, etc., should by the same token provide similar truth conditions for similar sentences. The truth conditions assigned to two sentences containing quantifiers should reflect in relevantly similar ways the contribution made by the quantifiers. Any departure from a theory thus homogeneous would have to be strongly motivated to be worth considering. Such a departure, for example, might manifest itself in a theory that gave an account of the contribution of quantifiers in mathematical reasoning different from that in normal everyday reasoning about pencils, elephants, and vice-presidents. David Hilbert urged such an account in "On the Infinite" [reprinted in this volume] which is discussed briefly below. Later on, I will try to say more about what conditions I would expect a satisfactory general theory of truth for our language to meet, as well as more about how such an account is to mesh with what I take to be a reasonable account of knowledge. Suffice it to say here that, although it

---

[2] I shall in fact have nothing to say about meaning in this paper. I believe that the concept is in much deserved disrepute, but I don't dismiss it for all that. Recent work, most notably by Kripke, suggests that what passed for a long time for meaning – namely the Fregean "sense" – has less to do with truth than Frege or his immediate followers thought it had. Reference is what is presumably most closely connected with truth, and it is for *this* reason that I will limit my attention to reference. If it is granted that change of reference can take place without a corresponding change in meaning, and that truth is a matter of reference, then talk of meaning is largely beside the point of the cluster of problems that concern us in this paper. These comments are not meant as arguments, but only as explanation.

will often be convenient to present my discussion in terms of theories of mathematical truth, we should always bear in mind that what is really at issue is our over-all philosophical view. I will argue that, *as an over-all view,* it is unsatisfactory – not so much because we lack a seemingly satisfactory account of mathematical truth or because we lack a seemingly satisfactory account of mathematical knowledge – as because we lack any account that satisfactorily brings the two together. I hope that it is possible ultimately to produce such an account; I hope further that this paper will help to bring one about by bringing into sharper focus some of the obstacles that stand in its way.

I. Two kinds of account

Consider the following two sentences:

(1)   There are at least three large cities older than New York.
(2)   There are at least three perfect numbers greater than 17.

Do they have the same logicogrammatical form? More specifically, are they both of the form

(3)   There are at least three *FG*'s that bear *R* to *a*.

where 'There are at least three' is a numerical quantifier eliminable in the usual way in favor of existential quantifiers, variables, and identity; '*F*' and '*G*' are to be replaced by one-place predicates, '*R*' by a two-place predicate, and '*a*' by the name of an element of the universe of discourse of the quantifiers? What are the truth conditions of (1) and (2)? Are they relevantly parallel? Let us ignore both the vagueness of 'large' and 'older than' and the peculiarities of attributive-adjective constructions in English which make a large city not something large and a city but more (although not exactly) like something large *for* a city. With those complications set aside, it seems clear that (3) accurately reflects the form of (1) and thus that (1) will be true if and only if the thing named by the expression replacing '*a*' ('New York') bears the relation designated by the expression replacing '*R*' ('① is older than ②') to at least three elements (of the domain of discourse of the quantifiers) which satisfy the predicates replacing '*F*' and '*G*' ('large' and 'city', respectively). This, I gather, is what a suitable truth definition would tell us. And I think it's right. Thus, if (1) is true, it is because certain cities stand in a certain relation to each other, etc.

But what of (2)? May we use (3) in the same way as a matrix in spelling out the conditions of *its* truth? That sounds like a silly question to which the obvious answer is "Of course." Yet the history of the subject (the philosophy of mathematics) has seen many other answers. Some (including one of my past and present selves[3]), reluctant to face the consequences of combining what I shall dub such a "standard" semantical account with a platonistic view of the nature of numbers, have shied away from supposing that numerals are names and thus, by implication, that (2) is of the form (3). David Hilbert (1926) chose a different but equally divergent approach, in his case in an attempt to arrive at a satisfactory account of the use of the notion of infinity in mathematics. On one construal, Hilbert can be seen as segregating a class of statements and methods, those of "intuitive" mathematics, as those which needed no further justification. Let us suppose that these are all "finitely verifiable" in some sense that is not precisely specified. Statements of arithmetic that do not share this property – typically, certain statements containing quantifiers – are seen by Hilbert as instrumental devices for going from "real" or "finitely verifiable" statements to "real" statements, much as an instrumentalist regards theories in natural science as a way of going from observation sentences to observation sentences. These mathematically "theoretical" statements Hilbert called "ideal elements," likening their introduction to the introduction of points "at infinity" in projective geometry: they are introduced as a convenience to make simpler and more elegant the theory of the things you really care about. If their introduction does not lead to contradiction and if they have these other uses, then it is justified: hence the search for a consistency proof for the full system of first-order arithmetic.

If this is a reasonable, if sketchy, account of Hilbert's view, it indicates that he did not regard all quantified statements semantically on a par with one another. A semantics for arithmetic as he viewed it would be very hard to give. But hard or not, it would certainly not treat the quantifier in (2) in the same way as the quantifier in (1). Hilbert's view as outlined represents a flat denial that (3) is the model according to which (2) is constructed.

On other such accounts, the truth conditions for arithmetic sentences are given as their formal derivability from specified sets of axioms. When coupled with the desire to attribute a truth value to each closed sentence of arithmetic, these views were torpedoed by the incompleteness theorems. They could be restored at least to internal consistency either by the liberalization of what counts as derivability (e.g., by including the application of an ω-rule in permissible derivations) or by abandoning the

[3]See my "What Numbers Could Not Be," 1965 [reprinted in this volume].

desire for completeness. For lack of a better term and because they almost invariably key on the syntactic (combinatorial) features of sentences, I will call such views "combinatorial" views of the determinants of mathematical truth. The leading idea of combinatorial views is that of assigning truth values to arithmetic sentences on the basis of certain (usually proof-theoretic) syntactic facts about them. Often, truth is defined as (formal) derivability from certain axioms. (Frequently a more modest claim is made – the claim to truth-in-*S*, where *S* is the particular system in question.) In any event, in such cases truth is conspicuously not explained in terms of reference, denotation, or satisfaction. The "truth" predicate is syntactically defined.

Similarly, certain views of truth in arithmetic on which the Peano axioms are claimed to be "analytic" of the concept of number are also "combinatorial" in my sense. And so are conventionalist accounts, since what marks them as conventionalist is the contrast between them and the "realist" account that analyzes (2) by assimilating it to (1), via (3).

Finally, to make one further distinction, a view is not automatically "combinatorial" if it interprets mathematical propositions as being about combinatorial matters, either self-referentially or otherwise. For such a view might analyze mathematical propositions in a "standard" way in terms of the names and quantifiers they might contain and in terms of the properties they ascribe to the objects within their domains of discourse – which is to say that the underlying concept of *truth* is essentially Tarski's. The difference is that its proponents, although realists in their analysis of mathematical language, part ways with the platonists by construing the mathematical universe as consisting exclusively of mathematically unorthodox objects: Mathematics for them is limited to metamathematics, and that to syntax.

I will defer to later sections my assessment of the relative merits of these various approaches to the truth of such sentences as (2). At this point I wish only to introduce the distinction between, on the one hand, those views which attribute the obvious syntax (and the obvious semantics) to mathematical statements, and, on the other, those which, ignoring the apparent syntax and semantics, attempt to state truth conditions (or to specify and account for the existing distribution of truth values) on the basis of what are evidently non-semantic syntactic considerations. Ultimately I will argue that each kind of account has its merits and defects: each addresses itself to an important component of a coherent over-all philosophic account of truth and knowledge.

But what are these components, and how do they relate to one another?

II. Two conditions

A. The first component of such an over-all view is more directly concerned with the concept of truth. For present purposes we can state it as the requirement that there be an over-all theory of truth in terms of which it can be certified that the account of mathematical truth is indeed an account of mathematical *truth*. The account should imply truth conditions for mathematical propositions that are evidently conditions of their truth (and not *simply,* say, of their theoremhood in some formal system). This is not to *deny* that being a theorem of some system can be a truth condition for a given proposition or class of propositions. It is rather to require that any theory that proffers theoremhood as a condition of truth also *explain the connection between truth and theoremhood.*

Another way of putting this first requirement is to demand that any theory of mathematical truth be in conformity with a general theory of truth – a theory of truth theories, if you like – which certifies that the property of sentences that the account calls "truth" is indeed truth. This, it seems to me, can be done only on the basis of some general theory for at least the language as a whole (I assume that we skirt paradoxes in some suitable fashion). Perhaps the applicability of this requirement to the present case amounts only to a plea that the semantical apparatus of mathematics be seen as part and parcel of that of the natural language in which it is done, and thus that whatever *semantical* account we are inclined to give of names or, more generally, of singular terms, predicates, and quantifiers in the mother tongue include those parts of the mother tongue which we classify as mathematese.

I suggest that, if we are to meet this requirement, we shouldn't be satisfied with an account that fails to treat (1) and (2) in parallel fashion, on the model of (3). There may well be *differences,* but I expect these to emerge at the level of the analysis of the reference of the singular terms and predicates. I take it that we have only one such account: Tarski's, and that its essential feature is to define truth in terms of reference (or satisfaction) on the basis of a particular kind of syntactico-semantical analysis of the language, and thus that any putative analysis of mathematical truth must be an analysis of a concept which is a truth concept at least in Tarski's sense. Suitably elaborated, I believe this requirement to be inconsistent with all the accounts that I have termed "combinatorial." On the other hand, the account that assimilates (2) above to (1) and (3) obviously meets this condition, as do many variants of it.

408

B. My second condition on an over-all view presupposes that we have mathematical knowledge and that such knowledge is no less knowledge for being mathematical. Since our knowledge is of truths, or can be so construed, an account of mathematical truth, to be acceptable, must be consistent with the possibility of having mathematical knowledge: the conditions of the truth of mathematical propositions cannot make it impossible for us to know that they are satisfied. This is not to argue that there cannot be unknowable truths – only that not all truths can be unknowable, for we know some. The minimal requirement, then, is that a satisfactory account of mathematical truth must be consistent with the possibility that some such truths be knowable. To put it more strongly, the concept of mathematical truth, as explicated, must fit into an over-all account of knowledge in a way that makes it intelligible how we have the mathematical knowledge that we have. An acceptable semantics for mathematics must fit an acceptable epistemology. For example, if I know that Cleveland is between New York and Chicago, it is because there exists a certain relation between the truth conditions for that statement and my present "subjective" state of belief (whatever may be our accounts of truth and knowledge, they must connect with each other in this way). Similarly, in mathematics, it must be possible to link up what it is for *p* to be true with my belief that *p*. Though this is extremely vague, I think one can see how the second condition tends to rule out accounts that satisfy the first, and to admit many of those which do not. For a typical "standard" account (at least in the case of number theory or set theory) will depict truth conditions in terms of conditions on objects whose nature, as normally conceived, places them beyond the reach of the better understood means of human cognition (e.g., sense perception and the like). The "combinatorial" accounts, on the other hand, usually arise from a sensitivity to precisely this fact and are hence almost always motivated by epistemological concerns. Their virtue lies in providing an account of mathematical propositions based on the procedures we follow in justifying truth claims in mathematics: namely, proof. It is not surprising that *modulo* such accounts of mathematical truth, there is little mystery about how we can obtain mathematical knowledge. We need only account for our ability to produce and survey formal proofs.[4] However, squeezing the balloon at that point apparently makes it bulge on the side of truth: the more nicely we tie up the concept of proof, the more

[4]Properly done, this is of course an enormous task. Nevertheless it sets to one side accounting for the burden that is borne by the semantics of the system and by our understanding of it, concentrating instead on our ability to determine that certain formal objects have certain syntactically defined properties.

closely we link the definition of proof to combinatorial (rather than semantic) features, the more difficult it is to connect it up with the truth of what is being thus "proved" – or so it would appear.

These then are the two requirements. Separately, they seem innocuous enough. In the balance of this paper I will both defend them further and flesh out the argument that jointly they seem to rule out almost every account of mathematical truth that has been proposed. I will consider in turn the two basic approaches to mathematical truth that I mentioned above, weighing their relative advantages in light of the two fundamental principles that I am advancing. I hope that the principles themselves will receive some illumination and support as I do so.

III. The standard view

I call the "platonistic" account that analyzes (2) as being of the form (3) "the standard view." Its virtues are many, and it is worth enumerating them in some detail before passing to a consideration of its defects.

As I have already pointed out, this account assimilates the logical form of mathematical propositions to that of apparently similar empirical ones: empirical and mathematical propositions alike contain predicates, singular terms, quantifiers, etc.

But what of sentences that are not composed (or correctly analyzable as being composed of) names, predicates, and quantifiers? More directly to the point, what of sentences that do not belong to the kind of language for which Tarski has shown us how to define truth? I would say that we need for such languages (if there are any) an account of truth of the sort that Tarski supplied for "referential" languages. I assume that the truth conditions for the language (e.g., English) to which mathematese appears to belong are to be elaborated much along the lines that Tarski articulated. So, to some extent, the question posed in the previous section – How are truth conditions for (2) to be explained? – may be interpreted as asking whether the sublanguage of English in which mathematics is done is to receive the same sort of analysis as I am assuming is appropriate for much of the rest of English. If so, then the qualms I shall sketch in the next section concerning how to fit mathematical knowledge into an overall epistemology clearly apply – though they can perhaps be laid to rest by a suitable modification of theory. If, on the other hand, mathematese is not to be analyzed along referential lines, then we are clearly in need not only of an account of truth (i.e., a semantics) for this new kind of language, but also of a new *theory of truth theories* that relates truth for referential (quantificational) languages to truth for these new (newly analyzed) languages. Given such an account, the task of accounting for mathematical knowledge would still remain; but it would presumably be an easier task, since the new semantical picture of mathematese would in most cases have been prompted by epistemological considerations. However, I do not give this alternative serious consideration in this paper because I don't think that anyone has ever actually chosen it. For to choose it is explicitly to consider *and reject* the "standard" interpretation of mathematical language, despite its superficial and initial plausibility, and then to provide an alternative semantics as a substitute.[5] The "combinatorial" theorists whom I discuss or refer to have usually wanted to have their cake and eat it too: they have not realized that the truth conditions that their account supplies for mathematical language have not been connected to the referential semantics which they assume is *also* appropriate for that language. Perhaps the closest candidate for an exception is Hilbert in the view I sketched briefly in the opening pages of this paper. But to pursue this further here would take us too far afield. Let us return, therefore, to our praise of the "standard view."

One of its primary advantages is that the truth definitions for individual mathematical theories thus construed will have the same recursion clauses as those employed for their less lofty empirical cousins. Or to put it another way, they can all be taken as parts of the same language for which we provide a single account for quantifiers regardless of the subdiscipline under consideration. Mathematical and empirical disciplines will not be distinguished in point of logical grammar. I have already underscored the importance of this advantage: it means that the logicogrammatical theory we employ in less recondite and more tractable domains will serve us well here. We can do with one, uniform, account and need not invent another for mathematics. This should hold true on virtually any grammatical theory coupled with semantics adequate to account for truth. My bias for what I call a Tarskian theory stems simply from the fact that he has given us the only viable systematic general account we have of truth. So, one consequence of the economy attending the standard view is that logical relations are subject to uniform treatment: they are invariant with subject matter. Indeed, they help define the concept of "subject matter." The same rules of inference may be used and their use accounted for by the same theory which provides us with our ordinary account of inference, thus avoiding a double standard. If we reject the standard view, mathematical inference will need a new and special account. As it is, standard uses of quantifier inferences are

[5] I sometimes think this is one of the things that Hilary Putnam wants to do in his stimulating article "Mathematics without Foundations," 1967a [reprinted in this volume].

justified by some sort of soundness proof. The formalization of theories in first-order logic requires for *its* justification the assurance (provided by the Completeness theorem) that all the logical consequences of the postulates will be forthcoming as theorems. The standard account delivers these guarantees. The obvious answers seem to work. To reject the standard view is to discard these answers. New ones would have to be found.

So much for the obvious virtues of this account. What are its faults?

As I suggested above, the principal defect of the standard account is that it appears to violate the requirement that our account of mathematical truth be susceptible to integration into our over-all account of knowledge. Quite obviously, to make out a persuasive case to this effect it would be necessary to sketch the epistemology I take to be at least roughly correct and on the basis of which mathematical truths, standardly construed, do not seem to constitute knowledge. This would require a lengthy detour through the general problems of epistemology. I will leave that to another time and content myself here with presenting a brief summary of the salient features of that view which bear most immediately on our problem.

IV. Knowledge

I favor a causal account of knowledge on which for X to know that S is true requires some causal relation to obtain between X and the referents of the names, predicates, and quantifiers of S. I believe in addition in a causal theory of *reference,* thus making the link to my saying knowingly that S *doubly* causal. I hope that what follows will dispel some of the fog which surrounds this formulation.

For Hermione to know that the black object she is holding is a truffle is for her (or at least requires her) to be in a certain (perhaps psychological) state.[6] It also requires the cooperation of the rest of the world, at least to the extent of permitting the object she is holding to be a truffle. Further – and this is the part I would emphasize – in the normal case, that the black object she is holding is a truffle must figure in a suitable way in a causal explanation of her belief that the black object she is holding is a

[6] If possible, I would like to avoid taking any stand on the cluster of issues in the philosophy of mind or psychology concerning the nature of psychological states. Any view on which Hermione can learn that the cat is on the mat by looking at a real cat on a real mat will do for my purposes. If looking at a cat on a mat puts Hermione into a state and you wish to call that state a physical, or psychological, or even physiological state, I will not object so long as it is understood that such a state, if it is her state of knowledge, is causally related in an appropriate way to the cat's having been on the mat when she looked. If there is no such state, then so much the worse for my view.

truffle. But what is a "suitable way"? I will not try to say. A number of authors have published views that seem to point in this direction,[7] and, despite differences among them, there seems to be a core intuition which they share and which I think is correct although very difficult to pin down.

That some such view must be correct and underlies our conception of knowledge is indicated by what we would say under the following circumstances. It is claimed that X knows that p. We think that X could not know that p. What reasons can we offer in support of our view? If we are satisfied that X has normal inferential powers, that p is indeed true, etc., we are often thrown back on arguing that X could not have come into possession of the relevant evidence or reasons: that X's four-dimensional space-time worm does not make the necessary (causal) contact with the grounds of the truth of the proposition for X to be in possession of evidence adequate to support the inference (if an inference was relevant). The proposition p places restrictions on what the world can be like. Our knowledge of the world, combined with our understanding of the restrictions placed by p, given by the truth conditions of p, will often tell us that a given individual could not have come into possession of evidence sufficient to come to know p, and we will thus deny his claim to the knowledge.

As an account of our knowledge about medium-sized objects, in the present, this is along the right lines. It will involve, causally, some direct reference to the facts known, and, through that, reference to these objects themselves. Furthermore, such knowledge (of houses, trees, truffles, dogs, and bread boxes) presents the clearest case and the easiest to deal with.

Other cases of knowledge can be explained as being based on inferences based on cases such as these, although there must evidently be interdependencies. This is meant to include our knowledge of general laws and theories, and, through them, our knowledge of the future and much of the past. This account follows closely the lines that have been proposed by empiricists, but with the crucial modification introduced by the explicitly causal condition mentioned above – but often left out of modern accounts, largely because of attempts to draw a careful distinction between "discovery" and "justification."

In brief, in conjunction with our other knowledge, we use p to determine the range of possible relevant evidence. We use what we know of X (the putative knower) to determine whether there could have been an appropriate kind of interaction, whether X's current belief that p is

[7]To cite but a few: Harman 1973; Goldman 1967: 357–72; Skyrms 1967: 373–89.

causally related in a suitable way with what is the case because *p* is true – whether his evidence is drawn from the range determined by *p*. If not, then *X* could not know that *p*. The connection between what must be the case if *p* is true and the causes of *X*'s belief can vary widely. But there is always *some* connection, and the connection relates the grounds of *X*'s belief to the subject matter of *p*.

It must be possible to establish an appropriate sort of connection between the truth conditions of *p* (as given by an adequate truth definition for the language in which *p* is expressed) and the grounds on which *p* is said to be known, at least for propositions that one must *come to know* – that are not innate. In the absence of this, no connection has been established between *having those grounds* and *believing a proposition which is true*. Having those grounds cannot be fitted into an explanation of *knowing p*. The link between *p* and justifying a belief in *p on those grounds* cannot be made. But for that knowledge which is properly regarded as some form of justified true belief, then the link *must* be made. (Of course not *all* knowledge need be justified true belief for the point to be a sound one.)

It will come as no surprise that this has been a preamble to pointing out that combining *this* view of knowledge with the "standard" view of mathematical truth makes it difficult to see how mathematical knowledge is possible. If, for example, numbers are the kinds of entities they are normally taken to be, then the connection between the truth conditions for the statements of number theory and any relevant events connected with the people who are supposed to have mathematical knowledge cannot be made out.[8] It will be impossible to account for how anyone knows any properly number-theoretical propositions. This second condition on an account of mathematical truth will not be satisfied, because we have no account of how we know that the truth conditions for mathematical propositions obtain. One obvious answer – that some of these propositions are true if and only if they are derivable from certain axioms via certain rules – will not help here. For, to be sure, we can ascertain that *those* conditions obtain. But in such a case, what we lack is the link between truth and proof, when truth is directly defined in the standard way. In short, although it may be a truth condition of certain number-theoretic propositions that they be derivable from certain axioms according to certain rules, *that* this is a truth condition must also follow from the account of *truth* if the condition referred to is to help connect truth and knowledge, if it is by their proofs that we know mathematical truths.

[8]For an expression of healthy skepticism concerning this and related points, see Steiner 1973: 57–66.

Of course, given some set-theoretical account of arithmetic, both the syntax and the semantics of *arithmetic* can be set out so as superficially to meet the conditions we have laid down. But the regress that this invites is transparent, for the same questions must then be asked about the set theory in terms of which the answers are couched.

V. Two examples

There are many accounts of mathematical truth and mathematical knowledge. The theses I have been defending are intended to apply to them all. Rather than try to be comprehensive, however, I will devote these last few pages to the examination of two representative cases: one "standard" view and one "combinatorial" view. First the standard account, as expressed by one of its most explicit and lucid proponents, Kurt Gödel.

Gödel is thoroughly aware that on a realist (i.e., standard) account of mathematical truth our explanation of how we know the basic postulates must be suitably connected with how we interpret the referential apparatus of the theory. Thus, in discussing how we can resolve the continuum problem, once it has been shown to be undecidable by the accepted axioms, he paints the following picture:

…the objects of transfinite set theory…clearly do not belong to the physical world and even their indirect connection with physical experience is very loose…

But, despite their remoteness from sense experience, we do have a perception also of the objects of set theory, as is seen from the fact that the axioms force themselves upon us as being true. I don't see why we should have less confidence in this kind of perception, i.e., in mathematical intuition, than in sense perception, which induces us to build up physical theories and to expect that future sense perceptions will agree with them and, moreover, to believe that a question not decidable now has meaning and may be decided in the future. [Gödel 1964; pp. 483–4 in this volume]

I find this picture both encouraging and troubling. What troubles me is that without an account of *how* the axioms "force themselves upon us as being true," the analogy with sense perception and physical science is without much content. For what is missing is *precisely* what my second principle demands: an account of the link between our cognitive faculties and the objects known. In physical science we have at least a start on such an account, and it is causal. We accept as knowledge only those beliefs which we can appropriately relate to our cognitive faculties. Quite appropriately, our conception of knowledge goes hand in hand with our conception of ourselves as knowers. To be sure, there is a *superficial* analogy. For, as Gödel points out, we "verify" axioms by deducing consequences from them concerning areas in which we seem to have more direct "perception" (clearer intuitions). But we are never told how we know even these, clearer, propositions. For example, the "verifiable" consequences of axioms of higher infinity are (otherwise undecidable) number-theoretical propositions which themselves are "verifiable" by computation up to any given integer. But the story, to be helpful anywhere, must tell us how we know statements of computational arithmetic – *if they mean what the standard account would have them mean.* And *that* we are not told. So the analogy is at best superficial.

So much for the troubling aspects. More important perhaps and what I find encouraging is the evident basic agreement which motivates Gödel's attempt to draw a parallel between mathematics and empirical science. He sees, I think, that something must be said to bridge the chasm, created by his realistic and platonistic interpretation of mathematical propositions, between the entities that form the subject matter of mathematics and the human knower. Instead of tinkering with the logical form of mathematical propositions or with the nature of the objects known, he postulates a special faculty through which we "interact" with these objects. We seem to agree on the analysis of the fundamental problem, but clearly disagree about the epistemological issue – about what avenues are open to us through which we may come to know things.

If our account of empirical knowledge is acceptable, it must be in part because it tries to make the connection evident in the case of our theoretical knowledge, where it is not *prima facie* clear how the causal account is to be filled in. Thus, when we come to mathematics, the absence of a coherent account of how our mathematical intuition is connected with the truth of mathematical propositions renders the over-all account unsatisfactory.

To introduce a speculative historical note, with some foundation in the texts, it might not be unreasonable to suppose that Plato had recourse to the concept of *anamnesis* at least in part to explain how, given the nature of the forms as he depicted them, one could ever have knowledge of them.[9]

The "combinatorial" view of mathematical truth has epistemological roots. It starts from the proposition that, whatever may be the "objects" of mathematics, our knowledge is obtained from proofs. Proofs are or can be (for some, must be) written down or spoken; mathematicians can survey them and come to agree that they *are* proofs. It is largely through these proofs that mathematical knowledge is obtained and transmitted.

---

[9]"The soul, then, as being immortal, and having been born again many times, and having seen all things that exist, whether in this world or in the world below, has knowledge of them all" (Plato, *Meno,* 81).

In short, this aspect of mathematical knowledge – its (essentially linguistic) means of production and transmission gives their impetus to the class of views that I call "combinatorial."

Noticing the role of proofs in the production of knowledge, it seeks the grounds of truth in the proofs themselves. Combinatorial views receive additional impetus from the realization that the platonist casts a shroud of mystery over how knowledge can be obtained at all. Add that realization to the belief that mathematics is a child of our own begetting (mathematical discovery, on these views, is seldom discovery about an independent reality), and it is not surprising that one looks for acts of conception to account for the birth. Many accounts of mathematical truth fall under this rubric. Perhaps almost all. I have mentioned several in passing, and I discussed Hilbert's view in "On the Infinite" very briefly. The final example I wish to consider is that of conventionalist accounts – the cluster of views that the truths of logic and mathematics are true (or can be made true) in virtue of explicit conventions where the conventions in question are usually the postulates of the theory. Once more, I will probably do them all an injustice by lumping together a number of views which their proponents would most certainly like to keep apart.

Quine, in his classic paper on this subject (1964, reprinted in this volume), has dealt clearly, convincingly, and decisively with the view that the truths of *logic* are to be accounted for as the products of convention – far better than I could hope to do here. He pointed out that, since we must account for infinitely many truths, the characterization of the eligible sentences as truths must be wholesale rather than retail. But wholesale characterization can proceed only via general principles – and, if we are supposed not to understand any logic at all, we cannot extract the individual instances from the general principles: we would need logic for such a task.

Persuasive as this may be, I wish to add another argument – not because I think this dead horse needs further flogging, but both because Quine's argument is limited to the case of logic and because the principal points I wish to bring out do not emerge sufficiently from it. Indeed, Quine grants the conventionalist certain principles I should like to deny him. In resting his case against conventionalism on the need for a wholesale characterization of infinitely many truths, Quine concedes that were there only finitely many truths to be reckoned with, the conventionalist might have a chance to make out his case. He says:

If truth assignments could be made one by one, rather than an infinite number at a time, the above difficulty would disappear; truths of logic ... would simply be asserted severally by fiat, and the problem of inferring them from more general conventions would not arise. [p. 353 in this volume]

Thus, if some way could be found to make sentences of logic wear their truth values upon their sleeves, the objections to the conventionalist account of truth would disappear – for we would have determined truth values for all the sentences, which is all that one could ask.

I wonder, however, what such a sprinkling of the word 'true' would accomplish. Surely it cannot suffice in order to determine a concept of truth to assign values to each and every sentence of the language [suppose now that the language is set theory, in some first-order formalization] (let those with an even number of horseshoes be "true").

What would make such an assignment of the predicate 'true' the determination of *the concept of truth?* Simply the use of that monosyllable? Tarski has suggested that satisfaction of Convention T is a necessary and sufficient condition on a definition of truth for a particular language.[10] A mere (recursive) distribution of truth values can be parlayed into a truth theory that satisfies convention T. We can rest with that provided we are prepared to beg what I think is the main question and ignore the concept of translation that occurs in its (Convention T's) formulation. What would be missing, hard as it is to state, is the theoretical apparatus employed by Tarski in providing truth definitions, i.e., the analysis of truth in terms of the "referential" concepts of naming, predication, satisfaction, and quantification. A definition that does not proceed by the customary recursion clauses for the customary grammatical forms may not be adequate, even if it satisfies Convention T. The explanation must proceed through reference and satisfaction and, furthermore, must be supplemented with an account of reference itself. But the defense of this last claim is too involved a matter to take up here.[11]

The Quine of "Truth by Convention" felt that to determine the truth values of all the contexts that contain a word suffices to determine its reference. That *might* be so, if we already had the concept of truth and

---

[10]"The Concept of Truth in Formalized Languages," reprinted in Tarski 1956. Convention T is stated on pp. 187–8 as follows:

CONVENTION T. A formally correct definition of the symbol 'Tr', formulated in the metalanguage, will be called *an adequate definition of truth* if it has the following consequences:

(α) all sentences which are obtained from the expression '$x \in$ Tr if and only if p' by substituting for the symbol 'x' a structural-descriptive name of any sentence of the language in question and for the symbol 'p' the expression which forms the translation of this sentence into the metalanguage;

(β) the sentence 'for any x, if $x \in$ Tr then $x \in$ S' (in other words 'Tr $\subseteq$ S').

[11]For an excellent presentation of a similar view, see Field, 1972: 347–5.

chased the reference of the term that interested us down through the truth definition. But there seems to be something patently wrong with trying to fix the concept of truth *itself* in this way. In so doing, we throw away the very crutch which enables that method to work for other concepts. Truth and reference go hand in hand. Our concept of truth, insofar as we have one, proceeds through the mediation of the concepts Tarski has used to define it for the class of languages he has considered – the essence of Tarski's contribution goes much further than Convention T, but includes the schemata for the actual definition as well: an analysis of truth for a language that did not proceed through the familiar devices of predication, quantification, etc., should not give us satisfaction.

If this is at all near the mark, then it should be clear why "combinatorial" views of the nature of mathematical truth fail on my account. They avoid what seems to me to be the necessary route to an account of truth: through the subject matter of the propositions whose truth is being defined. Motivated by epistemological considerations, they come up with truth conditions whose satisfaction or nonsatisfaction mere mortals can ascertain; but the price they pay is their inability to connect these so-called "truth conditions" with the truth of the propositions for which they are conditions.

Even if it is granted that the truths of first-order logic do not stem from conventions, it might still be claimed that the rest of mathematics (set theory, for logicists; set theory, number theory, and other things for nonlogicists) consists of conventions formalized in first-order logic. This view too is subject to the objection that such a concept of convention need not bring *truth* along with it.[12] Indeed it is clear that it does not. For, even ignoring more general objections, once the logic is fixed, it becomes possible that the conventions thus stipulated turn out to be inconsistent. Hence it cannot be maintained that setting down conventions *guarantees* truth. But if it does not *guarantee* truth, what distinguishes those cases in which it provides for it from those in which it does not? Consistency cannot be the answer. To urge it as such is to *mis*construe the significance of the fact that *in*consistency is *proof* that truth has not been attained. The deeper reason once more is that postulational stipulation makes no connection between the propositions and their subject matter – stipulation does not provide for truth. At best, it limits the class of truth definitions (interpretations) consistent with the stipulations. But that is not enough.

[12] Identical arguments will apply to the view, perhaps indistinguishable from this one, that the postulates constitute *implicit definitions* of existing concepts (as opposed to stipulating how new ones are to be understood), if that is advanced to explain how we know the axioms to be true (we learned the language by learning *these* postulates).

To clarify the point, consider Russell's oft-cited dictum: "The method of 'postulating' what we want has many advantages; they are the same as the advantages of theft over honest toil" (Russell 1919: 71). On the view I am advancing, that's false. For with theft at least you come away with the loot, whereas implicit definition, conventional postulation, and their cousins are incapable of bringing truth. They are not only morally but practically deficient as well.

Models and reality

HILARY PUTNAM

In 1922 Skolem delivered an address before the Fifth Congress of Scandinavian Mathematics in which he pointed out what he called a "relativity of set-theoretic notions". This "relativity" has frequently been regarded as paradoxical; but today, although one hears the expression "the Löwenheim-Skolem Paradox", it seems to be thought of as only an *apparent* paradox, something the cognoscenti enjoy but are not seriously troubled by. Thus van Heijenoort writes, "The existence of such a 'relativity' is sometimes referred to as the Löwenheim-Skolem Paradox, but, of course, it is not a paradox in the sense of an antinomy; it is a novel and unexpected feature of formal systems." In this address I want to take up Skolem's arguments, not with the aim of refuting them but with the aim of extending them in somewhat the direction he seemed to be indicating. It is not my claim that the "Löwenheim-Skolem Paradox" is an antinomy *in formal logic;* but I shall argue that it *is* an antinomy, or something close to it, in *philosophy of language*. Moreover, I shall argue that the resolution of the antinomy – the only resolution that I myself can see as making sense – has profound implications for the great metaphysical dispute about realism which has always been the central dispute in the philosophy of language.

The structure of my argument will be as follows: I shall point out that in many different areas there are three main positions on reference and truth: there is the extreme Platonist position, which posits nonnatural mental powers of directly "grasping" forms (it is characteristic of this position that "understanding" or "grasping" is itself an irreducible and unexplicated notion); there is the verificationist position which replaces the classical notion of truth with the notion of verification or proof, at least when it comes to describing how the language is understood; and there is the moderate realist position which seeks to preserve the central-

Presidential Address delivered before the Winter Meeting of the Association for Symbolic Logic in Washington, D.C., December 29, 1977. I wish to thank bas van Fraassen for valuable comments on and criticisms of an earlier version.

Reprinted with the kind permission of the editors and the Association for Symbolic Logic, Inc., from *The Journal of Symbolic Logic* 45.3 (September 1980): 464–82.

ity of the classical notions of truth and reference without postulating nonnatural mental powers. I shall argue that it is, unfortunately, the *moderate* realist position which is put into deep trouble by the Löwenheim-Skolem Theorem and related model-theoretic results. Finally I will opt for verificationism as a way of preserving the outlook of scientific or empirical realism, which is totally jetisoned by Platonism, even though this means giving up *metaphysical* realism.

The Löwenheim-Skolem Theorem says that a satisfiable first-order theory (in a countable language) has a countable model. Consider the sentence:

(i) $-(\exists R)(R$ is *one-to-one. The domain of* $R \subset N$. *The range of values of R is S*)

where '*N*' is a formal term for the set of all whole numbers and the three conjuncts in the matrix have the obvious first-order definitions.

Replace '*S*' with the formal term for the set of all real numbers in your favorite formalized set theory. The (i) will be a *theorem* (proved by Cantor's celebrated "diagonal argument"). So your formalized set theory *says* that a certain set (call it "*S*") is nondenumerable. So *S* must *be* nondenumerable in all *models* of your set theory. So your set theory – say ZF (Zermelo-Fraenkel set theory) has only nondenumerable models. But this is impossible! For, by the Löwenheim-Skolem Theorem, *no* theory can have *only* nondenumerable models; if a theory has a nondenumerable model, it must have denumerably infinite ones as well. Contradiction.

The resolution of this apparent contradiction is not hard, as Skolem points out (and it is not this apparent contradiction that I referred to as an antinomy, or close to an antinomy). For (i) only "says" that S is nondenumerable when the quantifier $(\exists R)$ is interpreted as ranging over *all* relations on $N \times S$. But when we pick a *denumerable* model for the language of set theory, "$(\exists R)$" does not range over *all* relations; it ranges only over relations *in the model*. (i) only "says" that S is nondenumerable in a *relative* sense: the sense that the members of S cannot be put in one-to-one correspondence with a subset of N by any R *in the model*. A set S can be "nondenumerable" in this *relative* sense and yet be denumerable "in reality". This happens when there *are* one-to-one correspondences between S and N but all of them lie outside the given model. What is a "countable" set from the point of view of one model may be an uncountable set from the point of view of another model. As Skolem sums it up, "even the notions 'finite', 'infinite', 'simply infinite sequence' and so forth turn out to be merely relative within axiomatic set theory".

The philosophical problem. Up to a point all commentators agree on the significance of the existence of "unintended" interpretations, e.g., models in which what are "supposed to be" nondenumerable sets are "in reality" denumerable. All commentators agree that the existence of such models shows that the "intended" interpretation, or, as some prefer to speak, the "intuitive notion of a set", is not "captured" by the formal system. But if *axioms* cannot capture the "intuitive notion of a set", what possibly could?

A technical fact is of relevance here. The Löwenheim-Skolem Theorem has a strong form (the so-called "downward Löwenheim-Skolem Theorem"), which requires the axiom of choice to prove, and which tells us that a satisfiable first-order theory (in a countable language) has a countable model which is a submodel of any given model. In other words if we are given a nondenumerable model M for a theory, then we can find a countable model M' of that same theory in which the predicate symbols stand for the same relations (restricted to the smaller universe in the obvious way) as they did in the original model. The only difference between M and M' is that the "universe of M' – i.e., the totality that the variables of quantification range over – is a proper subset of the "universe" of M.

Now the argument that Skolem gave, and that shows that "the intuitive notion of a set" (if there is such a thing) is not "captured" by any formal system, shows that even a *formalization of total science* (if one could construct such a thing), or even a *formalization of all our beliefs* (whether they count as "science" or not), could not rule out denumerable interpretations, and, *a fortiori,* such a formalization could not rule out *unintended* interpretations of this notion.

This shows that "theoretical constraints", whether they come from set theory itself or from "total science", cannot fix the interpretation of the notion *set* in the "intended" way. What of "operational constraints"?

Even if we allow that there might be a *denumerable infinity* of measurable "magnitudes", and that each of them might be measured to *arbitrary rational accuracy* (which certainly seems a utopian assumption), it would not help. For, by the "downward Löwenheim-Skolem Theorem", we can find a countable submodel of the "standard" model (if there is such a thing) in which countably many predicates (each of which may have countably many things in its extension) have their extensions preserved. In particular, we can fix the values of countable many magnitudes at all rational space-time points, and still find a countable submodel which meets all the constraints. In short, there certainly seems to be a *countable* model of our *entire body of belief* which meets all operational constraints.

The philosophical problem appears at just this point. If we are told, "axiomatic set theory does not capture the intuitive notion of a set", then it is natural to think that *something else* - our "understanding" - does capture it. But what can our "understanding" come to, at least for a naturalistically minded philosopher, which is more than *the way we use our language?* The Skolem argument can be extended, as we have just seen, to show that the *total use of the language* (operational plus theoretical constraints) does not "fix" a unique "intended interpretation" any more than axiomatic set theory by itself does.

This observation can push a philosopher of mathematics in two different ways. If he is inclined to Platonism, he will take this as evidence that the mind has mysterious faculties of "grasping concepts" (or "perceiving mathematical objects") which the naturalistically minded philosopher will never succeed in giving an account of. But if he is inclined to some species of verificationism (i.e., to indentifying truth with verifiability, rather than with some classical "correspondence with reality") he will say, "Nonsense! All the 'paradox' shows is that our understanding of 'The real numbers are nondenumerable' consists in our knowing *what it is for this to be proved,* and not in our 'grasp' of a 'model'." In short, the extreme positions - Platonism and verificationism - seem to receive comfort from the Löwenheim-Skolem Paradox; it is only the "moderate" position (which tries to avoid mysterious "perceptions" of "mathematical objects" while retaining a classical notion of truth) which is in deep trouble.

An epistemological/logical digression. The problem just pointed out is a serious problem for any philosopher or philosophically minded logician who wishes to view set theory as the description of a determinate independently existing reality. But from a mathematical point of view, it may appear immaterial: what does it matter if there are many different models of set theory, and not a unique "intended model" *if they all satisfy the same sentences?* What we want to know as mathematicians is what sentences of set theory are true; we do not want to have the sets themselves in our hands.

Unfortunately, the argument can be extended. First of all, the theoretical constraints we have been speaking of must, on a naturalistic view, come from only two sources: they must come from something like human decision or convention, whatever the source of the "naturalness" of the decisions or conventions may be, or from human experience, both experience with nature (which is undoubtedly the source of our most basic "mathematical intuitions", even if it be unfashionable to say so), and experience with "doing mathematics". It is hard to believe that either or both of these sources together can ever give us a *complete* set of axioms for set theory (since, for one thing, a complete set of axioms would have to be nonrecursive, and it is hard to envisage coming to have a nonrecursive set of axioms in the literature or in our heads even in the unlikely event that the human race went on forever doing set theory); and if a complete set of axioms is impossible, and the intended model*s* (in the plural) are singled out only by theoretical plus operational constraints then sentences which are independent of the axioms which we will arrive at in the limit of set-theoretic inquiry really have *no* determinate truth value; they are just true in some intended models and false in others.

To show what bearing this fact may have on actual set-theoretic inquiry, I will have to digress for a moment into technical logic. In 1938 Gödel put forward a new axiom for set theory: the axiom "$V=L$". Here L is the class of all constructible sets, that is, the class of all sets which can be defined by a certain constructive procedure if we pretend to have names available for all the ordinals, however large. (Of course, this sense of "constructible" would be anathema to constructive mathematicians.) V is the universe of all sets. So "$V=L$" just says *all sets are constructible.* By considering the inner model for set theory in which "$V=L$" is true, Gödel was able to prove the relative consistency of ZF and ZF *plus* the axiom of choice and the generalized continuum hypothesis.

"$V=L$" is certainly an important sentence, mathematically speaking. Is it *true?*

Gödel briefly considered proposing that we *add* "$V=L$" to the accepted axioms for set theory, as a sort of meaning stipulation, but he soon changed his mind. His later view was that "$V=L$" is *really* false, even though it is consistent with set theory, if set theory is itself consistent.

Gödel's intuition is widely shared among working set theorists. But does this "intuition" make sense?

Let *MAG* be a countable set of physical magnitudes which includes all magnitudes that sentient beings in this physical universe can actually measure (it certainly seems plausible that we cannot hope to measure more than a countable number of physical magnitudes). Let *OP* be the "correct" assignment of values; that is, the assignment which assigns to each member of *MAG* the value that that magnitude actually has at each rational space-time point. Then all the information "operational constraints" might give us (and, in fact, infinitely more) is coded into *OP*.

One technical term: an *ω-model* for a set theory is a model in which the *natural numbers* are ordered as they are "supposed to be"; that is, the sequence of "natural numbers" of the model is an ω-sequence.

425

Now for a small theorem.[1]

THEOREM. *ZF plus V = L has an ω-model which contains any given countable set of real numbers.*

PROOF. Since a countable set of reals can be coded as a single real by well-known techniques, it suffices to prove that *for every real s, there is an M such that M is an ω-model for ZF plus V = L and s is represented in M.*

By the "downward Löwenheim-Skolem Theorem", this statement is true if and only if the following statement is:

For every real s, there is a countable M such that M is an ω-model for ZF plus V = L and s is represented in M.

Countable structures with the property that the "natural numbers" of the structure form an ω-sequence can be coded as reals by standard techniques. When this is properly done, the predicate "*M is an ω-model for ZF plus V = L and s is represented in M*" becomes a two-place *arithmetical* predicate of *reals M, s*. The above sentence thus has the logical form (*for every real s*) (*there is a real M*) $(\cdots M, s, \cdots)$. In short, the sentence is a Π_2-sentence.

Now, consider this sentence *in the inner model V = L*. For every *s in the inner model* – that is, for every *s* in *L* – there is a model – namely *L* itself – which satisfies "*V = L*" and contains *s*. By the downward Löwenheim-Skolem Theorem, there is a countable submodel which is elementary equivalent to *L* and contains *s*. (Strictly speaking, we need here not just the downward Löwenheim-Skolem Theorem, but the "Skolem hull" construction which is used to prove that theorem.) By Gödel's work, this countable submodel itself lies in *L*, and as is easily verified, so does the real that codes it. So the above Π_2-sentences is true in the inner model *V = L*.

But Schoenfield has proved that Π_2-sentences are *absolute:* if a Π_2-sentence is true in *L*, then it must be true in *V*. So the above sentence is true in *V*.   □

What makes this theorem startling is the following reflection: suppose that Gödel is right, and "*V = L*" is *false* ("in reality"). Suppose that there is, in fact, a *non-constructible real number* (as Gödel also believes). Since the predicate "is constructible" is absolute in *β-models* – that is, in

---

[1]Barwise (1971) has proved the much stronger theorem that every countable model of ZF has a proper end extension which is a model of ZF + V = L. The theorem in the text was proved by me before 1963.

models in which the "wellorderings" *relative to the model* are well-orderings "in reality" (recall Skolem's "relativity of set-theoretic notions"!), no model containing such a nonconstructible s can satisfy "s is constructible" and be a β-*model*. But, by the above theorem, a model containing s *can* satisfy "s is constructible" (because it satisfies "$V=L$", and "$V=L$" says *everything* is constructible) and be an ω-*model*.

Now, suppose we formalize *the entire language of science* within the set theory ZF *plus* $V=L$. Any model for ZF which contains an abstract set isomorphic to *OP* can be extended to a model for this formalized language of science which is *standard with respect to OP* – hence, even if *OP* is nonconstructible "in reality", we can find a model *for the entire language of science* which satisfies *everything is constructible* and which assigns the correct values to all the physical magnitudes in *MAG* at all rational space-time points.

The claim Gödel makes is that "$V=L$" is false "in reality". But what on earth can this mean? It must mean, at the very least, that in the case just envisaged, the model we have described in which "$V=L$" holds would not be *the intended model*. But why not? It satisfies all theoretical constraints; and we have gone to great length to make sure it satisfies all operational constraints as well.

Perhaps someone will say that "$V \neq L$" (or something which implies that V does not equal L) should be added to the axioms of ZF as an additional "theoretical constraint". (Gödel often speaks of new axioms someday becoming evident.) But, while this may be acceptable from a nonrealist standpoint, it can hardly be acceptable from a realist standpoint. For the realist standpoint is that there is *a fact of the matter* – a fact independent of our legislation – as to whether $V=L$ or not. A realist like Gödel holds that we have access to an "intended interpretation" of ZF, where the access is not simply by linguistic stipulation.

What the above argument shows is that if the "intended interpretation" is fixed only by theoretical plus operational constraints, then if "$V \neq L$" does not follow from those constraints – if we do not *decide* to make $V=L$ true or to make $V=L$ false – then there will be "intended" models in which $V=L$ is *true*. If I am right, then the "relativity of set-theoretic notions" extends to a *relativity of the truth value of* "$V=L$" (and, by similar arguments, of the axiom of choice and the continuum hypothesis as well).

Operational constraints and counterfactuals. It may seem to some that there is a major equivocation in the notion of what *can* be measured, or observed, which endangers the apparently crucial claim that the evidence we *could* have amounts to at most denumerably many facts. Imagine a measuring apparatus that simply detects the presence of a particle within a finite volume *dv* around its own geometric center during each full minute on its clock. Certainly it comes up with at most denumerably many reports (each *yes* or *no*) even if it is left to run forever. But how many are the facts it *could* report? Well, if it were jiggled a little, by chance let us say, its geometric center would shift *r* centimeters in a given direction. It would then report totally different facts. Since for each number *r* it could be jiggled that way, the number of reports it could produce is nondenumerable – and it does not matter to this that we, and the apparatus itself, are incapable of distinguishing every real number *r* from every other one. The problem is simply one of scope for the modal word "can". In my argument, I must be identifying what I call observational constraints, not with the totality of facts that could be registered by observation – i.e., ones that either will be registered, or would be registered if certain chance perturbations occurred – but with the totality of facts that will in actuality be registered or observed, whatever those be.

In reply, I would point out that even if the measuring apparatus *were* jiggled *r* centimeters in a given direction, we could only know the real number *r* to some rational approximation. Now, if the intervals involved are all rational, there are only *countably* many facts of the form: *if action A* (an action described with respect to place, time, and character up to some finite "tolerance") were performed, then the result $r \pm \epsilon$ (a result described up to some rational tolerance) *would be obtained with probability in the interval a, b*. To know all facts of this form would be to know the *probability distribution* of all possible observable results of all possible actions. Our argument shows that a model could be constructed which agrees with all of these facts.

There is a deeper point to be made about this objection, however. Suppose we "first orderize" counterfactual talk, say, by including *events* in the ontology of our theory and introducing a predicate ("subjunctively necessitates") for the counter-factual connection between unactualized event types at a given place-time. Then our argument shows that a model exists which fits all the facts that will actually be registered or observed and fits our theoretical constraints, and this model *induces* an interpretation of the counterfactual idiom (a "similarity metric on possible worlds", in David Lewis' theory) which renders true just the counterfactuals that are true according to some completion of our theory. Thus appeal to counterfactual observations cannot rule out any models at all unless the interpretation of the counterfactual idiom itself is *already* fixed by something beyond operational and theoretical constraints.

Models and reality

(A related point is made by Wittgenstein in his *Philosophical Investigations:* talk about what an ideal machine – or God – could compute is talk *within* mathematics – in disguise – and cannot serve to fix the interpretation of mathematics. "God", too, has many interpretations.)

"Decision" and "convention". I have used the word "decision" in connection with open questions in set theory, and obviously this is a poor word. One cannot simply sit down in one's study and "decide" that "$V = L$" is to be true, or that the axiom of choice is to be true. Nor would it be appropriate for the mathematical community to call an international convention and legislate these matters. Yet, it seems to me that if we encountered an extra-terrestrial species of intelligent beings who had developed a high level of mathematics, and it turned out that they *rejected* the axiom of choice (perhaps because of the Tarski-Banach Theorem[2]), it would be wrong to regard them as simply making a *mistake.* To do *that* would, on my view, amount to saying that acceptance of the axiom of choice is built into our notion of rationality itself; that does not seem to me to be the case. To be sure, our acceptance of choice is not arbitrary; all kinds of "intuitions" (based, most likely, on experience with the finite) support it; its mathematical fertility supports it; but none of this is *so* strong that we could say that an equally successful culture which based *its* mathematics on principles *incompatible* with choice (e.g., on the so-called axiom of determinacy[4]) was *irrational.*

But if both systems of set theory – ours and the extra-terrestrials' – count as *rational,* what sense does it make to call one *true* and the others *false?* From the Platonist's point of view there is no trouble in answering this question. "The axiom of choice is true – true in *the* model", he will say (if he believes the axiom of choice). "We are right and the extra-terrestrials are wrong." But what is *the* model? If the intended model is singled out by theoretical and operational constraints, then, first, "the" intended model is plural not singular (so the "the" is inappropriate – our theoretical and operational constraints fit many models, not just one,

---

[2]This is a very counterintuitive consequence of the axiom of choice. Call two objects A, B "congruent by finite decomposition" if they can be divided into finitely many disjoint point sets $A_1, \ldots, A_n, B_1, \ldots B_n$, such that $A = A_1 \cup A_2 \cup \cdots \cup A_n$, $B = B_1 \cup B_2 \cup \cdots \cup B_n$, and (for $i = 1, 2, \ldots, n$) A_i is congruent to B_i. Then Tarski and Banach showed that *all spheres are congruent by finite decomposition.*

[3]This axiom, first studied by J. Mycielski (1964), asserts that infinite games with perfect information are determined, i.e. there is a winning strategy for either the first or second player. AD (the axiom of determinacy) implies the existence of a nontrivial countably additive two-valued measure on the real numbers, contradicting a well-known consequence of the axiom of choice.

and so do those of the extra-terrestrials as we saw before). Secondly, the intended models for us do satisfy the axiom of choice and the extra-terrestrially intended models do not; we are not talking about the same models, so there is no question of a "mistake" on one side or the other.

The Platonist will reply that what this really shows is that we have some mysterious faculty of "grasping concepts" (or "intuiting mathematical objects") and it is *this* that enables us to fix a model as *the* model, and not just operational and theoretical constraints; but this appeal to mysterious faculties seems both unhelpful as epistemology and unpersuasive as science. What neural process, after all, could be described as the perception of a mathematical object? Why of *one* mathematical object rather than another? I do not doubt that *some* mathematical axioms are built in to our notion of rationality ("every number has a successor"); but, if the axiom of choice and the continuum hypothesis are not, then, I am suggesting, Skolem's argument, or the foregoing extension of it, casts doubt on the view that these statements have a truth value independent of the theory in which they are embedded.

Now, suppose this is right and the axiom of choice is true when taken in the sense that it receives from *our* embedding theory and false when taken in the sense that it receives from extra-terrestrial theory. Urging this relativism is not advocating *unbridled* relativism; I do not doubt that there are some objective (if evolving) canons of rationality; I simply doubt that we would regard them as settling this sort of question, let alone as singling out *one* unique "rationally acceptable set theory". If this is right, then one is inclined to say that the extra-terrestrials have decided to let the axiom of choice be false and we have decided to let it be true; or that we have different "conventions"; but, of course, none of these words is literally right. It may well be the case that the idea that statements have their truth values *independent* of embedding theory is so deeply built into our ways of talking that there is simply no "ordinary language" word or short phrase which refers to the theory-dependence of meaning and truth. Perhaps this is why Poincaré was driven to exclaim "Convention, yes! Arbitrary, no!" when he was trying to express a similar idea in another context.

Is the problem a problem with the notion of a "set"? It would be natural to suppose that the problem Skolem points out, the problem of a surprising "relativity" of our notions, has to do with the notion of a "set", given the various problems which are *known* to surround *that* notion, or, at least, has to do with the problem of reference to "mathematical objects". But this is not so.

To see why it is not so, let us consider briefly the vexed problem of reference to theoretical entities in physical science. Although this may seem to be a problem more for philosophers of science or philosophers of language than for logicians, it is a problem whose logical aspects have frequently been of interest to logicians, as is witnessed by the expressions "Ramsey sentence", "Craig translation", etc. Here again, the realist – or, at least, the hard-core metaphysical realist – wishes it to be the case that *truth* and *rational acceptability* should be *independent* notions. He wishes it to be the case that what, e.g., electrons *are* should be distinct (and possibly different from) what we believe them to be or even what we would believe them to be given the best experiments and the epistemically best theory. Once again, the realist – the hard-core metaphysical realist – holds that our intentions single out "the" model, and that our beliefs are then either true or false in "the" model *whether we can find out their truth values or not.*

To see the bearing of the Löwenheim-Skolem Theorem (or of the intimately related Gödel Completeness Theorem and its model-theoretic generalizations) on this problem, let us again do a bit of model construction. This time the operational constraints have to be handled a little more delicately, since we have need to distinguish operational concepts (concepts that describe what we see, feel, hear, etc., as we perform various experiments, and also concepts that describe our acts of picking up, pushing, pulling, twisting, looking at, sniffing, listening to, etc.) from nonoperational concepts.

To describe our operational constraints we shall need three things. First, we shall have to fix a sufficiently large "observational vocabulary". Like the "observational vocabulary" of the logical empiricists, we will want to include in this set – call it the set of "O-terms" – such words as "red", "touches", "hard", "push", "look at", etc. Second, we shall assume that there *exists* (whether we can define it or not) a set of *S* which can be taken to be the set of macroscopically observable things and events (observable with the human sensorium, that means). The notion of an observable thing or event is surely vague; so we shall want *S* to be a generous set, that is, God is to err in the direction of counting too many things and events as "observable for humans" when He defines the set *S*, if it is necessary to err in either direction, rather than to err in the direction of leaving out some things that might be counted as borderline "observables". If one is a realist, then such a set *S* must exist, of course, even if our knowledge of the world and the human sensorium does not permit *us* to define it at the present time. The reason we allow *S* to contain events (and not just things) is that, as Richard Boyd has pointed out, some of the entities we can directly observe are *forces* – we can *feel* forces – and forces are not objects. But I assume that forces can be construed as predicates of either objects, e.g., our bodies, or of suitable events.

The third thing we shall assume given is a valuation (call it, once again '*OP*') which assigns the correct truth value to each n-place O-term (for $n = 1, 2, 3, \ldots$) on each n-tuple of elements of S on which it is defined. O-terms are in general also defined on things not in S; for example, two molecules too small to see with the naked eye may touch, a dust-mote too small to see may be black, etc. Thus *OP* is a *partial* valuation in a double sense; it is defined on only a subset of the predicates of the language, namely the O-terms, and even on these it only fixes a part of the extension, namely the extension of $T \upharpoonright S$ (the restriction of T to S), for each O-term T.

Once again, it is the valuation *OP* that captures our "operational constraints". Indeed, it captures these "from above", since it may well contain *more* information than we could actually get by using our bodies and our senses in the world.

What shall we do about "theoretical constraints"? Let us assume that there exists a possible formalization of present-day total science, call it 'T', and also that there exists a possible formalization of *ideal* scientific theory, call it 'T_I'. T_I is to be "ideal" in the sense of being *epistemically* ideal *for humans*. Ideality, in this sense, is a rather vague notion; but we shall assume that, when God makes up T_I, He constructs a theory which it would be rational for scientists to accept, or which is a limit of theories that it would be rational to accept, as more and more evidence accumulates, and also that he makes up a theory which is compatible with the valuation *OP*.

Now, the theory T is, we may suppose, well confirmed at the present time, and hence rationally acceptable on the evidence we *now* have; but there is a clear sense in which it may be false. Indeed, it may well lead to false predictions, and thus conflict with *OP*. But T_I, by hypothesis, does not lead to any false predictions. Still, the metaphysical realist claims – and it is just this claim that makes him a *metaphysical* as opposed to an empirical realist – that T_I may be, in reality, false. What is not knowable as true may nonetheless be true; what is epistemically most justifiable to believe may nonetheless be false, on this kind of realist view. The striking connection between issues and debates in the philosophy of science and issues and debates in the philosophy of mathematics is that this sort of realism runs into *precisely* the same difficulties that we saw Platonism run into. Let us pause to verify this.

Since the ideal theory T_I must, whatever other properties it may or may not have, have the property of being *consistent,* it follows from the Gödel Completeness Theorem (whose proof, as all logicians know, is intimately related to one of Skolem's proofs of the Löwenheim-Skolem Theorem), that T_I has models. We shall assume that T_I contains a primitive or defined term denoting each member of S, the set of "observable things and events". The assumption that we made, that T_I agrees with *OP*, means that all those sentences about members of S which *OP* requires to be true are theorems of T_I. Thus if M is any model of T_I, M has to have a member corresponding to each member of S. We can even replace each member of M which corresponds to a member of S by that member of S itself, modifying the interpretation of the predicate letters accordingly, and obtain a model M' in which each term denoting a member of S in the "intended" interpretation does denote that member of S. Then the extension of each O-term in that model will be partially correct to the extent determined by *OP*: that is, everything that *OP* "says" is in the extension of $\underline{P}$ is in the extension of $\underline{P}$, and everything that *OP* "says" is in the extension of the complement of $\underline{P}$ is in the extension of the complement of $\underline{P}$, for each O-term, in any such model. In short, such a model is standard with respect to $P \restriction S$ (P restricted to S) for each O-term P.

Now, such a model satisfies all operational constraints, since it agrees with *OP*. It satisfies those theoretical constraints we would impose in the ideal limit of inquiry. So, once again, it looks as if any such model is "intended" – for what else could single out a model as "intended" than this? But if this is what it *is* to be an "intended model", T_I must be *true* – true in all intended models! The metaphysical realist's claim that even the ideal theory T_I might be false "in reality" seems to collapse into unintelligibility.

Of course, it might be contended that "true" does not follow from "true in all intended models". But "true" is the same as "true in the intended *interpretation*" (or "in *all* intended interpretations", if there may be more than one interpretation intended – or permitted – by the speaker), on any view. So to follow this line – which is, indeed, the right one, in my view – one needs to develop a theory on which interpretations are specified *other* than by specifying models.

Once again, an appeal to mysterious powers of the mind is made by some. Chisholm (following the tradition of Brentano) contends that the mind has a faculty of *referring to external objects* (or perhaps to external properties) which he calls by the good old name "intentionality". Once again most naturalistically minded philosophers (and, of course, psychologists), find the postulation of unexplained mental faculties unhelpful epistemology and almost certainly bad science as well.

There are two main tendencies in the philosophy of science (I hesitate to call them "views", because each tendency is represented by many different detailed views) about the way in which the reference of theoretical terms gets fixed. According to one tendency, which we may call the Ramsey tendency, and whose various versions constituted the received view for many years, theoretical terms come in batches or clumps. Each clump – for example, the clump consisting of the primitives of electromagnetic theory – is defined by a theory, in the sense that all the models of that theory which are standard on the observation terms count as intended models. The theory is "true" just in case it has such a model. (The "Ramsey sentence" of the theory is just the second-order sentence that asserts the existence of such a model.) A sophisticated version of this view, which amounts to relativizing the Ramsey sentence to an open set of "intended applications", has recently been advanced by Joseph Sneed.

The other tendency is the realist tendency. While realists differ among themselves even more than proponents of the (former) received view do, realists unite in agreeing that a theory may have a true Ramsey sentence and not be (in reality) true.

The first of the two tendencies I described, the Ramsey tendency, represented in the United States by the school of Rudolf Carnap, accepted the "relativity of theoretical notions", and abandoned the realist intuitions. The second tendency is more complex. Its, so to speak, conservative wing, represented by Chisholm, joins Plato and the ancients in postulating mysterious powers wherewith the mind "grasps" concepts, as we have already said. If we have more available with which to fix the intended model than merely theoretical and operational constraints, then the problem disappears. The radical pragmatist wing, represented, perhaps, by Quine, is willing to give up the intuition that T_1 might be false "in reality". This radical wing is "realist" in the sense of being willing to assert that present-day science, taken more or less at face value (i.e., without philosophical reinterpretation) is at least approximately true; "realist" in the sense of regarding reference as trans-theoretic (a theory with a true Ramsey sentence may be false, because later inquiry may establish an incompatible theory as better); but not *metaphysical* realist. It is the moderate "center" of the realist tendency, the center that would like to hold on to metaphysical realism *without* postulating mysterious powers of the mind that is once again in deep trouble.

Pushing the problem back: the Skolemization of absolutely everything.
We have seen that issues in the philosophy of science having to do with reference of theoretical terms and issues in the philosophy of mathe-

matics having to do with the problem of singling out a unique "intended model" for set theory are both connected with the Löwenheim-Skolem Theorem and its near relative, the Gödel Completeness Theorem. Issues having to do with reference also arise in philosophy in connection with sense data and material objects and, once again, these connect with the model-theoretic problems we have been discussing. (In some way, it really seems that the Skolem Paradox underlies the *characteristic* problems of 20th century philosophy.)

Although the philosopher John Austin and the psychologist Fred Skinner both tried to drive sense data out of existence, it seems to me that most philosophers and psychologists think that there are such things as *sensations,* or *qualia.* They may not be objects of perception, as was once thought (it is becoming increasingly fashionable to view them as states or conditions of the sentient subject, as Reichenbach long ago urged we should); we may not have incorrigible knowledge concerning them; they may be somewhat ill-defined entities rather than the perfectly sharp particulars they were once taken to be; but it seems reasonable to hold that they are part of the legitimate subject matter of cognitive psychology and philosophy and not mere pseudo-entities invented by bad psychology and bad philosophy.

Accepting this, and taking the operational constraint this time to be that we wish the ideal theory to correctly predict all sense data, it is easily seen that the previous argument can be repeated here, this time to show that (if the "intended" models are the ones which satisfy the operational and theoretical constraints we now have, or even the operational and theoretical constraints we would impose in some limit) then, either the present theory is "true", in the sense of being "true in all intended models", provided it leads to no false predictions about sense data, or else the ideal theory is "true". The first alternative corresponds to taking the theoretical constraints to be represented by current theory; the second alternative corresponds to taking the theoretical constraints to be represented by the ideal theory. This time, however, it will be the case that even terms referring to ordinary material objects – terms like 'cat' and 'dog' – get differently interpreted in the different "intended" models. It seems, this time, as if we cannot even refer to ordinary middle sized physical objects except as formal constructs variously interpreted in various models.

Moreover, if we agree with Wittgenstein that the *similarity relation* between sense data we have at different times is not itself something present to my mind – that "fixing one's attention" on a sense datum and thinking "by 'red' I mean whatever is like *this*" does not really pick out any relation of similarity at all – and make the natural move of supposing that the intended models of my language when I now and in the future talk of the sense data I had at some past time t_0 are singled out by operational and theoretical constraints, then, again, it will turn out that my *past* sense data are mere formal constructs which get differently interpreted in various models. If we further agree with Wittgenstein that the notion of truth requires a *public* language (or requires at least states of the self at more than one time – that a "private language for one specious present" makes no sense), then even my *present* sense data are in this same boat.... In short, one can "Skolemize" absolutely everything. It seems to be absolutely impossible to fix a determinate reference (without appeal to nonnatural mental powers) for *any* term at all. If we apply the argument to the very metalanguage we use to talk about the predicament...?

The same problem has even surfaced recently in the field of cognitive psychology. The standard model for the brain/mind in this field is the modern computing machine. This computing machine is thought of as having something analogous to a formalized language in which it computes. (This hypothetical brain language has even received a name – "mentalese".) What makes the model of cognitive psychology a *cognitive* model is that "mentalese" is thought to be a medium whereby the brain constructs an *internal representation* of the external world. This idea runs immediately into the following problem: if "mentalese" is to be a vehicle for describing the external world, then the various predicate letters must have extensions which are sets of external things (or sets of *n*tuples of external things). But if the way "mentalese" is "understood" by the deep structures in the brain that compute, record, etc. in this "language" is *via* what artificial intelligence people call "procedural semantics" – that is, if the brain's *program for using* "mentalese" comprises its entire "understanding" of "mentalese" – where the program for using "mentalese", like any program, refers only to what is *inside* the computer – then how do *extensions* ever come into the picture at all? In the terminology I have been employing in this address, the problem is this: if the extension of predicates in "mentalese" is fixed by the theoretical and operational constraints "hard wired in" to the brain, or even by theoretical and operational constraints that it evolves in the course of inquiry, then these will not fix a *determinate* extension for any predicate. If thinking is ultimately done in "mentalese", then *no concept we have will have a determinate extension.* Or so it seems.

The bearing of causal theories of reference. The term "causal theory of reference" was originally applied to my theory of the reference of natural kind terms and Kripke's theory of the reference of proper names. These theories did not attempt to *define* reference, but rather attempted to say something about how reference is fixed, if it is not fixed by associating definite descriptions with the terms and names in question. Kripke and I argued that the intention to preserve reference through a historical chain of uses and the intention to cooperate socially in the fixing of reference make it possible to use terms successfully to refer although no one definite description is associated with any term by all speakers who use that term. These theories assume that individuals can be singled out for the purpose of a "naming ceremony" and that inferences to the existence of definite theoretical entities (to which names can then be attached) can be successfully made. Thus these theories did not address the question as to how any term can acquire a determinate reference (or any gesture, e.g., pointing – of course, the "reference" of gestures is just as problematic as the reference of terms, if not more so). Recently, however, it has been suggested by various authors that some account can be given of how at least some basic sorts of terms refer in terms of the notion of a "causal chain". In one version (cf. Evans 1973: 187–208), a version strikingly reminiscent of the theories of Ockham and other 14th century logicians, it is held that a term refers to "the dominant source" of the beliefs that contain the term. Assuming we can circumvent the problem that the dominant cause of our beliefs concerning *electrons* may well be *textbooks,*[4] it is important to notice that even if a *correct* view of this kind can be elaborated, it will do nothing to resolve the problem we have been discussing.

The problem is that adding to our hypothetical formalized language of science a body of theory titled "causal theory of reference" *is* just adding more *theory*. But Skolem's argument, and our extensions of it, are not affected by enlarging the theory. Indeed, you can even take the theory to consist of *all true sentences,* and there will be many models – models differing on the extension of every term not fixed by *OP* (or whatever you take *OP* to be in a given context) – which satisfy the entire theory. If "refers" can be defined in terms of some causal predicate or predicates in the metalanguage of our theory, then, since each model of the object language extends in an obvious way to a corresponding model of the metalanguage, it will turn out that, *in each model M, reference$_M$ is definable in terms of causes$_M$*; but, unless the word 'causes' (or whatever the causal predicate or predicates may be) is already glued to one definite relation with metaphysical glue, this does not fix a determinate extension for 'refers' at all.

---

[4]Evans handles this case by saying that there are appropriateness conditions on the type of causal chain which must exist between the item referred to and the speaker's body of information.

This is not to say that the construction of such a theory would be worthless as philosophy or as natural science. The program of cognitive psychology already alluded to – the program of describing our brains as computers which construct an "internal representation of the environment" seems to require that "mentalese" utterances be, in some cases at least, describable as the causal product of devices in the brain and nervous system which "transduce" information from the environment, and such a description might well be what the causal theorists are looking for. The program of realism in the philosophy of science – of *empirical* realism, not metaphysical realism – is to show that scientific theories can be regarded as better and better representations of an objective world with which we are interacting; if such a view is to be part of science itself, as empirical realists contend it should be, then the interactions with the world by means of which this representation is formed and modified must themselves be part of the subject matter of the representation. But the problem as to how the *whole representation,* including the empirical theory of knowledge that is a part of it, can determinately refer is not a problem that can be solved by developing more and better empirical theory.

Ideal theories and truth. One reaction to the problem I have posed would be to say: there are many ideal theories in the sense of theories which satisfy the operational constraints, and in addition have all the virtues (simplicity, coherence, containing the axiom of choice, whatever) that humans like to demand. But there are no "facts of the matter" not reflected in constraints on ideal theories in this sense. Therefore, what is really true is what is common to all such ideal theories; what is really false is what they all deny; all other statements are neither true nor false.

Such a reaction would lead to too few truths, however. It may well be that there are rational beings – even rational human species – which do not employ our color predicates, or who do not employ the predicate "person", or who do not employ the predicate "earthquake" (cf. Wiggins 1977). I see no reason to conclude from this that *our* talk of red things, or of persons, or of earthquakes, lacks truth value. If there are many ideal theories (and if "ideal" is itself a somewhat interest-relative notion), if there are many theories which (given appropriate circumstances) it is perfectly rational to accept, then it seems better to say that, insofar as these theories say different (and sometimes, apparently incompatible) things, that some facts are "soft" in the sense of depending for their truth value on the speaker, the circumstances of utterance, etc. This is what we have to say in any case about cases of ordinary vagueness, about ordinary causal talk, etc. It is what we say about apparently

Models and reality

incompatible statements of simultaneity in the special theory of relativity. To grant that there is more than one true version of reality is not to deny that some versions are false.

It may be, of course, that there *are* some truths that *any* species of rational inquirers would eventually acknowledge. (On the other hand, the set of these may be empty, or almost empty.) But to say that *by definition* these are all the truths there are is to redefine the notion in a highly restrictive way. (It also assumes that the notion of an "ideal theory" is perfectly clear; an assumption which seems plainly false.)

Intuitionism. It is a striking fact that this entire problem does *not* arise for the standpoint of mathematical intuitionism. This would not be a surprise to Skolem: it was precisely his conclusion that "most mathematicians want mathematics to deal, ultimately, with performable computing operations and not to consist of formal propositions about objects called this or that."

In intuitionism, knowing the meaning of a sentence or predicate consists in associating the sentence or predicate with a procedure which enables one to recognize when one has a proof that the sentence is constructively true (i.e., that it is possible to carry out the constructions that the sentence asserts can be carried out) or that the predicate applies to a certain entity (i.e., that a certain full sentence of the predicate is constructively true). The most striking thing about this standpoint is that the *classical notion of truth is nowhere used* – the semantics is entirely given in terms of the notion of "constructive proof", *including the semantics of "constructive proof" itself.*

Of course, the intuitionists do not think that "constructive proof" can be formalized, or that "mental constructions" can be identified with operations in our *brains.* Generally, they assume a strongly intentionalist and *a prioristic* posture in philosophy – that is, they assume the existence of mental entities called "meanings" and of a special faculty of intuiting constructive relations between these entities. These are not the aspects of intuitionism I shall be concerned with. Rather I wish to look on intuitionism as an example of what Michael Dummett has called "non-realist semantics" – that is, a semantic theory which holds that *a language is completely understood when a verification procedure is suitably mastered,* and not when truth conditions (in the classical sense) are learned.

The problem with realist semantics – truth-conditional semantics – as Dummett has emphasized, is that if we hold that the understanding of the sentences of, say, set theory consists in our knowledge of their "truth conditions", then how can we possibly say what *that* knowledge in turn consists in? (It cannot, as we have just seen, consist in the use of lan-

439

guage or "mentalese" under the control of operational plus theoretical constraints, be they fixed or evolving, since such constraints are too weak to provide a determinate extension for the terms, and it is this that the realist wants.)

If, however, the understanding of the sentences of a mathematical theory consists in the mastery of verification procedures (which need not be fixed once and for all – we can allow a certain amount of "creativity"), then a mathematical theory can be completely understood, and this understanding does not presuppose the notion of a "model" at all, let alone an "intended model".

Nor does the intuitionist (or, more generally, the "nonrealist" semanticist) have to foreswear *forever* the notion of a model. He has to foreswear reference to models in his account of *understanding*; but, once he has succeeded in understanding a rich enough language to serve as a metalanguage for some theory T (which may itself be simply a sublanguage of the metalanguage, in the familiar way), he can define 'true in T' à la Tarski, he can talk about "models" for T, etc. He can even define 'reference' or ('satisfaction') exactly as Tarski did.

Does the whole "Skolem Paradox" arise again to plague him at this stage? The answer is that it does not. To see why it does not, one has to realize what the "existence of a model" means in *constructive* mathematics.

"Objects" in constructive mathematics are *given through descriptions.* Those descriptions do not have to be mysteriously attached to those objects by some nonnatural process (or by metaphysical glue). Rather the possibility of *proving* that a certain construction (the "sense", so to speak, of the description of the model) has certain constructive properties is what is asserted and *all* that is asserted by saying the model "exists". In short, *reference is given through sense, and sense is given through verification-procedures and not through truth-conditions.* The "gap" between our theory and the "objects" simply disappears – or, rather, it never appears in the first place.

Intuitionism liberalized. It is not my aim, however, to try to convert my audience to intuitionism. Set theory may not be the "paradise" Cantor thought it was, but it is not such a bad neighborhood that I want to leave of my own accord, either. Can we separate the philosophical idea behind intuitionism, the idea of "nonrealist" semantics, from the restrictions and prohibitions that the historic intuitionists wished to impose upon mathematics?

The answer is that we can. First, as to set theory: the objection to *impredicativity,* which is the intuitionist ground for rejecting much of classical set theory, has little or no connection with the insistence upon verificationism itself. Indeed, intuitionist mathematics is itself "impredicative", inasmuch as the intuitionist notion of constructive proof presupposes constructive proofs which refer to the totality of *all* constructive proofs.

Second, as to the propositional calculus: it is well known that the classical connectives can be reintroduced into an intuitionist theory by reinterpretation. The important thing is not whether one uses "classical propositional calculus" or not, but how one *understands* the logic if one does use it. Using classical logic as an intuitionist would understand it, means, for example, keeping track of when a disjunction is selective (i.e., one of the disjuncts is constructively provable), and when it is nonselective; but this does not seem like too bad an idea.

In short, while intuitionism may go with a greater interest in constructive mathematics, a liberalized version of the intuitionist standpoint need not rule out "classical" mathematics as either illegitimate or unintelligible. What about the language of empirical science? Here there are greater difficulties. Intuitionist logic is given in terms of a notion of *proof,* and proof is supposed to be a *permanent* feature of statements. Moreover, proof is nonholistic; there is such a thing as the proof (in either the classical or the constructive sense) of an isolated mathematical statement. But verification in empirical science is a matter of degree, not a "yes-or-no" affair; even if we made it a "yes-or-no" affair in some arbitrary way, verification is a property of empirical sentences that can be *lost;* in general the "unit of verification" in empirical science is the theory and not the isolated statement.

These difficulties show that sticking to the intuitionist standpoint, however liberalized, would be a bad idea in the context of formalizing empirical science. But they are not incompatible with "nonrealist" semantics. The crucial question is this: do we think of the *understanding* of the language as consisting in the fact that speakers possess (collectively if not individually) an evolving network of verification procedures, or as consisting in their possession of a set of "truth conditions"? If we choose the first alternative, the alternative of "nonrealist" semantics, then the "gap" between words and world, between our *use* of the language and its "objects", never appears.[5] Moreover, the "nonrealist"

---

[5]To the suggestion that we identify truth with being verified, or accepted, or accepted in the long run, it may be objected that a person could reasonably, and possibly truly, make the assertion:

A; but it could have been the case that A and our scientific development differ in such a way to make $\bar{A}$ part of the ideal theory accepted in the long run; in that circumstance, it would have been the case that A but it was not true that A.

semantics is not *inconsistent* with realist semantics; it is simply *prior* to it, in the sense that it is the "nonrealist" semantics that must be internalized if the language is to be understood.

Even if it is not inconsistent with realist semantics, taking the non-realist semantics as our picture of how the language is understood undoubtedly will affect the way we view questions about reality and truth. For one thing, verification in empirical science (and, to a lesser extent, in mathematics as well, perhaps) sometimes depends on what we before called "decision" or "convention". Thus facts may, on this picture, depend on our interests, saliencies and decisions. There will be many "soft facts". (Perhaps whether $V = L$ or not is a "soft fact".) I cannot, myself, regret this. If appearance and reality end up being end-points on a continuum rather than being the two halves of a monster Dedekind cut in all we conceive and do not conceive, it seems to me that philosophy will be much better off. The search for the "furniture of the Universe" will have ended with the discovery that the Universe is not a furnished room.

Where did we go wrong? – The problem solved. What Skolem really pointed out is this: no interesting theory (in the sense of first-order theory) can, in and of itself, determine its own objects up to isomorphism. Skolem's argument can be extended as we saw, to show that if theoretical constraints do not determine reference, then the addition of operational constraints will not do it either. It is at this point that reference itself begins to seem "occult"; that it begins to seem that one cannot be any kind of a realist without being a believer in nonnatural mental powers. Many moves have been made in response to this predicament, as we noted above. Some have proposed that *second-order* formalizations are the solution, at least for mathematics; but the "intended" interpretation of the second-order formalism is not fixed by the use of the formalism (the formalism itself admits so-called "Henkin models", i.e., models in which the second-order variables fail to range over the *full* power set of the universe of individuals), and it becomes necessary to

This argument is fallacious, however, because the different "scientific development" means here the choice of a different version; we cannot assume the *sentence* $\ulcorner A \urcorner$ has a fixed meaning independent of what version we accept.

In fact the same problem can confront a metaphysical realist. Realists also have to recognize that there are cases in which the reference of a term depends on which theory one accepts, so that A can be a true sentence if T_1 is accepted and a false one if T_2 is accepted, where T_1 and T_2 are both true theories. But then imagine someone saying,

A; but it could have been the case that our A and our scientific development differ in such a way that T_2 was accepted. In that case, it would have been the case that A but A would not have been true.

attribute to the mind special powers of "grasping second-order notions". Some have proposed to accept the conclusion that mathematical language is only partially interpreted, and likewise for the language we use to speak of "theoretical entities" in empirical science; but then are "ordinary material objects" any better off? Are sense data better off? Both Platonism and phenomenalism have run rampant at different times and in different places in response to this predicament.

The problem, however, lies with the predicament itself. The predicament only *is* a predicament because we did two things: first, we gave an account of understanding the language in terms of programs and procedures for *using* the language (what else?); then, secondly, we asked what the possible "models" for the language were, thinking of the models as existing "out there" *independent of any description.* At this point, something really weird had already happened, had we stopped to notice. On any view, the understanding of the language must determine the reference of the terms, or, rather, must determine the reference given the context of use. If the use, even in a fixed context, does not determine reference, then use is not understanding. The language, on the perspective we talked ourselves into, has a full program of use; but it still lacks an *interpretation.*

This is the fatal step. To adopt a theory of meaning according to which a language whose whole use is specified still lacks something – viz. its "interpretation" – is to accept a problem which *can* only have crazy solutions. To speak as if *this* were my problem, "I know how to use my language, but, now, how shall I single out an interpretation?" is to speak nonsense. Either the use *already* fixes the "interpretation" or *nothing* can.

Nor do "causal theories of reference", etc., help. Basically, trying to get out of this predicament by *these* means is hoping that the *world* will pick one definite extension for each of our terms even if *we* cannot. But the world does not pick models or interpret languages. *We* interpret our languages or nothing does.

We need, therefore, a standpoint which links use and reference in just the way that the metaphysical realist standpoint refuses to do. The standpoint of "non-realist semantics" is precisely that standpoint. From that standpoint, it is trivial to say that a model in which, as it might be, the set of cats and the set of dogs are permuted (i.e., 'cat' is assigned the set of dogs as its extension, and 'dog' is assigned the set of cats) is "unintended" even if corresponding adjustments in the extensions of all the other predicates make it end up that the operational and theoretical constraints of total science or total belief are all "preserved". Such a model would be unintended *because we do not intend the word 'cat' to refer to dogs*. From the metaphysical realist standpoint, this answer does not work; it just pushes the question back to the metalanguage. The axiom of the metalanguage, " 'cat' refers to cats" cannot rule out such an unintended interpretation of the object language, unless the metalanguage itself already has had *its* intended interpretation singled out; but we are in the same predicament with respect to the metalanguage that we are in with respect to the object language, from that standpoint, so all is in vain. However, from the viewpoint of "nonrealist" semantics, the metalanguage is completely understood, and so is the object language. So we can *say and understand,* " 'cat' refers to cats". Even though the model referred to satisfies the theory, etc., it is "unintended"; we recognize that it is unintended *from the description through which it is given* (as in the intuitionist case). Models are not lost noumenal waifs looking for someone to name them; they are constructions within our theory itself, and they have names from birth.

PART IV
The concept of set

Russell's mathematical logic

KURT GÖDEL

Mathematical logic, which is nothing else but a precise and complete formulation of formal logic, has two quite different aspects. On the one hand, it is a section of Mathematics treating of classes, relations, combinations of symbols, etc., instead of numbers, functions, geometric figures, etc. On the other hand, it is a science prior to all others, which contains the ideas and principles underlying all sciences. It was in this second sense that Mathematical Logic was first conceived by Leibniz in his *Characteristica universalis,* of which it would have formed a central part. But it was almost two centuries after his death before his idea of a logical calculus really sufficient for the kind of reasoning occurring in the exact sciences was put into effect (in some form at least, if not the one Leibniz had in mind) by Frege and Peano.[1] Frege was chiefly interested in the analysis of thought and used his calculus in the first place for deriving arithmetic from pure logic. Peano, on the other hand, was more interested in its applications within mathematics and created an elegant and flexible symbolism, which permits expressing even the most complicated mathematical theorems in a perfectly precise and often very concise manner by single formulas.

It was in this line of thought of Frege and Peano that Russell's work set in. Frege, in consequence of his painstaking analysis of the proofs, had not gotten beyond the most elementary properties of the series of integers, while Peano had accomplished a big collection of mathematical theorems expressed in the new symbolism, but without proofs. It was

Reprinted with the kind permission of the author, editor, and publisher from Paul A. Schilpp, ed., *The Philosophy of Bertrand Russell,* The Library of Living Philosophers, Evanston, Ill. (Evanston & Chicago: Northwestern University, 1944), pp. 125–53. The author asked to note (1) that since the original publication of this paper advances have been made in some of the problems discussed and that the formulations given could be improved in several places, and (2) that the term "constructivistic" in this paper is used for a strictly anti-realistic kind of constructivism. Its meaning, therefore, is not identical with that used in current discussions on the foundations of mathematics. If applied to the actual development of logic and mathematics it is equivalent with a certain kind of "predicativity" and hence different both from "intuitionistically admissible" and from "constructive" in the sense of the Hilbert School.

[1]Frege has doubtless the priority, since his first publication about the subject, which already contains all the essentials, appeared ten years before Peano's.

only in *Principia Mathematica* that full use was made of the new method for actually deriving large parts of mathematics from a very few logical concepts and axioms. In addition, the young science was enriched by a new instrument, the abstract theory of relations. The calculus of relations had been developed before by Peirce and Schröder, but only with certain restrictions and in too close analogy with the algebra of numbers. In *Principia* not only Cantor's set theory but also ordinary arithmetic and the theory of measurement are treated from this abstract relational standpoint.

It is to be regretted that this first comprehensive and thorough going presentation of a mathematical logic and the derivation of Mathematics from it is so greatly lacking in formal precision in the foundations (contained in *1–*21 of *Principia*), that it presents in this respect a considerable step backwards as compared with Frege. What is missing, above all, is a precise statement of the syntax of the formalism. Syntactical considerations are omitted even in cases where they are necessary for the cogency of the proofs, in particular in connection with the "incomplete symbols." These are introduced not by explicit definitions, but by rules describing how sentences containing them are to be translated into sentences not containing them. In order to be sure, however, that (or for what expressions) this translation is possible and uniquely determined and that (or to what extent) the rules of inference apply also to the new kind of expressions, it is necessary to have a survey of all possible expressions, and this can be furnished only by syntactical considerations. The matter is especially doubtful for the rule of substitution and of replacing defined symbols by their *definiens*. If this latter rule is applied to expressions containing other defined symbols it requires that the order of elimination of these be indifferent. This however is by no means always the case ($\varphi!\hat{u} = \hat{u}[\varphi!u]$, e.g., is a counter-example). In *Principia* such eliminations are always carried out by substitutions in the theorems corresponding to the definitions, so that it is chiefly the rule of substitution which would have to be proved.

I do not want, however, to go into any more details about either the formalism or the mathematical content of *Principia*,[2] but want to devote the subsequent portion of this essay to Russell's work concerning the analysis of the concepts and axioms underlying Mathematical Logic. In this field Russell has produced a great number of interesting ideas some of which are presented most clearly (or are contained only) in his earlier writings. I shall therefore frequently refer also to these earlier writings,

[2]Cf. in this respect Quine 1941.

although their content may partly disagree with Russell's present stand-point.

What strikes one as surprising in this field is Russell's pronouncedly realistic attitude, which manifests itself in many passages of his writings. "Logic is concerned with the real world just as truly as zoology, though with its more abstract and general features," he says, e.g., in his *Introduction to Mathematical Philosophy* (edition of 1920, p. 169). It is true, however, that this attitude has been gradually decreasing in the course of time[3] and also that it always was stronger in theory than in practice. When he started on a concrete problem, the objects to be analyzed, (e.g., the classes or propositions) soon for the most part turned into "logical fictions." Though perhaps this need not necessarily mean [according to the sense in which Russell uses this term] that these things do not exist, but only that we have no direct perception of them.

The analogy between mathematics and a natural science is enlarged upon by Russell also in another respect (in one of his earlier writings). He compares the axioms of logic and mathematics with the laws of nature and logical evidence with sense perception, so that the axioms need not necessarily be evident in themselves, but rather their justification lies (exactly as in physics) in the fact that they make it possible for these "sense perceptions" to be deduced; which of course would not exclude that they also have a kind of intrinsic plausibility similar to that in physics. I think that (provided "evidence" is understood in a sufficiently strict sense) this view has been largely justified by subsequent developments, and it is to be expected that it will be still more so in the future. It has turned out that (under the assumption that modern mathematics is consistent) the solution of certain arithmetical problems requires the use of assumptions essentially transcending arithmetic, i.e., the domain of the kind of elementary indisputable evidence that may be most fittingly compared with sense perception. Furthermore it seems likely that for deciding certain questions of abstract set theory and even for certain related questions of the theory of real numbers new axioms based on some hitherto unknown idea will be necessary. Perhaps also the apparently unsurmountable difficulties which some other mathematical problems have been presenting for many years are due to the fact that the necessary axioms have not yet been found. Of course, under these circumstances mathematics may lose a good deal of its "absolute certainty;" but, under the influence of the modern criticism of the foundations, this has already happened to a large extent. There is some resemblance

---

[3]The above quoted passage was left out in the later editions of the *Introduction*.

between this conception of Russell and Hilbert's "supplementing the data of mathematical intuition" by such axioms as, e.g., the law of excluded middle which are not given by intuition according to Hilbert's view; the borderline however between data and assumptions would seem to lie in different places according to whether we follow Hilbert or Russell.

An interesting example of Russell's analysis of the fundamental logical concepts is his treatment of the definite article "the". The problem is: what do the so-called descriptive phrases (i.e., phrases as, e.g., "the author of *Waverley*" or "the king of England") denote or signify[4] and what is the meaning of sentences in which they occur? The apparently obvious answer that, e.g., "the author of *Waverley*" signifies Walter Scott, leads to unexpected difficulties. For, if we admit the further apparently obvious axiom, that the signification of a composite expression, containing constituents which have themselves a signification, depends only on the signification of these constituents (not on the manner in which this signification is expressed), then it follows that the sentence "Scott is the author of *Waverley*" signifies the same thing as "Scott is Scott"; and this again leads almost inevitably to the conclusion that all true sentences have the same signification (as well as all false ones).[5] Frege actually drew this conclusion; and he meant it in an almost metaphysical sense, reminding one somewhat of the Eleatic doctrine of the "One." "The True" – according to Frege's view – is analyzed by us in different ways in different propositions; "the True" being the name he uses for the common signification of all true propositions (cf. 1892b: 35).

Now according to Russell, what corresponds to sentences in the outer world is facts. However, he avoids the term "signify" or "denote" and uses "indicate" instead (in his earlier papers he uses "express" or "being a symbol for"), because he holds that the relation between a sentence and a fact is quite different from that of a name to the thing named. Furthermore, he uses "denote" (instead of "signify") for the relation between things and names, so that "denote" and "indicate" together would correspond to Frege's "*bedeuten*". So, according to Russell's

---

[4]I use the term "signify" in the sequel because it corresponds to the German word "*bedeuten*" which Frege, who first treated the question under consideration, used in this connection.

[5]The only further assumptions one would need in order to obtain a rigorous proof would be: (1) that "$\varphi(a)$" and the proposition "a is the object which has the property φ and is identical with a" mean the same thing and (2) that every proposition "speaks about something," i.e., can be brought to the form $\varphi(a)$. Furthermore one would have to use the fact that for any two objects a, b, there exists a true proposition of the form $\varphi(a, b)$ as, e.g., $a \neq b$ or $a = a \cdot b = b$.

terminology and view, true sentences "indicate" facts and, correspondingly, false ones indicate nothing.[6] Hence Frege's theory would in a sense apply to false sentences, since they all indicate the same thing, namely nothing. But different true sentences may indicate many different things. Therefore this view concerning sentences makes it necessary either to drop the above-mentioned principle about the signification (i.e., in Russell's terminology the corresponding one about the denotation and indication) of composite expressions or to deny that a descriptive phrase denotes the object described. Russell did the latter[7] by taking the viewpoint that a descriptive phrase denotes nothing at all but has meaning only in context; for example, the sentence "the author of *Waverley* is Scotch", is defined to mean: "There exists exactly one entity who wrote *Waverley* and whoever wrote *Waverley* is Scotch." This means that a sentence involving the phrase "the author of *Waverley*" does not (strictly speaking) assert anything about Scott (since it contains no constituent denoting Scott), but is only a roundabout way of asserting something about the concepts occurring in the descriptive phrase. Russell adduces chiefly two arguments in favor of this view, namely (1) that a descriptive phrase may be meaningfully employed even if the object described does not exist (e.g., in the sentence: "The present king of France does not exist"). (2) That one may very well understand a sentence containing a descriptive phrase without being acquainted with the object described; whereas it seems impossible to understand a sentence without being acquainted with the objects about which something is being asserted. The fact that Russell does not consider this whole question of the interpretation of descriptions as a matter of mere linguistic conventions, but rather as a question of right and wrong, is another example of his realistic attitude, unless perhaps he was aiming at a merely psychological investigation of the actual processes of thought. As to the question in the logical sense, I cannot help feeling that the problem raised by Frege's puzzling conclusion has only been evaded by Russell's theory of descriptions and that there is something behind it which is not yet completely understood.

There seems to be one purely formal respect in which one may give preference to Russell's theory of descriptions. By defining the meaning

[6]From the indication (*Bedeutung*) of a sentence is to be distinguished what Frege called its meaning (*Sinn*) which is the conceptual correlate of the objectively existing fact (or "the True"). This one should expect to be in Russell's theory a possible fact (or rather the possibility of a fact), which would exist also in the case of a false proposition. But Russell, as he says, could never believe that such "curious shadowy" things really exist. Thirdly, there is also the psychological correlate of the fact which is called "signification" and understood to be the corresponding belief in Russell's latest book. "Sentence" in contradistinction to "proposition" is used to denote the mere combination of symbols.

[7]He made no explicit statement about the former; but it seems it would hold for the logical system of *Principia,* though perhaps more or less vacuously.

of sentences involving descriptions in the above manner, he avoids in his logical system any axioms about the particle "the", i.e., the analyticity of the theorems about "the" is made explicit; they can be shown to follow from the explicit definition of the meaning of sentences involving "the". Frege, on the contrary, has to assume an axiom about "the", which of course is also analytic, but only in the implicit sense that it follows from the meaning of the undefined terms. Closer examination, however, shows that this advantage of Russell's theory over Frege's subsists only as long as one interprets definitions as mere typographical abbreviations, not as introducing names for objects described by the definitions, a feature which is common to Frege and Russell.

I pass now to the most important of Russell's investigations in the field of the analysis of the concepts of formal logic, namely those concerning the logical paradoxes and their solution. By analyzing the paradoxes to which Cantor's set theory had led, he freed them from all mathematical technicalities, thus bringing to light the amazing fact that our logical intuitions (i.e., intuitions concerning such notions as: truth, concept, being, class, etc.) are self-contradictory. He then investigated where and how these common-sense assumptions of logic are to be corrected and came to the conclusion that the erroneous axiom consists in assuming that for every propositional function there exists the class of objects satisfying it, or that every propositional function exists "as a separate entity;"[8] by which is meant something separable from the argument (the idea being that propositional functions are abstracted from propositions which are primarily given) and also something distinct from the combination of symbols expressing the propositional function; it is then what one may call the notion or concept defined by it.[9] The existence of this concept already suffices for the paradoxes in their "intensional" form, where the concept of "not applying to itself" takes the place of Russell's paradoxical class.

Rejecting the existence of a class or concept in general, it remains to determine under what further hypotheses (concerning the propositional function) these entities do exist. Russell pointed out (1907: 29) two possible directions in which one may look for such a criterion, which he

[8]In Russell 1907: 29. If one wants to bring such paradoxes as "the liar" under this viewpoint, universal (and existential) propositions must be considered to involve the class of objects to which they refer.

[9]"Propositional function" (without the clause "as a separate entity") may be understood to mean a proposition in which one or several constituents are designated as arguments. One might think that the pair consisting of the proposition and the argument could then for all purposes play the rôle of the "propositional function as a separate entity," but it is to be noted that this pair (as one entity) is again a set or a concept and therefore need not exist.

called the zig-zag theory and the theory of limitation of size, respectively, and which might perhaps more significantly be called the intensional and the extensional theory. The second one would make the existence of a class or concept depend on the extension of the propositional function (requiring that it be not too big), the first one on its content or meaning (requiring a certain kind of "simplicity," the precise formulation of which would be the problem).

The most characteristic feature of the second (as opposed to the first) would consist in the non-existence of the universal class or (in the intensional interpretation) of the notion of "something" in an unrestricted sense. Axiomatic set theory as later developed by Zermelo and others can be considered as an elaboration of this idea as far as classes are concerned.[10] In particular the phrase "not too big" can be specified (as was shown by J. v. Neumann 1929: 227) to mean: not equivalent with the universe of all things, or, to be more exact, a propositional function can be assumed to determine a class when and only when there exists no relation (in intension, i.e., a propositional function with two variables) which associates in a one-to-one manner with each object, an object satisfying the propositional function and vice versa. This criterion, however, does not appear as the basis of the theory but as a consequence of the axioms and inversely can replace two of the axioms (the axiom of replacement and that of choice).

For the second of Russell's suggestions too, i.e., for the zig-zag theory, there has recently been set up a logical system which shares some essential features with this scheme, namely Quine's system (cf. 1937: 70). It is, moreover, not unlikely that there are other interesting possibilities along these lines.

Russell's own subsequent work concerning the solution of the paradoxes did not go in either of the two afore-mentioned directions pointed out by himself, but was largely based on a more radical idea, the "no-class theory," according to which classes or concepts *never* exist as real objects, and sentences containing these terms are meaningful only to such an extent as they can be interpreted as a *façon de parler*, a manner of speaking about other things (cf. p. [460]). Since in *Principia* and elsewhere, however, he formulated certain principles discovered in the course of the development of this theory as general logical principles without mentioning any longer their dependence on the no-class theory, I am going to treat of these principles first.

I mean in particular the vicious circle principle, which forbids a certain

---

[10]The intensional paradoxes can be dealt with, e.g., by the theory of simple types or the ramified hierarchy, which do not involve any undesirable restrictions if applied to concepts only and not to sets.

kind of "circularity" which is made responsible for the paradoxes. The fallacy in these, so it is contended, consists in the circumstance that one defines (or tacitly assumes) totalities, whose existence would entail the existence of certain new elements of the same totality, namely elements definable only in terms of the whole totality. This led to the formulation of a principle which says that no totality can contain members definable only in terms of this totality, or members involving or presupposing this totality [vicious circle principle]. In order to make this principle applicable to the intensional paradoxes, still another principle had to be assumed, namely that "every propositional function presupposes the totality of its values" and therefore evidently also the totality of its possible arguments (cf. Whitehead and Russell 1910-13, 2: 39). [Otherwise the concept of "not applying to itself" would presuppose no totality (since it involves no quantifications),[11] and the vicious circle principle would not prevent its application to itself.] A corresponding vicious circle principle for propositional functions which says that nothing defined in terms of a propositional function can be a possible argument of this function is then a consequence (cf. Whitehead and Russell 1910-13, 1: 47, section 4). The logical system to which one is led on the basis of these principles is the theory of orders in the form adopted, e.g., in the first edition of *Principia,* according to which a propositional function which either contains quantifications referring to propositional functions of order n or can be meaningfully asserted of propositional functions of order n is at least of order $n+1$, and the range of significance of a propositional function as well as the range of a quantifier must always be confined to a definite order.

In the second edition of *Principia,* however, it is stated in the Introduction (pp. XI and XII) that "in a limited sense" also functions of a higher order than the predicate itself (therefore also functions defined in terms of the predicate as, e.g., in $p\lq\kappa \in \kappa$) can appear as arguments of a predicate of functions; and in appendix B such things occur constantly. This means that the vicious circle principle for propositional functions is virtually dropped. This change is connected with the new axiom that functions can occur in propositions only "through their values," i.e., extensionally, which has the consequence that any propositional function can take as an argument any function of appropriate type, whose extension is defined (no matter what order of quantifiers is used in the definition of this extension). There is no doubt that these things are quite unobjection-

---

[11]Quantifiers are the two symbols ($\exists x$) and (x) meaning respectively, "there exists an object x" and "for all objects x." The totality of objects x to which they refer is called their range.

able even from the constructive standpoint (see below and p. [456]), provided that quantifiers are always restricted to definite orders. The paradoxes are avoided by the theory of simple types,[12] which in *Principia* is combined with the theory of orders (giving as a result the "ramified hierarchy") but is entirely independent of it and has nothing to do with the vicious circle principle (cf. pp. [464–5]).

Now as to the vicious circle principle proper, as formulated on p. [454], it is first to be remarked that, corresponding to the phrases "definable only in terms of," "involving," and "presupposing," we have really three different principles, the second and third being much more plausible than the first. It is the first form which is of particular interest, because only this one makes impredicative definitions[13] impossible and thereby destroys the derivation of mathematics from logic, effected by Dedekind and Frege, and a good deal of modern mathematics itself. It is demonstrable that the formalism of classical mathematics does not satisfy the vicious circle principle in its first form, since the axioms imply the existence of real numbers definable in this formalism only by reference to all real numbers. Since classical mathematics can be built up on the basis of *Principia* (including the axiom of reducibility), it follows that even *Principia* (in the first edition) does not satisfy the vicious circle principle in the first form, if "definable" means "definable within the system" and no methods of defining outside the system (or outside other systems of classical mathematics) are known except such as involve still more comprehensive totalities than those occurring in the systems.

I would consider this rather as a proof that the vicious circle principle is false than that classical mathematics is false, and this is indeed plausible also on its own account. For, first of all one may, on good grounds, deny that reference to a totality necessarily implies reference to all single elements of it or, in other words, that "all" means the same as an infinite logical conjunction. One may, e.g., follow Langford's (1927: 599) and Carnap's (1931: 103 [51 in this volume], and 1937: 162) suggestion to

---

[12]By the theory of simple types I mean the doctrine which says that the objects of thought (or, in another interpretation, the symbolic expressions) are divided into types, namely: individuals, properties of individuals, relations between individuals, properties of such relations, etc. (with a similar hierarchy for extensions), and that sentences of the form: "*a* has the property φ," "*b* bears the Relation R to *c*," etc. are meaningless, if *a*, *b*, *c*, *R*, φ are not of types fitting together. Mixed types (such as the class of all classes of finite types) are excluded. That the theory of simple types suffices for avoiding also the epistemological paradoxes is shown by a closer analysis of these. (Cf. Ramsey 1926a and Tarski 1935b: 399.)

[13]These are definitions of an object α by reference to a totality to which α itself (and perhaps also things definable only in terms of α) belong. As, e.g., if one defines a class α as the intersection of all classes satisfying a certain condition φ and then concludes that α is a subset also of such classes *u* as are defined in terms of α (provided they satisfy φ).

interpret "all" as meaning analyticity or necessity or demonstrability. There are difficulties in this view; but there is no doubt that in this way the circularity of impredicative definitions disappears.

Secondly, however, even if "all" means an infinite conjunction, it seems that the vicious circle principle in its first form applies only if the entities involved are constructed by ourselves. In this case there must clearly exist a definition (namely the description of the construction) which does not refer to a totality to which the object defined belongs, because the construction of a thing can certainly not be based on a totality of things to which the thing to be constructed itself belongs. If, however, it is a question of objects that exist independently of our constructions, there is nothing in the least absurd in the existence of totalities containing members, which can be described (i.e., uniquely characterized)[14] only by reference to this totality (cf. Ramsey 1926a: 338 or 1931: 1). Such a state of affairs would not even contradict the second form of the vicious circle principle, since one cannot say that an object described by reference to a totality "involves" this totality, although the description itself does; nor would it contradict the third form, if "presuppose" means "presuppose for the existence" not "for the knowability."

So it seems that the vicious circle principle in its first form applies only if one takes the constructivistic (or nominalistic) standpoint[15] toward the objects of logic and mathematics, in particular toward propositions, classes and notions, e.g., if one understands by a notion a symbol together with a rule for translating sentences containing the symbol into such sentences as do not contain it, so that a separate object denoted by the symbol appears as a mere fiction.[16]

Classes and concepts may, however, also be conceived as real objects, namely classes as "pluralities of things" or as structures consisting of a plurality of things and concepts as the properties and relations of things existing independently of our definitions and constructions.

It seems to me that the assumption of such objects is quite as legitimate as the assumption of physical bodies and there is quite as much reason to believe in their existence. They are in the same sense necessary to obtain a satisfactory system of mathematics as physical bodies are necessary for a

[14]An object a is said to be described by a propositional function $\varphi(x)$ is $\varphi(x)$ is true for $x=a$ and for no other object.

[15]I shall use in the sequel "constructivism" as a general term comprising both these standpoints and also such tendencies as are embodied in Russell's "no class" theory.

[16]One might think that this conception of notions is impossible, because the sentences into which one translates must also contain notions so that one would get into an infinite regress. This, however, does not preclude the possibility of maintaining the above viewpoint for all the more abstract notions, such as those of the second and higher types, or in fact for all notions except the primitive terms which might be only a very few.

satisfactory theory of our sense perceptions and in both cases it is impossible to interpret the propositions one wants to assert about these entities as propositions about the "data," i.e., in the latter case the actually occurring sense perceptions. Russell himself concludes in the last chapter of his book on *Meaning and Truth* (1940), though "with hesitation," that there exist "universals," but apparently he wants to confine this statement to concepts of sense perceptions, which does not help the logician. I shall use the term "concept" in the sequel exclusively in this objective sense. One formal difference between the two conceptions of notions would be that any two different definitions of the form $\alpha(x) = \varphi(x)$ can be assumed to define two different notions α in the constructivistic sense. (In particular this would be the case for the nominalistic interpretation of the term "notion" suggested above, since two such definitions give different rules of translation for propositions containing α.) For concepts, on the contrary, this is by no means the case, since the same thing may be described in different ways. It might even be that the axiom of extensionality[17] or at least something near to it holds for concepts. The difference may be illustrated by the following definition of the number two: "Two is the notion under which fall all pairs and nothing else." There is certainly more than one notion in the constructivistic sense satisfying this condition, but there might be one common "form" or "nature" of all pairs.

Since the vicious circle principle, in its first form does apply to constructed entities, impredicative definitions and the totality of all notions or classes or propositions are inadmissible in constructivistic logic. What an impredicative definition would require is to construct a notion by a combination of a set of notions to which the notion to be formed itself belongs. Hence if one tries to effect a retranslation of a sentence containing a symbol for such an impredicatively defined notion it turns out that what one obtains will again contain a symbol for the notion in question (cf. Carnap 1931: 103 [51 in this volume] and 1937: 162). At least this is so if "all" means an infinite conjunction; but Carnap's and Langford's idea (mentioned on pp. [455–6]) would not help in this connection, because "demonstrability," if introduced in a manner compatible with the constructivistic standpoint towards notions, would have to be split into a hierarchy of orders, which would prevent one from obtaining the desired results.[18] As Chwistek (1933: 367) has shown, it is even possible under

---

[17]I.e., that no two different properties belong to exactly the same things, which, in a sense, is a counterpart to Leibniz's *Principium identitatis indiscernibilium,* which says no two different things have exactly the same properties.

[18]Nevertheless the scheme is interesting because it again shows the constructibility of notions which can be meaningfully asserted of notions of arbitrarily high order.

certain assumptions admissible within constructivistic logic to derive an actual contradiction from the unrestricted admission of impredicative definitions. To be more specific, he has shown that the system of simple types becomes contradictory if one adds the "axiom of intensionality" which says (roughly speaking) that to different definitions belong different notions. This axiom, however, as has just been pointed out, can be assumed to hold for notions in the constructivistic sense.

Speaking of concepts, the aspect of the question is changed completely. Since concepts are supposed to exist objectively, there seems to be objection neither to speaking of all of them (cf. p. [461]) nor to describing some of them by reference to all (or at least all of a given type). But, one may ask, isn't this view refutable also for concepts because it leads to the "absurdity" that there will exist properties φ such that $\varphi(a)$ consists in a certain state of affairs involving all properties (including φ itself and properties defined in terms of φ), which would mean that the vicious circle principle does not hold even in its second form for concepts or propositions? There is no doubt that the totality of all properties (or of all those of a given type) does lead to situations of this kind, but I don't think they contain any absurdity.[19] It is true that such properties φ [or such propositions $\varphi(a)$] will have to contain themselves as constituents of their content [or of their meaning], and in fact in many ways, because of the properties defined in terms of φ; but this only makes it impossible to construct their meaning (i.e., explain it as an assertion about sense perceptions or any other non-conceptual entities), which is no objection for one who takes the realistic standpoint. Nor is it self-contradictory that a proper part should be identical (not merely equal) to the whole, as is seen in the case of structures in the abstract sense. The structure of the series of integers, e.g., contains itself as a proper part and it is easily seen that there exist also structures containing infinitely many different parts, each containing the whole structure as a part. In addition there exist, even within the domain of constructivistic logic, certain approximations to this self-reflexivity of impredicative properties, namely propositions which contain as parts of their meaning not themselves but their own formal demonstrability (cf. Gödel 1931: 173 or Carnap 1937, §35). Now formal demonstrability of a proposition (in case the axioms and rules of inference are correct) implies this proposition

[19]The formal system corresponding to this view would have, instead of the axiom of reducibility, the rule of substitution for functions described, e.g., in Hilbert-Bernays 1934–9, 1: 90, applied to variables of any type, together with certain axioms of intensionality required by the concept of property which, however, would be weaker than Chwistek's. It should be noted that this view does not necessarily imply the existence of concepts which cannot be expressed in the system, if combined with a solution of the paradoxes along the lines indicated on p. [466].

and in many cases is equivalent to it. Furthermore, there doubtlessly exist sentences referring to a totality of sentences to which they themselves belong as, e.g., the sentence: "Every sentence (of a given language) contains at least one relation word."

Of course this view concerning the impredicative properties makes it necessary to look for another solution of the paradoxes, according to which the fallacy (i.e., the underlying erroneous axiom) does not consist in the assumption of certain self-reflexivities of the primitive terms but in other assumptions about these. Such a solution may be found for the present in the simple theory of types and in the future perhaps in the development of the ideas sketched on pp. [452-3 and 466]. Of course, all this refers only to concepts. As to notions in the constructivistic sense there is no doubt that the paradoxes are due to a vicious circle. It is not surprising that the paradoxes should have different solutions for different interpretations of the terms occurring.

As to classes in the sense of pluralities or totalities it would seem that they are likewise not created but merely described by their definitions and that therefore the vicious circle principle in the first form does not apply. I even think there exist interpretations of the term "class" (namely as a certain kind of structures), where it does not apply in the second form either.[20] But for the development of all contemporary mathematics one may even assume that it does apply in the second form, which for classes as mere pluralities is, indeed, a very plausible assumption. One is then led to something like Zermelo's axiom system for set theory, i.e., the sets are split up into "levels" in such a manner that only sets of lower levels can be elements of sets of higher levels (i.e., $x \in y$ is always false if x belongs to a higher level than y). There is no reason for classes in this sense to exclude mixtures of levels in one set and transfinite levels. The place of the axiom of reducibility is now taken by the axiom of classes [Zermelo's *Aussonderungsaxiom*] which says that for each level there exists for an arbitrary propositional function $\varphi(x)$ the set of those x of this level for which $\varphi(x)$ is true, and this seems to be implied by the concept of classes as pluralities.

Russell adduces two reasons against the extensional view of classes, namely the existence of (1) the null class, which cannot very well be a collection, and (2) the unit classes, which would have to be identical with their single elements. But it seems to me that these arguments could, if anything, at most prove that the null class and the unit classes (as distinct from their only element) are fictions (introduced to simplify the calculus like the points at infinity in geometry), not that all classes are fictions.

[20]Ideas tending in this direction are contained in Mirimanoff 1917a: 37-52, 1917b: 209-17, 1920: 29-52.

But in Russell the paradoxes had produced a pronounced tendency to build up logic as far as possible without the assumption of the objective existence of such entities as classes and concepts. This led to the formulation of the aforementioned "no class theory," according to which classes and concepts were to be introduced as a *façon de parler*. But propositions, too, (in particular those involving quantifications; Russell 1906a: 627) were later on largely included in this scheme, which is but a logical consequence of this standpoint, since, e.g., universal propositions as objectively existing entities evidently belong to the same category of idealistic objects as classes and concepts and lead to the same kind of paradoxes, if admitted without restrictions. As regards classes this program was actually carried out, i.e., the rules for translating sentences containing class names or the term "class" into such as do not contain them were stated explicitly; and the basis of the theory, i.e., the domain of sentences into which one has to translate is clear, so that classes can be dispensed with (within the system *Principia*), but only if one assumes the existence of a concept whenever one wants to construct a class. When it comes to concepts and the interpretation of sentences containing this or some synonymous term, the state of affairs is by no means as clear. First of all, some of them (the primitive predicates and relations such as "red" or "colder") must apparently be considered as real objects;[21] the rest of them (in particular according to the second edition of *Principia,* all notions of a type higher than the first and therewith all logically interesting ones) appear as something constructed (i.e., as something not belonging to the "inventory" of the world); but neither the basic domain of propositions in terms of which finally everything is to be interpreted, nor the method of interpretation is as clear as in the case of classes (see below).

This whole scheme of the no-class theory is of great interest as one of the few examples, carried out in detail, of the tendency to eliminate assumptions about the existence of objects outside the "data" and to replace them by constructions on the basis of these data.[22] The result has been in this case essentially negative; i.e., the classes and concepts introduced in this way do not have all the properties required for their use in mathematics, unless one either introduces special axioms about the data (e.g., the axiom of reducibility), which in essence already mean the existence in the data of the kind of objects to be constructed, or makes the fiction that one can form propositions of infinite (and even non-

[21] In Appendix C of *Principia* a way is sketched by which these also could be constructed by means of certain similarity relations between atomic propositions, so that these latter would be the only ones remaining as real objects.

[22] The "data" are to be understood in a relative sense here, i.e., in our case as logic without the assumption of the existence of classes and concepts.

denumerable) length (cf. Ramsey 1926a: 338 or 1931: 1), i.e., operates with truth-functions of infinitely many arguments, regardless of whether or not one can construct them. But what else is such an infinite truth-function but a special kind of an infinite extension (or structure) and even a more complicated one than a class, endowed in addition with a hypothetical meaning, which can be understood only by an infinite mind? All this is only a verification of the view defended above that logic and mathematics (just as physics) are built up on axioms with a real content which cannot be "explained away."

What one can obtain on the basis of the constructivistic attitude is the theory of orders (cf. p. [454]); only now (and this is the strong point of the theory) the restrictions involved do not appear as *ad hoc* hypotheses for avoiding the paradoxes, but as unavoidable consequences of the thesis that classes, concepts, and quantified propositions do not exist as real objects. It is not as if the universe of things were divided into orders and then one were prohibited to speak of all orders; but, on the contrary, it is possible to speak of all existing things; only, classes and concepts are not among them; and if they are introduced as a *façon de parler,* it turns out that this very extension of the symbolism gives rise to the possibility of introducing them in a more comprehensive way, and so on indefinitely. In order to carry out this scheme one must, however, presuppose arithmetic (or something equivalent) which only proves that not even this restricted logic can be built up on nothing.

In the first edition of *Principia,* where it was a question of actually building up logic and mathematics, the constructivistic attitude was, for the most part, abandoned, since the axiom of reducibility for types higher than the first together with the axiom of infinity makes it absolutely necessary that there exist primitive predicates of arbitrarily high types. What is left of the constructive attitude is only: (1) The introduction of classes as a *façon de parler;* (2) the definition of $\sim$, $\vee$, $\cdot$, etc., as applied to propositions containing quantifiers (which incidentally proved its fecundity in a consistency proof for arithmetic); (3) the step-by-step construction of functions of orders higher than 1, which, however, is superfluous owing to the axiom of reducibility; (4) the interpretation of definitions as mere typographical abbreviations, which makes every symbol introduced by definition an incomplete symbol (not one naming an object described by the definition). But the last item is largely an illusion, because, owing to the axiom of reducibility, there always exist real objects in the form of primitive predicates, or combinations of such, corresponding to each defined symbol. Finally also Russell's theory of descriptions is something belonging to the constructivistic order of ideas.

In the second edition of *Principia* (or to be more exact, in the introduction to it) the constructivistic attitude is resumed again. The axiom of reducibility is dropped and it is stated explicitly that all primitive predicates belong to the lowest type and that the only purpose of variables (and evidently also of constants) of higher orders and types is to make it possible to assert more complicated truth-functions of atomic propositions,[23] which is only another way of saying that the higher types and orders are solely a *façon de parler*. This statement at the same time informs us of what kind of propositions the basis of the theory is to consist, namely of truth-functions of atomic propositions.

This, however, is without difficulty only if the number of individuals and primitive predicates is finite. For the opposite case (which is chiefly of interest for the purpose of deriving mathematics), Ramsey (cf. Ramsey 1926a: 338 or 1931: 1) took the course of considering our inability to form propositions of infinite length as a "mere accident," to be neglected by the logician. This of course solves (or rather cuts through) the difficulties; but it is to be noted that, if one disregards the difference between finite and infinite in this respect, there exists a simpler and at the same time more far-reaching interpretation of set theory (and therewith of mathematics). Namely, in case of a finite number of individuals, Russell's *aperçu* that propositions about classes can be interpreted as propositions about their elements becomes literally true, since, e.g., "$x \in m$" is equivalent to "$x = a_1 \lor x = a_2 \lor \ldots \lor x = a_k$" where the a_i are the elements of m; and "there exists a class such that ..." is equivalent to "there exist individuals $x_1, x_2, \ldots, x_n$ such that ...,"[24] provided n is the number of individuals in the world and provided we neglect for the moment the null class which would have to be taken care of by an additional clause. Of course, by an iteration of this procedure one can obtain classes of classes, etc., so that the logical system obtained would resemble the theory of simple types except for the circumstance that mixture of types would be possible. Axiomatic set theory appears, then, as an extrapolation of this scheme for the case of infinitely many individuals or an infinite iteration of the process of forming sets.

Ramsey's viewpoint is, of course, everything but constructivistic, unless one means constructions of an infinite mind. Russell, in the second edition of *Principia,* took a less metaphysical course by confining himself to such truth-functions as can actually be constructed. In this way one is again led to the theory of orders, which, however, appears now in a new light, namely as a method of constructing more and more complicated

---

[23]I.e., propositions of the form $S(a), R(a, b)$, etc., where S, R are primitive predicates and a, b individuals.

[24]The x_i may, of course, as always, be partly or wholly identical with each other.

truth-functions of atomic propositions. But this procedure seems to pre-suppose arithmetic in some form or other (see next paragraph).

As to the question of how far mathematics can be built up on this basis (without any assumptions about the data – i.e., about the primitive pred-icates and individuals – except, as far as necessary, the axiom of infinity), it is clear that the theory of real numbers in its present form cannot be obtained.[25] As to the theory of integers, it is contended in the second edition of *Principia* that it can be obtained. The difficulty to be over-come is that in the definition of the integers as "those cardinals which belong to every class containing 0 and containing $x+1$ if containing x," the phrase "every class" must refer to a given order. So one obtains integers of different orders, and complete induction can be applied to integers of order n only for properties of order n; whereas it frequently happens that the notion of integer itself occurs in the property to which induction is applied. This notion, however, is of order $n+1$ for the inte-gers of order n. Now, in Appendix B of the second edition of *Principia,* a proof is offered that the integers of any order higher than 5 are the same as those of order 5, which of course would settle all difficulties. The proof as it stands, however, is certainly not conclusive. In the proof of the main lemma *89.16, which says that every subset α (of arbitrary high order)[26] of an inductive class β of order 3 is itself an inductive class of order 3, induction is applied to a property of β involving α [namely $\alpha-\beta\neq\Lambda$, which, however, should read $\alpha-\beta\sim\in\text{Induct}_2$ because (3) is evidently false]. This property, however, is of an order >3 if α is of an order >3. So the question whether (or to what extent) the theory of inte-gers can be obtained on the basis of the ramified hierarchy must be con-sidered as unsolved at the present time. It is to be noted, however, that, even in case this question should have a positive answer, this would be of no value for the problem whether arithmetic follows from logic, if propositional functions of order n are defined (as in the second edition of *Principia*) to be certain finite (though arbitrarily complex) combinations (of quantifiers, propositional connectives, etc.), because then the notion of finiteness has to be presupposed, which fact is concealed only by taking such complicated notions as "propositional function of order n" in an unanalyzed form as primitive terms of the formalism and giving their definition only in ordinary language. The reply may perhaps be

[25]As to the question how far it is possible to build up the theory of real numbers, pre-supposing the integers, cf. Weyl 1918.
[26]That the variable α is intended to be of undetermined order is seen from the later appli-cations of *89.17 and from the note to *89.17. The main application is in line (2) of the proof of *89.24, where the lemma under consideration is needed for α's of arbitrarily high orders.

offered that in *Principia* the notion of a propositional function of order n is neither taken as primitive nor defined in terms of the notion of a finite combination, but rather quantifiers referring to propositional functions of order n (which is all one needs) are defined as certain infinite conjunctions and disjunctions. But then one must ask: Why doesn't one define the integers by the infinite disjunction: $x=0 \lor x=0+1 \lor x=0+1+1 \lor \dots$ *ad infinitum,* saving in this way all the trouble connected with the notion of inductiveness? This whole objection would not apply if one understands by a propositional function of order n one "obtainable from such truth-functions of atomic propositions as presuppose for their definition no totalities except those of the propositional functions of order $<n$ and of individuals"; this notion, however, is somewhat lacking in precision.

The theory of orders proves more fruitful if considered from a purely mathematical standpoint, independently of the philosophical question whether impredicative definitions are admissible. Viewed in this manner, i.e., as a theory built up within the framework of ordinary mathematics, where impredicative definitions are admitted, there is no objection to extending it to arbitrarily high transfinite orders. Even if one rejects impredicative definitions, there would, I think, be no objection to extend it to such transfinite ordinals as can be constructed within the framework of finite orders. The theory in itself seems to demand such an extension since it leads automatically to the consideration of functions in whose definition one refers to all functions of finite orders, and these would be functions of order ω. Admitting transfinite orders, an axiom of reducibility can be proved. This, however, offers no help to the original purpose of the theory, because the ordinal α – such that every propositional function is extensionally equivalent to a function of order α – is so great, that it presupposes impredicative totalities. Nevertheless, so much can be accomplished in this way, that all impredicativities are reduced to one special kind, namely the existence of certain large ordinal numbers (or, well-ordered sets) and the validity of recursive reasoning for them. In particular, the existence of a well-ordered set, of order type ω_1 already suffices for the theory of real numbers. In addition this transfinite theorem of reducibility permits the proof of the consistency of the Axiom of Choice, of Cantor's Continuum-Hypothesis and even of the generalized Continuum-Hypothesis (which says that there exists no cardinal number between the power of any arbitrary set and the power of the set of its subsets) with the axioms of set theory as well as of *Principia.*

I now come in somewhat more detail to the theory of simple types which appears in *Principia* as combined with the theory of orders; the former is, however, (as remarked above) quite independent of the latter, since mixed types evidently do not contradict the vicious circle principle in any way. Accordingly, Russell also based the theory of simple types on entirely different reasons. The reason adduced (in addition to its "consonance with common sense") is very similar to Frege's, who, in his system, already had assumed the theory of simple types for functions, but failed to avoid the paradoxes, because he operated with classes (or rather functions in extension) without any restriction. This reason is that (owing to the variable it contains) a propositional function is something ambiguous (or, as Frege says, something unsaturated, wanting supplementation) and therefore can occur in a meaningful proposition only in such a way that this ambiguity is eliminated (e.g., by substituting a constant for the variable or applying quantification to it). The consequences are that a function cannot replace an individual in a proposition, because the latter has no ambiguity to be removed, and that functions with different kinds of arguments (i.e., different ambiguities) cannot replace each other; which is the essence of the theory of simple types. Taking a more nominalistic viewpoint (such as suggested in the second edition of *Principia* and in *Meaning and Truth*) one would have to replace "proposition" by "sentence" in the foregoing considerations (with corresponding additional changes). But in both cases, this argument clearly belongs to the order of ideas of the "no class" theory, since it considers the notions (or propositional functions) as something constructed out of propositions or sentences by leaving one or several constituents of them undetermined. Propositional functions in this sense are so to speak "fragments" of propositions, which have no meaning in themselves, but only insofar as one can use them for forming propositions by combining several of them, which is possible only if they "fit together," i.e., if they are of appropriate types. But, it should be noted that the theory of simple types (in contradistinction to the vicious circle principle) cannot in a strict sense follow from the constructive standpoint, because one might construct notions and classes in another way, e.g., as indicated on p. [462], where mixtures of types are possible. If on the other hand one considers concepts as real objects, the theory of simple types is not very plausible, since what one would expect to be a concept (such as, e.g., "transitivity" or the number two) would seem to be something behind all its various "realizations" on the different levels and therefore does not exist according to the theory of types. Nevertheless, there seems to be some truth behind this idea of realizations of the same concept on various levels, and one might, therefore, expect the theory of simple types to prove useful or necessary at least as a stepping-stone for a more satisfactory system, a way in which it has already been used by Quine (cf. 1937: 70). Also

Russell's "typical ambiguity" is a step in this direction. Since, however, it only adds certain simplifying symbolic conventions to the theory of types, it does not *de facto* go beyond this theory.

It should be noted that the theory of types brings in a new idea for the solution of the paradoxes, especially suited to their intensional form. It consists in blaming the paradoxes not on the axiom that every propositional function defines a concept or class, but on the assumption that every concept gives a meaningful proposition, if asserted for any arbitrary object or objects as arguments. The obvious objection that every concept can be extended to all arguments, by defining another one which gives a false proposition whenever the original one was meaningless, can easily be dealt with by pointing out that the concept "meaningfully applicable" need not itself be always meaningfully applicable.

The theory of simple types (in its realistic interpretation) can be considered as a carrying through of this scheme, based, however, on the following additional assumption concerning meaningfulness: "Whenever an object x can replace another object y in one meaningful proposition, it can do so in every meaningful proposition."[27] This of course has the consequence that the objects are divided into mutually exclusive ranges of significance, each range consisting of those objects which can replace each other; and that therefore each concept is significant only for arguments belonging to one of these ranges, i.e., for an infinitely small portion of all objects. What makes the above principle particularly suspect, however, is that its very assumption makes its formulation as a meaningful proposition impossible,[28] because x and y must then be confined to definite ranges of significance which are either the same or different, and in both cases the statement does not express the principle or even part of it. Another consequence is that the fact that an object x is (or is not) of a given type also cannot be expressed by a meaningful proposition.

It is not impossible that the idea of limited ranges of significance could be carried out without the above restrictive principle. It might even turn out that it is possible to assume every concept to be significant everywhere except for certain "singular points" or "limiting points," so that the paradoxes would appear as something analogous to dividing by zero. Such a system would be most satisfactory in the following respect: our logical intuitions would then remain correct up to certain minor correc-

---

[27]Russell formulates a somewhat different principle with the same effect (Whitehead and Russell 1910–13, 1: 95).

[28]This objection does not apply to the symbolic interpretation of the theory of types, spoken of on p. [465], because there one does not have objects but only symbols of different types.

tions, i.e., they could then be considered to give an essentially correct, only somewhat "blurred," picture of the real state of affairs. Unfortunately the attempts made in this direction have failed so far;[29] on the other hand, the impossibility of this scheme has not been proved either, in spite of the strong inconsistency theorems of Kleene and Rosser (1935: 630).

In conclusion I want to say a few words about the question whether (and in which sense) the axioms of *Principia* can be considered to be analytic. As to this problem it is to be remarked that analyticity may be understood in two senses. First, it may have the purely formal sense that the terms occurring can be defined (either explicitly or by rules for eliminating them from sentences containing them) in such a way that the axioms and theorems become special cases of the law of identity and disprovable propositions become negations of this law. In this sense even the theory of integers is demonstrably non-analytic, provided that one requires of the rules of elimination that they allow one actually to carry out the elimination in a finite number of steps in each case.[30] Leaving out this condition by admitting, e.g., sentences of infinite (and nondenumerable) length as intermediate steps of the process of reduction, all axioms of *Principia* (including the axioms of choice, infinity and reducibility) could be proved to be analytic for certain interpretations (by considerations similar to those referred to on p. [462]).[31] But this observation is of doubtful value, because the whole of mathematics as applied to sentences of infinite length has to be presupposed in order to prove this analyticity, e.g., the axiom of choice can be proved to be analytic only if it is assumed to be true.

In a second sense a proposition is called analytic if it holds, "owing to the meaning of the concepts occurring in it," where this meaning may perhaps be undefinable (i.e., irreducible to anything more fundamental).[32] It would seem that all axioms of *Principia,* in the first edition, (except the axiom of infinity) are in this sense analytic for certain interpretations of the primitive terms, namely if the term "predicative function" is replaced either by "class" (in the extensional sense) or (leaving out the axiom of choice) by "concept," since nothing can express better the

[29]A formal system along these lines is Church's "A Set of Postulates for the Foundation of Logic" (1932: 346; 1933: 839), where, however, the underlying idea is expressed by the somewhat misleading statement that the law of excluded middle is abandoned. However, this system has been proved to be inconsistent. See Kleene and Rosser 1935.

[30]Because this would imply the existence of a decision-procedure for all arithmetical propositions. Cf. Turing 1937: 230.

[31]Cf. also Ramsey (1926a: 338 or 1931: 1), where, however, the axiom of infinity cannot be obtained, because it is interpreted to refer to the individuals in the world.

[32]The two significations of the term *analytic* might perhaps be distinguished as tautological and analytic.

meaning of the term "class" than the axiom of the classes (cf. p. [459]) and the axiom of choice, and since, on the other hand, the meaning of the term "concept" seems to imply that every propositional function defines a concept.[33] The difficulty is only that we don't perceive the concepts of "concept" and of "class" with sufficient distinctness, as is shown by the paradoxes. In view of this situation, Russell took the course of considering both classes and concepts (except the logically uninteresting primitive predicates) as non-existent and of replacing them by constructions of our own. It cannot be denied that this procedure has led to interesting ideas and to results valuable also for one taking the opposite viewpoint. On the whole, however, the outcome has been that only fragments of Mathematical Logic remain, unless the things condemned are reintroduced in the form of infinite propositions or by such axioms as the axiom of reducibility which (in case of infinitely many individuals) is demonstrably false unless one assumes either the existence of classes or of infinitely many "*qualitates occultae.*" This seems to be an indication that one should take a more conservative course, such as would consist in trying to make the meaning of the terms "class" and "concept" clearer, and to set up a consistent theory of classes and concepts as objectively existing entities. This is the course which the actual development of Mathematical Logic has been taking and which Russell himself has been forced to enter upon in the more constructive parts of his work. Major among the attempts in this direction (some of which have been quoted in this essay) are the simple theory of types (which is the system of the first edition of *Principia* in an appropriate interpretation) and axiomatic set theory, both of which have been successful at least to this extent, that they permit the derivation of modern mathematics and at the same time avoid all known paradoxes. Many symptoms show only too clearly, however, that the primitive concepts need further elucidation.

It seems reasonable to suspect that it is this incomplete understanding of the foundations which is responsible for the fact that Mathematical Logic has up to now remained so far behind the high expectations of Peano and others who (in accordance with Leibniz's claims) had hoped

---

[33]This view does not contradict the opinion defended above that mathematics is based on axioms with a real content, because the very existence of the concept of, e.g., "class" constitutes already such an axiom; since, if one defined, e.g., "class" and "∈" to be "the concepts satisfying the axioms," one would be unable to prove their existence. "Concept" could perhaps be defined in terms of "proposition" (cf. p. [465]) (although I don't think that this would be a natural procedure); but then certain axioms about propositions, justifiable only with reference to the undefined meaning of this term, will have to be assumed. It is to be noted that this view about analyticity makes it again possible that every mathematical proposition could perhaps be reduced to a special case of $a = a$, namely if the reduction is effected not in virtue of the definitions of the terms occurring, but in virtue of their meaning, which can never be completely expressed in a set of formal rules.

that it would facilitate theoretical mathematics to the same extent as the decimal system of numbers has facilitated numerical computations. For how can one expect to solve mathematical problems systematically by mere analysis of the concepts occurring, if our analysis so far does not even suffice to set up the axioms? But there is no need to give up hope. Leibniz did not in his writings about the *Characteristica universalis* speak of a utopian project; if we are to believe his words he had developed this calculus of reasoning to a large extent, but was waiting with its publication till the seed could fall on fertile ground (1875–90, 7: 12; Vacca 1903: 72; Leibniz 1923–, 1: preface). He went even so far (1875–90, 7: 187) as to estimate the time which would be necessary for his calculus to be developed by a few select scientists to such an extent "that humanity would have a new kind of an instrument increasing the powers of reason far more than any optical instrument has ever aided the power of vision." The time he names is five years, and he claims that his method is not any more difficult to learn than the mathematics or philosophy of his time. Furthermore, he said repeatedly that, even in the rudimentary state to which he had developed the theory himself, it was responsible for all his mathematical discoveries; which, one should expect, even Poincaré would acknowledge as a sufficient proof of its fecundity.[34]

---

[34] I wish to express my thanks to Professor Alonzo Church, of Princeton University, who helped me find the correct English expressions in a number of places.

What is Cantor's continuum problem?

KURT GÖDEL

1. The concept of cardinal number

Cantor's continuum problem is simply the question: How many points are there on a straight line in euclidean space? An equivalent question is: How many different sets of integers do there exist?

This question, of course, could arise only after the concept of "number" had been extended to infinite sets; hence it might be doubted if this extension can be effected in a uniquely determined manner and if, therefore, the statement of the problem in the simple terms used above is justified. Closer examination, however, shows that Cantor's definition of infinite numbers really has this character of uniqueness. For whatever "number" as applied to infinite sets may mean, we certainly want it to have the property that the number of objects belonging to some class does not change if, leaving the objects the same, one changes in any way whatsoever their properties or mutual relations (e.g., their colors or their distribution in space). From this, however, it follows at once that two sets (at least two sets of changeable objects of the space-time world) will have the same cardinal number if their elements can be brought into a one-to-one correspondence, which is Cantor's definition of equality between numbers. For if there exists such a correspondence for two sets A and B it is possible (at least theoretically) to change the properties and relations of each element of A into those of the corresponding element of B, whereby A is transformed into a set completely indistinguishable from B, hence of the same cardinal number. For example, assuming a square and a line segment both completely filled with mass points (so that at each point of them exactly one mass point is situated), it follows, owing to the demonstrable fact that there exists a one-to-one correspondence between the points of a square and of a line segment and, therefore, also between the corresponding mass points, that the mass points of the square can be so rearranged as exactly to fill out the line segment, and

This is a revised and expanded version of a paper of the same title that appeared in *The American Mathematical Monthly* 54 (1947): 515–25. It is printed here with the kind permission of The Mathematical Association of America.

470

vice versa. Such considerations, it is true, apply directly only to physical objects, but a definition of the concept of "number" which would depend on the kind of objects that are numbered could hardly be considered to be satisfactory.

So there is hardly any choice left but to accept Cantor's definition of equality between numbers, which can easily be extended to a definition of "greater" and "less" for infinite numbers by stipulating that the cardinal number M of a set A is to be called less than the cardinal number N of a set B if M is different from N but equal to the cardinal number of some subset of B. That a cardinal number having a certain property exists is defined to mean that a set of such a cardinal number exists. On the basis of these definitions, it becomes possible to prove that there exist infinitely many different infinite cardinal numbers or "powers," and that, in particular, the number of subsets of a set is always greater than the number of its elements; furthermore, it becomes possible to extend (again without any arbitrariness) the arithmetical operations to infinite numbers (including sums and products with any infinite number of terms or factors) and to prove practically all ordinary rules of computation.

But, even after that, the problem of identifying the cardinal number of an individual set, such as the linear continuum, would not be well-defined if there did not exist some systematic representation of the infinite cardinal numbers, comparable to the decimal notation of the integers. Such a systematic representation, however, does exist, owing to the theorem that for each cardinal number and each set of cardinal numbers[1] there exists exactly one cardinal number immediately succeeding in magnitude and that the cardinal number of every set occurs in the series thus obtained.[2] This theorem makes it possible to denote the cardinal number immediately succeeding the set of finite numbers by $\aleph_0$ (which is the power of the "denumerably infinite" sets), the next one by $\aleph_1$, etc.; the one immediately succeeding all $\aleph_i$ where i is an integer, by $\aleph_\omega$, the next one by $\aleph_{\omega+1}$, etc. The theory of ordinal numbers provides the means for extending this series further and further.

---

[1] As to the question of why there does not exist a set of all cardinal numbers, see footnote 12.

[2] The axiom of choice is needed for the proof of this theorem (see Fraenkel and Bar-Hillel 1958). But it may be said that this axiom, from almost every possible point of view, is as well-founded today as the other axioms of set theory. It has been proved consistent with the other axioms of set theory which are usually assumed, provided that these other axioms are consistent (see Gödel 1940). Moreover, it is possible to define in terms of any system of objects satisfying the other axioms a system of objects satisfying those axioms *and* the axiom of choice. Finally, the axiom of choice is just as evident as the other set-theoretical axioms for the "pure" concept of set explained in footnote 11.

2. The continuum problem, the continuum hypothesis,
and the partial results concerning its truth obtained so far

So the analysis of the phrase "how many" unambiguously leads to a definite meaning for the question stated in the second line of this paper: The problem is to find out which one of the $\aleph$'s is the number of points of a straight line or (which is the same) of any other continuum (of any number of dimensions) in a euclidean space. Cantor, after having proved that this number is greater than $\aleph_0$, conjectured that it is $\aleph_1$. An equivalent proposition is this: Any infinite subset of the continuum has the power either of the set of integers or of the whole continuum. This is Cantor's continuum hypothesis.

But, although Cantor's set theory now has had a development of more than seventy years and the problem evidently is of great importance for it, nothing has been proved so far about the question what the power of the continuum is or whether its subsets satisfy the condition just stated, except (1) that the power of the continuum is not a cardinal number of a certain special kind, namely, not a limit of denumerably many smaller cardinal numbers,[3] and (2) that the proposition just mentioned about the subsets of the continuum is true for a certain infinitesimal fraction of these subsets, the analytic[4] sets.[5] Not even an upper bound, however large, can be assigned for the power of the continuum. Nor is the quality of the cardinal number of the continuum known any better than its quantity. It is undecided whether this number is regular or singular, accessible or inaccessible, and (except for König's negative result) what its character of confinality (see footnote 4) is. The only thing that is known, in addition to the results just mentioned, is a great number of consequences of, and some propositions equivalent to, Cantor's conjecture (Sierpiński 1934a).

This pronounced failure becomes still more striking if the problem is considered in its connection with general questions of cardinal arithmetic. It is easily proved that the power of the continuum is equal to $2^{\aleph_0}$. So the continuum problem turns out to be a question from the "multiplication table" of cardinal numbers, namely, the problem of evaluating a certain infinite product (in fact the simplest non-trivial one that can be formed). There is, however, not one infinite product (of factors >1) for

[3]See Hausdorff, *Mengenlehre* 1914: 68, or Bachmann 1955: 167. The discoverer of this theorem, J. König, asserted more than he had actually proved (1905: 177).
[4]See the list of definitions pp. 480–1.
[5]See Hausdorff 1914, 3rd ed.: 32. Even for complements of analytic sets the question is undecided at present, and it can be proved only that they either have the power $\aleph_0$ or $\aleph_1$ or continuum or are finite (see Kuratowski 1933–50, 1: 246).

472

which so much as an upper bound for its value can be assigned. All one knows about the evaluation of infinite products are two lower bounds due to Cantor and König (the latter of which implies the aforementioned negative theorem on the power of the continuum), and some theorems concerning the reduction of products with different factors to exponentiations and of exponentiations to exponentiations with smaller bases or exponents. These theorems reduce[6] the whole problem of computing infinite products to the evaluation of $\aleph_\alpha^{cf(\aleph_\alpha)}$ and the performance of certain fundamental operations on ordinal numbers, such as determining the limit of a series of them. All products and powers, can easily be computed[7] if the "generalized continuum hypothesis" is assumed; i.e., if it is assumed that $2^{\aleph_\alpha} = \aleph_{\alpha+1}$ for every α, or, in other terms, that the number of subsets of a set of power $\aleph_\alpha$ is $\aleph_{\alpha+1}$. But, without making any undemonstrated assumption, it is not even known whether or not $m < n$ implies $2^m < 2^n$ (although it is trivial that it implies $2^m \leqslant 2^n$), nor even whether $2^{\aleph_0} < 2^{\aleph_1}$.

3. Restatement of the problem on the basis of an analysis of the foundations of set theory and results obtained along these lines

This scarcity of results, even as to the most fundamental questions in this field, to some extent may be due to purely mathematical difficulties; it seems, however (see Section 4), that there are also deeper reasons involved and that a complete solution of these problems can be obtained only by a more profound analysis (than mathematics is accustomed to giving) of the meanings of the terms occurring in them (such as "set", "one-to-one correspondence", etc.) and of the axioms underlying their use. Several such analyses have already been proposed. Let us see then what they give for our problem.

First of all there is Brouwer's intuitionism, which is utterly destructive in its results. The whole theory of the $\aleph$'s greater than $\aleph_1$ is rejected as meaningless (Brouwer 1909: 569). Cantor's conjecture itself receives several different meanings, all of which, though very interesting in themselves, are quite different from the original problem. They lead partly to affirmative, partly to negative answers (Brouwer 1907, I: 9; III: 2). Not everything in this field, however, has been sufficiently clarified. The

---

[6]This reduction can be effected, owing to the results and methods of a paper by Tarski (1925: 1).

[7]For regular numbers $\aleph_\alpha$, one obtains immediately:

$$\aleph_\alpha^{cf(\aleph_\alpha)} = \aleph_\alpha^{\aleph_\alpha} = 2^{\aleph_\alpha} = \aleph_{\alpha+1}.$$

"semi-intuitionistic" standpoint along the lines of H. Poincaré and H. Weyl[8] would hardly preserve substantially more of set theory.

However, this negative attitude toward Cantor's set theory, and toward classical mathematics, of which it is a natural generalization, is by no means a necessary outcome of a closer examination of their foundations, but only the result of a certain philosophical conception of the nature of mathematics, which admits mathematical objects only to the extent to which they are interpretable as our own constructions or, at least, can be completely given in mathematical intuition. For someone who considers mathematical objects to exist independently of our constructions and of our having an intuition of them individually, and who requires only that the general mathematical concepts must be sufficiently clear for us to be able to recognize their soundness and the truth of the axioms concerning them, there exists, I believe, a satisfactory foundation of Cantor's set theory in its whole original extent and meaning, namely axiomatics of set theory interpreted in the way sketched below.

It might seem at first that the set-theoretical paradoxes would doom to failure such an undertaking, but closer examination shows that they cause no trouble at all. They are a very serious problem, not for mathematics, however, but rather for logic and epistemology. As far as sets occur in mathematics (at least in the mathematics of today, including all of Cantor's set theory), they are sets of integers, or of rational numbers (i.e., of pairs of integers), or of real numbers (i.e., of sets of rational numbers), or of functions of real numbers (i.e., of sets of pairs of real numbers), etc. When theorems about all sets (or the existence of sets in general) are asserted, they can always be interpreted without any difficulty to mean that they hold for sets of integers as well as for sets of sets of integers, etc. (respectively, that there either exist sets of integers, or sets of sets of integers, or ... etc., which have the asserted property). This concept of set,[9] however, according to which a set is something obtainable from the integers (or some other well-defined objects) by iterated

---

[8]See Weyl 1918 (1932 ed.). If the procedure of construction of sets described there (p. 20) is iterated, a sufficiently large (transfinite) number of times, one gets exactly the real numbers of the model for set theory mentioned in Section 4, in which the continuum hypothesis is true. But this iteration is not possible within the limits of the semi-intuitionistic standpoint.

[9]It must be admitted that the spirit of the modern abstract disciplines of mathematics, in particular of the theory of categories, transcends this concept of set, as becomes apparent, e.g., by the self-applicability of categories (see MacLane 1961). It does not seem, however, that anything is lost from the mathematical content of the theory if categories of different levels are distinguished. If there existed mathematically interesting proofs that would not go through under this interpretation, then the paradoxes of set theory would become a serious problem for mathematics.

application[10] of the operation "set of",[11] not something obtained by dividing the totality of all existing things into two categories, has never led to any antinomy whatsoever; that is, the perfectly "naïve" and uncritical working with this concept of set has so far proved completely self-consistent.[12]

But, furthermore, the axioms underlying the unrestricted use of this concept of set or, at least, a subset of them which suffices for all mathematical proofs devised up to now (except for theorems about the existence of extremely large cardinal numbers, see footnote 16), have been formulated so precisely in axiomatic set theory[13] that the question of whether some given proposition follows from them can be transformed, by means of mathematical logic, into a purely combinatorial problem concerning the manipulation of symbols which even the most radical intuitionist must acknowledge as meaningful. So Cantor's continuum problem, no matter what philosophical standpoint is taken, undeniably retains at least this meaning: to find out whether an answer, and if so which answer, can be derived from the axioms of set theory as formulated in the systems cited.

Of course, if it is interpreted in this way, there are (assuming the consistency of the axioms) *a priori* three possibilities for Cantor's conjecture: It may be demonstrable, disprovable, or undecidable.[14] The third alternative (which is only a precise formulation of the foregoing conjecture, that the difficulties of the problem are probably not purely mathematical), is the most likely. To seek a proof for it is, at present, perhaps the most promising way of attacking the problem. One result along these

[10]This phrase is meant to include transfinite iteration; i.e., the totality of sets obtained by finite iteration is considered to be itself a set and a basis for further applications of the operation "set of".

[11]The operation "set of x's" (where the variable "x" ranges over some given kind of objects) cannot be defined satisfactorily (at least not in the present state of knowledge), but can only be paraphrased by other expressions involving again the concept of set, such as: "multitude" ("combination", "part") is conceived of as something which exists in itself no matter whether we can define it in a finite number of words (so that random sets are not excluded).

[12]It follows at once from this explanation of the term "set" that a set of all sets or other sets of a similar extension cannot exist, since every set obtained in this way immediately gives rise to further applications of the operation "set of" and, therefore, to the existence of larger sets.

[13]See, e.g., Bernays, 1937–54, 2: 65; 6: 1; 7: 65, 133; 8: 89. Von Neumann 1925: 219; cf. also von Neumann 1929: 227; and 1928: 669; Gödel 1940; Bernays 1958. By including very strong axioms of infinity, much more elegant axiomatizations have recently become possible. (See Bernays 1961.)

[14]In case the axioms were inconsistent the last one of the four *a priori* possible alternatives for Cantor's conjecture would occur, namely, it would then be both demonstrable and disprovable by the axioms of set theory.

lines has been obtained already, namely, that Cantor's conjecture is not disprovable from the axioms of set theory, provided that these axioms are consistent (see Section 4).

It is to be noted, however, that on the basis of the point of view here adopted, a proof of the undecidability of Cantor's conjecture from the accepted axioms of set theory (in contradistinction, e.g., to the proof of the transcendency of π) would by no means solve the problem. For if the meanings of the primitive terms of set theory as explained on pages [474-5] and in footnote 11 are accepted as sound, it follows that the set-theoretical concepts and theorems describe some well-determined reality, in which Cantor's conjecture must be either true or false. Hence its undecidability from the axioms being assumed today can only mean that these axioms do not contain a complete description of that reality. Such a belief is by no means chimerical, since it is possible to point out ways in which the decision of a question, which is undecidable from the usual axioms, might nevertheless be obtained.

First of all the axioms of set theory by no means form a system closed in itself, but, quite on the contrary, the very concept of set[15] on which they are based suggests their extension by new axioms which assert the existence of still further iterations of the operation "set of". These axioms can be formulated also as propositions asserting the existence of very great cardinal numbers (i.e., of sets having these cardinal numbers). The simplest of these strong "axioms of infinity" asserts the existence of inaccessible numbers (in the weaker or stronger sense) $> \aleph_0$. The latter axiom, roughly speaking, means nothing else but that the totality of sets obtainable by use of the procedures of formation of sets expressed in the other axioms forms again a set (and, therefore, a new basis for further applications of these procedures) (Zermelo 1930: 29). Other axioms of infinity have first been formulated by P. Mahlo.[16] These axioms show clearly, not only that the axiomatic system of set theory as used today is

---

[15]Similarly the concept "property of set" (the second of the primitive terms of set theory) suggests continued extensions of the axioms referring to it. Furthermore, concepts of "property of property of set" etc. can be introduced. The new axioms thus obtained, however, as to their consequences for propositions referring to limited domains of sets (such as the continuum hypothesis) are contained (as far as they are known today) in the axioms about sets.

[16]See Mahlo 1911: 190–200; 1913: 269–76. From Mahlo's presentation of the subject, however, it does not appear that the numbers he defines actually exist. In recent years considerable progress has been made as to the axioms of infinity. In particular, some have been formulated that are based on principles entirely different from those of Mahlo, and Dana Scott has proved that one of them implies the negation of proposition A (mentioned on p. [478]). So the consistency proof for the continuum hypothesis explained on p. [479] does *not* go through if this axiom is added. However, that these axioms are implied by the

incomplete, but also that it can be supplemented without arbitrariness by new axioms which only unfold the content of the concept of set explained above.

It can be proved that these axioms also have consequences far outside the domain of very great transfinite numbers, which is their immediate subject matter: each of them, under the assumption of its consistency, can be shown to increase the number of decidable propositions even in the field of Diophantine equations. As for the continuum problem, there is little hope of solving it by means of those axioms of infinity which can be set up on the basis of Mahlo's principles (the aforementioned proof for the undisprovability of the continuum hypothesis goes through for all of them without any change). But there exist others based on different principles (see footnote 16); also there may exist, besides the usual axioms, the axioms of infinity, and the axioms mentioned in footnote 15, other (hitherto unknown) axioms of set theory which a more profound understanding of the concepts underlying logic and mathematics would enable us to recognize as implied by these concepts (see, e.g., footnote 19).

Secondly, however, even disregarding the intrinsic necessity of some new axiom, and even in case it has no intrinsic necessity at all, a probable decision about its truth is possible also in another way, namely, inductively by studying its "success." Success here means fruitfulness in consequences, in particular in "verifiable" consequences, i.e., consequences demonstrable without the new axiom, whose proofs with the help of the new axiom, however, are considerably simpler and easier to discover, and make it possible to contract into one proof many different proofs. The axioms for the system of real numbers, rejected by the intuitionists, have in this sense been verified to some extent, owing to the fact that analytical number theory frequently allows one to prove number-theoretical theorems which, in a more cumbersome way, can subsequently be verified by elementary methods. A much higher degree of verification than that, however, is conceivable. There might exist axioms so abundant in their verifiable consequences, shedding so much light upon a whole field, and yielding such powerful methods for solving problems (and even solving them constructively, as far as that is possible) that, no matter whether or not they are intrinsically necessary, they would have to be accepted at least in the same sense as any well-established physical theory.

general concept of set in the same sense as Mahlo's has not been made clear yet. See Tarski 1962: 134; Scott 1961; Hanf and Scott 1961: 445. Mahlo's axioms have been derived by Azriel Lévy from a general principle about the system of all sets (1960: 233). See also Bernays 1961: 1, where almost all set-theoretical axioms are derived from Lévy's principle.

4. Some observations about the question: In what sense and in which direction may a solution of the continuum problem be expected?

But are such considerations appropriate for the continuum problem? Are there really any clear indications for its unsolvability by the accepted axioms? I think there are at least two:

The first results from the fact that there are two quite differently defined classes of objects both of which satisfy all axioms of set theory that have been set up so far. One class consists of the sets definable in a certain manner by properties of their elements;[17] the other of the sets in the sense of arbitrary multitudes, regardless of if, or how, they can be defined. Now, before it has been settled what objects are to be numbered, and on the basis of what one-to-one correspondences, one can hardly expect to be able to determine their number, expect perhaps in the case of some fortunate coincidence. If, however, one believes that it is meaningless to speak of sets except in the sense of extensions of definable properties, then, too, he can hardly expect more than a small fraction of the problems of set theory to be solvable without making use of this, in his opinion essential, characteristic of sets, namely, that they are extensions of definable properties. This characteristic of sets, however, is neither formulated explicitly nor contained implicitly in the accepted axioms of set theory. So from either point of view, if in addition one takes into account what was said in Section 2, it may be conjectured that the continuum problem cannot be solved on the basis of the axioms set up so far, but, on the other hand, may be solvable with the help of some new axiom which would state or imply something about the definability of sets.[18]

The latter half of this conjecture has already been verified; namely, the concept of definability mentioned in footnote 17 (which itself is definable in axiomatic set theory) makes it possible to derive, in axiomatic set theory, the generalized continuum hypothesis from the axiom that every set is definable in this sense.[19] Since this axiom (let us call it "A") turns out to be demonstrably consistent with the other axioms, under the

---

[17] Namely, definable by certain procedures, "*in terms* of ordinal numbers" (i.e., roughly speaking, under the assumption that for each ordinal number a symbol denoting it is given). See Gödel 1940 and 1939. The paradox of Richard, of course, does not apply to this kind of definability, since the totality of ordinals is certainly not denumerable.

[18] D. Hilbert's program for a solution of the continuum problem (see Hilbert 1926: 161), which, however, has never been carried through, also was based on a consideration of all possible definitions of real numbers.

[19] On the other hand, from an axiom in some sense opposite to this one, the negation of Cantor's conjecture could perhaps be derived. I am thinking of an axiom which (similar to Hilbert's completeness axiom in geometry) would state some maximum property of the

assumption of the consistency of these other axioms, this result (regardless of the philosophical position taken toward definability) shows the consistency of the continuum hypothesis with the axioms of set theory, provided that these axioms themselves are consistent.[20] The proof in its structure is similar to the consistency-proof of non-euclidean geometry by means of a model within euclidean geometry. Namely, it follows from the axioms of set theory that the sets definable in the aforementioned sense form a model of set theory in which the proposition A and, therefore, the generalized continuum hypothesis is true.

A second argument in favor of the unsolvability of the continuum problem on the basis of the usual axioms can be based on certain facts (not known at Cantor's time) which seem to indicate that Cantor's conjecture will turn out to be wrong,[21] while, on the other hand, a disproof of it is demonstrably impossible on the basis of the axioms being assumed today.

One such fact is the existence of certain properties of point sets (asserting an extreme rareness of the sets concerned) for which one has succeeded in proving the existence of non-denumerable sets having these properties, but no way is apparent in which one could expect to prove the existence of examples of the power of the continuum. Properties of this type (of subsets of a straight line) are: (1) being of the first category on every perfect set (Sierpiński 1934b: 270, and Kuratowski 1933–50, 1: 269ff), (2) being carried into a zero set by every continuous one-to-one mapping of the line onto itself (Lusin and Sierpiński 1918: 35, and Sierpiński 1934b: 270). Another property of a similar nature is that of being coverable by infinitely many intervals of any given lengths. But in this case one has so far not even succeeded in proving the existence of non-denumerable examples. From the continuum hypothesis, however, it follows in all three cases that there exist, not only examples of the power of the continuum,[22] but even such as are carried into themselves (up to denumerably many points) by *every* translation of the straight line (Sierpiński 1935a: 43).

Other highly implausible consequences of the continuum hypothesis are that there exist: (1) subsets of a straight line of the power of the con-

system of all sets, whereas axiom A states a minimum property. Note that only a maximum property would seem to harmonize with the concept of set explained in footnote 11.

[20]See Gödel 1940 and 1939: 220. I take this opportunity to correct a mistake in the notation and a misprint which occurred in this paper: in lines 25–29, p. 221; 4–6 and 10, p. 222; 11–19, p. 223, the letter α should be replaced (at all places where it occurs) by μ. Also, in Theorem 6, p. 222, the symbol "$\equiv$" should be inserted between $\phi_\alpha(x)$ and $\phi_\alpha(x')$. For a carrying through of the proof in all details, Gödel 1940 is to be consulted.

[21]Views tending in this direction have been expressed also by Lusin 1935: 129ff. See also Sierpiński 1934a: 132.

[22]For the third case see Sierpiński 1934a (1st ed.): 39, Theorem 1.

tinuum which are covered (up to denumerably many points) by *every* dense set of intervals (Lusin 1914: 1259); (2) infinite dimensional subsets of Hilbert space which contain no non-denumerable finite-dimensional subset (in the sense of Menger-Urysohn) (Hurewicz 1932: 8); (3) an infinite sequence A^i of decompositions of any set M of the power of the continuum into continuum many mutually exclusive sets A_x^i such that, in whichever way a set $A_{x_i}^i$ is chosen for each i, $\prod_{i=0}^{\infty}(M - A_{x_i}^i)$ is denumerable.[23] (1) and (3) are very implausible even if "power of the continuum" is replaced by "$\aleph_1$".

One may say that many results of point-set theory obtained without using the continuum hypothesis also are highly unexpected and implausible (Blumenthal 1940: 346). But, true as that may be, still the situation is different there, in that, in most of those instances (such as, e.g., Peano's curves), the appearance to the contrary can be explained by a lack of agreement between our intuitive geometrical concepts and the set-theoretical ones occurring in the theorems. Also, it is very suspicious that, as against the numerous plausible propositions which imply the negation of the continuum hypothesis, not one plausible proposition is known which would imply the continuum hypothesis. I believe that adding up all that has been said one has good reason for suspecting that the role of the continuum problem in set theory will be to lead to the discovery of new axioms which will make it possible to disprove Cantor's conjecture.

Definitions of some of the technical terms

Definitions 4–15 refer to subsets of a straight line, but can be literally transferred to subsets of euclidean spaces of any number of dimensions if "interval" is identified with "interior of a parallelepipedon."

1. I call *the character of confinality* of a cardinal number m (abbreviated by "$cf(m)$") the smallest number n such that m is the sum of n numbers $< m$.

2. A cardinal number m is *regular* if $cf(m) = m$, otherwise singular.

3. An infinite cardinal number m is *inaccessible* if it is regular and has no immediate predecessor (i.e., if, although it is a limit of numbers $< m$, it is not a limit of fewer than m such numbers); it is *strongly inaccessible* if each product (and, therefore, also each sum) of fewer than m numbers $< m$ is $< m$. (See Sierpiński and Tarski 1930: 292; Tarski 1938: 68.)

---

[23]See Braun and Sierpiński 1932: 1, proposition (Q). This proposition is equivalent with the continuum hypothesis.

It follows from the generalized continuum hypothesis that these two concepts are equivalent. $\aleph_0$ is evidently inaccessible, and also strongly inaccessible. As for finite numbers, 0 and 2 and no others are strongly inaccessible. A definition of inaccessibility, applicable to finite numbers, is this: m is inaccessible if (1) any sum of fewer than m numbers $< m$ is $< m$, and (2) the number of numbers $< m$ is m. This definition, for transfinite numbers, agrees with that given above, and for finite numbers yields $0, 1, 2$ as inaccessible. So inaccessibility and strong inaccessibility turn out not to be equivalent for finite numbers. This casts some doubt on their equivalence for transfinite numbers, which follows from the generalized continuum hypothesis.

4. A set of intervals is *dense* if every interval has points in common with some interval of the set. (The end-points of an interval are not considered as points of the interval.)

5. A *zero-set* is a set which can be covered by infinite sets of intervals with arbitrarily small lengths-sum.

6. A *neighborhood* of a point P is an interval containing P.

7. A subset A of B is *dense in B* if every neighborhood of any point of B contains points of A.

8. A point P is in the *exterior* of A if it has a neighborhood containing no point of A.

9. A subset A of B is *nowhere dense in B* if those points of B which are in the exterior of A are dense in B, or (which is equivalent) if for no interval I the intersection IA is dense in IB.

10. A subset A of B is *of the first category in B* if it is the sum of denumerably many sets nowhere dense in B.

11. A set A is *of the first category on B* if the intersection AB is of the first category in B.

12. A point P is called *a limit point* of a set A if any neighborhood of P contains infinitely many points of A.

13. A set A is called *closed* if it contains all its limit points.

14. A set is *perfect* if it is closed and has no isolated point (i.e., no point with a neighborhood containing no other point of the set).

15. *Borel-sets* are defined as the smallest system of sets satisfying the postulates:
    (1) The closed sets are Borel-sets.
    (2) The complement of a Borel-set is a Borel-set.
    (3) The sum of denumerably many Borel-sets is a Borel-set.

16. A set is *analytic* if it is the orthogonal projection of some Borel-set of a space of next higher dimension. (Every Borel-set therefore is, of course, analytic.)

481

Supplement to the second edition [1963]

Since the publication of the preceding paper, a number of new results have been obtained; I would like to mention those that are of special interest in connection with the foregoing discussions.

1. A. Hajnal has proved that, if $2^{\aleph_0} \neq \aleph_2$ could be derived from the axioms of set theory, so could $2^{\aleph_0} = \aleph_1$ (1956: 131). This surprising result could greatly facilitate the solution of the continuum problem, should Cantor's continuum hypothesis be demonstrable from the axioms of set theory, which, however, probably is not the case.

2. Some new consequences of, and propositions equivalent with, Cantor's hypothesis can be found in the new edition of W. Sierpiński's book (1934a, 2nd ed.). In the first edition, it had been proved that the continuum hypothesis is equivalent with the proposition that the euclidean plane is the sum of denumerably many "generalized curves" (where a generalized curve is a point set definable by an equation $y = f(x)$ in some cartesian coordinate system). In the second edition (p. 207)[24], it is pointed out that the euclidean plane can be proved to be the sum of fewer than continuum many generalized curves under the much weaker assumption that the power of the continuum is not an inaccessible number. A proof of the converse of this theorem would give some plausibility to the hypothesis $2^{\aleph_0} =$ the smallest inaccessible number $> \aleph_0$. However, great caution is called for with regard to this inference, because the paradoxical appearance in this case (like in Peano's "curves") is due (at least in part) to a transference of our geometrical intuition of curves to something which has only some of the characteristics of curves. Note that nothing of this kind is involved in the counterintuitive consequences of the continuum hypothesis mentioned on pp. [479–80].

3. C. Kuratowski has formulated a strengthening of the continuum hypothesis (1948: 131), whose consistency follows from the consistency-proof mentioned in Section 4. He then drew various consequences from this new hypothesis.

4. Very interesting new results about the axioms of infinity have been obtained in recent years (see footnotes 13 and 16).

In opposition to the viewpoint advocated in Section 4 it has been suggested (Errera 1953: 176–83) that, in case Cantor's continuum problem should turn out to be undecidable from the accepted axioms of set theory, the question of its truth would lose its meaning, exactly as the question of the truth of Euclid's fifth postulate by the proof of the consistency of non-euclidean geometry became meaningless for the mathe-

---

[24]Or Sierpiński 1951: 9. See related results in Kuratowski (1951: 15) and Sikorski 1951: 18.

matician. I therefore would like to point out that the situation in set theory is very different from that in geometry, both from the mathematical and from the epistemological point of view.

In the case of the axiom of the existence of inaccessible numbers, e.g., (which can be proved to be undecidable from the von Neumann-Bernays axioms of set theory provided that it is consistent with them) there is a striking asymmetry, mathematically, between the system in which it is asserted and the one in which it is negated.[25]

Namely, the latter (but not the former) has a model which can be defined and proved to be a model in the original (unextended) system. This means that the former is an extension in a much stronger sense. A closely related fact is that the assertion (but not the negation) of the axiom implies new theorems about integers (the individual instances of which can be verified by computation). So the criterion of truth explained on p. [476] is satisfied, to some extent, for the assertion, but not for the negation. Briefly speaking, only the assertion yields a "fruitful" extension, while the negation is sterile outside its own very limited domain. Cantor's continuum hypothesis, too, can be shown to be sterile for number theory and to be true in a model constructible in the original system, whereas for some other assumption about the power of the continuum this perhaps is not so. On the other hand neither one of those asymmetries applies to Euclid's fifth postulate. To be more precise, both it and its negation are extensions in the weak sense.

As far as the epistemological situation is concerned, it is to be said that by a proof of undecidability a question loses its meaning only if the system of axioms under consideration is interpreted as a hypothetico-deductive system; i.e., if the meanings of the primitive terms are left undetermined. In geometry, e.g., the question as to whether Euclid's fifth postulate is true retains its meaning if the primitive terms are taken in a definite sense, i.e., as referring to the behavior of rigid bodies, rays of light, etc. The situation in set theory is similar, the difference is only that, in geometry, the meaning usually adopted today refers to physics rather than to mathematical intuition and that, therefore, a decision falls outside the range of mathematics. On the other hand, the objects of transfinite set theory, conceived in the manner explained on pp. [474–5] and in footnote 11, clearly do not belong to the physical world and even their indirect connection with physical experience is very loose (owing primarily to the fact that set-theoretical concepts play only a minor role in the physical theories of today).

But, despite their remoteness from sense experience, we do have some-

---

[25]The same asymmetry also occurs on the lowest levels of set theory, where the consistency of the axioms in question is less subject to being doubted by skeptics.

thing like a perception also of the objects of set theory, as is seen from the fact that the axioms force themselves upon us as being true. I don't see any reason why we should have less confidence in this kind of perception, i.e., in mathematical intuition, than in sense perception, which induces us to build up physical theories and to expect that future sense perceptions will agree with them and, moreover, to believe that a question not decidable now has meaning and may be decided in the future. The set-theoretical paradoxes are hardly any more troublesome for mathematics than deceptions of the senses are for physics. That new mathematical intuitions leading to a decision of such problems as Cantor's continuum hypothesis are perfectly possible was pointed out earlier (pp. [476-7]).

It should be noted that mathematical intuition need not be conceived of as a faculty giving an *immediate* knowledge of the objects concerned. Rather it seems that, as in the case of physical experience, we *form* our ideas also of those objects on the basis of something else which *is* immediately given. Only this something else here is *not,* or not primarily, the sensations. That something besides the sensations actually is immediately given follows (independently of mathematics) from the fact that even our ideas referring to physical objects contain constituents qualitatively different from sensations or mere combinations of sensations, e.g., the idea of object itself, whereas, on the other hand, by our thinking we cannot create any qualitatively new elements, but only reproduce and combine those that are given. Evidently the "given" underlying mathematics is closely related to the abstract elements contained in our empirical ideas.[26] It by no means follows, however, that the data of this second kind, because they cannot be associated with actions of certain things upon our sense organs, are something purely subjective, as Kant asserted. Rather they, too, may represent an aspect of objective reality, but, as opposed to the sensations, their presence in us may be due to another kind of relationship between ourselves and reality.

However, the question of the objective existence of the objects of mathematical intuition (which, incidentally, is an exact replica of the question of the objective existence of the outer world) is not decisive for the problem under discussion here. The mere psychological fact of the existence of an intuition which is sufficiently clear to produce the axioms of set theory and an open series of extensions of them suffices to give meaning to the question of the truth or falsity of propositions like

[26]Note that there is a close relationship between the concept of set explained in footnote 14 and the categories of pure understanding in Kant's sense. Namely, the function of both is "synthesis," i.e., the generating of unities out of manifolds (e.g., in Kant, of the idea of *one* object out of its various aspects).

Cantor's continuum hypothesis. What, however, perhaps more than anything else, justifies the acceptance of this criterion of truth in set theory is the fact that continued appeals to mathematical intuition are necessary not only for obtaining unambiguous answers to the questions of transfinite set theory, but also for the solution of the problems of finitary number theory[27] (of the type of Goldbach's conjecture),[28] where the meaningfulness and unambiguity of the concepts entering into them can hardly be doubted. This follows from the fact that for every axiomatic system there are infinitely many undecidable propositions of this type.

It was pointed out earlier (p. [477]) that, besides mathematical intuition, there exists another (though only probable) criterion of the truth of mathematical axioms, namely their fruitfulness in mathematics and, one may add, possibly also in physics. This criterion, however, though it may become decisive in the future, cannot yet be applied to the specifically set-theoretical axioms (such as those referring to great cardinal numbers), because very little is known about their consequences in other fields. The simplest case of an application of the criterion under discussion arises when some set-theoretical axiom has number-theoretical consequences verifiable by computation up to any given integer. On the basis of what is known today, however, it is not possible to make the truth of any set-theoretical axiom reasonably probable in this manner.

Postscript

Shortly after the completion of the manuscript of this paper the question of whether Cantor's Continuum Hypothesis is provable from the von Neumann-Bernays axioms of set theory (the axiom of choice included) was settled in the negative by Paul J. Cohen (1963a, 1964). It turns out that for a wide range of $\aleph_\tau$, the equality $2^{\aleph_0} = \aleph_\tau$ is consistent and an extension in the weak sense (that is, it implies no new number-theoretical theorems). Whether for a suitable concept of "standard" definition there exist definable $\aleph_\tau$ not excluded by König's theorem (see p. [472] above) for which this is not so is still an open question (of course, it must be assumed that the existence of the $\aleph_\tau$ in question is either demonstrable or has been postulated).

---

[27]Unless one is satisfied with inductive (probable) decisions, such as verifying the theorem up to very great numbers, or more indirect inductive procedures (see pp. [478, 485]).

[28]I.e., universal propositions about integers which can be decided in each individual instance.

485

The iterative conception of set

GEORGE BOOLOS

A set, according to Cantor, is "any collection ... into a whole of definite, well-distinguished objects ... of our intuition or thought."[1] Cantor also defined a set as a "many, which can be thought of as one, i.e., a totality of definite elements that can be combined into a whole by a law."[2] One might object to the first definition on the grounds that it uses the concepts of *collection* and *whole,* which are notions no better understood than that of *set,* that there ought to be sets of objects that are not objects of our thought, that 'intuition' is a term laden with a theory of knowledge that no one should believe, that *any* object is "definite," that there should be sets of ill-distinguished objects, such as waves and trains, etc., etc. And one might object to the second on the grounds that 'a many' is ungrammatical, that if something is "a many" it should hardly be thought of as one, that *totality* is as obscure as *set,* that it is far from clear how laws can combine anything into a whole, that there ought to be other combinations into a whole than those effected by "laws," etc., etc. But it cannot be denied that Cantor's definitions could be used by a person to identify and gain some understanding of the sort of object of which Cantor wished to treat. Moreover, they do suggest – although, it must be conceded, only very faintly – two important characteristics of sets: that a set is "determined" by its elements in the sense that sets with exactly the same elements are identical, and that, in a sense, the clarification of which is one of the principal objects of the theory whose rationale we shall give, the elements of a set are "prior to" it.

It is not to be presumed that the concepts of *set* and *member of* can be explained or defined by means of notions that are simpler or conceptually more basic. However, as a theory about sets might itself provide the sort of elucidation about sets and membership that good definitions

Reprinted with the kind permission of the author and the editors from the *Journal of Philosophy* 68 (1971): 215–32.

[1] "Unter einer 'Menge' verstehen wir jede Zusammenfassung M von bestimmten wohl-unterschiedenen Objekten m unserer Anschauung oder unseres Denkens (welche die 'Elemente' von M genannt werden) zu einem Ganzen" (Cantor 1932: 282).

[2] "...jedes Viele, welches sich als Eines denken lässt, d.h. jeden Inbegriff bestimmter Elemente, welcher durch ein Gesetz zu einem Ganzen verbunden werden kann" (Cantor 1932: 204).

might be hoped to offer, there is no reason for such a theory to begin with, or even contain, a definition of 'set'. That we are unable to give informative definitions of *not* or *for some* does not and should not prevent the development of quantificational logic, which provides us with significant information about these concepts.

I. Naive set theory

Here is an idea about sets that might occur to us quite naturally, and is perhaps suggested by Cantor's definition of a set as a totality of definite elements that can be combined into a whole *by a law*.

By the law of excluded middle, any (one-place) predicate in any language either applies to a given object or does not. So, it would seem, to any predicate there correspond two sorts of thing: the sort of thing to which the predicate applies (of which it is true) and the sort of thing to which it does not apply. So, it would seem, for any predicate there is a set of all and only those things to which it applies (as well as a set of just those things to which it does not apply). Any set whose members are exactly the things to which the predicate applies – by the axiom of extensionality, there cannot be *two* such sets – is called the *extension* of the predicate. Our thought might therefore be put: "Any predicate has an extension." We shall call this proposition, together with the argument for it, the *naive conception of set*.

The argument has great force. How could there *not* be a collection, or set, of just those things to which any given predicate applied? Isn't anything to which a predicate applies similar to all other things to which it applies in precisely the respect that it applies to them; and how could there fail to be a set of all things similar to one another in this respect? Wouldn't it be extremely implausible to say, of any particular predicate one might consider, that there weren't two kinds of thing it determined, namely, a kind of thing of which it is true, and a kind of thing of which it is not true? And why should one not take these kinds of things to be sets? Aren't kinds sets? If not, what is the difference?

Let us denote by '$\mathcal{K}$' a certain standardly formalized first-order language, whose variables range over all sets and individuals ($=$ non-sets), and whose nonlogical constants are a one-place predicate letter 'S' abbreviating 'is a set', and a two-place predicate letter '$\in$', abbreviating 'is a member of'. Which sentences of this language, together with their consequences, do we believe state truths about sets? Otherwise put, which formulas of $\mathcal{K}$ should we take as axioms of a set theory on the strength of our beliefs about sets?

If the naive conception of set is correct, there should (at least) be a set of just those things to which ϕ applies, if ϕ is a formula of $\mathcal{K}$. So (the universal closure of) $\ulcorner(\exists y)(Sy \;\&\; (x)(x \in y \leftrightarrow \phi))\urcorner$ should express a truth about sets (if no occurrence of 'y' in ϕ is free).

We call the theory whose axioms are the axiom of extensionality (to which we later recur), i.e., the sentence

$$(x)(y)(Sx \;\&\; Sy \;\&\; (z)(z \in x \leftrightarrow z \in y) \rightarrow x = y)$$

and all formulas $\ulcorner(\exists y)(Sy \;\&\; (x)(x \in y \leftrightarrow \phi))\urcorner$ (where 'y' does not occur free in ϕ) *naive set theory*.

Some of the axioms of naive set theory are the formulas

$$(\exists y)(Sy \;\&\; (x)(x \in y \leftrightarrow x \neq x))$$
$$(\exists y)(Sy \;\&\; (x)(x \in y \leftrightarrow (x = z \lor x = w)))$$
$$(\exists y)(Sy \;\&\; (x)(x \in y \leftrightarrow (\exists w)(x \in w \;\&\; w \in z)))$$
$$(\exists y)(Sy \;\&\; (x)(x \in y \leftrightarrow (Sx \;\&\; x = x)))$$

The first of these formulas states that there is a set that contains no members. By the axiom of extensionality, there can be at most one such set. The second states that there is a set whose sole members are z and w; the third, that there is a set whose members are just the members of members of z.

The last, which states that there is a set that contains all sets whatsoever, is rather anomalous; for if there is a set that contains all sets, a universal set, that set contains itself, and perhaps the mind ought to boggle at the idea of something's *containing* itself. Nevertheless, naive set theory is simple to state, elegant, initially quite credible, and natural in that it articulates a view about sets that might occur to one quite naturally.

Alas, it is inconsistent.

Proof of the inconsistency of naive set theory
(Russell's paradox)

No set can contain all and only those sets which do not contain themselves. For if any such set existed, if it contained itself, then, as it contains *only* those sets which do not contain themselves, it would not contain itself; but if it did not contain itself, then, as it contains *all* those sets which do not contain themselves, it would contain itself. Thus any such set would have to contain itself if and only if it did not contain itself. Consequently, there is no set that contains all and only those sets which do not contain themselves.

This argument, which uses no axioms of naive set theory, or any other set theory, shows that the sentence

$$\sim(\exists y)(Sy \,\&\, (x)(x \in y \leftrightarrow (Sx \,\&\, \sim x \in x)))$$

is *logically valid* and, hence, is a theorem of any theory that is expressed in $\mathcal{K}$. But one of the axioms and, hence, one of the theorems, of naive set theory is the sentence

$$(\exists y)(Sy \,\&\, (x)(x \in y \leftrightarrow (Sx \,\&\, \sim x \in x)))$$

Naive set theory is therefore inconsistent.

II. The iterative conception of set

Faced with the inconsistency of naive set theory, one might come to believe that any decision to adopt a system of axioms about sets would be *arbitrary* in that no explanation could be given why the particular system adopted had any greater claim to describe what we conceive sets and the membership relation to be like than some other system, perhaps incompatible with the one chosen. One might think that no answer could be given to the question: why adopt *this* particular system rather than that or this other one? One might suppose that any apparently consistent theory of sets would have to be unnatural in some way or fragmentary, and that, if consistent, its consistency would be due to certain provisions that were laid down for the express purpose of avoiding the paradoxes that show naive set theory inconsistent, but that lack any independent motivation.

One might imagine all this; but there is another view of sets: the *iterative conception of set,* as it is sometimes called, which often strikes people as entirely natural, free from artificiality, not at all ad hoc, and one they might perhaps have formulated themselves.

It is, perhaps, no more natural a conception than the naive conception, and certainly not quite so simple to describe. On the other hand, it is, as far as we know, consistent: not only are the sets whose existence would lead to contradiction not assumed to exist in the axioms of the theories that express the iterative conception, but the more than fifty years of experience that practicing set theorists have had with this conception have yielded a good understanding of what can and what cannot be proved in these theories, and at present there just is no suspicion at all that they are inconsistent.[3]

[3]The conception is well known among logicians; a rather different version of it is sketched in Shoenfield (1967: chap. 10). I learned of it principally from Putnam, Kripke, and Donald Martin. Authors of set-theory texts either omit it or relegate it to back pages; philosophers, in the main, seem to be unaware of it, or of the preeminence of ZF, which may be said to embody it. It is due primarily to Zermelo and Russell.

The standard, first-order theory that expresses the iterative conception of set as fully as a first-order theory in the language $\mathcal{L}$ of set theory[4] can, is known as *Zermelo-Fraenkel set theory,* or 'ZF' for short. There are other theories besides ZF that embody the iterative conception: one of them, Zermelo set theory, or "Z", which will occupy us shortly, is a *subsystem* of ZF in the sense that any theorem of Z is also a theorem of ZF; two others, von-Neumann-Bernays-Gödel set theory and Morse-Kelley set theory, are supersystems (or extensions) of ZF, but they are most commonly formulated as second-order theories.

Other theories of sets, incompatible with ZF, have been proposed.[5] These theories appear to lack a motivation that is independent of the paradoxes in the following sense: they are not, as Russell has written, "such as even the cleverest logician would have thought of if he had not known of the contradictions" (1959: 80). A final and satisfying resolution to the set-theoretical paradoxes cannot be embodied in a theory that blocks their derivation by artificial technical restrictions on the set of axioms that are imposed *only because* paradox would otherwise ensue; these other theories survive only through such artificial devices. ZF alone (together with its extensions and subsystems) is not only a consistent (apparently) but also an independently motivated theory of sets: there is, so to speak, a "thought behind it" about the nature of sets which might have been put forth even if, impossibly, naive set theory had been consistent. The thought, moreover, can be described in a rough, but informative way without first stating the theory the thought is behind.

In order to see why a conception of set other than the naive conception might be desired even if the naive conception were consistent, let us take another look at naive set theory and the anomalousness of its axiom, '$(\exists y)(Sy \ \& \ (x)(x \in y \leftrightarrow (Sx \ \& \ x=x)))$'.

According to this axiom there is a set that contains all sets, and therefore there is a set that contains itself. It is important to realize how odd the idea of something's containing itself is. Of course a set can and must *include* itself (as a subset). But *contain* itself? Whatever tenuous hold on the concepts of *set* and *member* were given one by Cantor's definitions of 'set' and one's ordinary understanding of 'element', 'set', 'collection', etc. is altogether lost if one is to suppose that some sets are members of themselves. The idea is paradoxical not in the sense that it is contradictory to suppose that some set is a member of itself, for, after all, '$(\exists x)(Sx \ \& \ x \in x)$' is obviously consistent, but that if one understands '$\in$' as meaning 'is a member of', it is very, very peculiar to suppose it

<hr/>

[4]$\mathcal{L}$ contains (countably many) variables, ranging over (pure) sets, '=', and '$\in$', which is its sole nonlogical constant.
[5]For example, Quine's systems NF and ML.

true. For when one is told that a set is a collection into a whole of definite elements of our thought, one thinks: Here are some things. Now we bind them up into a whole.[6] *Now* we have a set. We don't suppose that what we come up with after combining some elements into a whole could have been one of the very things we combined (not, at least, if we are combining two or more elements).

If $(\exists x)(Sx \& x \in x)$, then $(\exists x)(\exists y)(Sx \& Sy \& x \in y \& y \in x)$. The supposition that there are sets x and y each of which belongs to the other is almost as strange as the supposition that some set is a self-member. There is of course an infinite sequence of such cyclical pathologies: $(\exists x)(\exists y)(\exists z)(Sx \& Sy \& Sz \& x \in y \& y \in z \& z \in x)$, etc. Only slightly less pathological are the suppositions that there is an ungrounded set,[7] or that there is an infinite sequence of sets $x_0, x_1, x_2, \ldots$, each term of which belongs to the previous one.

There does not seem to be any argument that is guaranteed to persuade someone who really does not see the peculiarity of a set's belonging to itself, or to one of its members, etc., that these states of affairs are peculiar. But it is in part the sense of their oddity that has led set-theorists to favor conceptions of set, such as the iterative conception, according to which what they find odd does not occur.

We describe this conception now. Our description will have three parts. The first is a rough statement of the idea. It contains such expressions as 'stage', 'is formed at', 'earlier than', 'keep on going', which must be exorcised from any formal theory of sets. From the rough description it sounds as if sets were continually being created, which is not the case. In the second part, we present an axiomatic theory which partially formalizes the idea roughly stated in the first part. For reference, let us call this theory the *stage theory*. The third part consists in a derivation from the stage theory of the axioms of a theory of sets. These axioms are formulas of $\mathcal{L}$, the language of set theory, and contain none of the metaphorical expressions which are employed in the rough statement and of which abbreviations are found in the language in which the stage theory is expressed.

Here is the idea, roughly stated:

A set is any collection that is formed at some stage of the following process: Begin with individuals (if there are any). An individual is an object that is not a set; individuals do not contain members. At stage zero (we count from zero instead of one) form all possible collections of

---

[6]We put a "lasso" around them, in a figure of Kripke's.

[7]x is *ungrounded* if x belongs to some set z, each of whose members has a member in common with z.

individuals. If there are no individuals, only one collection, the null set, which contains no members, is formed at this 0th stage. If there is only one individual, two sets are formed: the null set and the set containing just that one individual. If there are two individuals, four sets are formed; and in general, if there are n individuals, 2^n sets are formed. Perhaps there are infinitely many individuals. Still, we assume that one of the collections formed at stage zero is the collection of all individuals, however many of them there may be.

At stage one, form all possible collections of individuals and sets formed at stage zero. If there are any individuals, at stage one some sets are formed that contain both individuals and sets formed at stage zero. Of course some sets are formed that contain only sets formed at stage zero. At stage two, form all possible collections of individuals, sets formed at stage zero, and sets formed at stage one. At stage three, form all possible collections of individuals and sets formed at stages zero, one, and two. At stage four, form all possible collections of individuals and sets formed at stages zero, one, two, and three. Keep on going in this way, at each stage forming all possible collections of individuals and sets formed at earlier stages.

Immediately after all of stages zero, one, two, three,..., there is a stage; call it stage omega. At stage omega, form all possible collections of individuals formed at stages zero, one, two,.... One of these collections will be the set of *all* sets formed at stages zero, one, two,....

After stage omega there is a stage omega plus one. At stage omega plus one form all possible collections of individuals and sets formed at stages zero, one, two..., and omega. At stage omega plus two form all possible collections of individuals and sets formed at stages zero, one, two,..., omega, and omega plus one. At stage omega plus three form all possible collections of individuals and sets formed at earlier stages. Keep on going in this way.

Immediately after all of stages zero, one, two,..., omega, omega plus one, omega plus two,..., there is a stage, call it stage omega plus omega (or omega times two). At stage omega plus omega form all possible collections of individuals and sets formed at earlier stages. At stage omega plus omega plus one ....

...omega plus omega plus omega (or omega times three)...

...(omega times four)...

...omega times omega... ...

Keep on going in this way....

According to this description, sets are formed over and over again: in fact, according to it, a set is formed at every stage later than that at which it is first formed. We could continue to say this if we liked; instead we shall say that a set is formed only once, namely, at the earliest stage at which, on our old way of speaking, it would have been said to be formed.

That is a rough statement of the iterative conception of set. According to this conception, no set belongs to itself, and hence there is no set of all sets; for every set is formed at some earliest stage, and has as members only individuals or sets formed at still earlier stages. Moreover, there are not two sets x and y, each of which belongs to the other. For if y belonged to x, y would have had to be formed at an earlier stage than the earliest stage at which x was formed, and if x belonged to y, x would have had to be formed at an earlier stage than the earliest stage at which y was formed. So x would have had to be formed at an earlier stage than the earliest stage at which it was formed, which is impossible. Similarly, there are no sets x, y, and z such that x belongs to y, y to z, and z to x. And in general, there are no sets $x_0, x_1, x_2, \ldots, x_n$ such that x_0 belongs to x_1, x_1 to $x_2, \ldots, x_{n-1}$ to x_n, and x_n to x_0. Furthermore it would appear that there is no sequence of sets $x_0, x_1, x_2, x_3, \ldots$ such that x_1 belongs to x_0, x_2 belongs to x_1, x_3 belongs to x_2, and so forth. Thus, if sets are as the iterative conception has them, the anomalous situations do not arise in which sets belong to themselves or to others that in turn belong to them.

The sets of which ZF in its usual formulation speaks ("quantifies over") are not all the sets there are, if we assume that there are some individuals, but only those which are formed at some stage under the assumption that there are no individuals. These sets are called *pure* sets. All members of a pure set are pure sets, and any set, all of whose members are pure, is itself pure. It may not be obvious that any pure sets are ever formed, but the set Λ, which contains no members at all, is pure, and is formed at stage 0. $\{\Lambda\}$ and $\{\{\Lambda\}\}$ are also both pure and are formed at stages 1 and 2, respectively. There are many others. From now on, we shall use the word 'set' to mean 'pure set'.

Let us now try to state a theory, the stage theory, that precisely expresses much, but not all, of the content of the iterative conception. We shall use a language, $\mathcal{J}$, in which there are two sorts of variables: variables 'x', 'y', 'z', 'w', ..., which range over sets, and variables 'r', 's', 't', which range over stages. In addition to the predicate letters '$\in$' and '$=$' of $\mathcal{L}$, $\mathcal{J}$ also contains two new two-place predicate letters 'E', read 'is earlier than', and 'F', read 'is formed at'. The rules of formation of $\mathcal{J}$ are perfectly standard.

Here are some axioms governing the sequence of stages:

(I) $(x) \sim sEs$ (No stage is earlier than itself.)

(II) $(r)(s)(t)((rEs\ \&\ sEt) \rightarrow rEt)$ (*Earlier than* is transitive.)

(III) $(s)(t)(sEt \lor s=t \lor tEs)$ (*Earlier than* is connected.)

(IV) $(\exists s)(t)(t \neq s \to sEt)$ (There is an earliest stage.)

(V) $(s)(\exists t)(sEt \, \& \, (r)(rEt \to (rEs \lor r=s)))$ (*Immediately* after any stage there is another.)

Here are some axioms describing when sets and their members are formed:

(VI) $(\exists s)((\exists t)tEs \, \& \, (t)(tEs \to (\exists r)(tEr \, \& \, rEs)))$ (There is a stage, not the earliest one, which is not immediately after any one stage. In the rough description, stage omega was such a stage.)

(VII) $(x)(\exists s)(xFs \, \& \, (t)(xFt \to t=s))$ (Every set is formed at some unique stage.)

(VIII) $(x)(y)(s)(t)((y \in x \, \& \, xFs \, \& \, yFt) \to tEs)$ (Every member of a set is formed *before,* i.e., at an earlier stage than, the set.)

(IX) $(x)(s)(t)(xFs \, \& \, tEs \to (\exists y)(\exists r)(y \in x \, \& \, yFr \, \& \, (t=r \lor tEr)))$ (If a set is formed at a stage, then, at or after any earlier stage, at least one of its members is formed. So it never happens that all members of a set are formed before some stage, but the set is not formed at that stage, if it has not been formed already.)

We may capture part of the content of the idea that at any stage every *possible* collection (or set) of sets formed at earlier stages is formed (if it has not yet been formed) by taking as axioms all formulas $\ulcorner(s)(\exists y)(x)(x \in y \leftrightarrow (\chi \, \& \, (\exists t)(tEs \, \& \, xFt)))\urcorner$, where χ is a formula of the language $\mathcal{J}$ in which no occurrence of 'y' is free. Any such axiom will say that for any stage there is a set of just those sets to which χ applies that are formed before that stage. Let us call these axioms *specification axioms*.

There is still one important feature contained in our rough description that has not yet been expressed in the stage theory: the analogy between the way sets are *inductively generated* by the procedure described in the rough statement and the way the natural numbers $0, 1, 2, \ldots$ are inductively generated from 0 by the repeated application of the successor operation. One way to characterize this feature is to assert a suitable induction principle concerning sets and stages; for, as Frege, Dedekind, Peano, and others have enabled us to see, the content of the idea that objects of a certain kind are inductively generated in a certain way is just the proposition than an appropriate induction principle holds of those objects.

The principle of mathematical induction, the induction principle governing the natural numbers, has two forms, which are interderivable on certain assumptions about the natural numbers. The first version of the principle is the statement

$$(P)[(P0 \mathbin{\&} (n)[Pn \rightarrow PSn]) \rightarrow (n)Pn]$$

which may be read, 'If 0 has a property and if whenever a natural number has the property its successor does, then every natural number has the property'. The second version is the statement

$$(P)[(n)((m)[m<n \rightarrow Pm] \rightarrow Pn) \rightarrow (n)Pn]$$

It may be read, '*If each natural number has a property provided that all smaller natural numbers have it, then every natural number has the property*'.

The induction principle about sets and stages that we should like to assert is modeled after the second form of the principle of mathematical induction. Let us say that a stage *s is covered by* a predicate if the predicate applies to every set formed at *s*. Our analogue for sets and stages of the second form of mathematical induction says that *if each stage is covered by a predicate provided that all earlier stages are covered by it, then every stage is covered by the predicate*. The full force of this assertion can be expressed only with a second-order quantifier. However, we can capture some of its content by taking as axioms all formulas

$$\ulcorner(s)((t)(tEs \rightarrow (x)(xFt \rightarrow \theta)) \rightarrow (x)(xFs \rightarrow \chi)) \rightarrow (s)(x)(xFs \rightarrow \chi)\urcorner$$

where χ is a formula of $\mathscr{J}$ that contains no occurrences of '*t*' and θ is just like χ except for containing a free occurrence of '*t*' wherever χ contains a free occurrence of '*s*'. [Observe that '$(x)(xFs \rightarrow \chi)$' says that χ applies to every set formed at stage *s* and, hence, that *s* is covered by χ.] We call these axioms *induction axioms*.

III. Zermelo set theory

We complete the description of the iterative conception of set by showing how to derive the axioms of a theory of sets from the stage theory. The axioms we derive speak only about sets and membership: they are formulas of $\mathscr{L}$.

The axiom of the null set: $(\exists y)(x) \sim x \in y$. (There is a set with no members.)

Derivation. Let $\chi = $ '$x=x$'. Then

$$(s)(\exists y)(x)(x \in y \leftrightarrow (x=x \mathbin{\&} (\exists t)(tEs \mathbin{\&} xFt)))$$

495

is a specification axiom, according to which, for any stage, there is a set of all sets formed at earlier stages. As there is an earliest stage, stage 0, before which no sets are formed, there is a set that contains no members. Note that, by axiom (IX) of the stage theory, any set with no members is formed at stage 0; for if it were formed later, it would have to have a member (that was formed at or after stage 0).

The axiom of pairs: $(z)(w)(\exists y)(x)(x \in y \leftrightarrow (x=z \lor x=w))$. (For any sets z and w, not necessarily distinct, there is a set whose sole members are z and w.)

Derivation. Let $\chi = {}^{\prime}(x=z \lor x=w){}^{\prime}$. Then

$$(s)(\exists y)(x)(x \in y \leftrightarrow ((x=z \lor x=w) \, \& \, (\exists t)(tEs \, \& \, xFt)))$$

is a specification axiom, according to which, for any stage, there is a set of all sets formed at earlier stages that are identical with either z or w. Any set is formed at some stage. Let r be the stage at which z is formed; s, the stage at which w is formed. Let t be a stage later than both r and s. Then there is a set of all sets formed at stages earlier than t that are identical with z or w. So there is a set containing just z and w.

The axiom of unions: $(z)(\exists y)(x)(x \in y \leftrightarrow (\exists w)(x \in w \, \& \, w \in z))$. (For any set z, there is a set whose members are just the members of members of z.)

Derivation. '$(s)(\exists y)(x)(x \in y \leftrightarrow ((\exists w)(x \in w \, \& \, w \in z) \, \& \, (\exists t)(tEs \, \& \, xFt)))$' is a specification axiom, according to which, for any stage, there is a set of all members of members of z formed at earlier stages. Let s be the stage at which z is formed. Every member of z is formed before s, and hence every member of a member of z is also formed before s. Thus there is a set of all members of members of z.

The power-set axiom: $(z)(\exists y)(x)(x \in y \leftrightarrow (w)(w \in x \rightarrow w \in z))$. (For any set z, there is a set whose members are just the subsets of z.)

Derivation. '$(s)(\exists y)(x)(x \in y \leftrightarrow ((w)(w \in x \rightarrow w \in z) \, \& \, (\exists t)(tEs \, \& \, xFt)))$' is a specification axiom, according to which, for any stage, there is a set of all subsets of z formed at earlier stages. Let t be the stage at which z is formed and let s be the next later stage. If x is a subset of z, then x is formed before s. For otherwise, by axiom (IX), there would be a member of x that was formed at or after t and, hence, that was not a member of z.

So there is a set of all subsets of z formed before s, and hence a set of all subsets of z.

The axiom of infinity:

$$(\exists y)((\exists x)(x \in y \,\&\, (z) \sim z \in x)$$
$$\&\, (x)(x \in y \rightarrow (\exists z)(z \in y \,\&\, (w)(w \in z \leftrightarrow (w \in x \lor w = x)))))$$

(Call a set null if it has no members. Call z a successor of x if the members of z are just those of x and x itself. Then there is a set which contains a null set and which contains a successor of any set it contains.)

Derivation. Let us first observe that every set x has a successor. For let y be a set containing just x and x (axiom of pairs), and let w be a set containing just x and y (axiom of pairs again), and let z contain just the members of members of w (axiom of unions). Then z is a successor of x, for its members are just x and x's members. Next, note that if z is a successor of x, x is formed at r, and t is the next stage after r, then z is formed at t. For every member of z is formed before t. So z is formed at or before t, by axiom (IX). But x, which is in z, is formed at r. So z cannot be formed at or before r. So z cannot be formed before t. Now, by axiom (VI), there is a stage s, not the earliest one, which is not immediately after any stage. '$(s)(\exists y)(x)(x \in y \leftrightarrow (x = x \,\&\, (\exists t)(t\text{E}s \,\&\, x\text{F}t)))$' is a specification axiom, according to which, for any stage, there is a set of all sets formed at earlier stages. So there is a set y of all sets formed before s. y thus contains all sets formed at stage 0, and hence contains a null set. And if y contains x, y contains all successors of x (and there are some), for all these are formed at stages immediately after stages before s and, hence, at stages themselves before s.

Axioms of separation (Aussonderungsaxioms): All formulas

$$\ulcorner (z)(\exists y)(x)(x \in y \leftrightarrow (x \in z \,\&\, \phi)) \urcorner$$

where ϕ is a formula of L in which no occurrence of 'y' is free.

Derivation. If ϕ is a formula of $\mathcal{L}$ in which no occurrence of 'y' is free, then $\ulcorner (s)(\exists y)(x)(x \in y \leftrightarrow ((x \in z \,\&\, \phi) \,\&\, (\exists t)(t\text{E}s \,\&\, x\text{F}t))) \urcorner$ is a specification axiom, which we may read, 'for any stage s, there is a set of all sets formed at earlier stages, which belong to z and to which ϕ applies. Let s be the stage at which z is formed. All members of z are formed before s. So, for any z, there is a set of just those members of z to which ϕ applies, which we would write, $\ulcorner (z)(\exists y)(x)(x \in y \leftrightarrow (x \in z \,\&\, \phi)) \urcorner$. A formal derivation of an Aussonderungsaxiom would use the specification axiom described and axioms (VII) and (VIII) of the stage theory.

Axioms of regularity: All formulas

$$\ulcorner(\exists x)\phi \rightarrow (\exists x)(\phi \& (y)(y \in x \rightarrow \sim\psi))\urcorner,$$

where ϕ does not contain 'y' at all and ψ is just like ϕ except for containing an occurrence of 'y' wherever ϕ contains a free occurrence of 'x'.

Derivation. The idea: Suppose ϕ applies to some set x'. x' is formed at some stage. That stage is therefore not covered by $\ulcorner\sim\phi\urcorner$. By an induction axiom, there is then a stage s not covered by $\ulcorner\sim\phi\urcorner$, although all stages earlier than s are covered by $\ulcorner\sim\phi\urcorner$. Since s is not covered by $\ulcorner\sim\phi\urcorner$, there is an x, formed at s, to which $\ulcorner\sim\phi\urcorner$ does not apply, i.e., to which ϕ applies. If y is in x, however, y is formed before s, and hence the stage at which it is formed is covered by $\ulcorner\sim\phi\urcorner$. So $\ulcorner\sim\phi\urcorner$ applies to y (which is what $\ulcorner\sim\psi\urcorner$ says).

For a formal derivation, contrapose, reletter, and simplify the induction axiom

$$\ulcorner(s)((t)(tEs \rightarrow (x)(xFt \rightarrow \sim\phi)) \rightarrow (x)(xFs \rightarrow \sim\phi))$$
$$\rightarrow (s)(x)(xFs \rightarrow \sim\phi)\urcorner$$

so as to obtain

$$\ulcorner(\exists s)(\exists x)(xFs \& \phi) \rightarrow (\exists s)(\exists x)(xFs \& \phi \& (y)(t)(tEs \& yFt \rightarrow \sim\psi))\urcorner$$

Assume $\ulcorner(\exists x)\phi\urcorner$. Use axiom (VII) and modus ponens to obtain

$$\ulcorner(\exists s)(\exists x)(xFs \& \phi \& (y)(t)(tEs \& yFt \rightarrow \sim\psi))\urcorner$$

Use axioms (VII) and (VIII) to obtain $\ulcorner(\exists x)(\phi \& (y)(y \in x \rightarrow \sim\psi))\urcorner$ from this.

The axioms of regularity (partially) express the analogue for sets of the version of mathematical induction called the *least-number principle:* if there is a number that has a property, then there is a least number with that property. The analogue itself has been called the *principle of set-theoretical induction.*[8] Here is an application of set-theoretical induction.

Theorem: No set belongs to itself.
Proof. Suppose that some set belongs to itself, i.e., that $(\exists x)x \in x$.

$$(\exists x)x \in x \rightarrow (\exists x)(x \in x \& (y)(y \in x \rightarrow \sim y \in y))$$

[8] By Tarski, among others.

498

is an axiom of regularity. By modus ponens, then, some set x belongs to itself though no member of x (not even x) belongs to itself. This is a contradiction.

The axioms whose derivations we have given are those statements which are often taken as axioms of ZF and which are deducible from all (sufficiently strong[9]) theories that can fairly be called formalizations of the iterative conception, as roughly described. (The axiom of extensionality has a special status, which we discuss below.) Other axioms than those we have given could have been taken as axioms of the stage theory. For example, we could have fairly taken as an axiom a statement asserting the existence of a stage, not immediately later than any stage, but later than some stage that is itself neither the earliest stage nor immediately later than any stage. Such an axiom would have enabled us to deduce a stronger axiom of infinity than the one whose derivation we have given, but this stronger statement is not commonly taken as an axiom of ZF. We could also have derived other statements from the stage theory, such as the statement that no set belongs to any of its members, but this statement is never taken as an axiom of ZF. We do not believe that the axioms of replacement or choice can be inferred from the iterative conception.

One of the axioms of regularity,

$$(z)((\exists x)x \in z \rightarrow (\exists x)(x \in z \mathrel{\&} (y)(y \in x \rightarrow \sim y \in z)))$$

is sometimes called *the* axiom of regularity; in the presence of other axioms of ZF, all the other axioms of regularity follow from it. The name 'Zermelo set theory' is perhaps most commonly given to the theory whose axioms are '$(x)(y)((z)(z \in x \leftrightarrow x \in y) \rightarrow x=y)$', i.e., the axiom of extensionality, and the axioms of the null set, pairs, and unions, the power-set axiom, the axiom of infinity, all the Aussonderungsaxioms, and the axiom of regularity.[10] With the exception of the axiom of extensionality, then, all the axioms of Zermelo set theory follow from the stage theory.

IV. Zermelo-Fraenkel set theory

The axioms of replacement. ZF is the theory whose axioms are those of Zermelo set theory and all axioms of replacement.[11] A formula of $\mathcal{L}$ is an

---

[9]'Sufficiently strong' may here be taken to mean "at least as strong as the stage theory."
[10]Zermelo (1908) took as axioms versions of the axioms of extensionality, the null set, pairs (and unit set), unions, the power-set axiom, the axiom of infinity, the Aussonderungsaxioms, and the axiom of choice.
[11]Sometimes the axiom of choice is also considered one of the axioms of ZF.

axiom of replacement if it is the translation into $\mathcal{L}$ of the result "substituting" a formula of $\mathcal{L}$ for 'F' in

$$F \text{ is a function} \rightarrow (z)(\exists y)(x)(x \in y \leftrightarrow (\exists w)(w \in z \;\&\; F(w) = x))$$

There is an extension of the stage theory from which the axioms of replacement could have been derived. We could have taken as axioms all instances (that can be expressed in $\mathcal{J}$) of a principle which may be put, 'If each set is correlated with at least one stage (no matter how), then for any set z there is a stage s such that for each member w of z, s is later than some stage with which w is correlated'. This *bounding* or *cofinality* principle is an attractive further thought about the interrelation of sets and stages, but it does seem to us to be a *further* thought, and not one that can be said to have been meant in the rough description of the iterative conception. For that there are exactly ω_1 stages does not seem to be excluded by anything said in the rough description; it would seem that $R\omega_1$ (see below) is a model for any statement of $\mathcal{L}$ that can (fairly) be said to have been implied by the rough description, and not all of the axioms of replacement hold in $R\omega_1$.[12] Thus the axioms of replacement do not seem to us to follow from the iterative conception.

Adding the axioms of replacement to those of Zermelo set theory enables us to define a sequence of sets, $\{R_\alpha\}$, with which the stages of the stage theory may be identified. Suppose we put $R_0 =$ the null set; $R_{\alpha+1} = R_\alpha \cup$ the power set of R_α, and $R_\lambda = \bigcup_{\beta < \lambda} R_\beta$ (λ a limit ordinal) – axioms of replacement ensure that the operation R is well-defined – and say that s is a stage if $(\exists \alpha)s = R_\alpha$, that x is formed at s if x is subset but not a member of s, and that s is earlier than t if, for some α, β, $s = R_\alpha$, $t = R_\beta$, and $\alpha < \beta$. Then we can prove as theorems of ZF not only the translations into the language of set theory of the axioms of the stage theory, but also those of all those stronger axioms asserting the existence of stages further and further "out" that might have been suggested by the rough description (and those of the instances of the bounding principle which are expressible in $\mathcal{J}$ as well). ZF thus enables us to describe and assert the full first-order content of the iterative conception within the language of set theory.

Although they are not derived from the iterative conception, the reason for adopting the axioms of replacement is quite simple: they have many desirable consequences and (apparently) no undesirable ones. In addition to theorems about the iterative conception, the consequences of replacement include a satisfactory if not ideal[13] theory of infinite numbers,

---

[12]Worse yet, R_{δ_1} would also seem to be such a model. (δ_1 is the first nonrecursive ordinal.)
[13]An ideal theory would decide the continuum hypothesis, at least.

and a highly desirable result that justifies inductive definitions on well-founded relations.

The axiom of extensionality. The axiom of extensionality enjoys a special epistemological status shared by none of the other axioms of ZF. Were someone to deny another of the axioms of ZF, we would be rather more inclined to suppose, on the basis of his denial alone, that he believed that axiom false than we would if he denied the axiom of extensionality. Although 'there are unmarried bachelors' and 'there are no bachelors' are equally preposterous things to say, if someone were to say the former, he would far more invite the suspicion that he did not mean what he said than someone who said the latter. Similarly, if someone were to say, "there are distinct sets with the same members," he would thereby justify us in thinking his usage nonstandard far more than someone who asserted the denial of some other axiom. Because of this difference, one might be tempted to call the axiom of extensionality "analytic," true by virtue of the meanings of the words contained in it, but not to consider the other axioms analytic.

It has been persuasively argued, by Quine and others, however, that until we have an acceptable explanation of how a sentence (or what it says) can be true in virtue of meanings, we should refrain from calling *anything* analytic. It seems probable, nevertheless, that whatever justification for accepting the axiom of extensionality there may be, it is more likely to resemble the justification for accepting most of the classical examples of analytic sentences, such as 'all bachelors are unmarried' or 'siblings have siblings' than is the justification for accepting the other axioms of set theory. That the concepts of *set* and *being a member of* obey the axiom of extensionality is a far more central feature of our use of them than is the fact that they obey any other axiom. A theory that denied, or even failed to affirm, some of the other axioms of ZF might still be called a set theory, albeit a deviant or fragmentary one. But a theory that did not affirm that the objects with which it dealt were identical if they had the same members would only by charity be called a theory of *sets* alone.

The axiom of choice. One form of the axiom of choice, sometimes called the "multiplicative axiom," is the statement, 'For any x, if x is a set of nonempty disjoint sets (two sets are disjoint if nothing is a member of both), then there is a set, called a *choice set* for x, that contains exactly one member of each of the members of x'.

It seems that, unfortunately, the iterative conception is neutral with respect to the axiom of choice. It is easy to show that, since, as is now known, neither the axiom of choice nor its negation is a theorem of ZF, neither the axiom nor its negation can be derived from the stage theory. Of course the stage theory, which is supposed to formalize the rough description, could be extended so as to decide the axiom. But it seems that no additional axiom, which would decide choice, can be inferred from the rough description without the assumption of the axiom of choice itself, or some equally uncertain principle, in the inference. The difficulty with the axiom of choice is that the decision whether to regard the rough description as implying a principle about sets and stages from which the axiom could be derived is as difficult a decision, because essentially the same decision, as the decision whether to accept the axiom.

Suppose that we tried to derive the axiom by arguing in this manner: Let x be a set of nonempty disjoint sets. x is formed at some stage s. The members of members of x are formed at earlier stages than s. Hence, at s, if not earlier, there is a set formed that contains exactly one member of each member of x. But to assert this is to beg the question. How do we know that such a choice set *is* formed? If a choice set is formed, it is indeed formed at or before s. But how do we know that one is formed at all? To argue that at s we can choose one member from each member of x and so form a choice set for x is also to beg the question: "we *can't* choose" one member from each member of x if there is no choice set for x.

To say this is not to say that the axiom of choice is not both obvious and indispensable. It is only to say that the justification for its acceptance is not to be found in the iterative conception of set.

What is the iterative conception of set?

CHARLES PARSONS

I intend to raise here some questions about what is nowadays called the 'iterative conception of set'. Examination of the literature will show that it is not so clear as it should be what this conception *is*.

Some expositions of the iterative conception rest on a 'genetic' or 'constructive' conception of the existence of sets. An example is the subtle and interesting treatment of Professor Wang.[1] This conception is more metaphysical, and in particular more idealistic, than I would expect most set theorists to be comfortable with. In my discussion I shall raise some difficulties for it.

In the last part of the paper I introduce an alternative based on some hints of Cantor and on the Russellian idea of typical ambiguity. This is not less metaphysical though it is intended to be less idealistic. I see no way to obtain philosophical understanding of set theory while avoiding metaphysics; the only alternative I can see is a positivistic conception of set theory. Perhaps the latter would attract some who agree with the critical part of my argument.

However, the positive part of the paper will concentrate on the notion of proper class and the meaning of unrestricted quantifiers in set theory. That these issues are closely related is evident since in Zermelo-type set theories the universe is a proper class.

The concept of set is also intimately related to that of *ordinal*. Although this relation will be remarked on in several places, a more complete account of it, and thus of the more properly *iterative* aspect of the iterative conception, will have to be postponed until another occasion.

I

One can state in approximately neutral fashion what is essential to the 'iterative' conception: sets form a well-founded hierarchy in which the

Reprinted with the kind permission of the author, the editors and the publisher from *Proceedings of the 5th International Congress of Logic, Methodology and Philosophy of Science 1975, Part I: Logic, Foundations of Mathematics, and Computability Theory*, Robert E. Butts and Jaakko Hintikka, eds., D. Reidel 1977, pp. 335–67.

[1]Wang (1974, chap. 6, which is reprinted in this volume). A widely cited writer whose viewpoint I would also describe as genetic is Shoenfield (1967: 238–40).

elements of a set precede the set itself. In axiomatic set theory, this idea is most directly expressed by the axiom of foundation, which says that any non-empty set has an '∈-minimal' element.[2] But what makes it possible to use such an assumption in *motivating* the axioms of set theory is that other evident or persuasive principles of set existence are compatible with it and even suggest it, as is indicated by the von Neumann relative consistency proof for the axiom of foundation.

On the 'genetic' conception that I will discuss shortly, the hierarchy arises because sets are taken to be 'formed' or 'constituted' from previously given objects, sets of individuals [see footnote 2]. But one can speak more abstractly and generally of the elements of a set as being *prior* to the set. In axiomatic set theory with foundation, this receives a mathematically explicit formulation, in which the relation of priority is assumed to be well-founded.

For motivation and justification of set theory, it is important to ask in what this 'priority' consists. However, for the practice of set theory from there on, only the abstract structure of the relation matters. Here we should recall that the hierarchy of sets can be 'linearized' in that each set can be assigned an ordinal as its *rank*. Individuals, and for smoothness of theory the empty set,[3] obtain rank 0. In general, the rank of a set is the least ordinal greater than the ranks of all its elements.[4]

It should be observed that the notion of well-foundedness is prima facie second-order and thus is not totally captured by the first-order

[2] I shall consider throughout set theories which allow individuals (*Urelemente*); this requires trivial modifications of the most usual axioms, but the choice among possible ways of doing this is of no importance for us. In extensionality and foundation, the main parameters are restricted to sets (or at least nonindividuals, if classes are admitted).

The literature on the foundations of set theory does not sufficiently emphasize that the exclusion of individuals in the standard axiomatizations of set theory is a rather artificial step, taken for the convenience of pure mathematics. An applied set theory would normally have to have individuals. What is more relevant for the present discussion is that some of the intuitions about sets with which set theory starts concern sets of individuals. First-order set theory with individuals is compatible with the assumption that there are no individuals and therefore with the usual individual-free set theory.

[3] From the genetic point of view, this is an artifice: individuals are presumably given prior to any sets, even the empty set, so that if rank directly reflects order of construction, the empty set should have rank 1. The same holds for the alternative viewpoint I present below.

[4] In a set theory with individuals, some usual theorems about ranks, for example that for every ordinal α there is a set R_α of all sets of rank $<\alpha$, require the assumption that there is a set of all individuals. It follows that there cannot be too many individuals; for example, ordinals cannot all be construed as individuals. The plausibility of this assumption depends on the intended application. It would be a piece of highly dubious metaphysics to assume there is a set of *absolutely* all individuals, if for no other reason because it is not settled once and for all what *is* an individual. Pure mathematics should be independent of this question; for it the individuals can be an arbitrary set, class, or sometimes structure. However, below I shall assume that the individuals constitute a set.

axiom of foundation.[5] ZF with foundation has models in which the relation representing membership is not well-founded. However, it can be seen that in such a model there is a (not first-order definable) binary relation on the universe for which *replacement* fails. The axioms of separation and replacement are also prima facie second-order, and the fault for such failure of well-foundedness lies in the fact that their full content is not captured by the first-order schemata. But then the problem of stating clearly the iterative conception of set is bound up with the problem of the relation of set theory to second-order conceptions. This problem was already present at the historical beginning of axiomatic set theory with Zermelo's use of the notion of 'definite property'.

The idea that the elements of a set are prior to the set is highly persuasive as an approach to the paradoxes. If we suppose that the elements of a set must be 'given' before the set, then no set can be an element of itself, and there can be no universal set. The reasoning leading to the Russell and Cantor paradoxes is cut off.

However, one does not deal so directly with the Burali-Forti paradox. Why should it not be that all ordinals are individuals and therefore 'prior' to all sets, so that there is no obstacle of this kind to the existence of a set of all ordinals? To be sure, once we look at things in this way it becomes persuasive to view the Burali-Forti argument as just a proof that there is no set of all ordinals. Moreover, the conception of ordinals as order types of well-ordered sets would suggest that for any ordinal there is at least one set of that order type to which it is not prior, so that the existence of a set of all ordinals would imply that later in the priority ordering no new order types could arise. But if there were a set of all ordinals, W, then $W \cup \{W\}$ would have just such an order type.

That ordinals need to fit into a priority ordering with sets and indeed be 'cofinal' with them seems to have been neglected in discussion of iterative set theory, perhaps because in the formal theory ordinals are construed as sets, so that this happens automatically.

One would like to maintain that the requirement of priority is the *only* principle limiting the existence of sets, so that at a given position 'arbitrary multitudes' of objects which are at earlier positions form sets. Although it is difficult to make sense of this, it at least should imply a comprehension principle: given a predicate 'F' which is definitely true or false of each object prior to the position in question, there is *at* the position a set whose elements are just the prior objects satisfying 'F'. In particular, the axiom of separation follows: since the elements of x are prior

[5] However, in set theories with classes foundation for classes follows from its assumption for sets.

to x, $\{z: z \in x \wedge Fz\}$ exists and is not posterior to x. But to apply this idea more generally, some way of marking positions is needed. The genetic approach in effect assumes such to be given. Conversely, if set theory is assumed, the ordinals offer such a marking.

II

We have now gone about as far as we can without explaining what I have called the genetic approach. Put most generally, it supposes that sets are 'formed', 'constructed', or 'collected' from their elements in a succession of stages. The first part of this idea has some plausibility as an interpretation of some of Cantor's preliminary remarks about what a set is. Thus Cantor's famous 'definition' of 1895:

By a 'set' we understand any collection M into a whole of definite, well-distinguished objects of our intuition or our thought (which will be called the 'elements' of M).[6]

If we were to take 'collection into a whole' quite literally as an operation, then the priority of the elements of a set to the set would simply be priority in order of construction. Cantor's language suggests rather that 'collection' (*Zusammenfassung*) is an operation of the *mind;* in this case the requirement would be that the objects be represented to the *mind* before the operation of collection is performed. However, it will be clear that as it stands this temporal reading is too crude.

It may seem that these notions belong only to the early history of set theory, and in particular that they would have disappeared with the discrediting of logical psychologism at the end of the last century. But the fact is that they are to be found in the contemporary literature. Thus Schoenfield writes (1967: 238):

A closer examination of the paradox [Russell's] shows that it does not really contradict the intuitive notion of a set. According to this notion, a set A is formed by gathering together certain objects to form a single object, which is the set A. Thus before the set A is formed, we must have available all of the objects which are to be members of A.

Although Schoenfield says that we form sets 'in successive stages', he does not offer an interpretation, temporal or otherwise, of the stages, although he does use 'earlier' to express their order.

[6]Cantor 1932: 282. Wang calls this definition "genetic" (1974: 188) and speculates that the difference between this and previous ones (in particular, Cantor 1932: 204) may be due to Cantor's awareness of the Burali-Forti paradox. It should be remarked that a genetic conception of *ordinals* is intimated in (Cantor 1932: 195-6), a text from 1883.

What is the iterative conception of set?

Wang writes, "The set is a single object formed by collecting the members together" (1974: 181) [530 in this volume]. He recognizes that the concept of collecting is highly problematic and makes an interesting attempt to explain it; he interprets it as an operation of the mind. We shall discuss his views shortly.

We now have the familiar conception of sets as formed in a well-ordered sequence of stages, where a set can be formed at a given stage only from sets formed at earlier stages and from whatever objects were available at the outset.

The language of Cantor, Shoenfield, and Wang invites regarding the intuitive concept of set as analogous to the concepts of constructive mathematics, where one also uses the idea of mathematical objects as constructed in successive stages, and where there is no stage at which all constructions are complete. An immediately obvious limitation of the analogy is that in the typical constructive case (e.g. orthodox intuitionism) the succession of stages is simply succession in *time,* and incompletability arises from the fact that the theory is a theory of an idealized finite mind which is located at some point in time and has available only what it has constructed in the past and its intentional attitudes toward the future. The same interpretation of iterative set theory would require that the stages be thought of as a kind of 'super-time' of a structure richer than can be represented in time on any intelligible account of construction in time. It is hard to see what the conception of an idealized mind is that would fit here; it would differ not only from finite minds but also from the divine mind as conceived in philosophical theology, for the latter is thought of either as in time, and therefore as doing things in an order with the same structure as that in which finite beings operate, or its eternity is interpreted as complete liberation from succession.

It may seem that there is a much more obvious conflict between iterative set theory and a constructive interpretation of it: set theory is the very paradigm of a *platonistic* theory. As is customary in discussing the foundations of mathematics, platonism means here not just accepting abstract entities or universals but epistemological or metaphysical realism with respect to them. Thus a platonistic interpretation of a theory of mathematical objects takes the truth or falsity of statements of the theory, in particular statements of existence, to be objectively determined independently of the possibilities of our knowing this truth or falsity. Contrast, for example, the traditional intuitionist conception of a mathematical statement as an indication of a 'mental construction' that constitutes a proof of the statement.

Perhaps it would be rash to rule out an interpretation of set theory that

507

would not be platonistic.[7] But in any case it seems that a platonistic interpretation is flatly incompatible with viewing the 'formation' of sets as an operation of the mind. However, that there is not a direct contradiction should be evident when we observe that we are concerned in set theory with what formations of sets are *possible*. In contrast to the situation with intuitionism, we do not require that a statement to the effect that it is possible to 'construct' a set satisfying a certain condition should be itself an indication of a construction. Even if we construe the formation of sets as a mental operation, what is possible with respect to such formation can be viewed independently of our knowledge. Thus there is a *prima facie* resolution of the difficulty posed by platonism.

However, we have not reckoned yet with the actual content of the set-theoretic principles that seem to require a platonistic interpretation, such as the combination of classical logic with the postulation of a set of all sets of integers. In an iterative account, the individual steps of iteration are in Wang's word "maximum" (1974: 183 [532 in this volume]). Namely we regard as available at any given stage any set that *could* have been formed earlier. We could represent this assumption as that if a set *can* be formed at a given stage, then it *is* formed (or at least that it *exists* at that stage). This of course has effects on what can be formed later, since every *possibility* of set formation at stage α is such that its result is available at later stages and can therefore enter into further constructions.

We can illustrate this by the manner in which these ideas are used to justify the power set axiom. At a given stage α, any 'multitude' of available objects can be formed into a set. Let x be a set formed at stage α, and let y be a subset of x. Since the elements of y are all elements of x, they must have been available at stage α. Hence y *could have been formed* at stage α. x is available from stage $\alpha+1$, on; from our assumption it follows that y is available as well. Thus at stage $\alpha+1$, *every* subset x is available and $\mathfrak{P}(x)$ can be formed.

We should distinguish two principles that are playing a role here, and which can be confused with one another. One is the 'arbitrary' nature of sets, which, following Wang, we have expressed (provisionally) by saying that any 'multitude' of available objects can be formed into a set. The other is the principle that allows the transition from possibility of formation at stage α to availability at stage $\alpha+1$, perhaps by way of existence at stage α. Both principles may be taken to arise from the idea we

---

[7]Formalism apart, one ought not to rule out the possibility of an interpretation of set theory along constructivist lines, particularly in view of the broadening of the intuitionist outlook in recent years. ZF has recently been shown consistent relative to some set theories based on intuitionist logic. See Friedman (1973) and Powell (1975).

expressed above that the priority requirement, here interpreted to mean that sets are formed from *available* objects, is the *only* constraint on the existence of sets. But the second principle begins to undercut the idea of sets as *formed* from available objects, since the successive stages of formation are required *only* because a set must be formed from available objects, and not because of any successiveness in the process of formation itself. The question arises whether the interpretation of the priority of the elements of a set to the set in terms of order of construction does not reduce to viewing this priority as a matter of *constitution:* the elements are prior because they *constitute* the set (to use a more abstract phrase than they are its *parts,* which would invite inferences inappropriate to set theory). This view is close to what I advocate below, but it is quite different from the conception of set formation as an operation of the mind.

Wang makes an interesting attempt to develop the latter idea. He says that a multitude can be formed into a set only if its "range of variability" is "in some sense intuitive" (1974: 182) [531 in this volume]. I shall for the time being accept the notion of 'multitude'; the problems concerning it are related to the question of the notion of *class* in set theory. Wang indicates that to form a set is to "look through or run through or collect together" all the objects in the multitude.[8] Thus a condition for a multitude to form a set is that it should be possible thus to 'overview' it. This overviewing is a kind of intuition, presumably analogous to perception. Of course infinite multitudes can be 'overviewed'.

Clearly Wang does not maintain that human beings have the capacity to "run through" infinite collections. He speaks of overviewing "in an *idealized* sense" (1974: 182) [531 in this volume]. In other words, he has a highly abstract conception of the possibilities of intuition. In constructive conceptions of the arbitrary finite, we already disregard the actual bounds of human capacities, in the sense that if a sequence of steps has been performed we always *can* perform a further step, and any operation can be iterated. Wang's idealized overviewing carries such abstraction further in that finitude and even the limitations posed by the continuous structure of space-time (as the setting of the objects of perception and even of the mind itself) are disregarded.

The question arises what force it still has, on Wang's level of abstraction, to treat the possibilities involved in this kind of motivation of the axioms of set theory as possibilities of *intuition.* The analogy with sense-perception which is central to the constructive conception of intuition in Brouwer or to Hilbert's distinction between intuitive and formal mathe-

[8]Wang may be developing the remark of Gödel (1964: 272, no. 40) [484, n. 26, in this volume] that the function of the concept of set is "synthesis" in a sense close to Kant's.

matics seems to be almost totally lost. Consider Wang's remarks on the axioms of separation and power set. The former is stated thus: If a multitude A is included in a set x, then A is a set.

Since x is a given set, we can run through all members of x, and, therefore, we can do so with arbitrary omissions. In particular, we can in an idealized sense check against A and delete only those members of x which are not in A. In this way, we obtain an overview of all the objects in A and recognize A as a set (1974: 184) [533 in this volume].

The idealization seems to include something like omniscience: A may be given in some way that does not independently of the axiom assure us that it is a set, and yet we can use it in order to 'choose' the members of x that are in A. A may of course be given to us by a predicate containing quantifiers that do not range over a set; in deciding whether an element y of x is to be deleted, we cannot 'run through' the values of the bound variables as part of the process of checking y against A. It is not clear what more structured account of 'idealized checking' would yield the result Wang needs. An alternative would be to view subsets as run through not by verification but by arbitrary selection. But if a predicate is given, how are we to 'select' just those elements of x that satisfy it unless we can decide which ones do?[9]

More strain on the concept of intuition appears in Wang's treatment of the power set axiom (1974: 184) [534 in this volume]:

We have $\cdots$ an intuitive idea of running through with omissions. This general notion $\cdots$ provides us with an overview of all cases of AS [separation] as applied to x.

By saying that we have an "intuitive idea" of running through with

---

[9]Cf. the fact that in intuitionism the classical notion of set splits into those of spread and species.

It is of interest to look at a case, namely the hereditarily finite sets, where something like arithmetical intuition does yield the axiom of separation. Here we can argue by induction on n that there is a w such that

$$(*) \qquad (\forall z)(z \in w \leftrightarrow z \in \{x_1 \cdots x_n\} \wedge Fz).$$

For if $n = 0$, $w = \Lambda$. Suppose w satisfies (*) and consider $\{x_1 \cdots x_{n+1}\}$. If $\neg Fx_{n+1}$ then w satisfies

$$(\forall z)(z \in w \leftrightarrow z \in \{x_1 \cdots x_{n+1}\} \wedge Fz);$$

if Fx_{n+1}, then $w \cup \{x_{n+1}\}$ satisfies the condition. In effect this argument shows us that of the possible subsets of a finite set *one* satisfies the separation condition, without telling us which one.

One might consider interpreting the requirement that a set x can be 'overviewed' as meaning that it can be run through in a well-ordered way and then attempt a transfinite analogue of this inductive argument. It seems that to handle limit cases such an argument requires replacement, and of course separation can be deduced from replacement in a more trivial way without assuming x to be well-orderable.

omissions, he does not only mean that a *case* of such running through is intuitable, for that would not yield the result. Rather, for a given set x the concept of such runnings through is intuitive in the sense that 'we can' run through *all* cases of it. Something of the content of the idea of intuitive running through seems to be lost here. Clearly in the case of small finite sets of manageable objects, we really do see all the elements 'as a unity' in a way that preserves the articulation of the individual elements. Somewhat larger sets can be seen by a completable succession of steps of bringing one (or a few) objects under one's purview. If we consider arbitrary iterations of such steps, there is no longer any limit to how large a (finite) set can be thus intuited. We also have a simple and clear generative rule for sets such as the natural numbers, though the process of sensibly intuiting is in this case incompletable, so that the *givenness* to us of such a set depends in a more essential way on conception. But regarding the natural numbers as intuitable as a whole amounts just to abstracting from the above incompletability. There is another qualitative leap in dealing with all sets of integers, as has often been remarked on in the literature (as indeed Wang himself emphasizes when he stresses the importance of impredicativity).[10] Here, however we understand the notion of an 'arbitrary set' of integers, say by some picture of arbitrary selection, we do not have the conceptual grasp of what the totality contains that would be given by some method of generating them. The divorce from *sensible* intuition involved in treating this totality as 'intuitable' seems complete, unless perception is used only as a source of quite remote analogies. Two mathematical symptoms of this situation are the absence of a definable well-ordering of the continuum and our inability to solve the continuum problem.

I ought to make clear that I understand by 'intuition' a quasi-perceptual manner in which an object is presented to the mind. In this I follow Kant. The word 'intuition' is also used in the philosophy of mathematics and otherwise for any manner by which propositions can be known where this knowledge is not largely accounted for by deductive or inductive reasoning. There is a tendency to confuse these two senses. As for the appropriateness of my sense of intuition to Wang, I should point out that the other concept of intuition does not distinguish sets from 'multitudes' or other primitive notions that might enter into evident set-theoretic axioms. Moreover, intuition in the latter sense is purely *de dicto,* intuition *that* certain propositions are true, while Wang clearly requires intuition *de re,* intuition *of* sets.[11]

---

[10]For example the classic formulation of Bernays (1935: 275–6) [259–60 in this volume]. Cf. Wang (1974: 183) [532 in this volume].

[11]This distinction is explicitly made by Steiner (1975: 130–1). However, Steiner regards

I no longer understand Wang's talk of 'intuitively running through' where it is applied to the set of all sets of integers. In the above I have perhaps connected intuition more closely with the senses (more abstractly Kant's "sensibility") than Wang would find acceptable. But even quite abstract marks of sensibility, such as the structure of time, are lost in this case.

However, there might be an interpretation on which Wang's hypothesis that a 'multitude' is a set if and only if it is an object of intuition would be logical and ontological rather than perceptual and epistemic. To be an object of intuition would be simply to be an object rather than a Fregean "concept" or perhaps a property. In other words, Kant's contrast of intuitions as "singular representations" (*Logic*, §1) would give virtually the only essential mark of intuition.[12] Although this interpretation would bring Wang's hypothesis into accord with the views I express below, I shall not pursue it further in this paper.

I want now to turn to the axiom of replacement, about which Wang has most interesting things to say. Wang writes:

Once we adopt the viewpoint that we can in an idealized sense run through all members of a given set, the justification of *SAR*[13] is immediate. That is, if, for

only *de dicto* mathematical intuition as defensible. I have sketched in previous writing an account of arithmetical intuition on Kantian lines (1965: 201–3; 1969; 1971: 158–62, 166–7). Curiously, Steiner cites me and then says, "No one today, however, upholds hardcore intuition – the direct intuition of mathematical *objects*" (*ibid.*), although he then mentions Gödel as a possible exception. Since I was trying to elucidate the *forward* character of Kantian intuition, perhaps Steiner did not consider arithmetical intuition on my view to be "direct intuition of mathematical *objects*", particularly in (1971).

Some comment is in order on Gödel's view of mathematical intuition, particularly since he explicitly says, "We do have something like a perception also of the objects of set theory" (1964: 270) [483–4 in this volume]. This seems to commit Gödel to intuition *de re*. His immediately following remark does not give any argument for this; he says only that it "is seen from the fact that the axioms force themselves on us as being true", which implies only intuition *de dicto*.

However, it seems clear that Gödel holds (1964: 271 bottom) [484 in this volume], that our ideas of objects of certain kinds contain "constituents" which are *given* (not "created" by thinking) on the basis of which we "form our ideas" of these objects and postulate theories of them. "Evidently the 'given' underlying mathematics is closely related to the abstract elements contained in our empirical ideas" (1964: 272) [484 in this volume]. In the case of set theory, Gödel does not give any indication of wanting to distinguish sets as objects of intuition from other entities (such as "properties of sets" (1964: 264 n. 18) [476 n. 15 in this volume]) that the axioms might refer to.

Elsewhere Gödel, in contrast to the above passage, contrasts the intuitive with the abstract (1958: 281). There he seems to be using 'intuition' in a much narrower and more strictly Kantian sense. Of course there he is writing in German; possibly he would not use 'Anschauung' in the sense in which he uses 'intuition' in (1964).

[12]According to Hintikka, this is Kant's own view. See Hintikka 1969a and other essays reprinted in 1973 and 1974.

[13]SAR is the statement: if *b* is an operation and b_x is a set for every member *x* of a set *y*, then all these sets b_x form a set (Wang 1974: 186) [535 in this volume].

each element of the set, we put some other given object there, we are able to run through the resulting multitude as well. In this manner, we are justified in forming new sets by arbitrary replacements. If, however, one does not have this idea of running through all members of a given set, the justification of the replacement axiom is more complex. (1974: 186) [536 in this volume]

The picture here is marvelously persuasive; for me, it expresses very well why the axiom of replacement seems obvious. But something like the omniscience assumption of the discussion of separation is present in the remark that "*we put* some other given object there" and "are able to run through the resulting multitude." What is much more revealing is that the objects seem to have no relevant internal structure: Our ability to 'run through' a multitude is preserved if we replace its elements by *any* other objects, for example by much larger sets. It is as if the objects were given only as wholes, or at least that any internal structure would not affect the possibility of running through the totality. A model for this (conceptual rather than intuitive) is the case where the objects are given only by names. Wang seems to be making an hypothesis here, although I do not feel the same qualms as in the case of the power set about taking it as an hypothesis about what is intuitable. It is of course the *combination* of the replacement with the power set axiom that yields sets of very high ranks.

Wang expresses by his picture the idea, present in the earliest intimation of the axiom of replacement (Cantor 1932: 444; cf. Wang 1974: 211 [562 in this volume]) that whether a multitude forms a set depends only on its cardinality and not on the 'internal constitution' or relations of its elements. Put in this way, the axiom is not a principle of *iteration* of set formation, in line with the conclusion of Boolos (1971: 228–9 [500–1 in this volume]) that it does not follow from the iterative conception of set.[14] The most direct justification of replacement by appeal to ideas about stages seems to me somewhat circular.[15] That of Gödel cited by

[14]Boolos seems, however, to arrive at his conclusion too easily. He seems to assume that the 'stages' and their ordering have to be given independently of the concept of set, at least for the expression of *the* iterative conception. His actual axioms about stages (and a further possible one he mentions on p. 227 [499 in this volume]) would permit the stages to be ordered by a very simple recursive well-ordering. It seems to me that sets and 'stages' ought to be 'formed' together, so that the formation of certain sets should make possible going on to further stages.

However, the most obvious principle of this kind, that if a well-ordering has been constructed then there is a stage such that the earlier stages are ordered isomorphically to the given well-ordering, is weaker than the axiom of replacement.

[15]Thus Shoenfield (1967: 240) deduces replacement from a "cofinality principle": if "we have a set A, and $\cdots$ we have assigned a stage S_a to each element a of A. . . . There is to be a stage which follows all of the stages S_a" (p. 239). However, he justifies this by saying, "Since we can visualize the collection A as a single object (viz., the set A), we can also visualize the collection of stages S_a as a single object; so we can visualize a situation in which all

Wang (1974: 186 and 221, n. 5) [536 and n.4, 536 in this volume] I find less immediate and persuasive.

Although I admit that Wang's picture (apart from the question of omniscience) offers a plausible hypothesis about what is intuitable,[16] it seems to me to be equally plausible as an hypothesis about what can be thought or about what can *be,* and the latter interpretations fit better the case of power set. I want now to pursue the genetic conception of sets in this direction.

III

In the preceding section we saw a number of difficulties with the idea that sets are 'formed' from their elements, in particular by an activity of running through in intuition. I want now to suggest a more 'ontological' view of the hierarchy of sets.

The earliest attempt that we know to explain the paradoxes of set theory and to develop set theory in a way that avoids them is in Cantor's famous letter to Dedekind of July 28, 1899 (1932: 443-7). Cantor there presupposes his earlier 'many into one' characterizations of the notion of set, such as that of 1895 cited above. He begins (p. 443) with "the concept of a definite multiplicity (*Vielheit*)" What he calls an *inconsistent* multiplicity is one such that "the assumption of a 'being together' (*Zusammensein*) of *all* its elements leads to a contradiction". A consistent multiplicity or set is one whose "being collected together to 'one thing' is possible". It is noteworthy that Cantor here identifies the possibility of all the elements of a multiplicity *being together* with the possibility of their being collected together into one thing. This intimates the more recent conception that a 'multiplicity' that does not constitute a set is *merely potential,* according to which one can distinguish potential from actual being in some way so that it is impossible that *all* the elements of an inconsistent multiplicity should be actual.

I am here interpreting Cantor to mean that where there is an essential obstacle to a multiplicity's being collected into a unity, this is due to the fact that in a certain sense the multiplicity does not exist. It does not exist

the stages are completed" (*ibid.*). Here he is assuming that "visualizability as a single object" is preserved by replacement of a by S_a; but that is just the principle of replacement. Wang's picture seems more fundamental than the kind of argument Shoenfield gives.

Wang gives a similar argument (1974: 220 n. 4) [535-6 n.3 in this volume].

The argument does obtain general replacement from the special case where the range of the replacing function consists of stages.

[16]This plausibility is perhaps reinforced by the fact that replacement holds for the hereditarily finite and the hereditarily countable sets.

as a totality of its elements; if it did, they would form a unity or could at least be *collected* into a unity. But in the case of inconsistent multiplicities, this is impossible. The sense of this non-existence needs some further elucidation which Cantor does not supply. The language of potentiality and actuality is not in the text, though Cantor may have been suggesting it in calling an inconsistent multiplicity *absolutely infinite* (p. 443).

What seems to me of interest in the present connection in these hints of Cantor is that he seems to be trying to distinguish sets from "inconsistent multiplicities" without real use of any metaphor of *process* according to which sets are those multiplicities whose 'formation' can be 'completed'.[17] Such a metaphor makes the idea of an inconsistent multiplicity as a merely potential totality rather easy. I suggest interpreting Cantor by means of a modal language with quantifiers, where within a modal operator a quantifier always ranges over a set (not, however, one that is explicitly given or even that exists in the 'possible world' it might be taken to range over). Then it is not possible that all elements of, say, Russell's class exist, although for any element, it is possible that *it* exists. As it stands this conception requires it to be meaningful to talk of *any* set (or any object), even though the range of *this* quantifier does not constitute a unity; the elements of its range cannot all 'exist together'. However, at least some such talk can be replaced by ordinary quantification behind necessity.

What one would like to obtain from this conception is some interpretation of the stages of the iterative conception that also does not depend on the metaphor of process. However, I intend first to look at Cantor's conception of a multiplicity. Wang seems to use "multitude" in the same sense, although he does not use it to translate Cantor's *Vielheit* when he discusses Cantor's 1899 correspondence with Dedekind (1974: 211) [562 in this volume). These notions are among a number which occur in the literature on logic and set theory and which purport to be more comprehensive than the notion of set. The most respectable of these notions is that of (proper) class. We should also mention Frege's "concept", Zermelo's "definite property", and Shoenfield's "collection". Gödel's "property of sets" (1964, n. 18, p. 264) [n. 15, p. 476 in this volume] presumably also belongs on this list.

---

[17]However, this is not to say that Cantor now conceives sets as having no intrinsic relation to the mind. Hao Wang has pointed out to me that this would be questionable. For example he characterizes a consistent multiplicity as one the totality of whose elements "can be thought of without contradiction as 'being together', so that their being collected together (*Zusammengefasstwerden*) to 'one thing' is possible".

Of all these notions, perhaps the most developed from the philosophical side is Frege's notion of a concept; I shall use it for purposes of comparison. What is then striking is that neither Cantor's nor Wang's notion seems to be derived from predication as Frege's is. Since Cantor's notion is one of the prototypes of the notion of proper class, this fact seems to clash with the actual use of the notion of class in set theory (perhaps with exceptions; see below) according to which classes are derived from predication; Zermelo's "definite property" is a more immediate prototype than Cantor's *Vielheit*. I myself have suggested (1974c) that sets are not derived from predication while classes are.

Cantor in 1899 apparently thought of sets as a species of the genus multiplicity, and then perhaps the non-predicative (if not impredicative!) character of multiplicities in general was needed in order to preserve the 'arbitrariness' of sets against its being restricted by what we might express in language. Frege seems to have obtained the same freedom by his realism about concepts. However, for Frege the nature of a concept could apparently only be explained by appeal to predication (more generally, to 'unsaturated' expressions). The sharp distinction between concepts and objects is a shadow of the syntactical difference between expressions with and without argument places. This difference is then 'inherited' by concepts that are not denoted by expressions of any language we use or understand.

I want to suggest that predication plays a constitutive role in the explanation of Cantor's notion of multiplicity as well and that at least an "inconsistent multiplicity" must resemble a Fregean concept in not being straightforwardly an object. In the Cantorian context, predication seems to be essential in explaining how a multiplicity can be given to us not as a unity, that is as a set. Much the clearest case of this is understanding a predicate. Understanding 'x is an ordinal' is a kind of consciousness or knowledge of ordinals that does not so far 'take them as one' in such a way that they constitute an object. We might abstract from language and speak with Kant of knowledge through concepts, but whatever we make of this the predicational structure is still present.

The philosophy of Kant might suggest another way in which a multiplicity might be given not as a unity, namely as an 'unsynthesized manifold'. It seems clear that in the cases Kant actually envisaged, the objects involved would have the definiteness necessary to constitute a Cantorian 'multiplicity' only if they are a set. Even if we generalize the notion in some way, I do not see how such a 'manifold' can be taken up into explicit consciousness except perceptually (intuitively) or conceptually.

The idea that to be an *object* and to be a *unity* are the same thing is very tempting and has deep roots in the history of philosophy. An object is something whose identity with itself (represented in different ways) and difference from other objects can be meaningfully talked about; it is then subject to at least rudimentary application of *number*. This line of reasoning inclines us to identify Cantor's "multiplicity" with Frege's concept at least in that a multiplicity which is not a set is not an object. Some such assumption seems necessary to cut off the question why there are not multiplicities whose elements are not sets or individuals: multiplicities are multiplicities of *objects,* and under that condition there are no restrictions on the existence of multiplicities (although possibly on the use of quantifiers over them), but if a multiplicity is an object, then it is a set.

However, we have to deal with the fact that in Frege the gulf between concepts and objects comes from the structure of predication itself, so that a concept is irremediably not an object, even if only one object falls under it. Cantor evidently holds that some multiplicities just *are* sets, in particular those that are not too large. This may seem not a very essential difference: if a concept F is such that there is a set y of all x such that Fx, then the distinction between F and y is just the distinction between $(\ ) \in y$ and y.[18] For an inconsistent multiplicity there is no such 'reducibility'. In view of Russell's paradox, the idea of the predicative nature of the concept will motivate the idea that there should *be* inconsistent multiplicities, but it does not seem to motivate Cantor's particular principles as to what multiplicities are 'consistent'. For reasons which will become clear later, I do not think we have yet captured the sense in which an inconsistent multiplicity is not an object.

Let us look for a moment at the well-known difficulties of Frege's theory of concepts. The conception has the great attraction that it enables us to generalize predicate places without introducing nominalized predicates that purport to denote objects (classes or attributes) – something that has to be restricted on pain of Russell's paradox. But the temptation to nominalize is irresistible, as Frege himself discovered on two fronts. His construction of mathematics required an 'official' nominalization in postulating extensions. 'Unofficial' nominalizations cropped up repeatedly in his own informal talk about concepts and gave rise to the paradox that the concept *horse* is an object, not a concept.[19] At the end of his life Frege decided the temptation was to be resisted and that neither the expression "the concept F" nor the expression "the extension

---

[18] In a sense $(\ ) \in y$ *is F*, since coextensiveness for concepts is the analogue of identity for objects, but we cannot say that the concepts are *identical*.

[19] Frege 1892a. Only it might be a concept after all, since "is a concept" is syntactically such that it takes *object*-names as subjects, and is therefore a predicate of objects.

Of course a voluminous literature has grown up on this question.

of the concept F'' really denotes anything.[20] However, perhaps yielding to temptation can in one way or another be legitimized, even where the extensions postulated are not sets (cf. my, 1974c).

Does the Cantorian concept of ''multiplicity'' have to be understood realistically? To the extent that *sets* are understood realistically, of course ''consistent multiplicities'' are mind-independent in the corresponding sense. However, obviously it does not follow from the fact that we allow classical logic and impredicative reasoning about sets that we have to allow either about classes or other more general entities. The suggestion made below that such entities are at bottom *intensions* would imply, if we think of an intension in the traditional way as a meaning entertained by, and in some sense constructed by, the mind, that realism about them is inappropriate. However, in view of the interest of impredicative conceptions of classes for large cardinals, both predicative and impredicative conceptions should be pursued.[21]

[20]Frege (1969–76, 1: 288–9), a text written in 1924 or 1925. Cf. (1969–76, 1: 276–7), from 1919. The late evolution of Frege's thought on these matters is discussed in my (1976).

One can question whether the problem of generalizing predicate places is really solved by Frege's approach. Once we have generalizable variables in predicate places, we have new predicates that are not generalized by the variables in question – predicates which in Frege's semantics denote ''second-level concepts''. Hence the urge to extend the language by nominalization appears in Frege's context in another form. An ultimate Fregean canonical language would have to be a predicate calculus of order ω of which the semantics can no longer be expressed, unless we admit predicates of infinitely many arguments of different types. Surely *we* understand such a language by a means which from this Fregean point of view is a falsification, namely by a recursion in which *in general* variables with argument places of given types range over relations of these arguments, and each type is reached by finite iteration of the ascent from arguments to function. That involves a 'unification of universes' that Frege rejected, and which essentially contains nominalization.

Frege's logic contained bound variables only for objects and first-level functions, and free variables for one type of second-level function. He refers informally in at least one place to a third-level function (1893–1903, 1: 41), which would seem to be required by the *semantics* of his system. Formally, he thought higher levels dispensable because second-level functions could be replaced by first-level functions in which the function arguments were replaced by their *Wertverläufe* (1893–1903, 1: 42). This was untenable because it depended on the inconsistent axiom V. But of course in a less absolute way to replace functions by sets which are objects is just the procedure of set theory, which then does dispense with 'higher level functions' for most purposes. It is only quite recently, with the discussion of measurable and other very large cardinals, that higher than second-order concepts relative to the universe of sets have had any real application. See especially Reinhardt (1974a) and Wang (1977).

[21]Analogously to the theory of predicatively definable sets of natural numbers, one can explore mathematically the predicative definability of classes relative to the universe of sets. See Moschovakis (1971).

Wang's discussion of the axioms of separation and power set could lead one to think that impredicative reasoning about 'multitudes' is already involved in motivating the axiom of power set. Although this may be psychologically natural, what the power set axiom says is that given a set x there is a set of all sub*sets* of x, not that there is a set of all 'multitudes' whose elements are elements of x. Thus being an arbitrary subset of x has to be definite, but the 'multitude' of them is defined without quantifying over arbitrary *multitudes*. The

Let us now return to Cantor's suggestion that the elements of an "inconsistent multiplicity" cannot all exist together. I do not conclude that an inconsistent multiplicity does not exist in any sense; even the hypothesis that it is not an object will have to be qualified. However, one implication is clear: it is not a totality of its elements; it is not 'constituted' in a definite way by its elements. Its existence cannot require the prior existence of all its elements, because there is no such prior existence.

I wish to explicate the difference between sets and classes by means of some intensional principles about them. From the idea that a set is constituted by its elements, it is reasonable to conclude that it is *essential* to a set to have just the elements that it has and that the *existence* of a set requires that of each of its elements. Exactly how one states these principles depends on how one treats existence in modal languages. I shall assume that the truth of $x \in y$ requires that y exists (Ey). Then we have:

(1) $$x \in y \rightarrow Ex \wedge Ey$$
(2) $$x \in y \rightarrow \Box(Ey \rightarrow x \in y)$$
(3) $$x \notin y \wedge Ey \rightarrow \Box(x \notin y).^{22}$$

My proposal is that these principles should fail in some way if y is an "inconsistent multiplicity" or proper class. Indeed Reinhardt has suggested that proper classes differ from sets in that under counterfactual conditions they might have different elements (1974a: 196). I am endorsing this suggestion as an explication of the intuitions about "inconsistent multiplicities" that I have been discussing.

---

axiom of separation tells us that *any* 'multitude' of elements of x is a subset of x, so that the 'definiteness' of the property of being a subset of x implies that of being a submultitude of x. But we do not need to assume the definiteness of the latter property; indeed if we think of 'multitudes' intensionally (see below), it is only their *extensions* that become a definite totality by this reasoning.

[22] The most natural and elementary application of these principles is in relation to sets of ordinary objects that are the extensions of predicates contingently true of them. I intend to discuss these matters in a paper in preparation; cf. (1974d) and Tharp (1975).

(1)–(3) exactly parallel familiar principles of identity except that identity is usually treated as independent of existence.

(2) implies that set abstracts are not rigid designators. If 'F' is a predicate that holds of an object x, but not necessarily so, then $\Box(E\{z:Fz\} \rightarrow x \in \{z:Fz\})$ is true with the scope of the abstract outside the modal operator but false with the scope within. I assume that in any possible world $\{z:Fz\}$ is the set of existent z such that Fz in that world.

The free variables in (1)–(3) range over all possible objects, although for the present discussion the appropriate modal logic has *bound* variables ranging only over existing objects. If this treatment of free variables is thought to be too Meinongian, then (3) needs to be replaced by a schema

$$\forall x \Box(x \in y \rightarrow Fx) \rightarrow \Box\forall x(x \in y \rightarrow Fx),$$

or, in the second-order case, by the corresponding second-order axiom.

As before, for a given 'possible world' we should think of the bound variables as ranging over a set, perhaps an R_α (see note 4); but the sets that exist in that world are elements of the domain, while classes are arbitrary subsets of the domain.

Reinhardt does not use an intensional language; in his formulation the actual world V is part of a counterfactual 'projected universe' which is the domain of the quantifiers. He assumes a mapping j on $\mathfrak{P}(V)$ such that if $x \in V$, $jx = x$. We can think of jx as a 'counterpart' of x in the projected universe, in the sense of Lewis (1968). Thus sets are their own counterparts and can be strictly reidentified in alternative possible worlds. A set y can have no new elements in the projected universe and its only new non-elements are all x such that $x \notin V$. This accords with (1)–(3).

For a class P, jP can have additional elements in the projected universe, so that it violates (3), although jP agrees with P for 'actual' objects (elements of V).[23] Reinhardt's extensional language must distinguish jP from P; hence my thinking of jP as a 'counterpart'. A complication is that P itself occurs in the projected universe, though now as a set. Reinhardt himself suggests an alternative reading, by which a class x is an *intension*, so that P in the actual world and jP in the projected universe are the 'values' in two different possible worlds of the same intension. A formal language in which this reading might be formulated is the second-order modal language of Montague (1970), where the first-order variables range over objects and are interpreted (in the manner usual for modal logic) rigidly across possible worlds, and the second-order variables range over intensions, which in the semantics are functions from possible worlds to extensions of the appropriate type; but see pages [523–4].[24]

---

[23] Thus the relation of P and jP does not contradict (2). However, this is due to the special nature of the projected universe: j is an elementary embedding of the (sets and classes of) the actual world into it. (2) is presumably not an appropriate general principle about proper classes.

The explicit application to set theory of a modal conception of mathematical existence and the use of modal quantificational logic to explicate it seem to originate with Putnam (1967a) [reprinted in this volume]. Putnam does not address the questions about 'transworld identification' of sets that our principles (1)–(3) are meant to answer. However, it appears that his suggested translation of statements containing unrestricted quantifiers (p. 21) [310 in this volume] requires that a "standard concrete model of Zermelo set theory" should have a structure that is rigid, that is the relations are not changed when considered with respect to an alternative possible world. If this assumption is made, equivalents of (2) and (3) follow from the fact that a standard model is maximal for the ranks it contains (p. 20) [309 in this volume]. On "concreteness", cf. Parsons (1980, footnote 33).

Putnam seems to envisage a first-order formulation, which requires his "models" to be objects. The second-order formulation seems to us more appropriate not only for the set-class distinction but also for explicating the priority of the elements of a set to the set (Section V).

[24] A reformulation of Reinhardt's ideas in an intensional language would be desirable, in particular in order to eliminate the Meinongian ontology of possible non-actual sets and

No doubt what is most interesting about Reinhardt's idea is the impredicative use of proper classes that he combines with it, with the result that the ordinals in V are, in the projected universe, a measurable cardinal (1974b: 22; or Wang 1977: 327). However, in discussing the idea of proper classes as intensions I want to keep the predicative interpretation in mind as well.

IV

Cantor's conception suggests a more radical view than we have drawn from it so far, namely that one can in a sense not meaningfully quantify over absolutely all sets. In (1974a) and (1974c) I sketched all too briefly a 'relativistic' conception of quantifiers in set theory. The idea was that an interpretation that assigns to a sentence of set theory a definite sense would take its quantifiers to range over a set (presumably R_α for some large α), but that normally such a sentence would be so used as to be 'systematically ambiguous' as to *what* set the quantifiers ranged over.[25] A

classes that he uses, especially in (1974b). Montague's intensional logic would not be adequate as it stands for this purpose, since his first-order quantifiers range over all possible objects; however, there is no difficulty in reformulating it to fit an interpretation in which bound variables range over existing objects. If the only alternative possible worlds one wants to consider are those with more ranks, than the version of quantified modal logic of Schütte (1968) is applicable. This has the additional advantage that free variables also range over existing objects.

Pure modal logic, however, would not suffice to state Reinhardt's schema (S4) (1974a: 196), since it expresses a condition on a single 'possible world' for infinitely many formulae.

The question arises how a class P can recur 'in extension' in another possible world such as Reinhardt's projected universe. The answer is that it would be represented by its "rigidification", that is an attribute Q satisfying the condition.

$$\forall x(Px \leftrightarrow Qx) \wedge \Box\{\forall x(\Box Qx \vee \Box \neg Qx) \wedge$$
$$\forall R[\forall x \Box(Qx \to Rx) \to \Box\forall x(Qx \to Rx)]\}$$

I assume here that bound variables range over existing objects; otherwise Barcan's axiom would hold and the third conjunct would be unnecessary.

Wang (1977) formulates Reinhardt's ideas in the opposite direction, by eliminating the intensional motivation and thinking of V not as the 'actual world' but as a set which is an 'approximation' to the universe. What *mathematical* interest an intensional formulation would have is not clear; perhaps it would suggest 'intuitionistic' approaches to strong reflection axioms.

(*Added in proof.*) The statement that every attribute has a rigidification is just the formulation appropriate to this setting (without the Barcan formula) of the axiom E^σ (for $\sigma = (e)$) of Gallin (1975: 78). That the second conjunct above is equivalent to Gallin's $\forall x(\Box Qx \vee \Box \neg Qx)$ follows from Barcan's axiom. However, (1)-(3) (p. 519) and the comprehension axiom of p. 527 are inconsistent with the Barcan formula. Axiomatizations of set theory based on the ideas of this paper are described in Parsons 1981.

[25]Such a conception is hinted at in Zermelo (1930); see especially p. 47.

The conception of quantification over all sets advanced here is close to that of Putnam (1967a) [reprinted in this volume], except for the addition of the concept of systematic ambiguity.

merit of this view was (1974c: 8, 10–11) that it yielded a kind of reduction of classes to sets. Since I have here followed Cantor and Reinhardt in viewing classes as quite different from sets, here I can defend this relativistic position only in a modified form. I shall now present my understanding of the matter.

What *does* follow from the thesis that the elements of an inconsistent multiplicity cannot all exist together is that quantification over all sets does not obey the classical correspondence theory of truth. The totality of sets is not 'there' to constitute any 'fact' by virtue of which a sentence involving quantifiers over all sets would be true. The usual model-theoretic conception of logical validity thus leaves out the 'absolute' reading of quantifiers in set theory.

If the only constraint on an interpretation of a discourse in the language of set theory is that it should make statements proved in first-order logic from axioms accepted by the interpreter *true,* then the interpreter needs only a minimally stronger theory than that applied in the discourse to interpret it so that the quantifiers range over some R_α.[26] However, it seems clear that this condition is too weak, and it remains so even if the interpreter seeks to capture not just one particular discourse but what the set theorist he is interpreting might be taken to be *disposed* to assent to. This is the case envisaged in (1974c: 10), where we suppose that the interpreter takes the quantifiers to range over R_α for an α with an inaccessibility property undreamed of by the speaker.

Let us suppose this inaccessibility property to be P and that the interpreter chooses the least such α. One weakness of the reading is that it takes $(\exists\alpha)P\alpha$ to be false; although the supposition that the speaker is not disposed to assent to it is reasonable, the result is arbitrary in that we have no reason to suppose the speaker disposed to dissent from it.

A more decisive objection is that if the interpreter gets the speaker to understand P and convinces him that there is a cardinal satisfying it, then he must attribute to the speaker a meaning change brought about by this persuasion: previously his 'concept of set' excluded P-cardinals; now it admits them.

The speaker, however, can (going outside the language of set theory and talking of himself and his intentions) question this and say that P-cardinals are cardinals in just the sense in which he previously talked of cardinals; he will presumably reinforce this by assenting to a number

---

[26]The existence of an R_α such that V is an elementary extension of R_α is provable in ZF plus impredicative classes (Bernays-Morse set theory; see Drake (1974: 125)). What is essential, however, is not impredicative classes but allowing bound class variables in instances of replacement; one could use the system NB$^+$ mentioned in my (1974a, note 15).

of statements that follow directly from the existence of a P-cardinal by axioms or theorems of set theory he accepted previously.

The idea that quantifiers in set theory are systematically ambiguous was meant to meet this kind of objection by saying that the interpretation of the speaker's quantifiers as ranging over a single R_α cannot be an exactly correct interpretation, since it fixes the sense of statements whose sense is not fixed by their use to this degree. However, it still seems to imply that the speaker who is convinced of the existence of a P-cardinal undergoes a meaning change in a weaker sense, in that the ambiguity of his quantifiers is reduced by 'raising the ante' as to what degree of inaccessibility an α has to have so that R_α will 'do'.

We should observe that assertions in pure mathematics are made with a presumption of *necessity;* if we attribute this to our speaker we can see how P-cardinals are immediately captured by his previous set theory, since necessary generalizations are not limited in their force to what 'actually' exists. We can see the 'meaning change' in accepting P-cardinals as analogous to the speaker's considering a different possible world or range of possible worlds.

The force of this analogy is limited, as we can see by a little further reflection on the conception of 'inconsistent multiplicities' as intensions. It seems that we cannot consider a proper class as given even by an *intension* that is definite in the sense of, say, possible-world semantics, as a function from possible worlds to extensions. To begin with, it is only by an interpretation external to a discourse that one can speak in full generality of the range of its quantifiers and the extensions of its predicates. The systematic ambiguity of the language of set theory arises from the fact that such an interpretation can itself be mapped back into the language of set theory when stronger assumptions are made. Thus we should think of predicates whose 'extensions' are proper classes as really not having *fixed* extensions.[27]

This situation does not change if we enlarge the language of set theory to an intensional language. Here we are able to express the 'potentiality

---

[27]Cf. the remarks on the discomfort evinced by use of proper classes in Wang (1977, footnote 8). However, Wang does not make clear whether this discomfort would be removed if we confine ourselves to thinking "of these large classes as extensions or ranges of properties".

The point which I would emphasize is that if the language of set theory with quantifiers read as ranging over 'all sets' has a 'fixed' or 'definite' sense, then it is naturally extended by a satisfaction predicate, and definiteness of sense is preserved. But in the extended language one can of course construe the classes required by the Bernays-Gödel theory. Iteration of the procedure yields more classes.

In Wang's terms, this justification of classes no doubt falls within the conception of them as "extensions or ranges of properties". Still, such an enlargement of the language of set of the totality of sets' in that it is necessarily true that the domain of a bound variable *possibly* exists as a set. But however such an intensional language is formulated, it will still be possible to read it in a set-theoretic possible world semantics, and even if on the most straightforward reading the union of the domains for all possible worlds is all sets, the assumption that there is a set that realizes the properties of this union presumably has the same plausibility that other such reflection principles have. In such a model, of course a second-order intension will be represented by a set.[28]

We should not be surprised at this; it is really a consequence of the general nature of true 'systematic ambiguity', where there is no general concept of 'possible interpretation' which is not either inadequate or infected with the same difficulties as the language it interprets. Otherwise one could resolve the ambiguity as generality (meaning by *A*, '*A* on every possible interpretation') or indexically, by some contextual device or convention indicating which interpretation is meant. Russell's "typical ambiguity" was essential in that according to his theory of meaning there was no way of expressing by a single generalization *all* instances of a formula where the variables were understood as typically ambiguous. In (1974b) I handled semantical paradoxes by observing that paradoxical sentences could be taken to have no truth-value or not to express propositions on the interpretations presupposed by the semantic concepts occurring in them, while obtaining definite truth-values or coming to express propositions on interpretations 'from outside'. But at some point there must on this account be systematic ambiguity, or else one could generate 'super' paradoxes such as 'this sentence is not true on any interpretation'.

Thus although it is true to say that a proper class is given to us only 'in intension', this statement does not have quite its ordinary meaning. Obviously what is lacking is not just its being given to *the mind* 'in extension'; that is lacking for most sets as well. What is lacking has to do, one might say, with being, and moreover if the underlying intension had the

---

theory seems to be treated with reserve by many set theorists, although the reason could be just that in deductive power it is inferior to stronger axioms of infinity.

Locutions requiring either classes or satisfaction and truth are frequent in writings on set theory (cf. my, 1974c), but the characteristic informal use is very weak and could be captured by a free variable formalism for classes with very elementary operations on them, as George Boolos' comments on (1974c) reminded me.

[28]The 'straightforward reading' involves replacing 'set' by 'class' at certain points in the standard model-theoretic account of (modal) logical validity, just as in the case of ordinary logic in set theory. The same should be the case for set theories with intuitionistic logic, which are suggested by the same considerations as suggest the modal language. It should not be thought that changing to intuitionistic logic will remove the fundamental dilemmas about quantification over all sets.

fixed, completed existence a proper class lacks then the class would have it as well. However, some general ideas about intensional concepts do have application to this case. If we think of classes as given only by our understanding of the (perhaps indefinitely extendible) language of set theory, then the assumption that impredicative reasoning about classes is valid is rather arbitrary. This way of looking at classes corresponds to thinking of intensions as meanings and of meanings as constructions of the mind. This is the conception that is appropriate to applying intensional logic to propositional attitudes. Alternatively (and here we have a clearer theory) intensions are thought of as individuated by modal conditions, as 'functions from possible worlds to extensions'. This is the conception appropriate to modal logic. It seems neutral with respect to the question of impredicativity.

Let me make a final comment on the *predicative* conception of classes. If we understand a second-order language containing set theory in this way then set existence does for us the work of the axiom of reducibility in Russell's theory of types. For predicates which are high in a ramified hierarchy or which more generally are expressible only by 'logically complex' means, the existence of a set $\{x: Fx\}$ provides a simple equivalent $x \in a$ for a a *name* of $\{x: Fx\}$. Clearly it serves as an equivalent only *extensionally*. In the intensional situations envisaged above the equivalence of $x \in a$ and Fx will not be necessary even if the name a has been introduced by stipulation ('*a priori*' in Kripke's sense; cf. Tharp 1975), and therefore the two predicates will behave differently in intensional contexts. Thus the license for impredicativity given by assuming the existence of sets does not nullify the predicative conception of intensions even for intensions that have sets as extensions. Of course this 'reducibility' does not obtain for predicates that do not have sets as extensions.[29]

In conclusion, I would claim that the above discussion had added something to the explication of the idea that an 'inconsistent multiplicity' is not really an object, since even as an intension it is systematically ambiguous. The task remains to explain whether the ideas of the last two sections are helpful in understanding the 'stages' of the genetic conception and the underlying priority of the elements of a set to the set.

---

[29] It is commonly claimed that the axiom of reducibility nullifies Russell's ramification of his hierarchy of types. This claim depends on ignoring, presumably on the grounds that nonextensional features of functions are not significant for mathematics, the fact that Russell thought of propositional functions intensionally.

On the other hand it is hard to see what is left of Russell's no-class theory once the axiom of reducibility is admitted. Russell himself says that the axiom of reducibility accomplishes "what common sense effects by the admission of classes" (1908: 167), but he considers the axiom a weaker assumption than the existence of classes. The weakness must consist in the restrictions of the simple theory of types.

V

In the last two sections we sought to avoid using either epistemic concepts or the metaphor of process in trying to understand the conditions for the existence of sets. However, we concentrated largely on the distinction between sets and classes or 'multiplicities' and on discourse about absolutely all sets.

The idea that any available objects can be formed into a set is, I believe, correct, provided that it is expressed abstractly enough, so that 'availability' has neither the force of existence at a particular *time* nor of giveness to the human mind, and formation is not thought of as an action or Husserlian *Akt*. What we need to do is to replace the language of time and activity by the more bloodless language of potentiality and actuality.

Objects that exist together *can* constitute a set. However, we do have to distinguish between 'existing together' and 'constituting a set'. A multiplicity of objects that exist together *can* constitute a set, but it is not necessary that they *do*. Given the elements of a set, it is not necessary that the set exists together with them. If it is possible that there should be objects satisfying some condition, then the realization of this possibility is not as such the realization *also* of the possibility that there be a set of such objects. However, the converse does hold and is expressed by the principle that the existence of a set implies that of all its elements.

The same idea would be expressed in semantic terms by the supposition that we can use quantifiers and predicates in such a way that the range of the quantifiers and the objects satisfying any one of the predicates can constitute single objects, but these objects are not already captured by our discourse. However, this way of putting the matter might be taken to rule out too categorically an 'absolute' use of quantifiers and predicates. Without returning to an ontological characterization such as the Cantorian language of 'existing together', we can say that this is the condition under which quantifiers and predicates obtain definiteness of sense.

Above we suggested that the axiom of power set rests on a sort of principle of plenitude, according to which all the possible subsets of a given set are capable of existing 'at once'. Against what we have just said one might object that there is no intrinsic reason why the 'potentiality' of a set relative to its elements should not be nullified in our theory by a similar principle of plenitude.

The short answer to this objection is that such treatment would lead to contradictions, Russell's paradox in particular. We could apparently consistently assume (as in New Foundations) that the domain of discourse is a set in the domain, but then of course there will be other 'multiplicities' of elements of the domain that are not in it.[30]

A further point is that there seems to be an intrinsic ordering of 'relative possibility' in the element-set relation that is lacking for the arbitrary subsets of a given set. A set is an *immediate* possibility given its elements, the sets of which it is an element are at least at another remove. We do of course have conceptions of the 'simultaneous' realization even of infinite hierarchies in this ordering, but such a conception gives the possibility of sets that are still higher.

This observation should remind us that more is involved in the 'iterative conception' of set than the priority of element to set, since in Gödel's words we think of arbitrary sets as obtained by *iteration* of the "operation 'set of'" starting with individuals, and we have not yet dealt with the concept of iteration. To do so adequately would be beyond the scope of this paper. I shall make a few remarks.

First, our strategy has been to use modal concepts in order to save the idea that *any* multiplicity of objects can constitute a set; one makes only the proviso that they 'can exist together', and this proviso I take to be already given by the meaning of the quantifiers unless they are used in a 'systematically ambiguous' way. One saves thereby the universal comprehension axiom as well, though in a form that hardly seems 'naive' any more: In the second-order modal language it would have to be expressed by the statement that for every attribute P there is an attribute Q that is the rigidification of P (note 24 above) and such that

$$(4) \qquad \Diamond (\exists y)(\forall x)(x \in y \leftrightarrow Qx).[31]$$

However, even with the assumptions needed to obtain a version of the power set axiom we do not obtain greater power than that given by a much more traditional way of saving the comprehension axiom: the simple theory of types. It is clear that without some principle allowing for *transfinite* iteration of something like the above comprehension principle we will not obtain even the possible existence of sets of infinite rank, such as the usual axiom of infinity already requires. For the axiom of infinity, the principle needed is one allowing the conversion of a 'potential' infinity into an 'actual' infinity: we can easily show

---

[30]In the case of NF, these additional 'multiplicities' would correspond to the proper classes of ML.

If a model of NF is given as a set in the ordinary set-theoretic sense, the domain of the model and the V of the model will of course differ. The membership relation of the model will obviously not be the same as the membership relation 'from outside'.

[31]Thus if in some possible world $(\forall x)(x \in y \leftrightarrow Qx)$ holds, the elements of y are just the objects that *actually* have P. In many cases they will not be just the objects that have P in the world in question.

(5) $$\Box(\forall x)\Diamond(\exists y)(y=x\cup\{x\}),$$

but to use (4) to infer that ω possibly exists, we would need to get from (5) to

$$\Diamond(\forall x)(\exists y)(y=x\cup\{x\});$$

in terms of a set-theoretic semantics for the modal language, the possible worlds containing finite segments of ω need to be collected into a single one.

Second, it is clear that there has to be a priority of earlier to later ordinals, whether this is *sui generis* or derivative from the priority of element to set. One could of course assume a well-ordered structure of individuals, within which there would be no ontological priority of earlier to later elements. The axiom of infinity of *Principia* is such an assumption. To make it is natural enough, unless we assume a relation to the mind is essential to the natural numbers. Then it seems that smaller numbers are prior to larger ones by virtue of the order of time, as in Brouwer's (and apparently also Kant's) theory of intuition.

For reasons indicated above, no such structure can represent all ordinals. In fact larger ordinals seem conceivable to us only by characteristically set-theoretic means such as assuming that there is already a *set* closed under some operation on ordinals.

Third, it seems to me that the evidence of the axiom of foundation is more a matter of our not being able to understand how non-well-founded sets could be possible rather than in a stricter insight that they are *impossible*. We can understand starting with the immediately actual (individuals) and iterating the 'realization' of higher and higher possibilities. It seems that (at least as long as we hold to the priority of element to set) we do not understand how there could be sets that do not arise in this way. Non-well-founded $\in$-structures have been described (simple ones already in Mirimanoff 1917a), but we do not recognize them as structures of *sets* with $\in$ as the real membership relation, even when they satisfy the axioms of set theory.[32] We are at liberty to say that the *meaning* of 'set' is, in effect, 'well-founded set', but that does not exclude the possibility that someone might conceive a structure very like a 'real' $\in$-structure which violated foundation but which might be thought of as a structure of sets in a new sense closely related to the old.

I shall close with a rather speculative comment. The conception of 'inconsistent multiplicities' as indefinite or ambiguous raises a doubt

[32]Perhaps this could be said of trivial variants such as that resulting from identifying individuals with their unit classes. But here of course a slightly modified axiom of foundation holds.

about whether it is appropriate to talk of *the* cumulative hierarchy as most set theorists do. The definiteness of the power set is maintained even though the hope of deciding such questions as the continuum hypothesis and Souslin's hypothesis by means of convincing new axioms has not been realized. However, in this case the idea of the 'maximality' of the power set gives us some intuitive handle on the plausibility of the hypotheses or of 'axioms' such as $V = L$ that *do* decide them.

Maximality conceptions also contribute to the plausibility of large cardinal axioms. Here it seems conceivable in the abstract that we might see the possibility of a cardinal α with a 'structural property' P and of a cardinal β with such a property Q, where these properties are not 'compossible'; that is, we would see (perhaps even in ZF) that such α and β cannot both exist. That would yield two incompatible possibilities of cumulative hierarchies.

This has not happened with any of the types of large cardinals considered in recent years, where it has generally happened that of two such properties one (say P) implies the other, and indeed $P\alpha$ implies the existence of many smaller β such that $Q\beta$. That this is so has seemed rather remarkable; perhaps it is evidence against the views I have advanced.

However, one reason for thinking that 'incompatible large cardinals' will not arise is that by the Skolem-Löwenheim theorem both would reflect into the countable sets. If our confidence in the uniqueness of $\mathfrak{P}(\omega)$ is so great as to lead us to reject the possibility of incompatible large cardinals, one would still wish for some more direct reason for doing so.[33]

[33] I am indebted to Robert Bunn, William Craig, William C. Powell, Hilary Putnam, and Hao Wang for valuable discussions related to this paper. I regret that time did not permit me to follow up Mr. Bunn's remarks on Jourdain's attempt to develop the theory of inconsistent multiplicities.

The concept of set

HAO WANG

1. The (maximum) iterative concept

A set is a collection of previously given objects; the set is determined when it is determined for every given object x whether or not x belongs to it. The objects which belong to the set are its members, and the set is a single object formed by collecting the members together. The members may be objects of any sort: plants, animals, photons, numbers, functions, sets, etc.

According to the iterative concept, a set is something obtainable from some basic objects (such as the empty set, or the integers, or individuals, or some other well-defined urelements) by iterated applications of the rich operation 'set of' which permits the collecting together of any multitude of 'given' objects (in particular, sets) or any part thereof into a set. This process includes transfinite iterations. For example, the multitude of sets obtained by finite iteration is considered to be itself a set.

We understand this concept of set sufficiently well to see, after some deliberation, and in some cases even a great deal of deliberation, that the ordinary axioms of set theory are true for (or with respect to) this concept, and to be able to extend these axioms by proposing additional axioms and recognizing some of them to be true for (or with respect to) it.

The iterative concept involves at least four difficult ideas: the idea of 'given', the idea of collecting together, the idea of 'part' or subset, and the idea of iteration. The idea of iterations implies the potentiality of continuing to any stage (as indexed by a previously given ordinal number)[1] and adds an inductive element to the idea of 'given' (viz. all sets obtained at or before any given stage are viewed as given). The idea of urelement is not difficult for set theory, because we are in this context

Reprinted with the kind permission of the author and publisher from Hao Wang, *From Mathematics to Philosophy,* Routledge and Kegan Paul, 1974, pp. 181–223. Permission for publication in America kindly granted by Humanities Press, Inc.

[1] The reader who is not familiar with the technical concepts of set theory is referred to standard texts such as Kamke 1950 and Fraenkel 1953. For more specialized concepts and results, the reader may consult Hausdorff 1949, Gödel 1940, Bernays 1958, and Cohen 1966.

not interested in what an individual is but rather leave the question open. We do not attempt to determine what the correct urelements are.

It is a basic feature of reality that there are many things. When a multitude of given objects can be collected together, we arrive at a set. For example, there are two tables in this room. We are ready to view them as given both separately and as a unity, and justify this by pointing to them or looking at them or thinking about them either one after the other or simultaneously. Somehow the viewing of certain given objects together suggests a loose link which ties the objects together in our intuition, or a variable object which could be any one of them. In order that our mind may more effortlessly and unwaveringly fix our attention on this variable object, we, it could be suggested, concretize or reify the loosely linked bundle of objects and think of the more determinate range of variability. But then we seem to be forced by the surprising success of the reification to admit that there are certain objective grounds for our ostensively acquired intuition. It may be noted that Cantor discusses briefly the same phenomenon in connection with the move from a potentially infinite to an actually infinite.[2]

We can form a set from a multitude only in case the range of variability of this multitude is in some sense intuitive. This is the criterion for determining whether a multitude forms a set for us. The natural way of getting such intuitive ranges is by the use of intuitive concepts (defining properties). An intuitive concept, unlike an abstract concept such as that of mental illness or that of differentiable manifold, enables us to overview (or look through or run through or collect together), in an *idealized* sense, all the objects in the multitude which make up the extension of the concept, in such a way that there are no surprises as to the objects which fall under the concept. Hence, each intuitive concept determines an intuitive range of variability and therewith a set.

The overviewing of an infinite range of objects presupposes an infinite intuition which is an idealization. Strictly speaking, we can only run

---

[2]Cantor 1932. All references to Cantor are to this volume of his collected works.

Unterliegt es nämlich keinem Zweifel, dass wir die *veränderlichen* Grössen im Sinne des potentialen Unendlichen nicht missen können, so lässt sich daraus auch die Notwendigkeit des Aktual-Unendlichen folgendermassen beweisen: Damit eine solche veränderliche Grösse in einer mathematischen Betrachtung verwertbar sei, muss strenggenommen das 'Gebiet' ihrer Veränderlichkeit durch eine Definition vorher bekannt sein; dieses 'Gebiet' kann aber nicht selbst wieder etwas Veränderliches sein, da sonst jede feste Unterlage der Betrachtung fehlen würde; also ist dieses 'Gebiet' eine bestimmte aktualunendliche Wertmenge. So setzt jedes potentiale Unendliche, soll es streng mathematisch verwendar sein, ein Aktual-Unendliches voraus. Diese 'Gebiete der Veränderlichkeit' sind die eigentlichen Grundlagen der Analysis sowohl wie der Arithmetik und sie verdienen es daher in hohem Grade, selbst zum Gegenstand von Untersuchungen genommen zu werden, wie dies von mir in der 'Mengenlehre' (théorie des ensembles) geschehen ist (1886, pp. 410–11).

through finite ranges (and perhaps ones of rather limited size only). This idealization contains seeds for growth in itself. For example, not only are the infinitely many integers taken as given, but we also take as given the process of selecting integers from this unity of all integers, and therewith all possible ways of leaving integers out in the process. So we get a new intuitive idealization (viz. the set of all sets of integers) and then one goes on.

The concept of all subsets is often thought to be opaque because we envisage all possibilities independently of whether we can specify each in words; for example, just as there are 2^{10} subsets of a set with 10 members, we think of 2^a subsets of a set with a members when a is an infinite cardinal number. In particular, we do not concern ourselves over how a set is defined, e.g. whether by an impredicative definition. This is the sense in which the individual steps of iteration are 'maximum'. It is possible to get other iterative concepts by restricting the operation of going to the next stage, one familiar example being the constructible sets. The (maximum) iterative concept has been discussed by Bernays (1935) [reprinted in this volume] under the name of platonism.

The weakest 'platonistic' assumption introduced by arithmetic is that of the totality of integers... But analysis is not content with this modest variety of platonism; it reflects it to a stronger degree with respect to the following notions: set of numbers, sequence of numbers, and function. It abstracts from the possibility of giving definitions of sets, sequences, and functions. These notions are used in a 'quasi-combinatorial' sense, by which I mean: in the sense of an analogy of the infinite to the finite... In Cantor's theories, platonistic conceptions extend far beyond those of the theory of real numbers. This is done by iterating the use of the quasi-combinatorial concept of a function and adding methods of collection. This is the well-known method of set theory. [cf. pp. 259–60, this volume]

What is given at each stage depends on an orderly manner of iteration. Hence, the concept of ordinal number is essential to the iterative notion, in that we use ordinal numbers to index the stages of iteration. Thus, as we generate more and more sets according to the iterative concept, we encounter certain well-ordered sets among the sets generated. The order types of these well-ordered sets determine ordinal numbers which can be used to index (further) stages of the iterations. Given a totality of operations for generating sets, we can also survey all the ordinal numbers obtainable by these operations and introduce new ordinals. In general, for any ordinal number α, given by whatever means, we are permitted to carry the process of iteration to the α-th stage, and regard all the sets generated up to and including the α-th stage as given and proceed further.

The question of urelements involves us in the contrast between sets (or mathematical objects in general) and other objects. Philosophically it is important to realize that we are faced with a perfectly general situation and that we can begin initially with any collectable multitudes as urelements. There is nothing in the original iterative concept to rule out different kinds of urelements. For example, we can take all physical objects as the urelements, or all elementary particles, or all animals, or all integers, etc. In each case, if we are able to collect the urelements into a set x, we can carry out the process of iteration starting with x (or conceive of a hierarchy of transfinite types with x at the bottom). Since, however, the process of generating further sets from an initial set x of urelements is uniform, with respect to x, i.e. the process remains the same no matter what initial set x we might wish to choose, it is reasonable to consider just one typical general case as we have done in our first explanation of the iterative concept.

Moreover, for the abstract study of sets, it seems convenient to disregard nonsets altogether. This turns out to be feasible, because even if we start from nothing (i.e. neither urelements nor sets) initially, we can get the empty set 0. The use of this artificial special case of an empty set of urelements achieves a convenient purity. As a matter of fact, for the mathematical studies of sets it is customary to require that all members of sets are sets. This restriction excludes sets of tables and elephants, but does not exclude sets of numbers and functions which are identified with certain sets. Under this restriction, we say that a set is a collection of given sets.

On the basis of our explanations of the (maximum) iterative concept of set, we are able to see that the ordinary axioms of set theory (commonly referred to as *ZF* or *ZFC*) are true for the concept.

AE Axiom of extensionality. A set is completely determined by its members; i.e. two different sets may not contain the same members. If x and y have the same members, then $x=y$.

This may be viewed as a defining characteristic of sets (in contrast with properties).

AS Axiom of subset formation (axiom of comprehension). If a multitude A is included in a set x, then A is a set.

Since x is a given set, we can run through all members of x, and, therefore, we can do so with arbitrary omissions. In particular, we can in an idealized sense check against A and delete only those members of x which are not in A. In this way, we obtain an overview of all the objects in A and recognize A as a set.

AP Axiom of power set. All subsets of a set can be collected into a set.

For, if x is given, then all subsets of x are given individually by AS. We have, moreover, an intuitive idea of running through with omissions. This general notion, which is on a higher level than its application to each multitude A included in x, provides us with an overview of all cases of AS as applied to x. And the overview provides us with the basis of performing the collection to get the power set of x.

In our previous discussion about the urelements, we have reached the conclusion that for the abstract development of set theory we can conveniently disregard the diversity of urelements and, in fact, leave out nonsets altogether, taking an empty set of urelements. Now that we are justified in forming the power set of a given set, we are able to tidy up the iterative process in another direction. The operation of forming the power set of a given set eliminates the need to branch out from a given set x: there are different ways of forming subsets of x; we might otherwise be forced to distinguish between different kinds of subsets of x so that certain subsets of x are collected into one new set, and certain other subsets into another new set, and so on. By using the power set of x, we are able to pull together all subsets of x and summarize the formation of all possible subsets of x in a single new set, viz. the power set of x. In this way, we obtain a standard representation of all single applications of the rich operation 'set of' to any given totality of given objects. There are then no other obstacles against our construing the iterative conception in the sharper form of ranks or types or stages: every set is obtainable at some stage α (an ordinal number) and every stage R_α is obtained from the empty set (of urelements) by iterated applications of the operation 'set of,' which yields all members of the power set of R_β if $\alpha = \beta + 1$, and just gathers together all sets obtained at previous stages if α is a limit ordinal number. In other words, if R_α is the totality of sets obtained at all stages before the α-th, then $R_{\alpha+1}$ consists of all the subsets of R_α. For example, $R_0 = 0$, $R_1 = \{0\}$, $R_2 = \{0, \{0\}\}$, $R_3 = \{0, \{0\}, \{\{0\}\}, \{0, \{0\}\}\}$, and so on.

The iterative concept implies that we continue the iteration as far as possible; in particular, it implies that, for any given ordinal number α, there is an α-th stage. There is then the problem of getting ordinal numbers to index the stages. For example, we take for granted that we have the finite ordinal numbers to begin with. We are then led, as it happened to Cantor originally, to ω as the limit of all finite ordinal numbers (the natural numbers) and then to $\omega + 1$, and so on.

Thus, for each natural number n, we have a stage R_n. But there is no reason to stop there. So we have a further stage R_ω which collects together all the finite stages, as well as stages $R_{\omega+1}$, etc. From the way the stages are obtained, we see that for every set obtained, there is a first stage at which it appears, and that if there is at least one stage possessing a certain property, then there is a first stage possessing that property.

AF Axiom of foundation. Every set can be got at some stage; or, every nonempty set (or even multitude of sets) has a minimal member, i.e. a member x such that no member of x belongs to the set.

For there is a member x which is got at no later stage than any other member of the set. But all members of x are got at earlier stages and therefore cannot belong to the set.

AI Axiom of infinity. There is an infinite set (for example, R_ω).

AC Axiom of choice. Given any set x of nonempty sets, there is a set which contains exactly one member from each member of x.

Since every member of x is got at an earlier stage than x, all members of members of x are got earlier and any selection from these can be collected together to form a set.

AR Axiom of replacement. If b_x is a set for every member x of a set y, then the union of all these sets b_x is included in a set.

This form of *AR* differs from the more familiar form in two minor aspects: the use of the union and the weakening from being a set to being included in a set. The familiar form is *SAR*: if b is an operation and b_x is a set for every member x of a set y, then all these sets b_x form a set. The differences are introduced for certain esthetic reasons, which are not very relevant to our main interest here. We shall relegate a crude direct justification of *AR*, as well as an explanation of the relation between *AR* and *SAR*, to a footnote.[3] Here, we shall confine our attention to *SAR*.

[3] In very rough terms, we may directly justify the axiom *AR*, on the basis of the iterative concept, in the following manner.

In general, given any set y, we may consider the multitude of all stages $R_{\alpha(x)}$ where x is a member of y and $R_{\alpha(x)}$ is the first stage at which x appears. A reasonable principle for continuing the stages is to permit, for each given set y the collection or merging of all these stages $R_{\alpha(x)}$ into a new set. If, instead of $R_{\alpha(x)}$, we take any given set b_x, it is no less justifiable to collect or merge all the stages where these sets first appear, to get a new set. By this principle, if b_x appears at stage $R_{\beta(x)}$ then the result obtained by merging these stages, for all x belonging to y, contains the union of these sets b_x.

The minor additional procedure of forming the union of a set (i.e. merging the elements of a set) is conceptually a consequence of the intended process of iteration, since all members of members of a set a are given at earlier stages and therefore collected into a set b before the stage at which a emerges. A separate axiom for forming union is often included in formal systems of set theory because the defining properties do not mirror faithfully the intended extensional interpretation. The absorption of this feature into the replacement axiom (as stated above) is meant to render it less conspicuous. An inessential feature of the form of *AR* as stated is to get, instead of the image of y directly, a set which contains it. This amounts to taking, instead of b_x itself, its corresponding stage $R_{\beta(x)}$ so that (the union

Once we adopt the viewpoint that we can in an idealized sense run through all members of a given set, the justification of *SAR* is immediate. That is, if, for each element of the set, we put some other given object there, we are able to run through the resulting multitude as well. In this manner, we are justified in forming new sets by arbitrary replacements. If, however, one does not have this idea of running through all members of a given set, the justification of the replacement axiom is more complex.

Gödel points out that the axiom of replacement does not have the same kind of *immediate* evidence (previous to any closer analysis of the iterative concept of set) which the other axioms have. This is seen from the fact that it was not included in Zermelo's original system of axioms. He suggests that, heuristically, the best way of arriving at it from this standpoint is the following. From the very idea of the iterative concept of set it follows that if an ordinal number α has been obtained, the operation of power set (P) iterated α times leads to a set $P^{\alpha}(0)$. But, for the same reason, it would seem to follow that if, instead of P, one takes some larger jump in the hierarchy of types, e.g. the transition Q from x to $P^{|x|}(x)$ (where $|x|$ is the smallest ordinal of the well-orderings of x), $Q^{\alpha}(0)$ likewise is a set. Now, to assume this for any conceivable jump operation (even for those that are defined by reference to the universe of all sets or by use of the choice operation) is equivalent to the axiom of replacement.[4]

The seven axioms *ESPFICR* will be regarded as making up the ordinary system *ZF* (or *ZFC*) of set theory. The comments above about these axioms are intended to show that we can see them to be true for the iterative concept of set. Somewhat more formally, we can also recapitulate the hierarchy of sets resulting from the iterative concept, by assuming that the ordinal numbers are given initially, as follows.

of) the multitude of all the sets b_x, for x in y, is included in the set of all their corresponding stages $R_{\beta(x)}$. This serves the purpose of avoiding the somewhat inelegant situation of making the more basic axiom of comprehension a consequence of the replacement axiom.

It should perhaps be pointed out that these seven axioms *ESPFICR* are equivalent to other more commonly used sets of axioms taken as making up *ZF*. For a detailed proof, the reader may consult Shoenfield 1967: 240–3. The derivations making up the proof of equivalence go back at least to Zermelo (1930) and Bernays (1958); compare the references under notes 13 and 1.

[4]More explicitly, I would like to add as a supplement, it is a familiar fact that once we have replacement from sets of ordinals to get new sets of ordinals and we permit a stage R_{α} for each given ordinal α, we can get full replacement. And it is easily seen that replacement for sets of ordinals (i.e. given $f(\alpha)$, $\alpha < \beta$, there is γ, $f(\alpha) < \gamma$, for all $\alpha < \beta$) follows from the iteration of jumps (i.e. given f and β, there is γ^{α}, $f(0) < \gamma$, for all $\alpha < \beta$).

Gödel's explanation of the jump operation may also be viewed as a generalization of the way Cantor applies (in the development of his second number class) his second principle of generation according to which, if there is defined any definite succession of ordinal numbers of which there is no greatest, a new number is created which is defined as the next greater to them all (1883, p. 196).

R_0 = the empty set (or, sometimes, the set of integers)
$R_{\alpha+1}$ = the power set of R_α, i.e. the set of all subsets of R_α
R_λ = the union of all R_α, $\alpha < \lambda$, where λ is a limit ordinal
V = the union of all R_α, α any ordinal

In other words, the universe of all sets consists of all x such that x belongs to some R_α, α an ordinal. The smallest α such that x belongs to R_α is usually called the rank of x. Under this formulation, it is clear that the two difficult ideas are power set and ordinal number. In recent years, much effort has been devoted to finding more ordinals by introducing new cardinals to strengthen axiomatic set theory. In contrast, there has been little progress in efforts to enrich directly power sets (e.g. that of the set of integers) by new axioms. Both endeavors could be viewed as attempts to make our vague intuitive ideas more explicit.

The iterative concept seems close to Cantor's original idea,[5] and has been, in one form or another, developed and emphasized by Mirimanoff (1917a, b), von Neumann (1925), Zermelo (1930), Bernays (1935), and Gödel (1964).

This iterative concept of set is of course quite different from the dichotomy concept which regards each set as obtained by dividing the totality of all things into two categories (viz. those which have the property and those which do not). Following Gödel, one may speak of the two concepts as the mathematical versus the logical. To quote:

There exists, I believe, a satisfactory foundation of Cantor's set theory in its whole original extent and meaning, namely axiomatics of set theory interpreted in the way sketched below. It might seem at first sight that the set-theoretical paradoxes would doom to failure such an undertaking, but closer examination shows that they cause no trouble at all. They are a very serious problem, not for mathematics, however, but rather for logic and epistemology. [Gödel 1964: 262; 474 in this volume]

Many people have been puzzled by the fact that in an earlier paper on Russell, Gödel takes the paradoxes much more seriously (1944: 215–16 [452 in this volume]). 'By analyzing the paradoxes to which Cantor's set theory had led, he freed them from all mathematical technicalities, thus bringing to light the amazing fact that our logical intuitions (i.e. intuitions concerning such notions as: truth, concept, being, class, etc.) are self-contradictory.' The difference in emphasis, as Gödel explains, is due to a difference in the subject matter, because the whole paper on Russell is concerned with logic rather than mathematics. The full concept of class

[5]Compare the discussions to follow (in particular, notes 6, 8, 9). It may be said that not only the famous 1895 definition in terms of a collection of objects into a whole, but even also the 1883 definition in terms of one and many suggest strongly the iterative concept.

(truth, concept, being, etc.) is not used in mathematics, and the iterative concept, which is sufficient for mathematics, may or may not be the full concept of class. Therefore, the difficulties in these logical concepts do not contradict the fact that we have a satisfactory mathematical foundation of mathematics in terms of the iterative concept of set. In relation to logic as opposed to mathematics, Gödel believes that the unsolved difficulties are mainly in connection with the intensional paradoxes (such as the concept of not applying to itself) rather than with either the extensional or the semantic paradoxes. In terms of the contrast between bankruptcy and misunderstanding as considered below, Gödel's view is that the paradoxes in mathematics, which he identifies with set theory, are due to a misunderstanding, while logic, as far as its true principles are concerned, is bankrupt on account of the intensional paradoxes.[6]

One feels vaguely that the iterative concept corresponds pretty well to Cantor's 1895 'genetic' definition of set:[7] 'By a "set" we shall understand any collection into a whole M of definite, distinct objects m (which will be called the "elements" of M) of our intuition or our thought.' We are naturally curious to know a little more about the development of Cantor's concept and its relation to the iterative concept.

In 1882, Cantor explains that a set of elements is *well defined,* if by its definition and by the logical principle of excluded middle we must recognize as internally determined whether any object of the right kind belongs to the set or not.[8] One is inclined to think that the concept of set implicit in this context is closer to the logical concept rather than the mathematical one. In the next year, a set is defined, with references to Plato's notion of ideas and other related concepts, as[9] 'every Many, which can

---

[6]In order to prevent any misinterpretation of this remark, Professor Gödel suggests adding the following. 'This observation by no means intends to deny the fact that *some* of the principles of logic have been *formulated* quite satisfactorily, in particular all those that are used in the application of logic to the sciences including mathematics as it has just been defined.'

[7]'Unter einer "Menge" verstehen wir jede Zusammenfassung M von bestimmten wohlunterschieden Objekten m unserer anschauung oder unseres Denkens (welche die "Elemente" von M gennanten werden) zu einem Ganzen' (1895, p. 282).

[8]'Eine Mannigfaltigkeit (ein Inbegriff, eine Menge) von Elementen, die "irgendwelcher Begriffssphäre angehören, nenne ich *wohldefiniert,* wenn auf Grund ihrer Definition und infolge des logischen Prinzips vom ausgeschlossenen Dritten es als *intern bestimmt* angesehen werden muss, *sowohl* ob irgendein derselben Begriffssphäre angehöriges Objekt zu der gedachten Mannigfaltigkeit als Element gehört oder nicht, *wie auch,* ob zwei zur Menge gehörige Objekte, trotz formaler Unterschiede in der Art des Gegebenseins einander gleich sind oder nicht. Im allgemeinen werden die betreffenden Entscheidungen nicht mit den zu Gebote stehenden Methoden oder Fähigkeiten in Wirklichkeit sicher und genau ausführbar sein; darauf kommt es aber hier durchaus nicht an, sondern *allein* auf die *interne Determination,* welche in konkreten Fällen, wo es die Zwecke fordern, durch Vervollkommnung der Hilfsmittel zu einer *aktuellen (externen) Determination* auszubilden ist" (p. 150).

[9]'Unter einer "Mannigfaltigkeit" oder "Menge" verstehe ich nämlich allgemein jedes

be thought of as One, i.e. every totality of elements that can be united into a whole by a law.'

According to Fraenkel, Cantor had discovered the so-called Burali-Forti paradox no later than 1895, i.e. at least two years before Burali-Forti's publication, and had communicated it to, among others, Hilbert in 1896 (see p. 470 of Cantor's *Works*). This discovery may have something to do with the 'genetic' element in the famous 1895 definition. According to Zermelo (in Cantor 1932, p. 352, footnote 9), part of the reason why Cantor, in his treatise of 1895–7, deals extensively with the second number class rather than with all cardinal numbers was Cantor's awareness of the 'Burali-Forti paradox.' This may also explain why Cantor, in his 1895 paper, spoke of desiring to show that all cardinals form a well-ordered set 'in an extended sense' (p. 295). A concrete proposal along the line of distinguishing sets and classes was made in Cantor's letter to Dedekind in 1899 (pp. 443–4), not published until 1932.

There are also other differences between Cantor's outlook and the current one. But these seem to belong more appropriately to a footnote.[10]

Viele, welches sich als Eines denken lässt, d. h. jeden Inbegriff bestimmter Elemente, welcher durch ein Gesetz zu einem Ganzen verbunden werden kann,...' (p. 204). The parenthetical explanations of sets in the contexts of defining cardinality in 1884 (p. 387) and 1887 (p. 411) do not seem to add anything.

[10]Cantor does consider point sets (sets of real numbers) and sets of integers, as well as functions of point sets. But in his general development one sees more a theory of transfinite numbers than a set theory. He quickly extracts cardinal and ordinal numbers from sets and devotes most of his attention to these infinite numbers. The impression is that he believes there is a great variety of objects so that no neat structure can be imposed on all sets above and beyond that imposed by such basic notions as cardinal and ordinal numbers.

For Cantor, cardinals and ordinals are not sets but general concepts or universals abstracted from sets of equal cardinality and isomorphic well-ordered sets. For example, the cardinality of a set x is what is common to all sets 'equivalent' to x (p. 141, 1879; p. 387, 1887; p. 283, 1895; p. 444, 1899). Two sets are equivalent if there is a one-one correspondence between them. Cantor does work freely with numbers as objects and forms sets of them. For example, the first infinite cardinal is that of the set of finite cardinals.

On the other hand, we certainly cannot identify a number with its extension and take it as a set in our universe of sets. For example, the extension of the universal 1 is as large as the universe of all objects (including all sets) since, for each x, $\{x\}$ is a set of cardinality 1. Since the universe of objects consists of sets and urelements and numbers are objects but not sets, Cantor seems to treat them as urelements. This creates some problem with regard to the iterative concept of set. How do we assign ranks to the numbers? A natural suggestion is to give all urelements the rank 0 and the rank of a set would as in the current form, be determined inductively by the membership relation. But then R_0 which is a set would be too large and contain what Cantor calls an inconsistent manifold. One alternative would be distributing numbers into different ranks.

There is indeed a natural way of doing this, as has been carried out by Mirimanoff (1917a, b), Zermelo (unpublished work of 1915, see Bernays 1941: 6), and von Neumann (1923). Each ordinal number is identified with a canonical set representing it: take the empty set as 0, $\alpha \cup \{\alpha\}$ as $\alpha + 1$, the limit ordinal as the set of representatives of all preceding ordinals. Cantor did not make this convenient identification. But it seems likely that Cantor thinks of an open domain of objects which are not sets, such as physical objects,

With regard to the task of setting up the axioms of set theory (including the search for new axioms), we can distinguish two questions, viz. (1) what, roughly speaking, the principles are by which we introduce the axioms, (2) what their precise meaning is and why we accept such principles. The second question is incomparably more difficult. It is my impression that Gödel proposes to answer it by phenomenological investigations.

In connection with the first question, Gödel suggests the following summary of the principles which have actually been used for setting up axioms. It is understood that the same axiom can be justified by different principles which are nevertheless distinct in that they are based on different ideas; for example, inaccessible numbers are justified by either (2) or (3) below. The five principles to follow are illustrated by the discussion so far and the section below on new axioms and criteria of acceptability.

(1) Existence of sets representing intuitive ranges of variability, i.e. multitudes which, in some sense, can be 'overviewed' (see above).

(2) Closure principle: if the universe of sets is closed with respect to certain operations there exists a set which likewise is. This implies, e.g., the existence of inaccessible cardinals and of inaccessible cardinals equal to their index as inaccessible cardinals.

(3) Reflection principle: the universe of all sets is structurally undefinable. One possibility of making this statement precise is the following: The universe of sets cannot be uniquely characterized (i.e. distinguished from all its initial segments) by any internal structural property of the $\in$-relation in it, expressible in any logic of finite or transfinite type, including infinitary logics of any cardinal number. This principle may be considered as a generalization of (2). Further generalizations and other precisations are in the making in recent literature.

(4) Extensionalization: axioms such as comprehension and replacement are first formulated in terms of defining properties or relations. They are extensionalized as applying to arbitrary collections or extensional correlations. For example, we get the inaccessible

experiences, universals, properties, and whatnot. Hence, he would not have examined the idea that they can all either be put into R_0 or be given other appropriate ranks in a natural way. However that may be, Cantor's development of set theory as a mathematical subject can be embodied in a framework of pure sets, once the identification of numbers with suitable sets is made.

It is perhaps natural to think of the urelements as forming a set. If, on the other hand, one assumes that there are as many sets (or numbers) as urelements, one might perhaps modify the iterative model by defining $R_\alpha(a)$ relative to each set a of urelements such that $R_0(a)=a$ but $R_{\alpha+1}(a)$ and $R_\lambda(a)$ are defined as above.

numbers by (2) above only if we construe the axiom of replacement extensionally.

(5) Uniformity of the universe of sets (analogous to the uniformity of nature): the universe of sets does not change its character substantially as one goes over from smaller to larger sets or cardinals, i.e., the same or analogous states of affairs reappear again and again (perhaps in more complicated versions). In some cases it may be difficult to see what the analogous situations or properties are. But in cases of simple and, in some sense, 'meaningful' properties it is pretty clear that there is no analog except the property itself. This principle, e.g., makes the existence of strongly compact cardinals very plausible, due to the fact that there should exist generalizations of Stone's representation theorem for ordinary Boolean algebras to Boolean algebras with infinite sums and products.

2. Bankruptcy (contradiction) or misunderstanding (error)?

The reactions of Frege and Cantor to the paradoxes were sharply different and can be described as the bankruptcy theory versus the misunderstanding theory. The difference can undoubtedly be attributed completely to their different conceptions of set (the logical versus the mathematical notion). A related reason may perhaps be described as the difference between viewing sets from outside (Frege) and actually doing set theory (Cantor). Typically in philosophical discussions on the foundations of a subject, the emphasis of insiders and outsiders tends to differ. Even when the same statements are endorsed, quite different things could be intended. The meaning of methodological statements can be so indefinite that it is sometimes not easy to reconcile what a specialist says with what he does.

For example, Cantor, Zermelo, Mirimanoff, and von Neumann all seem to have basically the same conception of set, at least with regard to properties of sets which are implicit in the familiar axioms of today. Yet what they say sounds quite different. Cantor apparently thinks that the paradoxes are paradoxical only because the concept of set is not correctly understood (see, e.g., p. 470, letter of 1907). Zermelo construes the paradoxes as necessitating some restrictions on Cantor's 1895 definition of set:

It has not, however, been successfully replaced by one that is just as simple and does not give rise to such reservations. Under these circumstances there is at this point nothing left for us to do but to proceed in the opposite direction and, starting from set theory as it is historically given, to seek out the principles required for establishing the foundations of this mathematical discipline. In solving the problem we must, on the one hand, restrict these principles sufficiently to exclude all contradictions and, on the other, take them sufficiently wide to retain all that is valuable in this theory. (1908: 261)

According to Mirimanoff,

One believes and it appears evident, that the existence of individuals must imply the existence of sets of them; but Burali-Forti and Russell have shown by different examples that a set of individuals need not exist, even though the individuals exist. As we cannot accept this new fact, we are obliged to conclude that the proposition which appears evident to us and which we believe to be always true is inexact, or rather that it is only true under certain conditions. (1917a: 38)

In discussing attempts to axiomatize set theory, von Neumann (1925: 219-40 and 1961-3, vol. 1) emphasizes an arbitrary element:

Naturally, it can never be shown in this way that the antinomies are actually excluded; and much arbitrariness always attaches to the axioms. (There is, to be sure, a measure of justification of these axioms in that they turn into evident propositions of naive set theory, when the axiomatically meaningless word 'set' is taken in Cantor's sense. But what is deleted from naive set theory – and to avoid the antinomies it is essential to make some deletion – is absolutely arbitrary.) (1961-3, 1: 37)

In the extreme cases, the proponents of the misunderstanding theory propose to uncover flaws in seemingly correct arguments, while the bankruptcy theorists find our basic intuition proven to be contradictory and seek to reconstruct or salvage what they can, by ad hoc devices if necessary. The basic intuitive concept is often called naive set theory and identified with the belief in an absolute comprehension principle according to which any property defines a set. That some notion like this was actually seriously developed by Frege was a historical accident often advanced as evidence that we do have such a contradictory intuition. The principle, if correct, in fact appears to be the sort of thing which belongs to the domain of logic. Hence, it is much easier to understand Frege's enthusiasm over the thesis of reducibility of mathematics than that of his followers. Viewed in the light of Cantor's development of set theory, however, it is not at all clear that we do have such a contradictory intuition. It seems more appropriate to say that we have an inexact intuition which leads to the iterative conception as we notice the paradoxes and the flaws in them. It is, therefore, debatable whether we have such an intuition to begin with. But perhaps this could easily degenerate into a terminological debate.

Now, a stronger assertion is that the inconsistent concept is the only intuition of set we have. Hence, once the concept is seen to be wrong, we are left with nothing but the task of reconstruction as described in the above quotation from Zermelo. Taken literally, Zermelo's two constraints of not too narrow and not too broad are rather weak and leave room for much arbitrariness in that many mutually incompatible set theories are possible solutions. Moreover, there is implicitly an additional arbitrariness in deciding what results are to be taken as data for the reconstruction since the notion of 'valuable' (perhaps also implying 'reliable') is clearly ambiguous. Assuming that we have a good idea what the data are, the task as described sounds much like a combinatorial puzzle which in principle admits of diverse solutions. Even in the empirical sciences, such a situation does not satisfy the intellect. For example, the eightfold way theory of elementary particles has such a flavor but most people look for either a refutation of the theory or more basic principles from which the theory can be deduced. It will be said that, if in fact we do not have good intuitions about sets, then our wish for a stable solution of the paradoxes is futile. And it is not hard to see why the 'toughminded' position of crying bankruptcy has a certain appeal: it gives the impression of greater 'clarity,' a defiance of tradition, and independence from the slippery matter of intuition.

But the fact is we do arrive at a fairly stable iterative concept of set, whether or not we agree that this is the only original intuitive notion of set to begin with. And this concept was also implicit in the works of Zermelo and von Neumann who in the quoted contexts speak as bankruptcists, but used their good intuitions about sets in setting up their axiom systems. In any event, even if we agree that our intuition did once lead to contradictions, that fact does not justify the view that we run a high risk of self-contradiction whenever we use our intuition. The striking fact is that people do set theory by extensive appeals to their intuition and there is a practically universal agreement on the correctness or incorrectness of the results thus obtained, as results about sets. The iterative concept of set is an *intuitive* concept and *this* intuitive concept has led to no contradictions.

This is not to say that we have made the iterative concept fully exact and explicit: there remain problems about the indefiniteness of the concept of definite property and one-one correspondence, the range of ordinals we can envisage, the limitations of the axiomatic method. It is not even denied that, within this framework, there is room to experiment with new axioms and be open-minded as to the choice between alternatives. But the historical and conceptual matters sketched so far seem to discredit the bankruptcy view according to which, even today, the fundamental problem of the foundations of set theory remains the solution of the antinomies.

There is a related distinction between formalists and realists (or objectivists). As these positions get further refined, there is a certain convergence of views on matters regarding the correctness of results, even though there is a difference in choosing different problems to work on, for example, a preference by formalists for constructible sets and relative consistency results over speculations on and derivations from very large cardinals. Another difference is in the matter of working habits, so that one might be an avowed realist but think mainly formalistically, while an avowed formalist may use intuitions very efficiently in doing set theory and yet claim that set theory has only a formalistic model. In any case, any serious formalistic position does accept that we have perfectly reliable intuitions with regard to integers and some would claim intuitionistic reasoning as mostly evident. In other words, a formalist position on sets is given more content by contrasting set theory with other (usually more restricted) areas which do have more than a formal subject matter. One also thinks of degrees of reliability. The objectivistic position is a modification of realism with the goal of avoiding a number of extraneous difficulties with mathematical objects.

3. Objectivism and formalism in set theory

Different philosophical positions may be reached either by using the same data or by using different data. With the same data, the disagreement may often be apparent rather than real. More (relevant) data ought to be an advantage. It is not always easy to determine what is acceptable as data. For example, working informally with ZF is easier than with NF;[11] with ZF, while one pursues the argument without regard to formalization, the end results usually come out all right and can, if one wishes, be made into formal proofs from the axioms. This is at least in part due to our ability to think in terms of the intuitive models rather than the formal axioms. One might expect that the finite axiomatization of NF would yield fairly directly a contradiction, but the enumeration of all objects turns out to use an unstratified formula. Also, the ability to work with, e.g., a set which is a standard model of ZF, an assumption not formally provable in ZF, not only yields correct results but facilitates the flow of our arguments. Another point is the convergence of theories which at one stage were regarded as based on fundamentally different ideas.

[11]The system of Quine 1937: 70–80.

For this reason, it cannot be said that our belief in the superiority of one set theory is merely a result of sociological factors such as familiarity, conformity, and respect for authority. It seems to be unquestionable that we have come to accept axioms of extensionality, replacement, choice, and foundation. Perhaps more doubtful is the acceptance that the hypothesis of constructibility is false and that we suspend judgment on the truth or falsity of the hypothesis of measurable cardinals. Sometimes we accept an axiom once stated, sometimes it takes a fairly long time before an axiom is accepted (e.g. the axiom of choice), but at the end we reach an agreement. We do not have two camps of comparable force such that one accepts an axiom, the other rejects it. The agreement also persists in time, i.e. the community does not oscillate from one day to the next. There is also a pretty good agreement on when to suspend judgment. By empirical induction we expect similar agreement in the future, and similar persistence.

It is often not easy to give precise reasons why a certain axiom is accepted or rejected. And also, what is accepted need not always be reflected faithfully in a formal (statement of an) axiom. Hence, the possibility of refinement and modification is not excluded. And the search for more articulate explanations of these empirical facts is of philosophical interest. But it does not follow that unless clear reasons can be given, these surprising phenomena of agreement and coherence must be considered illusions. Here we tread on a thin line between passive acceptance of the fashion and capricious irreverence.

It appears that if mathematics deals with objects, then every mathematical proposition is true or false. There is a natural tendency to think of objects and models. On the other hand, we may wish to say that what is more basic is the successor or the membership relation and that they have certain properties. This does not confine us to any fixed formal systems. In the first place, the rule of induction, for example, is usually taken in an informal way or, in other words (what is really the same thing), taken as a second order statement. As is well known, we then have again the standard model. This is probably one way of upholding objectivism without relying on objects. In the second place, it is not implied that we know all the properties in advance. It is not excluded that we may in the process of studying the subject further come upon and accept new axioms. Perhaps this does leave room for the possibility that there is some yet undiscovered limitation which will show that, for example, the continuum hypothesis is undecidable in a certain stronger sense. The limiting case would be that there are certain absolutely undecidable propositions in set theory. But nobody knows how to work with the concept of absolute undecidability.

A very different position would be: since the continuum hypothesis is undecidable in *ZF, therefore,* the question of its truth loses meaning. In other words, axioms and theorems are true, but undecidable propositions can neither be true nor be false. Let us refer to this mixed position as *M.* A more radical and perhaps more consistent position says that it makes no sense to speak of propositions of set theory (or, according to another extreme viewpoint, any mathematical propositions) as having a truth value (or that they are not really propositions), axioms and hypotheses being in the same boat. This radical thesis depends either on a recommendation to use the word 'true' in a special way or on the contention that we have no reasonable intuitive concept of set at all. We shall not delay over it but confine our attention to the less radical mixed position *M.*

It is not easy to understand the position *M* in any coherent way. The axioms can only be true on account of an interpretation of the concepts involved. In order that the interpretation withhold judgment on undecidable propositions, the axioms would have to capture fully the 'interpretation.' This means, among other things, that we must not confine ourselves to interpretations in the ordinary sense of two-valued models because in such models every proposition is either true or false. Of course, with the usual axioms of number theory and set theory, we do believe that they do not capture completely our intended interpretations of the central concepts.

The historical origin of this curious position *M* is somewhat complicated. The desire to avoid occult qualities and operate with concrete material as much as possible leads to a delight in formal systems as syntactical objects. Perhaps a transfer of the dubious verifiability theory of meaning is made so that verifiability and falsifiability of propositions of set theory are identified with provability and refutability in a formal system. Apart from other difficulties with general verifiability theory, this viewpoint has its own problems: the limitations of formalization, and the unexplained source of the intrinsic meaningfulness of the axioms and theorems of a formal system.

A different line of defending this mixed position is to argue that propositions of set theory have no independent meaning but only derive their meaning from a superstructure, perhaps useful as a summary or for the economy of thought, which is not based on direct intuitions but on how efficiently one can get back ordinary mathematics from it. From this point of view, there is quite a bit of arbitrariness in our choice of formal systems of set theory, and if a system is adequate for ordinary mathematics, the meaning of the axioms is derived from its consequences in more meaningful areas as a sort of gift. Theorems of set theory which are derivable in the formal system and stay in the superstructure get in turn their meaning from the axioms. Hence, undecidable propositions of set theory cannot have meaning since the only possible source of meaning for them (viz. provability or refutability) is barred.

This viewpoint cannot account for the relatively stable iterative concept of set and makes the question of truth of the propositions of set theory quite thoroughly a relative matter, depending on which formal system one chooses to use. For example, the axiom of choice, the non-existence of a universal set or a complement set of every set, the existence of any subset of ω definable only by nonstratified formulas are each true in *ZF* but false in *NF*. Also, the uninhibited comparative study of different systems becomes somewhat of a mystery unless we have an intuitive set theory which we can use with no conscious regard toward formalization.

A favorite example against the pragmatic view that we accept an axiom because of its elegance (simplicity) and power (usefulness) is the constructibility hypothesis. It should be accepted according to the pragmatic view but is not generally accepted as true. Indeed, it is likely to be false according to the iterative concept of set. Basically, it is felt that the pragmatic view leaves out the criterion of intuitive plausibility. The constructibility hypothesis is not plausible in itself and, moreover, many of its consequences are not plausible. For example, it implies the existence of a definable well-ordering of the real numbers and fairly simple uncountable sets without perfect subsets; and these consequences are dubious – they have been said to be contrary to the intuition of ordinary mathematics. It implies a strange pattern of reduction theorems with regard to projective sets.[12] It is not a conceptually pure proposition because it allows ordinal numbers definable only by impredicative definitions or not definable at all, but proceeds to reject all further uses of impredicative definitions. The central argument is, perhaps, that by intention we view sets as arbitrary multiplicities regardless of how or if they can be defined. Hence, it is extremely unlikely that constructible sets, which are essentially the ordinal numbers only, give us all arbitrary sets.

Given this initial implausibility, one may be inclined to view with favor certain propositions which contradict the constructibility hypothesis. In particular, the existence of measurable cardinals is one such proposition, and it implies that there are only countably many constructible sets of integers. On account of the prior belief that the constructibility hypothesis is highly restrictive, this conclusion is seen as further evidence that it is false and as evidence that the measurable cardinal hypothesis has plausibility.

[12]Compare, e.g., D. A. Martin 1968: 687–9.

It has been suggested that possibly all sets are ordinal definable because we may have so many ordinal numbers that the collection turns out to be sufficiently rich. When this argument is applied to defend the constructibility hypothesis, we have a further difficulty in that higher ordinals give no more lower sets. For example, all constructible sets of integers are obtained at stage ω_1 and no large cardinals will change the situation. Here, one would perhaps wish to say that there are a lot of countable ordinals. If there are actually enough countable ordinals to make it true that all sets of integers are constructible, then, of course, the continuum hypothesis would also be true. On the other hand, there are familiar ways of foiling the constructibility axiom while retaining the continuum hypothesis. This is entirely in line with our belief, further substantiated by, though probably not completely dependent upon, the truth of the continuum hypothesis for constructible sets, that we are, relative to our present knowledge, more ready to deny the constructibility hypothesis than the continuum hypothesis.

There are different alternatives to the strong proposition that undecidability in *ZF* implies meaninglessness, or the related proposition that the *ZF* axioms constitute an 'implicit definition' of the concept of set. One alternative is to say that we can never know enough to conclude definitely whether the continuum hypothesis (or the constructibility hypothesis) is true or false. This position would permit our extension of *ZF* to include inaccessible and Mahlo numbers but exclude the possibility of finding clear axioms to decide the continuum hypothesis. A somewhat different alternative would be to say that at least ideas which we have today such as large cardinals cannot possibly lead to a decision on the continuum hypothesis. The position hardest to refute is perhaps that we have simply to withhold judgment: admittedly, as time goes on, we can discover new facts about sets, we can decide more propositions; but, for all we know, we may never be able to decide the continuum hypothesis. One feels uncomfortable if this is put forward as an empirical prediction. Otherwise we would like to see some general arguments. For example, taking into account the diverse possible ways in which languages can grow, we may feel that there are potentially uncountably many questions we can ask about sets. But, certainly we cannot answer uncountably many questions. Hence, why should the continuum hypothesis not be among the unanswerable ones? This can be answered by pointing out that even if we cannot answer all questions, we may be able to answer any question which is singled out as an object of special attention.

Since the continuum problem is to determine the number of sets of integers, it seems reasonable to expect that, barring surprising coincidence, we can only settle the question after we have determined what objects are to be numbered (what sets of integers are allowed) and on the basis of what one-one correspondences (compare Gödel 1964: 266) [478 in this volume]. But then we seem to be in a difficulty since thus far the determinations we can specify by precise axioms would tend to contradict the intended arbitrary character of sets and one-one correspondences. For example, this is the situation with the notion of constructible sets: we do not regard the continuum hypothesis as shown to be true because it follows from the constructibility hypothesis.

The general limitations of language and formal systems might also suggest that no plausible axiomatization of set theory is likely to be sufficiently refined to determine the exact size of the continuum. It may be the case, for example, that no plausible axioms of set theory will yield the continuum hypothesis, or determine a different specific cardinality for the continuum. But since we have no way of surveying all correct axioms of set theory, there is little likelihood that such a proposition will be established directly rather than approximated by scattered negative results, such as $2^{\aleph_0} \neq \aleph_1$, $2^{\aleph_0} \neq \aleph_4$, etc.

Reasons for believing the continuum hypothesis to be false have been put forward, and are regarded widely as unclear. If one believes the negation of the continuum hypothesis, then, of course, no formal system which includes only true axioms and is consistent with the continuum hypothesis can decide it. This had been used by Gödel as a reason for believing the continuum hypothesis undecidable in *ZF*, before P. J. Cohen established the fact.[13] According to Gödel (1964: 267) [479 in this volume], the continuum hypothesis has implausible consequences. For example, there are results which give uncountable sets which intuitively seem to contain very few members or are highly scattered (e.g. uncountable sets which are meager on every perfect set). But the continuum hypothesis implies that these sets are of the same size as the continuum. The uneasiness about such evidence is based on the feeling that most people do not have a well-developed intuition of large and small with regard to infinite sets apart from the actual development of set theory. On the other hand, it cannot be excluded that someone might have such intimate knowledge so that, for example, he can separate out the errors coming from using the preset-theoretical intuitive concept of largeness. With regard to the matter of intuition, Gödel notes a current fashion against the appeal to intuition and a consequent lack of practice in the conscious use of intuitions. He points out that intuition does not at all mean what first comes to mind but can and should be cultivated.

Some set theorist states that if $2^\omega = \omega_1$, then there must be a surprisingly

---

[13]See Gödel 1947 and Cohen 1966, 1963a: 1143-8, and 1964: 105-10.

delicate balance between the reals and the countable ordinals. But such a remark would be more forceful if it were used against $2^\omega = \omega_{17}$ say. As it is, one might say that, for all we know, 2^ω might be ω_2, or the first inaccessible number, or real-valued measurable, and that ω_1 is, for all we know, about as reasonable a candidate as any of these.

We do not argue for any strong sharp conclusions but rather try to apply what might be called the dialogue method to determine the limitations of one-sided views. For example, we are not able to establish in any clear sense the thesis that the set-theoretical concepts and theorems describe some well-determined reality, in which Cantor's conjecture must be either true or false. Yet the somewhat indeterminate meanings of the primitive terms of set theory as explained in the iterative notion are accepted as sound. According to Gödel (1964: 272) [484–5 in this volume]: 'The mere psychological fact of the existence of an intuition which is sufficiently clear to produce the axioms of set theory and an open series of extensions of them suffices to give meaning to the question of the truth or falsity of propositions like Cantor's continuum hypothesis.' By the phrase 'give meaning to the question,' Gödel means that there is a good chance of finding a unique answer to the question which will be accepted by all or most of those who are acquainted with the question.

The attraction of the dialogue method is perhaps due at least in part to the fact that what is most interesting in philosophy is not general conclusions but the meaning and limitation of these. Hence, we arrive at the content of these general statements by dialogues. For example, the fact that we have no inconsistency may be due to our limited range of activity relative to all formally possible proofs, and, in addition, our tendency to give proofs which can be interpreted in different frameworks.

It is more natural, certainly for most mathematicians, to deal with objects and models rather than formulas and formal systems. One might wish to claim this is just a shorthand way of doing things even though set theory is a formal game based on analogy and hasty generalizations. The central weakness of this position is of course its apparent inability to explain how the purely formal set theory can hang together so well.

It has been claimed that we have an informal consistency proof of set theory based on considerations about formulas. The point can be illustrated by thinking about the second order arithmetic. Let us try to find directly a countable model for the formal system. We do not have to worry about the fixed sets defined by conditions involving only integers. Now we consider the countably many formulas which contain variables over sets. For each statement $m \in x$ (i.e. $m \in \hat{m}\phi_n m$), we may attempt to try out the two possibilities of being true and being false, adding more set terms to satisfy the impredicatives in ϕ_n or $\neg\phi_n$. In this way, we would arrive at an intricate graph tree with countably many nodes. For each numeral m and each set term t, we have a formula $m \in t$ (a node) and two branches according as it is taken to be true or false. The truth and falsity of these countably many atomic formulas interact in a complicated way. The problem of consistency is to have a consistent selection of truth values for these atomic formulas, i.e. each formula gets a unique truth value so that all the defining conditions are satisfied.

Viewed in this way, we do not seem to have any good intuition that there must be such a model. In fact, we would find it very surprising if the combinatorial facts resulting from such a formalist outlook on the axioms come out right. In any case, it seems unreasonable to use such a picture with such apparently uncertain outcomes as a means of defending the formalist position.

Alternatively we may follow Gentzen and attempt to prove by transfinite induction that no proof can give a contradiction. In general, whether a formula $n \in t$ is true depends on whether certain other atomic formulas are true or false. If we could obtain an ordering of the degrees of impredicativities, we would be able to get the induction going. But the circular element in the impredicative definitions seems to suggest that we can only get such an ordering in some artificial way, perhaps by assuming what we wish to prove. In fact, it seems that this type of consideration, rather than increasing our belief in the formalist position, has the tendency of suggesting that the platonic picture is the only foothold, vague as it is, we can fall back on.

4. New axioms and criteria of acceptability

Consider first the conditions for accepting a hypothesis in set theory (axiom of choice, hypothesis of measurable cardinals, or some suitably restricted hypothesis of determinacy) as true. Two basic criteria are intrinsic necessity and pragmatic success. The former is related to but perhaps sharper than intuitive plausibility. The latter has various ramifications. One condition is to produce correct lower-order consequences, known (confirmation) and unknown (prediction), for example, about sets of reals, reals, and integers. Another condition is to supply powerful methods of solving problems and even methods which unify diverse results and go beyond them. It is also desirable that the hypothesis be easy to state and to understand. Briefly, we may speak of confirmation, prediction, power, unification (and therefore 'explanation'), and simplicity. The elements of power and unification also contain the component of elegance. Of course, these conditions are neither necessary nor sufficient, since the complicated notion of intrinsic necessity has to

dominate and since we may be willing to accept as true hypotheses satisfying only some of these conditions. Of course, there is nothing like a set
of quantitative measures by which we can calculate how well each hypothesis fares according to these criteria.

One might wish to view pragmatic success as merely an intermediate
criterion for screening candidates for new axioms and require that these
candidates eventually pass the test of intrinsic necessity. But there is then
the question how invariant the notion of intrinsic necessity is. The iterative concept is admittedly not perfectly clear. Some new axioms may be
seen as rendering more exact what we intend, while others may extend or
modify our notion in some natural way. If intrinsic necessity is the only
way to qualify a hypothesis as an axiom, then there would seem to be no
need of comparing set theory with physics. But if pragmatic success can
also make a proposition true, one might wonder whether the meaning of
the word 'true' is not stretched. To reply to this, we seem to be in a position to say that if pragmatic success is sufficient to make physical hypotheses true, why not also hypotheses in set theory?

There appears to be a sharp contrast between arbitrarily choosing to
call a hypothesis true and axioms being forced upon us. Looking backwards, it seems fair to say that axioms which we have accepted so far
were forced on us. Therefore, it seems reasonable to expect that we will
only accept a hypothesis as true in the future, if the evidence forces it
upon us. We do not have a clear idea how such forcing will take place.
Moreover, our axioms now all seem to be justified on intrinsic necessity
alone. This again suggests that we might choose to wait till existing and
future hypotheses achieve the state of intrinsic necessity relative to our
understanding before accepting any of them as axioms.

The same data can be interpreted in opposite ways. It has been claimed
that the axiom of choice was elevated to the status of an axiom only
because of repeated exposure and the psychological reluctance to tolerate
central undecidable propositions. One would then have to say that the
conceptual justification is no more than an ad hoc rationalization. In the
same vein, objects may be regarded as convenient metaphors for discussing formulas and properties.

In any case, it does not seem reasonable to call our choice 'arbitrary,'
since we feel we have good reasons for making certain choices and there
is a surprising degree of agreement among people who have thought
about alternative hypotheses. The two serious positions would seem to
be: (1) by accepting a new axiom, we change or extend our concept of set
(change the meaning of the word 'set'), and the meaning and truth of the
new axiom is determined by the changed concept; (2) we have all along
the same concept of set and we accept the new axiom because we have

discovered new facts about sets. There is a strong temptation to say that there is no genuine disagreement, because there is no sharp distinction between changing knowledge by changing meaning and changing knowledge by acquiring new information.

Nobody denies that our intuition of set develops. One should like to be open-minded and allow for the possibility of revoking or modifying our axioms if, for example, contradictions arise or the content of some axiom is rendered more precise by new findings. We cannot, however, disregard our experience so far which seems to indicate a measure of stability leaving little room for difficult choices. We may or may not be successful in explaining satisfactorily the apparently surprising degree of coherence and agreement in set theory, but it is important that we do not belittle this fact on the basis of preconceived philosophical ideas.

Intrinsic necessity depends on the concept of iterative model. In a general way, hypotheses which purport to enrich the content of power sets (say that of integers) or to introduce more ordinals conform to the intuitive model. We believe that the collection of all ordinals is very 'long' and each power set (of an infinite set) is very 'thick.' Hence, any axioms to such effects are in accordance with our intuitive concept. The difficulty is that in order to make specific assertions to increase the length or the thickness, we generally have to use propositions which imply other consequences as well. And then we can no longer justify all these propositions by appealing merely to our (maximum) iterative concept. In particular, there is no known positive principle that guides the search for new axioms to enrich power sets.

For example, consider the hypothesis that the cardinality of the continuum is real-valued measurable. This does deal with the power set of ω, and it decides the continuum hypothesis. But nobody is willing to take it as an axiom. It asserts a special relationship between cardinal numbers defined initially in different ways and is not the kind of proposition which one would be inclined to regard as directly justifiable on the basis of the intuitive iterative concept. Another example is the much more intensively studied axiom of determinacy (*AD*). This axiom *AD* does imply that there are a lot of real numbers (and comparatively few sets of real numbers). But, as it stands, *AD* embodies a generalization of DeMorgan's law with quantifiers (e.g. $\neg \forall x \exists y$ is equivalent to $\exists x \forall y \neg$) to the infinite case and asserts the existence of winning strategies for infinite games.

In its full generality, *AD* contradicts the axiom of choice. The attention is, therefore, mostly concentrated on restricted forms of *AD*. And according to the criterion of pragmatic success, the projective *AD* (say) performs very well indeed. It yields uniform and elegant proofs that all projective sets are Lebesgue measurable and have the Baire property, it yields pleasing new results on the reduction principles, it is easy to state and understand, etc. However, AD is not taken as an axiom in the sense that we can see directly from our intuitive concept that it or certain restricted forms of it are true. Rather it is generally viewed as an efficient hypothesis which yields elegant consequences and, in various restricted forms, may be derivable from more intuitive principles ('axioms of infinity' or large cardinal axioms) about the length of the collection of all ordinal numbers.

If we have somehow got hold of the 'real' power set of integers, CH should already enjoy a definite truth value even though we might not know what the value is. It is an empirical fact that we do not know how to enrich the power set directly by intuitively evident principles. Hence, the current search for new axioms (in particular, for the purpose of settling CH) centers around large cardinal axioms. Parallel to the imperfect information regarding the thickness of the power set of ω, our knowledge of the countable ordinals is also very incomplete. For example, the minimal α such that M_α is a model of ZF is a countable ordinal about which we have little to say without reference to the system ZF. Even if we look at ZF directly in an attempt to build a countable model, we are not clear whether and to what degree the circularity of assuming ZF to have a model can be avoided.

The fascination with axioms of infinity leads to the reaction: Why just this one jewel? This is undoubtedly connected with the impression that we can find axioms of infinity which mostly appear to be justifiable by an appeal to the inexact iterative concept of set. To begin with, it is commonly believed that the positive notion of continued iterations is sufficient to justify inaccessible numbers and Mahlo numbers. For example, the existence of (strong) inaccessible numbers means roughly just that the totality of sets obtainable by the procedures of set formation embodied in the axioms of ZF forms again a set. Hence, these same procedures are applicable to it to yield other new sets (Zermelo 1930: 29–47). Since the iterative concept permits unlimited extensions, the new axioms are seen to be introduced without arbitrariness. Moreover, each of these axioms, under the assumption of its consistency, can be shown to yield new number-theoretic theorems. Hence, they can be defended both on the ground of intrinsic necessity and, to some extent, on the ground of pragmatic success.

Another method of justifying axioms of infinity is by way of the reflection principles. The iterative concept implies that the universe of all sets is very large. When we have expressed certain properties of the universe, we can already find sets which have these properties. In other words, the reflection principle generalizes the relation of inaccessible numbers to the axioms of *ZF*. It says that, any time we try to capture the universe from what we positively possess (or can express), we fail the task and the characterization is satisfied by certain (large) sets. Such principles have been applied to justify (to derive) the existence of the inaccessible and the Mahlo numbers, as well as almost all axioms of *ZF*.[14] They have also been applied to justify larger cardinals. But, for example, reflection principles of diverse forms which are strong enough to justify measurable cardinals (by way of 1-extendible numbers) no longer appear to be clearly implied by the iterative concept of set.

There used to be a confused belief that axioms of infinity cannot refute the constructibility hypothesis (and therefore even less the continuum hypothesis) since *L* contains by definition all ordinals. For example, if there are measurable cardinals, they must be in *L*. However, in *L* they do not satisfy the condition of being measurable. This is no defect of these cardinals, unless one were of the opinion that *L* is the true universe. As is well known, all kinds of strange phenomena appear in nonstandard models. However, there does remain a feeling that the property of being a measurable cardinal says more than just largeness, although it implies largeness. It is often felt that the existence of measurable cardinals is more problematic than the existence of inaccessible numbers, even if we disregard the fact that the former is much stronger than the latter. In fact, there are different ways of introducing large cardinals. For example, sometimes we introduce large cardinals by singling out properties of ω in relation to the smaller ordinals and say that there exist cardinals greater than ω which have such properties.

Yet large cardinal hypotheses do occupy a preferred place among the candidates for new axioms about sets because in the majority of cases we expect to be able to show that they just make explicit that the iterative model contains ranks R_α for certain large α. Many of these hypotheses are linearly ordered in the sense that, for two hypotheses H_1 and H_2 we can either (1) derive H_1 from H_2 in *ZF* and find (by assuming H_2) a rank R_α which satisfies *ZF* plus H_1 but does not satisfy H_2, or (2) obtain the same results with H_1 and H_2 interchanged.

There are a number of different aspects of mathematics. In two senses, set theory is not sufficiently abstract to serve as foundations of mathematics. It might be said that we have real numbers as a basic datum, and it is less central how reasoning about real numbers is formalized. In another direction, mathematics is interested in abstract structures such as groups and fields which, though involving concepts like that of set, are independent of the detailed structures of our set theory.

[14]Compare Lévy 1960: 223–38; and Bernays 1961: 1–49.

The modern mathematical theory of categories suggests two rather distinct problems. One is whether the self-applicability of categories is essential so that mathematically interesting proofs would not go through under an interpretation of categories as sets or classes (perhaps of different levels or types). The other is whether such interpretation, even if successful in 'substance,' would not be too artificial as a codification of a type of natural mathematical practice.

5. Comparisons with geometry and physics

Set theory has been compared to geometry and to physics. There are different aspects with regard to which the comparison is made: objects of these disciplines (ontology), sources of our knowledge (epistemology), propositions (axioms or hypotheses) and their truth or acceptability (methodology).

The comparison between Euclid's fifth postulate and the continuum hypothesis is far fetched. Nobody proposed to call CH an axiom. There is a feeling that not only is the parallel postulate not evident, but the other postulates are also assumptions (together making up an implicit definition) of which we do not have sharp enough intuition to give justifications. The independence of CH, on the other hand, is not accompanied by doubts about the acceptability of the axioms of ZF. There is a sense of completeness of geometry with the parallel postulate or its alternatives either as first order theory or as second order theory (with completion coming from the different domain of sets). Hence, even if one takes the position that other postulates are evident or necessary, there is less reason to look for new axioms which would decide the parallel postulate.

Both the parallel postulate and its negation are extensions of 'absolute geometry' (determined by the remaining axioms of geometry) in the weak sense (in the sense of translatability or relative interpretability) (see Gödel 1964: 270–1) [483 in this volume]. This is equally true of CH at least relative to ZF. But axioms of infinity yield extensions in a stronger sense and there is an asymmetry between an axiom of infinity and its negation. Roughly speaking, an axiom of infinity is stronger and more fruitful than its negation. Epistemologically, there is of course also the difference that geometry is more directly connected to the physical world than set theory.

It seems clear that admitting space as a pure form of intuition need not commit us to the a priori character of Euclidean geometry. For example, we are willing to admit as a consequence of our form of intuition that all physical objects have spatial extension, but then it may be argued that

556

such a statement is analytic. The scepticism over the parallel postulate is often attributed to the difficulty of envisaging the infinite extension of the straight line. As is well known, there are various equivalent statements which do not mention the infinite extension and can be tested by experiments subject to inevitable inaccuracies in measuring continuous quantities. Two answers can be given to attempts to determine the truth of Euclidean geometry by empirical observations. One would be Poincaré's 'conventionalism.' The other would be to speak of a (local) 'space of intuition' which necessarily satisfies Euclidean geometry. It is true that this second reply is not refuted by experience. Only it is hard to give a convincing positive argument that we do have such exact intuitions, with or without the parallel postulate. Perhaps statements such as 'there are three noncollinear points' are evident and necessary.

It seems curious that while certain obvious things are proved in an elaborate manner, many other gaps are left wide open in Euclid. One explanation might be that the metaphorical definitions of points, lines, etc. are implicitly appealed to. On the other hand, there is a body of central theorems and proofs which are presumably proved quite exactly, and the foundations were accepted with a good deal of tolerance.

Another apparently puzzling feature is the relation of the fifth postulate to the rest. If one doubts the fifth postulate and it is shown to be independent, then the natural conclusion would seem to be that the fifth postulate is not a priori but the others are not affected. Instead, one began to question the necessity of all the postulates. This historical fact is perhaps a combination of two different factors. One is led to realize there is a difficulty in understanding the primitive concepts (or their definitions as originally given). Also, the existence of consistent alternatives shows that we do not have a 'complete' system that is necessary, and that, therefore, we probably do not have enough intuition to justify even parts of the system.

What is gained by comparing set theory with physics? One reason may be the suggestion of mathematical objects. In this comparison, set theory is quite different from arithmetic where, unlike evidence in physics, the general rules such as mathematical induction are perfectly evident.

It has been argued that just as physical objects are natural and necessary for organizing our physical experience, mathematical objects are natural and necessary for organizing our mathematical experience. Physical objects and not merely sensations are immediately given since they are not mere combinations of sensations and our thinking cannot create qualitatively new elements. Or perhaps what is given is not the physical objects but merely something different from sensations which generates the unity of one object out of the diversity of its many aspects. But then one is inclined to think of this something as contributed by the mind. The operation 'set of' is undoubtedly an instrument of synthesis. But there is always lurking somewhere the problem of infinity. To the extent the operation 'set of' suggests a synthesis, we seem to call up a picture of images which can possibly operate on infinite totalities and even permit infinite iterations. But the image of all subsets of a given infinite set (say ω) seems to involve an especially big jump. If somehow we have these subsets, we can apply the operation 'set of' to them. Yet to arrive at all subsets (say of ω), we seem to use something like an analogy.

Perhaps the fact that we can operate with such collections (e.g. in Cantor's diagonal argument) and even use them to prove theorems about natural numbers may be regarded as evidence. But such data are ambiguous and open to interpretations compatible with, e.g., predicative set theory. It seems reasonable to assert that not all data need be associated with certain things acting on our sense organs. But the only other possible source would seem to be either mind or a subtler form of objective reality which either is different from the physical world (e.g. remembering a platonic world) or is the same world but only acts on us in a different way (perhaps by an 'abstract perception').

It seems relatively easy to accept that axioms of set theory force themselves upon us as being true, or even that we feel we have an intuition that produces not only these axioms but an open series of extensions of them. It is perhaps reasonable to assume that all possible extensions will converge. But it seems a stronger assumption to say that the extensions will eventually yield in some sense a unique model so that the continuum hypothesis will be true or false in that model.

To say that mathematical objects exist objectively or even more that in some indeterminate sense we 'perceive' them seems to be stronger yet. Of course, if this is true, we are entitled to use the predicate calculus and reach the conclusion that every statement of set theory is meaningful. For our knowledge, we may still have the problem of recognizing whether a statement is true. This may be the reason why one could believe this strong position and yet not regard the criterion of pragmatic success as entirely superfluous.

The comparison with physics suggests that we look for evidence indirectly through consequences. While it is not necessary that we should be omniscient with regard to our intellectual creations, the difficulty in recognizing new axioms does appear more compatible with the view that mathematical objects exist independently of us. It would seem inevitable that applying this criterion would cost mathematical axioms much of their 'absolute certainty.' For it cannot be denied that success is a matter of degree and consequences (especially lower level consequences which are the more important for this criterion) do not at all determine the axioms in any unique manner. This does not necessarily obliterate the difference between physics and set theory (as with geometry), at least insofar as we do not at present envisage testing axioms of set theory by consequences in physics.

We may wish to compare the continuum hypothesis with a physical hypothesis which cannot be decided yet. But the analogy is certainly unclear since the latter is not only related to laws (axioms) of physics but depends on a good deal of empirical data. In fact, whatever undecidable propositions in physics may mean, they would seem to be of a radically different nature from those in formal mathematical systems.[15]

6. Digression on unbounded quantifications

The central problems of the iterative concept are: (1) the power set operation (i.e. what subsets are allowed); (2) ordinal numbers (i.e. what ordinals are allowed). Both involve an element of unlimited generality which cannot be rendered completely explicit. An explication of the concept of definite property in the principles of subset formation and replacement is relevant to approximations to this element of generality. In addition, there are principles for generating ordinals which depend on, besides iteration, also analogy and reflection.

Several people have questioned the legitimacy of using unbounded quantifiers (ranging over all sets) in defining new sets, even accepting the iterative interpretation of set theory. It has been suggested, among other things, that we are only justified in using all axioms of *ZF* if we combine it with the intuitionistic predicate calculus. This leads to an exact formal system and mathematical problems concerning the strength of such a system. The philosophical point at issue is, however, by no means equally clear.

If the totality of all sets is an unfinished totality, there is a problem in the use of unbounded quantifiers in the axiom of replacement and in the logical inferences. At any rate the logical notion of sentence is quite alien to Cantor's conception and there is a problem of comparing the sentential generating principle with the conceptual one (less correctly, the arithmetic one).

---

[15]Gödel points out that the hypothesis of measurable cardinals may imply more interesting (positive in some yet to be analyzed sense) universal number-theoretical statements beyond propositions such as the ordinary consistency statements such as, for instance, the equality of p_n (the n-th prime number) with some easily computable function. Such consequences can be rendered probable by verifying large numbers of numerical instances. Hence, the difference with the hypothesis of the expanding universe is not as great as we may think at first.

It does not seem unreasonable to regard at the same time ω as an unfinished totality and yet allow both quantifiers over all members of ω and the unrestricted law of excluded middle. Here it is clear that every member of ω is reachable from 0 by the successor function. Similarly, if we consider a theory of the second number class, we would be willing to do the same, even though we cannot explicitly give the operations for generating new countable ordinals. From this point of view, the current practice with the universe V of all sets seems entirely consistent.

It might be argued that we do look beyond ω_1 while V is a sort of absolute limit; ω_1 is a set but V is not a set. But the situation does not change, if we are only interested in countable ordinals and regard ω_1 as our universe. And the crucial point is, we do not anticipate incompatible alternative extensions of various approximations to ω_1 or V. Even though there are different ways of approximating to ω_1 and to V, we expect them to converge.

The genetic element is tied up with the distinction between viewing a class as one and as many. It is not excluded that we can talk about all sets and even form classes by conditions on all sets. Only in order that class be treated as one, we have to build it up from below. This imposes a requirement on its members (they must be 'given') but not on how one is to select sets from all given sets to form a new set. This is typically the situation with the replacement axiom: we make sure that all members of the new set are sets but choose them by arbitrary means, including the use of unbounded quantifiers. In particular, only a class as one can be a member of other classes or sets.

The concept of an unfinished or subjectively unfinishable totality is distinguishable from the idea that existence must agree with provable generation according to predetermined generating operations. Rather it permits the classical interpretation of quantifiers. We may also appeal to the reflection principles to argue that the unbounded quantifiers are not really unbounded. We are inclined to say that it is not problematic that we accept the classic logic with regard to propositions about all sets. There is no objection against viewing such propositions as determining definite properties.

If we adopt a constructive approach, then we do have a problem in allowing unlimited quantifiers to define other sets. Even then there remains the possibility of accepting the law of excluded middle. The difficulty is rather in establishing universal conclusions because we cannot survey all permissible operations.

Frege thinks of individuals, predicates, predicates of predicates, etc. He identifies sets with extensions of predicates and treats them as individuals (on the same level with individuals). This suggests immediately the idea of a type hierarchy of extensions, since extensions of predicates seem to be more closely related to predicates than to individuals. Alternatively, we may wish to think of all sets as objects (individuals) but distinguish them from extensions of predicates. In that case, we would be led to something like a finite type theory (perhaps ramified) based on the current set theory (a theory of individuals). But such a conception has little that is positive to say about how sets are to be generated.

7. Extracting axioms of set theory from Cantor's writings

As before, we discuss Cantor's views by reference to his collected works. It is relatively easy to be dissatisfied with Cantor's philosophical speculations on the transfinite. We can comfort ourselves that there is no need to take Cantor's flight into theology seriously, especially since we now possess more reasonable defenses of set theory.

One proof proceeds from the concept of God and concludes first from the highest perfection of God's essence the possibility of creating a transfinitum ordinatum. It then goes on to conclude from His divinity and glory the necessity of the actual successful creation of the transfinitum. Another proof shows a posteriori that the assumption of a transfinitum in natura naturata yields a better (because more complete) explanation than the opposite hypothesis, of the phenomena especially of the organisms and the psychical facts (1886, p. 400).

In one sense we may regard the integers as real insofar as they take up, by dint of definitions, a wholly determined place in our understanding, are well distinguished from all other ingredients of our thinking, but stand in definite relations to them and thereby modify the substance of our spirit in a definite way; I propose to call this kind of reality of the numbers their *intrasubjective* or *immanent reality*. The numbers can, however, also be ascribed reality insofar as they must be considered as an expression or picture of events and relations in the world outside our intellect, as further the different number classes I, II, III, etc. are representatives of cardinalities which actually appear in the physical and spiritual nature. I call this second kind of reality the *transsubjective* or *transient reality* of the integers...

This coherence of both kinds of reality has its proper root in the *unity* of *all, in which we ourselves participate*. The allusion to this coherence serves here the purpose of deriving a consequence for mathematics which seems very important to me, namely that in developing ideas in it we *only* have to account for the *immanent* reality of its concepts, and are not at all obliged to test their *transient* reality (1883, pp. 181–2).

This last special property of mathematics is for Cantor the ground for calling mathematics 'free.'

In the case of Cantor, what is more interesting for philosophy is perhaps not so much his metaphysical speculations as the conceptual framework revealed by his mathematical practice.

Cantor uses multiplicity or manifold (Vielheit) as a primitive concept which corresponds to class in current usage. According to him there are two kinds of (definite) multiplicity: the absolutely infinite or inconsistent multiplicities (proper classes in current usage) and the consistent ones which are called sets. He then states explicitly three axioms (pp. 443–4).

C1   Two equivalent classes are either both sets or both proper classes. (One-one replacement).[16]
C2   Every subclass of a set is a set. (Subset formation).
C3   For any set of sets, the elements of these sets again form a set; in other words, the union of a set of sets is a set. (Union).

There are undoubtedly axioms which appear too obvious (and pre-sumably too numerous) to Cantor to be stated explicitly. Such axioms can of course be added without violating Cantor's intention.

Like most mathematicians, Cantor uses implicitly the axiom of exten-sionality, for example, in establishing $P=Q$ for two point sets P and Q (but compare the introduction of $\equiv$ on p. 145). By the way, Dedekind mentions this axiom explicitly (1888, §2).

C4   If two classes A and B have the same elements, then $A=B$ (in par-ticular, two sets with the same extension are equal). (Extension-ality).

A more interesting case is the power set axiom. In at least two connec-tions, Cantor uses implicitly something like the power set axiom. In defining exponentiation of cardinals (pp. 287–9, 1895) Cantor considers the totality of functions from a set a into a set b and states explicitly that the totality is again a set. In particular, he shows that the cardinality of the linear continuum is $2^{\aleph_0}$. In presenting his diagonal argument to estab-lish $2^\lambda > \lambda$, for any cardinal λ, he considers for every set a (in particular, the set of all real numbers) with cardinality λ, the set b of all functions $f(x)$ which takes only 0 and 1 as values with x ranging over all members of a (in particular, all reals ≥ 0, ≤ 1) (pp. 279–80, 1890–1). Here, b is the power set of a.

C5   For every set a, all its subsets form a set. (Power set).

Of course, Cantor never doubts there are infinite sets. He freely uses the sets of all integers, all algebraic numbers, all reals, etc. (pp. 115, 126,

[16]Cantor does seem to apply implicitly the axiom of replacement to get ω_ω from $\omega, \omega_1, \dots$ in his 1895 paper (p. 296).

The concept of set

143, etc.). More specifically, he states explicitly (p. 293, 1895) that the totality of all finite cardinals forms a set.

C6 There is a set containing $0, 1, 2$, etc. (Infinity).

A basic indefiniteness in the above axioms is the notion of equivalence or one-one correspondence in C1: what language forms are permissible in specifying the one-one correspondence? The problem is illustrated by Zermelo's discussion (1908: §1.4, §1.6) of definite properties (which are presumably contrasted with things like poetic images and theological characterizations). One expects Cantor to accept a broad range of one-one correspondences so that, for example, any formula in the second order language with set and class variables would be permissible; or alternatively one might choose to exclude bound class variables.

As far as I know, Cantor does not discuss well-founded sets. But I believe he operates under the assumption that all sets are well-founded.

It is well known that Cantor does not consider the axiom of choice and often uses it implicitly. For example, on p. 293 (1895), he gives a 'proof' that every infinite set contains a subset of cardinality $\aleph_0$ but is not aware that an appeal to the axiom of choice is needed.

Cantor is much interested in the well-ordering theorem in the form that (1) the totality of all cardinal numbers can be well-ordered (p. 280, 1890–1; p. 295, 1895) or that (2) all infinite cardinal numbers are alephs (p. 447, 1899). It is perhaps worth remarking that on p. 280 and p. 295 he speaks of the totality of all cardinals as a 'well-ordered set' in quotation marks. Undoubtedly the later distinction between sets and classes is a way of removing these quotation marks. Actually, on p. 280 Cantor claims to have proved (1) in his paper of 1883 (pp. 165–208). It seems that this erroneous assertion is based on the implicit assumption of (2) as evident or proven. On p. 285 (1895), Cantor also asserts the comparability of any two cardinal numbers with a promise to return to a proof of this.

On p. 447, Cantor attempts to prove (2) by arguing that the cardinality of every set is an aleph. Suppose V is a class with no aleph as its cardinal number. Then, Cantor argues, the well-ordered class *On* of all ordinals is projectable into the class V and there must exist a subclass of V which is equivalent to *On*. Hence, by C1 and C2, V must be a proper class. The claim that labelling distinct members of V would use up all members of *On* appeals to some form of the axiom of choice.

From the above discussion, it would appear that what is currently called *ZF* or perhaps the second order theory of *ZF* would be a reasonable codification of Cantor's concept of set. This is under the assumption that we identify ordinal and cardinal numbers with sets in the now familiar manner and that one-one correspondences be specified more explicitly. Also, Zermelo's formulation of the axiom of choice and his use of it in proving the well-ordering theorem is a definite advance beyond Cantor.

It would seem a relatively simple matter to introduce the rank hierarchy and observe that every set has a rank. Cantor, however, seems to consider primarily only hierarchies of all numbers but not those of all sets. Mirimanoff seems to be the first to formulate explicitly the iterative concept with the rank hierarchy (1917a, b).

8. The hierarchies of Cantor and Mirimanoff

Cantor considers the well-ordered classes of all ordinals and all infinite cardinals (p. 444):

A $0, 1, 2, \ldots, \omega, \ldots$

B $\aleph_0, \aleph_1, \aleph_2, \ldots, \aleph_\omega, \ldots$

In the current treatment, each member of A or B is a set and, in general, $\aleph_\alpha$ is identified with ω_α ($\aleph_0 = \omega$). The consideration of the exponentiation of infinite cardinals (p. 288) and the cardinality of the power set of an infinite set (p. 280) also suggests another well-ordered class:

C $\aleph_0, 2^{\aleph_0}, 2^{2^{\aleph_0}}, \ldots$, ($C_0 = \aleph_0$, $C_{\alpha+1} = 2^{C_\alpha}$, $C_\lambda =$ the union of all C_β, $\beta > \lambda$, λ a limit number).

The generalized continuum hypothesis says in effect that each term of B is identical with the corresponding term of C. The continuum hypothesis asserts that $2^{\aleph_0} = \aleph_1$. It is well known that Cantor introduced the continuum hypothesis and spent much effort trying to prove it. As early as 1878, Cantor asked the question as to how many classes would result if we divide all infinite point sets into different classes according to their cardinalities, and stated (p. 132): 'By an inductive procedure, the presentation of which we do not enter upon here, one can prove the theorem that the number of classes of linear point sets yielded by this principle of partition is a finite one and, in fact, it is *two*.' And then in 1883, Cantor asked for the cardinality of the continuum and stated (p. 192): 'I hope to be able to answer soon with a strict proof that the cardinality sought is none other than that of our *second number class*.' This is followed by a footnote (number 10, see p. 207) stating in effect that $2^{\aleph_1} = \aleph_2$. At the end of his most important paper (published 1879–84), he promised again (p. 244, 1884) that the continuum hypothesis 'will be proved in later sections.'

Cantor's continuum problem is a sharp formulation of the simple and intrinsically interesting question: how many points are there on a line or how many sets of integers are there? It is remarkable that Cantor is able not only to give a stable extension of the concept of (cardinal) number to infinite sets without arbitrariness but also to give the well-ordered class of all alephs as a basis for comparison. The class of alephs uses the class of ordinal numbers and Cantor's concept of number classes to get the next aleph after each given single aleph or sequence of alephs indexed by an ordinal. That all infinite cardinals are alephs (a form of the well-ordering theorem) depends on the axiom of choice. Hence, Cantor's great interest in the well-ordering theorem is easily understandable, seeing that it is needed to give a definite shape to the collection of all cardinal numbers. The very fundamental character of the continuum problem as well as the impressive achievements of Cantor in arriving at a sharp formulation of the problem also explain both Cantor's organization of his ideas and his obsession with the continuum hypothesis. A solution of the continuum problem would have been the crowning event of his whole intellectual development.

Cantor had obtained the Burali-Forti paradox at least two years before Burali-Forti's publication in 1897 and communicated it to Hilbert in 1896 (p. 470). He evidently did not find the phenomenon shocking and his distinction between sets and inconsistent multiplicities does not strike one as an ad hoc device but rather like a sharpening of an incomplete intuition quite in the spirit of his general approach. But his description of the distinction is admittedly vague and all too brief. 'For a multiplicity can be such that the assumption of a "being together" of *all* of its elements leads to a contradiction, so that it is impossible to conceive of the multiplicity as a unity, as "one finished thing". Such multiplicities I call *absolutely infinite* or *inconsistent multiplicities*' (p. 443). These letters to Dedekind were not published until 1932.

In 1917, Mirimanoff published two papers[17] in which ideas similar to Cantor's are discussed in great detail and pursued further.

With regard to ideas familiar today, the concept of well-founded sets is introduced along with the rank function and employed to show (all in M1) that every well-founded set has a rank (the iterative model). Furthermore, the representation of ordinals commonly associated with the name of von Neumann is proposed in M1 and developed more extensively in M2. It is perhaps of some historical interest that von Neumann, in his paper of 1925 (219–40), did refer to M1 in connection with well-founded sets (1961–3, 1: 46n.) but apparently was not aware of the anticipation of his definition of ordinals.

[17]Mirimanoff 1917a: 37–52; 1917b: 207–17; 1920: 29–52. These will be referred to respectively as M1, M2, M3.

The fundamental problem of M1 is (p. 38): What are the necessary and sufficient conditions for a set of individuals to exist? The distinction between existent and nonexistent sets corresponds to that between sets and proper classes. For linguistic convenience, we shall translate Mirimanoff's terms into the familiar ones. As another example of different terminology, he speaks of 'ordinary sets' instead of 'well-founded classes (or sets).'

In M1, each class is associated with a membership tree that goes from the class to all of its members, and thence to all their members, etc. (p. 41). Each path in this tree is called a descent and a class is well founded if all the descents in its tree are finite (p. 42). The trees can only stop at indecomposable elements which are the noyaux (p. 43, nuclei, the urelements). In particular, the empty set is taken as an urelement and denoted by *e*. The well-founded sets are suggested by the paradox of sets which are not their own members. It is clear that the class of all well-founded sets is not a set (p. 43). This approach brings out a common feature of Russell's and Burali-Forti's paradoxes.

It is not assumed that we are only interested in well-founded sets, but rather a solution for the fundamental problem is only proposed for well-founded sets. Six axioms are stated in M1.

P1   Subset formation. Every subclass of a set is a set (p. 44. This is 'a property of sets which is far from being evident but which I regard as true, at least for the sets I consider in this work,' p. 43).

P2   A class equivalent to the class *On* of all ordinals is not a set (p. 45. By P1, any class containing a subclass equivalent to *On* is not a set).

P3   The urelements form a set which is regarded as given or known (p. 48).

P4   Power set. For every set *a* of well-founded sets, there is a set of all subsets of *a* (p. 49).

P5   Union. For every set *a* of well-founded sets, there is a set of all members of members of *a* (p. 49).

P6   One-one replacement. Given a set *a* and a one-one correspondence correlating each member of *a* with a well-founded set, there is a set of all the images of members of *a* (p. 49).

The concept of rank is introduced explicitly (p. 51): the rank of a well-founded set is the smallest ordinal number greater than the ranks of its members. The rank of an urelement (in particular, of *e*) is zero.

Two theorems are then proved (p. 51).

Th. 1 Every well-founded set has a rank. It is first observed that a set has a rank if all its members have ranks. This is proved by a lemma

566

to the effect that every set of ordinal numbers has a rank. Since ordinals under the Cantor conception are not sets, the von Neumann ordinals are introduced to represent them and P6 is applied to obtain the lemma (p. 50). If now a well-founded set x did not have a determinate rank, then there is at least one member x_1 of x having the same property; similarly x_1 has at least a member x_2 not having a determinate rank, and so forth – an absurd result, seeing that the whole sequence $x, x_1, x_2, \ldots$ stops at an urelement whose rank is zero.

This elegant proof makes an implicit use of the countable axiom of choice and is familiar nowadays.

Th. 2 For every α, the collection R_α of all well-founded sets of rank α forms a set.

First, if the theorem is true for all $\alpha < \pi$, then it is also true for π. Thus let Σ be the union of R_α for $\alpha < \pi$. By P6 and P5, this union is a set. But R_π is a collection of subsets of Σ and is therefore a set by P4 and P1. But the class of all ordinals is well-ordered and the class of ordinals for which Th. 2 is false, if not empty, must have a least number.

The two theorems together yield the result: there is a relation $S(\alpha, y) \equiv y = R_\alpha$, and for all well-founded sets x, $\exists \alpha \, \exists y \, (S(\alpha, y) \wedge x \in y)$. Let $H = \hat{x} (\exists \alpha)(S(\alpha, y) \wedge x \in y)$.

The solution for the fundamental problem thus obtained in M1 is the following (pp. 51–2): a collection of well-founded sets is a set if and only if the ranks of its members are bounded above by an ordinal number.

Clearly if we assume the axiom of foundation (that all sets are well founded), we can delete from the above discussion the qualification of well-foundedness, and obtain the result that the universe is the same as the union of all R_α (briefly $V = H$) and therewith reach the iterative concept of set described in §1.

Intuitively it seems very plausible that since we collect 'given' objects to form new sets, there should be enough ordinals to index the stages of this process of iterated expansion. The axiom of foundation sharpens the concept of iteration. Given the axiom, which says that every path in a membership tree leads back to an urelement in a finite number of steps, it becomes more plausible that we can, beginning with the set of all urelements, reach each set by steps indexed by ordinal numbers. The possibility of deriving $V = H$ from the axiom of foundation of course shows that no strong new axiom can contradict the iterative concept without refuting some of the basic axioms commonly accepted. In this respect, $V = H$ is very different from the much stronger 'axiom' of constructibility.

Of course, the basic axioms for getting larger sets are replacement and power set. By definition, power set only increases the rank by 1, and replacement has also this 'local' character because every set of ordinal numbers has an upper bound which is again an ordinal number. Replacement itself assures that the indices of the members of the image set again form a set. A justification of the axiom would be that proper classes are so large that a one-one correspondence never gets from a set to a proper class.

Once we have $V = H$, it seems reasonable to strengthen Cantor's axiom C1 to say also that (C1*) all proper classes are equivalent to V. Thus, given any class, either there is a bound on the ranks of all its members and then it is a set, or else the ranks are unbounded and then it forms a proper class C. If we use the members of C to count the members of V rank by rank, then we cannot stop at any R_α because we would then have a one-one correspondence between a proper class C and a set which is the union of $R_0, \ldots, R_\alpha$. To carry out this argument, we have to assume the global axiom of choice that the universe V can be well-ordered because the well-ordering of each R_α is not given explicitly by the local axiom of choice. Given $V = H$, this appears to be the natural generalization of the usual local axiom of choice by which each R_α can be well-ordered. The axiom C1* was first introduced by von Neumann (1925). The axiom of foundation and the rank model were rediscovered and treated more thoroughly by von Neumann (1925 and 1929) and also by Zermelo (1930). A particularly nice formulation of the axiom of foundation would seem to be: if every subset of a class X belongs to X, then $X = V$.

In M1, the ordinal numbers, $1, 2, 3$, etc. are represented by $\{e\}$, $\{e, \{e\}\}, \{e, \{e\}, \{e, \{e\}\}\}$, etc. These are obtained as follows. Let x be a well-ordered set, and let y be the set of all its segments, including the segment e. Replace these segments in y by the sets of the segments of these segments and apply an analogous transformation to the segments introduced in this manner, and so on (p. 45). A definition not appealing to given well-ordered sets is also given (p. 47): Definition of On: a set x represents an ordinal if: (1) x is a well-founded set based on the urelement e; (2) if y and z are two distinct elements of x, then $x \in y$ or $y \in x$ (connected); (3) $y \in x$, then $y \subseteq x$ (transitive).

If we adopt the axiom of foundation, this of course yields the currently employed definition of On which is usually regarded as an improvement over von Neumann's rediscovery of 1923 (199–208). Further properties of On are derived in M2. In the conclusion of M2, it is pointed out that On can be used in place of Cantor's ordinal numbers in dealing with well-ordered sets. 'I do not know whether this indirect method presents real advantages. In any case the classic theory of Cantor appears thus under a new aspect' (p. 217).

A totally new problem in the foundations of set theory is considered in M3, namely the nature of definite properties implicitly employed in stating the axioms of subset formation and replacement. The discussions are not nearly as conclusive as the extensional considerations in M1.[18]

Turning to the origins of Cantor's development of set theory we append some historical notes on a few anticipations and independent discoveries of some of Cantor's ideas in set theory. As is well known, Cantor began his mathematical career by works on the trigonometric series.[19] In trying to extend the uniqueness of representation in terms of these series to certain functions with infinitely many points of discontinuity, Cantor was faced with the problem of singling out suitable infinite sets of points on the line. This led to the notion of the derived set P' of a set P (viz. the set of limit points of P) which not only marked the beginning of Cantor's study of point set theory but paved the way for his construction of transfinite ordinal numbers later on.

In 1872, Cantor considered finite iterations of the operation of derived set and observed that there can be point sets P such that, for every n, $P^{(n)}$ is not empty and hence is infinite (p. 92). In 1880 (p. 145), Cantor introduced infinite iterations to ∞, $\infty + 1$, $n_0 \infty^m + \ldots + n_m$, ∞^∞, etc. In particular, $P^{(\infty)}$ is, for example, identified with the intersection of P', $P^{(2)}$, etc. 'Here we see a dialectical generation of concepts, which always leads yet farther, and remains both free from every arbitrariness and necessary and logical in itself' (p. 148). This was followed in 1883 by the first extensive development of transfinite numbers. Cantor observed that further progress of his investigations would depend on an extension of the concept of number which nobody had attempted yet, and that without this extension of the concept of number:

It would be impossible for me to advance freely a single step in the theory of sets; in this circumstance a justification or, if necessary, an excuse, may be found for my introduction of apparently strange ideas in my considerations. These concern an extension or continuation of the sequence of integers into the infinite; however daring these may appear, I can nevertheless express not only the hope but the firm conviction that this extension will in time be regarded as thoroughly simple, proper, and natural. At the same time, I by no means conceal from myself the fact that with this enterprise I place myself in a certain opposition to widespread views on the mathematical infinite and to oft-defended opinions on the essence of number (p. 165).

In a paper of 1880 (pp. 115–28), P. du Bois-Reymond claimed priority

---

[18]For related historical matters, compare also part II of the survey of Skolem's work in logic (1970: 35–40).

[19]A long discussion of the relation of Cantor's work in this area to his predecessors' is contained in a series of papers by Jourdain (1906–13).

on the concept of derived sets of infinite order. He considered a set D of intervals so distributed on the straight line that every interval contains some member of D as a part. Then he asserted: 'We are led to this kind of distribution of intervals, of which I have several examples, when we seek for points of accumulation of order ∞, whose existence I indicated to Mr. Cantor of Halle by letter years ago.' He also mentioned that he had introduced his notion of 'pentachic' before Cantor did his equivalent notion of 'everywhere dense.'

A more interesting anticipation was du Bois-Reymond's use of the diagonal method in his theory of growth[20] nearly twenty years before Cantor published in 1892 his famous diagonal proof of the theorem that every set has more subsets than elements. Consider increasing functions of one real variable x, for $x>0$. For two such functions $f(x)$ and $g(x)$, du Bois-Reymond stipulated that $f<g$ if $f(x)/g(x)$ tends to 0 as x increases indefinitely. The following theorem was proved. Let $f_1, f_2, \ldots$ be any sequence of increasing functions such that $f_1<f_2<f_3,\ldots$, then there exists an increasing function f such that $f_n<f$, for all n. He defined a new sequence $g_1, g_2, \ldots$ such that $g_1=f_1$, and $g_{n+1}(x)>g_n(x)$, for all x. Thus, by hypothesis, there exists x_1 such that $f_2(x)>g_1(x)$, for $x>x_1$. Let $g_2(x)=f_2(x)$, for $x>x_1$; $g_2(x)=g_1(x)+1$, for $x\leqslant x_1$. Similarly, let $g_{n+1}(x)=f_{n+1}(x)$, for $x>x_n$; $g_{n+1}(x)=g_n(x)+1$, for $x\leqslant x_n$. The desired function f can then be defined: for every n, $f(n)=g_n(n)$ and $f(x)=g_{n+1}(x)$; for $n<x<n+1$. This theorem is analogous to Cantor's theorem that every fundamental sequence of ordinal numbers defines a greater ordinal. An analog of the immediate successor would be to go from f to g such that $g(x)=xf(x)$.

[20]The relevant papers by P. du Bois-Reymond are: 1869: 10–45 (especially §7); 1873: 61–91 (especially the appendix); 1875: 363–414, and 1877: 149–67.

Bibliography

Ackermann, W. 1940. "Zur Widerspruchsfreiheit der reinen Zahlentheorie."
Mathematische Annalen, vol. 117.

Alston, W. P. 1958. "Ontological Commitments." *Philosophical Studies*, vol. 9.
Reprinted in Benacerraf and Putnam 1964.

Ambrose, A. 1935. "Finitism in Mathematics," parts 1 and 2. *Mind*, n.s. vol. 44.
  1955. "Wittgenstein on Some Questions in the Foundations of Mathematics."
Journal of Philosophy, vol. 52.
  1956. "On Entailment and Logical Necessity." *Proceedings of the Aristotelian
Society* (1955-6), n.s. vol. 56.

Anderson, A. R. 1958. "Mathematics and the 'Language Game.'" *Review of
Metaphysics*, vol. 11. Reprinted in Benacerraf and Putnam 1964.

Anscombe, G. E. M. 1959. *An Introduction to Wittgenstein's Tractatus*. London: Hutchinson.

Ayer, A. J. 1936a. "The *A Priori*." In Ayer 1936b. Excerpted in Benacerraf and
Putnam 1964 and reprinted in this volume.
  1936b. *Language, Truth and Logic*. London: Gollancz; New York: Oxford
University Press; 2nd ed., Gollancz, 1946; reprinted, New York: Dover,
1946.
  1951. "Symposium: On What There Is II." In *Freedom, Language, and
Reality*. Aristotelian Society Supplementary Volume 25.
  1959. Ed. *Logical Positivism*. New York: Free Press.

Bachmann, H. 1955. "Transfinite Zahlen." *Ergebnisse der Mathematik und ihrer
Grenzgebiete*, n.s. vol. 1.

Barwise, J. 1971. "Infinitary Methods in the Model Theory of Set Theory." In
*Logic Colloquium '69. Proceedings of the Summer School and Colloquium
in Mathematical Logic* (Manchester, England, 1969). Edited by R. O. Gandy
and C. M. E. Yates. *Studies in Logic and the Foundations of Mathematics*,
vol. 61. Amsterdam: North-Holland.
  1975. *Admissible Sets and Structures*. New York: Springer.
  1977. Ed. *Handbook of Mathematical Logic. Studies in Logic and the Foundations of Mathematics*, vol. 90. Amsterdam and New York: North-Holland.

Baumann, J. 1868-9. *Die Lehren von Raum, Zeit und Mathematik in der neueren
Philosophie nach ihrem ganzen Einfluss dargestellt und beurtheilt*, 2 vols.
Berlin: Reimer.

Becker, O. 1954. *Grundlagen der Mathematik in geschichtlicher Entwicklung*.
Freiburg: Alber.

Behmann, H. 1934. "Sind die mathematischen Urteile analytisch oder synthetisch?" *Erkenntnis*, vol. 4.

Benacerraf, P. 1965. "What Numbers Could Not Be." *Philosophical Review*, vol. 74. Reprinted in this volume.

1967. "God, the Devil, and Gödel." *The Monist*, vol. 51.

1973. "Mathematical Truth." *Journal of Philosophy*, vol. 70. Reprinted in this volume.

Benacerraf, P., and Putnam, H., eds. 1964. *Philosophy of Mathematics: Selected Readings*. Englewood Cliffs, N.J.: Prentice-Hall.

Bennett, J. 1959. "Analytic-Synthetic." *Proceedings of the Aristotelian Society* (1958-9), n.s. vol. 59.

1961a. "A Myth about Logical Necessity." *Analysis*, vol. 21.

1961b. "Symposium: On Being Forced to a Conclusion I." *Aristotelian Society Supplementary Volume 35*.

Bernays, P. 1935. "Sur le platonisme dans les mathématiques." *L'Enseignement mathématique*, 1st ser. vol. 34. Translated as "On Platonism in Mathematics" by C. D. Parsons in Benacerraf and Putnam 1964 and reprinted in this volume.

1937–54. "A System of Axiomatic Set Theory," parts 1-7. *Journal of Symbolic Logic*, vols. 2, 6, 7, 8, 13, and 19.

1941. "Sur les questions méthodologiques actuelles de la théorie hilbertienne de la démonstration." *Les Entretiens de Zurich sur les fondements et la méthode des sciences mathématiques* (1938). Edited by F. Gonseth. Zürich: Leemann.

1957. "Von der Syntax der Sprache zur Philosophie der Wissenschaften." *Dialectica*, vol. 11.

1958. *Axiomatic Set Theory*. With a historical introduction by Abraham A. Fraenkel. Amsterdam: North-Holland.

1959. "Comments on Ludwig Wittgenstein's *Remarks on the Foundations of Mathematics*" (Wittgenstein 1956). *Ratio*, vol. 2. Reprinted in Benacerraf and Putnam 1964.

1961. "Zur Frage der Unendlichkeitsschemata in der axiomatischen Mengenlehre." In *Essays on the Foundations of Mathematics, Dedicated to A. A. Fraenkel on His Seventieth Anniversary*. Edited by Y. Bar-Hillel, E. I. J. Poznanski, M. O. Rabin, and A. Robinson. Jerusalem: Magnes Press, Hebrew University.

Beth, E. W. 1956. "Semantic Construction of Intuitionistic Logic." *Mededelingen der Koninklijke Nederlandse Akademie van Wetenschappen, Afdeeling Letterkunde*, n.s. vol. 19.

1957. *La Crise de la raison et la logique*. Paris: Gauthier-Villars.

1959. *The Foundations of Mathematics*. Amsterdam: North-Holland.

Birkhoff, G., and MacLane, S. 1941. *A Survey of Modern Algebra*. New York: Macmillan.

Bishop, E. 1967. *Foundations of Constructive Analysis*. New York: McGraw-Hill.

Bibliography

Black, M. 1933. *The Nature of Mathematics: A Critical Survey*. London: K. Paul, Trench, Trubner; New York: Harcourt, Brace; reprinted, London: Routledge and Kegan Paul, 1950.

——— 1948. "The Semantic Definition of Truth." *Analysis*, vol. 8.

Blumenthal, L. 1940. "A Paradox, a Paradox, a Most Ingenious Paradox." *American Mathematical Monthly*, vol. 47.

Bolzano, B. 1851. *Paradoxien des Unendlichen*. Leipzig: Reclam. Translated as Bolzano 1950.

——— 1950. *Paradoxes of the Infinite*. A translation of Bolzano 1851 by F. Příhonský, with a historical introduction by D. A. Steele. London: Routledge and Kegan Paul; New Haven: Yale University Press.

Boole, G. 1847. *The Mathematical Analysis of Logic, Being an Essay Toward a Calculus of Deductive Reasoning*. Cambridge, England: Macmillan, Barclay, and Macmillan; reprinted, Oxford: Blackwell; New York: Philosophical Library, 1948.

——— 1854. *An Investigation of the Laws of Thought, on Which Are Founded the Mathematical Theories of Logic and Probabilities*. London: Walton and Maberly; reprinted, New York: Dover, 1951.

Boolos, G. 1971. "The Iterative Conception of Set." *Journal of Philosophy*, vol. 68. Reprinted in this volume.

——— 1975. "On Second-Order Logic." *Journal of Philosophy*, vol. 72.

Borel, E. 1912. "La Philosophie mathématique et l'infini." *Revue du mois*, vol. 14.

Braun, S., and Sierpiński, W. 1932. "Sur quelques propositions équivalentes à l'hypothèse du continu." *Fundamenta Mathematicae*, vol. 19.

Bridgman, P. W. 1934. "A Physicist's Second Reaction to Mengenlehre." *Scripta Mathematica*, vol. 2.

Brouwer, L. E. J. 1907. *Over de grondslagen der wiskunde*. Doctoral thesis, Municipal University of Amsterdam, 1907. Amsterdam and Leipzig: Mass and van Suchtelen; Groningen: Noordhoff.

——— 1908. "Die onbetrouwbaarheid der logische principes." *Tijdschrift voor wijsbegeerte*, vol. 2. Reprinted in Brouwer 1919.

——— 1909. "Die möglichen Mächtigkeiten." In *Atti del IV Congresso Internazionale dei Matematici* (Rome, 1908), 3 vols. Edited by G. Castelnuovo. Rome: Salviucci.

——— 1912. *Intuitionisme en Formalisme*. Groningen: Noordhoff. Reprinted in Brouwer 1919. Translated as Brouwer 1913.

——— 1913. "Intuitionism and Formalism." A translation of Brouwer 1912 by A. Dresden. *Bulletin of the American Mathematical Society*, vol. 20. Reprinted in Benacerraf and Putnam 1964 and in this volume.

——— 1919. *Wiskunde, waarheid, werkelijkheid*. Groningen: Noordhoff.

——— 1925-7. "Zur Begründung der intuitionistischen Mathematik," parts 1-3. *Mathematische Annalen*, vols. 93, 95, 96.

——— 1927. "Über Definitionsbereiche von Funktionen." *Mathematische Annalen*, vol. 97. Translated as "On the Domains of Definition of Functions" by S. Bauer-Mengelberg in van Heijenoort 1967.

573

Bibliography

1929. "Mathematik, Wissenschaft, und Sprache." *Monatshefte für Mathematik und Physik*, vol. 36.

1949. "Consciousness, Philosophy, and Mathematics." In *Proceedings of the Tenth International Congress of Philosophy* (Amsterdam, 1948), vol. 1, pt. 2. Edited by E. W. Beth, H. J. Pos, and J. H. A. Hollak. Amsterdam: North-Holland. Reprinted in Benacerraf and Putnam 1964 and in this volume.

1952. "Historical Background, Principles, and Methods of Intuitionism." *South African Journal of Science*, vol. 49.

Burali-Forti, C. 1897. "Una questione su i numeri transfiniti." *Rendiconti del Circolo Matematico di Palermo*, vol. 11. Translated as "A Question on Transfinite Numbers" by J. van Heijenoort in van Heijenoort 1967.

Butts, R. E., and Hintikka, J., eds. 1977. *Logic, Foundations of Mathematics, and Computability Theory. Proceedings of the Fifth International Congress of Logic, Methodology and Philosophy of Science* (London, Ontario, 1975), pt. 1. Dordrecht: Reidel.

Cantor, G. 1883a. "Über unendliche, lineare Punktmannichfaltigkeiten," part 5. *Mathematische Annalen*, vol. 21. Reprinted as Cantor 1883b.

1883b. *Grundlagen einer allgemeinen Mannichfaltigkeitslehre. Ein mathematischphilosophischer Versuch in der Lehre des Unendlichen.* Leipzig: Teubner. Reprinted in Cantor 1932.

1895-7. "Beiträge zur Begründung der transfiniten Mengenlehre," parts 1 and 2. *Mathematische Annalen*, vols. 46 and 49. Reprinted in Cantor 1932. Translated as Cantor 1915.

1915. *Contributions to the Founding of the Theory of Transfinite Numbers.* A translation of Cantor 1895-7 by P. E. B. Jourdain. Chicago and London: Open Court; reprinted, New York: Dover, 1952.

1932. *Gesammelte Abhandlungen mathematischen und philosophischen Inhalts.* Edited by E. Zermelo. Berlin: Springer; reprinted, Hildesheim: Olms, 1962.

Carnap, R. 1928a. *Der logische Aufbau der Welt.* Berlin: Weltkreis.

1928b. *Scheinprobleme in der Philosophie; das Fremdpsychische und der Realismusstreit.* Berlin: Weltkreis.

1930a. "Die alte und die neue Logik." *Erkenntnis*, vol. 1. Translated as "The Old and the New Logic" by I. Levi in Ayer 1959.

1930b. "Die Mathematik als Zweig der Logik." *Blätter für deutsche Philosophie*, vol. 4.

1931. "Die logizistische Grundlegung der Mathematik." *Erkenntnis*, vol. 2. Translated as "The Logicist Foundations of Mathematics" by E. Putnam and G. J. Massey in Benacerraf and Putnam 1964 and reprinted in this volume.

1934a. "Die Antinomien und die Unvollständigkeit der Mathematik." *Monatshefte für Mathematik und Physik*, vol. 41.

1934b. *Logische Syntax der Sprache.* Vienna: Springer. Translated as Carnap 1937.

1935a. "Formalwissenschaft und Realwissenschaft." *Erkenntnis*, vol. 5. Translated as "Formal and Factual Science" in Feigl and Brodbeck 1953.

Bibliography

1935b. "Ein Gültigkeitskriterium für die Sätze der klassischen Mathematik."
Monatshefte für Mathematik und Physik, vol. 42.

1935c. *Philosophy and Logical Syntax*. London: K. Paul, Trench, Trubner;
reprinted, New York: AMS Press, 1979.

1937. *The Logical Syntax of Language*. A translation of Carnap 1934b by
A. Smeaton. London: K. Paul, Trench, Trubner; New York: Harcourt,
Brace; reprinted several times.

1939. *Foundations of Logic and Mathematics. International Encyclopedia of
Unified Science*, vol. 1. Chicago: University of Chicago Press.

1942. *Introduction to Semantics*. Cambridge: Harvard University Press.

1943. *Formalization of Logic*. Cambridge: Harvard University Press.

1947. *Meaning and Necessity*. Chicago: University of Chicago Press; 2nd ed.,
enlarged, 1956. Sections 2 and 3 are reprinted in Feigl, Sellars, and Lehrer
1972.

1950. "Empiricism, Semantics, and Ontology." *Revue internationale de philosophie*, vol. 4. Reprinted in Carnap 1947 (2nd ed.), Linsky 1952, and Feigl,
Sellars, and Lehrer 1972. The slightly modified version in *Meaning and
Necessity* was published in Benacerraf and Putnam 1964 and is reprinted in
this volume.

1952. "Meaning Postulates." *Philosophical Studies*, vol. 3. Reprinted in
Carnap 1947 (2nd ed.) and Feigl, Sellars, and Lehrer 1972.

Carroll, L. See under Dodgson, C. L.

Castañeda, H.-N. 1959. "Arithmetic and Reality." *Australasian Journal of
Philosophy*, vol. 37. Reprinted in Benacerraf and Putnam 1964.

1961. "On Mathematical Proofs and Meaning." *Mind*, n.s. vol. 70.

Chihara, C. 1961. "Wittgenstein and Logical Compulsion." *Analysis*, vol. 21.

1963. "Mathematical Discovery and Concept Formation." *Philosophical
Review*, vol. 72.

1973. *Ontology and the Vicious-Circle Principle*. Ithaca, N.Y.: Cornell University Press.

Church, A. 1932. "A Set of Postulates for the Foundation of Logic," part 1.
Annals of Mathematics, 2nd ser. vol. 33.

1933. "A Set of Postulates for the Foundation of Logic," part 2. *Annals of
Mathematics*, 2nd ser. vol. 34.

1934. "The Richard Paradox." *American Mathematical Monthly*, vol. 41.

1936-8. "A Bibliography of Symbolic Logic," parts 1 and 2. *Journal of Symbolic Logic*, vols. 1 and 3.

1944. *Introduction to Mathematical Logic, Part I. Annals of Mathematics
Studies*, no. 13. Princeton: Princeton University Press; London: Oxford
University Press, Milford. Revised and enlarged, *Princeton Mathematical
Series*, no. 17 (1956).

1951. "Structure: A Formulation of the Logic of Sense and Denotation." In
Structure, Method, and Meaning: Essays in Honor of Henry M. Sheffer.
Edited by P. Henle, H. Kallen, and S. Langer. New York: Liberal Arts
Press.

1956. "Propositions and Sentences." In I. M. Bocheński, A. Church, and N. Goodman, *The Problem of Universals: A Symposium* (Aquinas Symposium, University of Notre Dame, 1956). Notre Dame, Ind.: University of Notre Dame Press.

1958. "Ontological Commitment." *Journal of Philosophy*, vol. 55.

1974. "Set Theory with a Universal Set." In *Proceedings of the Tarski Symposium* (University of California, Berkeley, 1971). *Proceedings of Symposia in Pure Mathematics*, vol. 25. Edited by L. Henkin, J. Addison, C. C. Chang, W. Craig, D. Scott, and R. Vaught. Providence, R.I.: American Mathematical Society.

Chwistek, L. 1933. "Die nominalistische Grundlegung der Mathematik." *Erkenntnis*, vol. 3.

Cohen, M. R., and Nagel, E. 1934. *An Introduction to Logic and Scientific Method.* New York: Harcourt.

Cohen, P. J. 1963a. "The Independence of the Continuum Hypothesis," part 1. *Proceedings of the National Academy of Sciences*, vol. 50.

1963b. "A Minimal Model for Set Theory." *Bulletin of the American Mathematical Society*, vol. 69.

1964. "The Independence of the Continuum Hypothesis," part 2. *Proceedings of the National Academy of Sciences*, vol. 51.

1966. *Set Theory and the Continuum Hypothesis.* New York: Benjamin.

Curry, H. B. 1951. *Outlines of a Formalist Philosophy of Mathematics.* Amsterdam: North-Holland.

1954. "Remarks on the Definition and Nature of Mathematics." *Dialectica*, vol. 8. Reprinted in Benacerraf and Putnam 1964 and in this volume.

1958. "Calculuses and Formal Systems." *Dialectica*, vol. 12.

Davis, M., ed. 1965. *The Undecidable: Basic Papers on Undecidable Propositions, Unsolvable Problems and Computable Functions.* Hewlett, N.Y.: Raven Press.

Dedekind, R. 1872. *Stetigkeit und irrationale Zahlen.* Brunswick: Vieweg. Reprinted in Dedekind 1930-2. Translated as "Continuity and Irrational Numbers" in Dedekind 1901.

1888. *Was sind und was sollen die Zahlen?* Brunswick: Vieweg. Reprinted in Dedekind 1920-2. Translated as "The Nature and Meaning of Numbers" in Dedekind 1901.

1901. *Essays on the Theory of Numbers.* Translations of Dedekind 1872 and 1888 by W. W. Beman. Chicago: Open Court; reprinted, New York: Dover, 1963.

1930-2. *Gesammelte mathematische Werke*, 3 vols. Edited by R. Fricke, E. Noether, and O. Ore. Brunswick: Vieweg.

Destouches, J.-L. 1951. "Sur la mécanique classique et l'intuitionisme." *Proceedings, Koninklijke Nederlandse Akademie van Wetenschappen*, ser. A vol 54, and *Indagationes Mathematicae*, vol. 13.

Dieudonné, J. 1951. "L'axiomatique dans les mathématiques modernes." *XVIIe Congrès international de philosophie des sciences: Philosophie Mathéma-*

tique/Mécanique (Paris, 1949). *Actualités scientifiques et industrielles*, no. 1137. Paris: Hermann.

Dodgson, C. L. [Carroll, L.]. 1895. "What the Tortoise Said to Achilles." *Mind*, n.s. vol. 4.

Drake, F. R. 1974. *Set Theory: An Introduction to Large Cardinals.* Amsterdam: North-Holland.

Dreben, B., Andrews, P., and Aanderaa, S. 1963. "False Lemmas in Herbrand." *Bulletin of the American Mathematical Society*, vol. 69.

Dubislav, W. 1925. "Über das Verhältnis der Logik zur Mathematik." *Annalen der Philosophie*, vol. 5.

——— 1930. "Über den sogennanten Gegenstand der Mathematik." *Erkenntnis*, vol. 1.

——— 1932. *Die Philosophie der Mathematik in der Gegenwart. Philosophische Forschungsberichte*, vol. 13. Berlin: Junker and Dünnhaupt.

Du Bois-Reymond, P. 1869. "Bemerkungen über die verschiedenen Werthe, welche eine Function zweier reellen Variabeln erhält, wenn man diese Variabeln entweder nach einander oder gewissen Beziehungen gemäss gleichzeitig verschwinden lässt." *Journal für die reine und angewandte Mathematik*, vol. 70.

——— 1873. "Eine neue Theorie der Convergenz und Divergenz von Reihen mit positiven Gliedern." *Journal für die reine und angewandte Mathematik*, vol. 76.

——— 1875. "Ueber asymptotische Werte, infinitäre Approximationen und infinitäre Auflösung von Gleichungen." *Mathematische Annalen*, vol. 8.

——— 1877. "Ueber die Paradoxen des Infinitärcalcüls." *Mathematische Annalen*, vol. 11.

——— 1880. "Der Beweis des Fundamentalsatzes der Integralrechnung: $\int_a^b F'(x)\,dx = F(b)-F(a)$." *Mathematische Annalen*, vol. 16.

Dummett, M. 1955-6. "Frege on Functions: A Reply" and "Note: Frege on Functions." *Philosophical Review*, vols. 64 and 65. Reprinted as "Frege on Functions" in Dummett 1978.

——— 1956. "Nominalism." *Philosophical Review*, vol. 65. Reprinted in Dummett 1978.

——— 1957. "Constructionalism." *Philosophical Review*, vol. 66. Reprinted in Dummett 1978.

——— 1959a. "Truth." *Proceedings of the Aristotelian Society* (1958-9), n.s. vol. 59. Reprinted in Dummett 1978.

——— 1959b. "Wittgenstein's Philosophy of Mathematics." *Philosophical Review*, vol. 68. Reprinted in Benacerraf and Putnam 1964 and Dummett 1978.

——— 1973. *Frege: Philosophy of Language.* London: Duckworth; New York: Harper and Row.

——— 1975. "The Philosophical Basis of Intuitionistic Logic." In *Logic Colloquium '73. Proceedings of the Logic Colloquium* (Bristol, 1973). *Studies of the Logic and the Foundations of Mathematics*, vol. 80. Edited by H. E. Rose and J. C. Shepherdson. Amsterdam: North-Holland; New York: American Elsevier. Reprinted in Dummett 1978 and in this volume.

Bibliography

1977. *Elements of Intuitionism.* Oxford: Clarendon Press.

1978. *Truth and Other Enigmas.* Cambridge: Harvard University Press.

Erdmann, B. 1877. *Die Axiome der Geometrie.* Leipzig: Voss.

Erdmann, J. E., ed. 1840. *God. Guil. Leibnitii Opera Philosophica.* Berlin: Eichler.

Errera, A. 1953. "Le Problème du continu." *Atti della Accademia Ligure di Scienze e Lettere* (Genova), vol. 9.

Evans, G. 1973. "The Causal Theory of Names I." *Aristotelian Society Supplementary Volume 47.* Reprinted in Schwartz 1977.

Feferman, S. 1960. "Arithmetization of Metamathematics in a General Setting." *Fundamenta Mathematicae,* vol. 49.

Feigl, H. 1950. "Existential Hypotheses." *Philosophy of Science,* vol. 17.

Feigl, H., and Brodbeck, M., eds. 1953. *Readings in the Philosophy of Science.* New York: Appleton-Century-Crofts.

Feigl, H., and Sellars, W., eds. 1949. *Readings in Philosophical Analysis.* New York: Appleton-Century-Crofts.

Feigl, H., Sellars, W., and Lehrer, K., eds. 1972. *New Readings in Philosophical Analysis.* New York: Appleton-Century-Crofts.

Field, H. H. 1972. "Tarski's Theory of Truth." *Journal of Philosophy,* vol. 69.

1980. *Science without Numbers.* Princeton: Princeton University Press.

Findlay, J. N. 1942. "Gödelian Sentences: A Non-numerical Approach." *Mind,* n.s. vol. 51.

Fraenkel, A. A. 1919. *Einleitung in die Mengenlehre.* Berlin: Springer; 2nd ed., 1923; 3rd ed., 1928; reprinted, New York: Dover, 1946.

1930. "Die heutigen Gegensätze in der Grundlegung der Mathematik." *Erkenntnis,* vol. 1.

1935. "Sur la notion d'existence dans les mathématiques." *L'Enseignement mathématique,* 1st ser. vol. 34.

1953. *Abstract Set Theory.* Amsterdam: North-Holland; 2nd rev. ed., 1961; 4th rev. ed. by A. Lévy, North-Holland and New York: American Elsevier, 1976.

Fraenkel, A. A., and Bar-Hillel, Y. 1958. *Foundations of Set Theory.* Amsterdam: North-Holland; 2nd rev. ed. by A. A. Fraenkel, Y. Bar-Hillel, and A. Lévy, *Studies in Logic and the Foundations of Mathematics,* vol. 67. Amsterdam: North-Holland, 1973.

Frege, G. 1879. *Begriffsschrift, eine der arithmetischen nachgebildete Formelsprache des reinen Denkens.* Halle: Nebert. Reprinted in Frege 1964b. Translated as "Begriffsschrift, a Formula Language, Modeled upon that of Arithmetic, for Pure Thought" by S. Bauer-Mengelberg in van Heijenoort 1967 and as Frege 1972. Chapter 1 is translated by P. T. Geach in Frege 1952.

1884. *Die Grundlagen der Arithmetik. Eine logisch-mathematische Untersuchung über den Begriff der Zahl.* Breslau: Koebner. Translated as Frege 1950; excerpted and translated as Frege 1964c.

1885. "Über formale Theorien der Arithmetik." *Sitzungsberichte der Jenaischen Gesellschaft für Medicin und Naturwissenschaft für das Jahr 1885.* Jena: Fischer. Reprinted in Frege 1967. Translated in Frege 1971.

Bibliography

1981. *Function und Begriff. Vortrag gehalten in der Sitzung vom 9. Januar 1891 der Jenaischen Gesellschaft für Medicin und Naturwissenschaft* (Jena). Reprinted in Frege 1962. Translated as "Function and Concept" by P. T. Geach in Frege 1952.

1892a. "Über Begriff und Gegenstand." *Vierteljahrsschrift für wissenschaftliche Philosophie*, vol. 16. Reprinted in Frege 1962 and 1967. Translated as "On Concept and Object" by P. T. Geach in Frege 1952 and reprinted in Feigl, Sellars, and Lehrer 1972.

1892b. "Über Sinn und Bedeutung." *Zeitschrift für Philosophie und philosophische Kritik*, n.s. vol. 100. Reprinted in Frege 1962. Translated as "On Sense and Nominatum" by H. Feigl in Feigl and Sellars 1949 and as "On Sense and Reference" by M. Black in Frege 1952.

1893–1903. *Grundgesetze der Arithmetik, begriffsschriftlich abgeleitet*, 2 vols. Jena: Pohle; reprinted, Hildesheim: Olms, 1962. Partial translations in Frege 1952 and 1964a.

1897. "Über die Begriffsschrift des Herrn Peano und meine eigene." *Berichte über die Verhandlungen der Königlich Sächsischen Gesellschaft der Wissenschaften zu Leipzig, Mathematische-Physische Klasse*, vol. 48. Translated as Frege 1969.

1903. "Über die Grundlagen der Geometrie." *Jahresberichte der Deutschen Mathematiker-Vereinigung*, vol. 12. Translated in Frege 1971.

1904. "Was ist eine Funktion?" *Festschrift Ludwig Boltzmann gewidmet zum sechzigsten Geburtstage 20 Februar 1904*. Leipzig: Barth. Translated as "What Is a Function?" by P. T. Geach in Frege 1952.

1919. "Der Gedanke – Die Verneinung: Zwei logische Untersuchungen." *Beiträge zur Philosophie des Deutschen Idealismus*, vol. 1. "Der Gedanke" is translated as Frege 1956. "Die Verneinung" is translated as "Negation" by P. T. Geach in Frege 1952. Both "Der Gedanke" ("The Thought") and "Die Verneinung" ("Negation") are translated in Frege 1977.

1950. *The Foundations of Arithmetic: A Logico-Mathematical Enquiry into the Concept of Numbers*. A translation, with German text, of Frege 1884 by J. L. Austin. Oxford: Blackwell; New York: Philosophical Library; 2nd rev. ed., 1953; reprinted, without German text, New York: Harper Torchbooks, 1960.

1952. *Translations from the Philosophical Writings of Gottlob Frege*. Edited and translated by P. T. Geach and M. Black. Oxford: Blackwell; 2nd ed., 1960. Includes Frege 1891, 1892a, 1892b, 1904, 1919, and selections from 1897 and 1893–1903.

1956. "The Thought: A Logical Inquiry." A translation of part 1 of Frege 1919 by A. Quinton and M. Quinton. *Mind*, n.s. vol. 65.

1962. *Funktion, Begriff, Bedeutung: Fünf logische Studien*. Edited by G. Patzig. Göttingen: Vandenhoeck and Ruprecht; 2nd rev. ed., 1965. Includes Frege 1891, 1892a, and 1892b.

1964a. *The Basic Laws of Arithmetic: Exposition of the System*. Translated and edited, with an introduction, by M. Furth. Berkeley and Los Angeles: University of California Press. Includes a partial translation of Frege 1893–1903.

Bibliography

1964b. *Begriffsschrift und andere Aufsätze.* Edited by I. Angelelli. Hildesheim: Olms. Includes Frege 1879.

1964c. "The Concept of Number." Excerpted from Frege 1884, and translated by M. S. Mahoney in Benacerraf and Putnam 1964 and reprinted in this volume.

1967. *Kleine Schriften.* Edited by I. Angelelli. Darmstadt and Hildesheim: Olms. Includes Frege 1885 and 1892a.

1969. "On Herr Peano's Begriffsschrift and My Own." A translation of Frege 1897 by V. H. Dudman. *Australasian Journal of Philosophy*, vol. 47.

1969–76. *Nachgelassene Schriften und Wissenschaftlicher Briefwechsel,* 2 vols. Vol. 1 edited by H. Hermes, F. Kambartel, and F. Kaulbach; vol. 2 edited by G. Gabriel, H. Hermes, K. Kambartel, C. Thiel, and A. Veraart. Hamburg: Meiner. Vol. 1 is translated as Frege 1979; vol. 2 is translated as Frege 1980.

1971. *On the Foundations of Geometry and Formal Theories of Arithmetic.* Translations of Frege 1885 and 1903 by E.-H. Kluge. London and New Haven: Yale University Press.

1972. *Conceptual Notation and Related Articles.* A translation of Frege 1879 by T. W. Bynum. Oxford: Clarendon Press.

1977. *Logical Investigations.* Translated by P. T. Geach and R. H. Stoothoff. Oxford: Blackwell; New York: Yale University Press. Includes a translation of Frege 1919.

1979. *Gottlob Frege: Posthumous Writings.* A translation of Fege 1969–76, vol. 1, by P. Long, R. White, and R. Hargreaves. London: Blackwell; Chicago: University of Chicago Press.

1980. *Gottlob Frege: Philosophical and Mathematical Correspondence.* A translation and abridgement of Frege 1969–76, vol. 2. Translated by H. Kaal; abridged by B. McGuinness. London: Blackwell; Chicago: University of Chicago Press.

Freudenthal, H. 1936. "Zur intuitionistischen Deutung logischer Formeln." *Compositio Mathematica*, vol. 4.

Friedman, H. 1973. "The Consistency of Classical Set Theory Relative to a Set Theory with Intuitionistic Logic." *Journal of Symbolic Logic*, vol. 38.

Furstenberg, H. 1977. "Ergodic Behavior of Diagonal Measures and a Theorem of Szemerédi on Arithmetic Progressions." *Journal d'analyse mathématique*, vol. 31.

Gallen, D. 1975. *Intensional and Higher-Order Modal Logic.* Amsterdam: North-Holland.

Gasking, D. 1940. "Mathematics and the World." *Australasian Journal of Psychology and Philosophy*, vol. 18. Reprinted in Benacerraf and Putnam 1964.

Geach, P. T. 1951a. "Frege's *Grundlagen.*" *Philosophical Review*, vol. 60.

1951b. "Symposium: On What There Is I." In *Freedom, Language, and Reality. Aristotelian Society Supplementary Volume 25.*

1956. "Discussion Note: On Frege's Way Out." *Mind*, n.s. vol. 65.

1959. "Russell on Meaning and Denoting." *Analysis*, vol. 19.

580

Bibliography

1961. "Frege." In G. E. M. Anscombe and P. T. Geach, *Three Philosophers.* Oxford: Blackwell; Ithaca, N.Y.: Cornell University Press.

Gentzen, G. 1938a. "Die gegenwärtige Lage in der mathematischen Grundlagenforschung." *Forschungen zur Logik und zur Grundlegung der exakten Wissenschaften*, n.s. vol. 4. Leipzig: Hirzel.

1938b. "Neue Fassung des Widerspruchsfreiheitbeweises für die reine Zahlentheorie." *Forschungen zur Logik und zur Grundlegung der exakten Wissenschaften*, n.s. vol. 4. Leipzig: Hirzel.

1969. *The Collected Papers of Gerhard Gentzen.* Edited by M. E. Szabo. Amsterdam: North-Holland.

Gergonne, J. D. 1819. "Essai sur la théorie des définitions." *Annales des mathématiques pures et appliquées*, vol. 9.

Gödel, K. 1931. "Über formal unentscheidbare Sätze der *Principia Mathematica* und verwandter Systeme I." *Monatshefte für Mathematik und Physik*, vol. 38. Translated as Gödel 1962 and as "On Formally Undecidable Propositions of *Principia Mathematica* and Related Systems" by J. van Heijenoort in van Heijenoort 1967.

1931-2a. "Eine Interpretation des intuitionistischen Aussagenkalküls." *Ergebnisse eines mathematischen Kolloquiums*, vol. 4.

1931-2b. "Zur intuitionistischen Arithmetik und Zahlentheorie." *Ergebnisse eines mathematischen Kolloquiums*, vol. 4.

1931-2c. "Zum intuitionistischen Aussagenkalkül." *Ergebnisse eines mathematischen Kolloquiums*, vol. 4.

1939. "Consistency-Proof for the Generalized Continuum-Hypothesis." *Proceedings of the National Academy of Sciences*, vol. 25.

1940. *The Consistency of the Axiom of Choice and of the Generalized Continuum-hypothesis with the Axioms of Set Theory. Annals of Mathematics Studies*, no. 3. Princeton: Princeton University Press; London: Oxford University Press, Milford; reprinted, Princeton University Press, 1951, 1953, and 1958.

1944. "Russell's Mathematical Logic." In *The Philosophy of Bertrand Russell.* Edited by P. A. Schilpp. Evanston and Chicago: Northwestern University; reprinted several times. Reprinted in Benacerraf and Putnam 1964 and in this volume.

1947. "What Is Cantor's Continuum Problem?" *American Mathematical Monthly*, vol. 54. Revised and expanded as Gödel 1964 in Benacerraf and Putnam 1964 and reprinted in this volume.

1958. "Über eine bisher noch nicht benützte Erweiterung des finiten Standpunktes." *Dialectica*, vol. 12.

1962. *On Formally Undecidable Propositions of* Principia Mathematica *and Related Systems.* A translation of Gödel 1931 by B. Meltzer, with an introduction by R. B. Braithwaite. Edinburgh: Oliver and Boyd.

1964. "What Is Cantor's Continuum Problem?" In Benacerraf and Putnam (1964) and reprinted in this volume.

Gogol, D. 1978. "The $\forall_n$ ∃-Completeness of Zermelo-Fraenkel Set Theory."

Zeitschrift für Mathematische Logik und Grundlagen der Mathematik, vol. 24.

Goldman, A. I. 1967. "A Causal Theory of Knowing." *Journal of Philosophy*, vol. 64.

Goodman, N. 1951. *The Structure of Appearance*. Cambridge: Harvard University Press.

1956. "A World of Individuals." In I. M. Bocheński, A. Church, and N. Goodman, *The Problem of Universals: A Symposium* (Aquinas Symposium, University of Notre Dame, 1956). Notre Dame, Ind.: University of Notre Dame Press. Reprinted in Benacerraf and Putnam 1964.

Goodman, N., and Quine, W. V. 1947. "Steps Toward a Constructive Nominalism." *Journal of Symbolic Logic*, vol. 12.

Goodstein, R. L. 1957. "Critical Notice: Wittgenstein's *Remarks on the Foundations of Mathematics.*" *Mind*, n.s. vol. 66.

1958. "On the Nature of Mathematical Systems." *Dialectica*, vol. 12.

1962. "Symposium: The Foundations of Mathematics I." *Aristotelian Society Supplementary Volume 36*.

Grelling, K. 1936. "The Logical Paradoxes." *Mind*, n.s. vol. 45.

Grice, H. P., and Strawson, P. F. 1956. "In Defense of a Dogma." *Philosophical Review*, vol. 65. Reprinted in Feigl, Sellars, and Lehrer 1972.

Griss, G. F. C. 1946a. *Idealistische Filosofie: een humanistische levens- en wereldbeschouwing*. Arnhem: Van Loghum Slaterus.

1946b. "Negationless Intuitionistic Mathematics." *Proceedings, Koninklijke Nederlandse Akademie van Wetenschappen*, vol. 49, and *Indagationes Mathematicae*, vol. 8.

Haack, S. 1974. *Deviant Logics: Some Philosophical Issues*. Cambridge University Press.

Hahn, H. 1933. *Logik, Mathematik und Naturerkennen. Einheitswissenschaft*, vol. 2. Vienna: Gerold. Translated as "Logic, Mathematics and Knowledge of Nature" by A. Pap in Ayer 1959.

Hajnal, A. 1956. "On a Consistency Theorem Connected with the Generalized Continuum Problem." *Zeitschrift für mathematische Logik und Grundlagen der Mathematik*, vol. 2.

Hanf, W. P., and Scott, D. 1961. "Classifying Inaccessible Cardinals." *Notices of the American Mathematical Society*, vol. 8.

Hardy, G. H. 1929. "Mathematical Proof." *Mind*, n.s. vol. 38.

Harman, G. H. 1973. *Thought*. Princeton: Princeton University Press.

Hart, W. P. 1970. "Skolem's Promises and Paradoxes." *Journal of Philosophy*, vol. 67.

1974. "On an Argument for Formalism." *Journal of Philosophy*, vol. 71.

Hausdorff, F. 1914. *Grundzüge der Mengenlehre*. Leipzig: Veit; 2nd rev. ed., Berlin and Leipzig: de Gruyter, 1927; 3rd ed., 1935. Translated as Hausdorff 1949.

1949. *Set Theory*. Translated from the 3rd German edition of Hausdorff 1914 by J. R. Auman. New York: Chelsea; reprinted, 1957, 1962.

Hempel, C. G. 1945a. "Geometry and Empirical Science." *American Mathematical Monthly*, vol. 52. Reprinted in Feigl and Sellars 1949.

1945b. "On the Nature of Mathematical Truth." *American Mathematical Monthly*, vol. 52. Reprinted in Feigl and Sellars 1949, Feigl and Brodbeck 1953, Benacerraf and Putnam 1964, and in this volume.

Hempel, C. G., and Oppenheim, P. 1936. *Der Typusbegriff im Lichte der neuen Logik*. Leiden: Sijthoff.

Henkin, L. 1949. "The Completeness of the First-Order Functional Calculus." *Journal of Symbolic Logic*, vol. 14. Reprinted in Hintikka 1969.

1950. "Completeness in the Theory of Types." *Journal of Symbolic Logic*, vol. 15. Reprinted in Hintikka 1969.

Herbrand, J. 1968. *Écrits logiques*. Edited by J. van Heijenoort. Paris: Presses Universitaires de France. Translated as Herbrand 1972.

1972. *Logical Writings*. A translation of Herbrand 1968. Edited by W. D. Goldfarb. Dordrecht: Reidel; Cambridge: Harvard University Press.

Heyting, A. 1931. "Die intuitionistische Grundlegung der Mathematik." *Erkenntnis*, vol. 2. Translated as "The Intuitionist Foundations of Mathematics" by E. Putnam and G. J. Massey in Benacerraf and Putnam 1964 and reprinted in this volume.

1934. *Mathematische Grundlagenforschung, Intuitionismus, Beweistheorie*. Berlin: Springer.

1948. "Formal Logic and Mathematics." *Synthèse*, vol. 6.

1956. *Intuitionism: an Introduction*. Amsterdam: North-Holland; 2nd rev. ed., 1966; 3rd rev. ed., 1971. Excerpted as "Disputation" in Benacerraf and Putnam 1964 and reprinted in this volume.

1958. "Blick von der intuitionistischen Warte." *Dialectica*, vol. 12.

1961. "Axiomatic Method and Intuitionism." In *Essays on the Foundations of Mathematics, Dedicated to A. A. Fraenkel on His Seventieth Anniversary*. Edited by Y. Bar-Hillel, E. I. J. Poznanski, M. O. Rabin, and A. Robinson. Jerusalem: Magnes Press, Hebrew University.

Hilbert, D. 1899a. "Grundlagen der Geometrie." *Festschrift zur Feier der Enthüllung des Gauss-Weber-Denkmals in Göttingen*. Leipzig: Teubner. Reprinted as Hilbert 1899b.

1899b. *Grundlagen der Geometrie*. Leipzig: Teubner; 2nd ed., 1903; 3rd rev. ed., 1909; 7th rev. and enlarged ed., 1930; 10th ed., with a supplement by P. Bernays, Stuttgart, 1968. Translated as Hilbert 1902 and 1971.

1902. *The Foundations of Geometry*. A translation of Hilbert 1899b by E. J. Townsend. Chicago: Open Court.

1905a. "Über die Grundlagen der Logik und der Arithmetik." *Verhandlungen des Dritten Internationalen Mathematiker-Kongresses in Heidelberg vom 8. bis 13. August 1904*. Lepizig: Teubner. Reprinted in Hilbert 1899b (3rd ed.). Translated as Hilbert 1905b and as "On the Foundations of Logic and Arithmetic" by B. Woodward in van Heijenoort 1967.

1905b. "On the Foundations of Logic and Arithmetic." A translation of Hilbert 1905a by G. B. Halsted. *Monist*, vol. 15.

1918. "Axiomatisches Denken." *Mathematische Annalen*, vol. 78. Reprinted in Hilbert 1932–5, vol. 3.

1922. "Neubegründung der Mathematik." *Abhandlungen aus dem mathematischen Seminar der Hamburgischen Universität*, vol. 1. Reprinted in Hilbert 1932–5, vol. 3.

1926. "Über das Unendliche." *Mathematische Annalen*, vol. 95. Translated as "On the Infinite" by E. Putnam and G. J. Massey in Benacerraf and Putnam 1964 and by S. Bauer-Mengelberg in van Heijenoort 1967. Reprinted in this volume.

1928. "Die Grundlagen der Mathematik." *Abhandlungen aus dem mathematischen Seminar der Hamburgischen Universität*, vol. 6. Reprinted in Hilbert 1899 (7th ed.) and in Hilbert, Weyl, and Bernays 1928. Translated as "The Foundations of Mathematics" by S. Bauer-Mengelberg and D. Føllesdal in van Heijenoort 1967.

1931. "Die Grundlegung der elementaren Zahlenlehre." *Mathematische Annalen*, vol. 104.

1932–5. *Gesammelte Abhandlungen*, 3 vols. Berlin: Springer; 2nd ed., 1970; 3rd ed., New York: Chelsea, 1981.

1971. *The Foundations of Geometry*. A translation of Hilbert 1899b (10th ed.) by L. Under. La Salle, Ill.: Open Court.

Hilbert, D., and Bernays, P. 1934–9. *Grundlagen der Mathematik*, 2 vols. Berlin: Springer; 2nd ed., 1968–70.

Hilbert, D., Weyl, H., and Bernays, P. 1928. *Die Grundlagen der Mathematik. Hamburger Mathematische Einzelschriften*, vol. 5. Leipzig: Teubner. Includes Hilbert 1928.

Hintikka, K. J. J. 1959. "Existential Presuppositions and Existential Commitments." *Journal of Philosophy*, vol. 56.

1969a. "On Kant's Concept of Intuition (*Anschauung*)." In *The First Critique: Reflections on Kant's Critique of Pure Reason*. Edited by T. Penelhum and J. J. MacIntosh. Belmont, Calif.: Wadsworth.

1969b. Ed. *The Philosophy of Mathematics*. Oxford: Oxford University Press.

1973. *Logic, Language-Games, and Information*. Oxford: Clarendon Press.

1974. *Knowledge and the Known: Historical Perspectives in Epistemology*. Dordrecht: Reidel.

Hook, S., ed. 1956. *American Philosophers at Work*. New York: Criterion.

Huntington, E. V. 1913. "A Set of Postulates for Abstract Geometry." *Mathematische Annalen*, vol. 73.

Hurewicz, W. 1932. "Une remarque sur l'hypothèse du continu." *Fundamenta Mathematicae*, vol. 19.

Jockusch, C. G., Jr. 1972. "Ramsey's Theorem and Recursion Theory." *Journal of Symbolic Logic*, vol. 37.

Jourdain, P. E. B. 1906–13. "The Development of the Theory of Transfinite Numbers," parts 1–4. *Archiv der Mathematik und Physik*, vols. 10, 14, 16, 22.

Jubien, M. 1977. "Ontology and Mathematical Truth." *Noûs*, vol. 11.

Kamke, E. 1928. *Mengenlehre*. Berlin and Leipzig: Gruyter; 2nd ed., 1947. Translated as Kamke 1950.

1950. *Theory of Sets*. A translation of Kamke 1928 (2nd ed.) by F. Bagemihl. New York: Dover.

Kant, I. 1781. *Kritik der reinen Vernunft*. Riga; 2nd ed., 1787. Translated as Kant 1881.

1800. *Immanuel Kants Logik, ein Handbuch zu Vorlesungen*. Edited by G. B. Jäsche. Königsberg: Nicolovius. Reprinted in Kant 1867–8, vol. 8.

1867–8. *Immanuel Kants Sämmtliche Werke*, 8 vols. Edited by G. Hartenstein. Leipzig: Voss.

1881. *Critique of Pure Reason*. A translation of Kant 1781 (2nd ed.) by F. M. Müller. London: Macmillan; 2nd rev. ed., 1896; 5th printing, 1953.

Kaufmann, F. 1930. *Das Unendliche in der Mathematik und seine Ausschaltung*. Leipzig and Vienna: Deuticke.

1931. "Bemerkungen zum Grundlagenstreit in Logik und Mathematik." *Erkenntnis*, vol. 2.

Kazemier, B. H., and Vuysje, D., eds. 1962. *Logic and Language: Studies Dedicated to Professor Rudolf Carnap on the Occasion of His Seventieth Birthday*. Dordrecht: Reidel.

Kino, A., Myhill, J., and Vesley, R., eds. 1970. *Intuitionism and Proof Theory. Proceedings of the Summer Conference at Buffalo, N.Y., 1968*. Amsterdam: North-Holland.

Kleene, S. C. 1949. "On the Intuitionistic Logic." In *Proceedings of the Tenth International Congress of Philosophy* (Amsterdam, 1948), vol. 1, pt. 2. Edited by E. W. Beth, H. J. Pos, and J. H. A. Hollak. Amsterdam: North-Holland.

1952. *Introduction to Metamathematics*. Amsterdam: North-Holland; Groningen: Noordhoff; New York: Van Nostrand.

Kleene, S. C. and Rosser, J. B. 1935. "The Inconsistency of Certain Formal Logics." *Annals of Mathematis*, vol. 36.

Kneale, W. 1946. "The Truths of Logic." *Proceedings of the Aristotelian Society* (1945–6), n.s. vol. 46.

Kneale, W., and Kneale, M. 1962. *The Development of Logic*. Oxford: Clarendon Press.

König, J. 1905. "Zum Kontinuum-Problem." *Mathematische Annalen*, vol. 60.

Körner, S. 1960. *The Philosophy of Mathematics: An Introduction*. London: Hutchinson; reprinted, New York: Harper and Brothers, 1962.

Kreisel, G. 1950. "Note on Arithmetic Models for Consistent Formulae of the Predicate Calculus." *Fundamenta Mathematicae*, vol. 37.

1951. "Some Remarks on the Foundations of Mathematics: An Expository Article." *Mathematical Gazette*, vol. 35.

1951–2. "On the Interpretation of Non-finitist Proofs," parts 1 and 2. *Journal of Symbolic Logic*, vols. 16 and 17.

1953a. "The Diagonal Method in Formalized Arithmetic." *British Journal for the Philosophy of Science*, vol. 3.

1953b. "On a Problem of Henkin's." *Indagationes Mathematicae*, vol. 15.

1953c. "A Variant to Hilbert's Theory of the Foundations of Arithmetic." *British Journal for the Philosophy of Science*, vol. 4.

585

1958a. "Hilbert's Programme." *Dialectica*, vol. 12. An expansion of the revised version published in Benacerraf and Putnam 1964 is reprinted in this volume.

1958b. "Mathematical Significance of Consistency Proofs." *Journal of Symbolic Logic*, vol. 23.

1958c. "Relative Consistency Proofs." *Journal of Symbolic Logic*, vol. 23.

1958d. "A Remark on Free Choice Sequences and the Topological Completeness Proofs." *Journal of Symbolic Logic*, vol. 23.

1959. "Interpretation of Analysis by Means of Constructive Functionals of Finite Types." In *Constructivity in Mathematics: Proceedings of the Colloquium* (Amsterdam, 1957). Edited by A. Heyting. Amsterdam: North-Holland.

1960a. "Ordinal Logics and the Characterization of Informal Concepts of Proof." *Proceedings of the International Congress of Mathematicians, 1958* (Edinburgh). Cambridge University Press.

1960b. "La Prédicativité." *Bulletin de la Société Mathématique de France*, vol. 88.

1962a. "Foundations of Intuitionistic Logic." In *Logic, Methodology and Philosophy of Science: Proceedings of the 1960 International Congress* (Stanford, Calif.). Edited by E. Nagel, P. Suppes, and A. Tarski. Stanford, Calif.: Stanford University Press.

1962b. Review of "Arithmetization of Metamathematics in a General Setting" (Feferman 1960). *Zentralblatt für Mathematik und ihre Grenzgebiete*, vol. 95.

1962c. Review of *Beweistheorie* (Schütte 1960). *Mathematical Gazette*, vol. 46.

1965. "Mathematical Logic." In *Lectures on Modern Mathematics*, vol. 3. Edited by T. L. Saaty. New York: Wiley.

1967. "Informal Rigour and Completeness Proofs." In *Problems in the Philosophy of Mathematics*. Edited by I. Lakatos. Amsterdam: North-Holland.

1968. "A Survey of Proof Theory." *Journal of Symbolic Logic*, vol. 33.

1976a. "Relative Consistency Proofs. II." *Journal of Symbolic Logic*, vol. 41.

1976b. "What Have We Learned from Hilbert's Second Problem?" In *Mathematical Developments Arising from Hilbert Problems. Proceedings of Symposia in Pure Mathematics* (Northern Illinois University, DeKalb, 1974), vol. 28. Edited by F. E. Browder. Providence, R.I.: American Mathematical Society.

1976c. "Wie die Beweistheorie zu ihren Ordinalzahlen kam und kommt." *Jahresberichte der Deutschen Mathematiker-Vereinigung*, vol. 78.

1980. Obituary of Kurt Gödel. *Biographical Memoirs of Fellows of the Royal Society*, vol. 28, with corrigenda and addenda in vols. 29 (1981) and 30 (1982).

1982. "Finiteness Theorems in Arithmetic: An Application of Herbrand's Theorem for Σ_2-Formulas." *Proceedings of the Herbrand Symposium, Logic Colloquium 1981*. Edited by J. Stern. Amsterdam: North-Holland.

Kreisel, G., and Takeuti, G. 1974. "Formally Self-referential Propositions for Cut-Free Classical Analysis and Related Systems." *Dissertationes Mathematicae*, vol. 118.

Kreisel, G., Mints, G. E., and Simpson, S. G. 1975. "The Use of Abstract Language in Elementary Metamathematics: Some Pedagogic Examples." In

Bibliography

Logic Colloquium (Boston, 1972–3). Edited by R. Parikh. *Lecture Notes in Mathematics*, no. 453. Edited by A. Dold and B. Eckmann. Berlin, Heidelberg, and New York: Springer.

Kripke, S. 1965. "Semantic Analysis of Intuitionistic Logic I." In *Formal Systems and Recursive Functions: Proceedings of the Eighth Logic Colloquium* (Oxford, 1963). Edited by J. N. Crossley and M. A. E. Dummett. Amsterdam: North-Holland.

1972. "Naming and Necessity." In *Semantics of Natural Language.* Edited by D. Davidson and G. Harman. Dordrecht: Reidel.

1975. "Outline of a Theory of Truth." *Journal of Philosophy*, vol. 72.

Kuratowski, C. 1933–50. *Topologie*, 2 vols. *Monografie Matematyczne*, vols. 3 and 21; 2nd ed., Warsaw: Państwowe Wydawnictwo Naukowe, 1948–52. Translated as Kuratowski 1966–8.

1948. "Ensembles projectifs et ensembles singuliers." *Fundamenta Mathematicae*, vol. 35.

1951. "Sur une charactérisation des alephs." *Fundamenta Mathematicae*, vol. 38.

1966–8. *Topology*, 2 vols. A translation of Kuratowski 1933–50 by J. Jaworowski. Warsaw: Pańtwowe Wydawnictwo Naukowe; London and New York: Academic Press.

Lakatos, I. 1962. "Symposium: The Foundations of Mathematics II." *Aristotelian Society Supplementary Volume 36.*

1967. Ed. *Problems in the Philosophy of Mathematics.* Amsterdam: North-Holland.

Langford, C. H. 1927. "On Inductive Relations." *Bulletin of the American Mathematical Society*, vol. 33.

Lear, J. 1977. "Sets and Semantics." *Journal of Philosophy*, vol. 74.

Leibniz, G. W. von. 1875–90. *Die Philosophischen Schriften von Gottfried Wilhelm Leibniz*, 7 vols. Edited by C. J. Gerhardt. Berlin: Weidmann; facsimile reprint, Hildesheim: Olms, 1962.

1923– *Sämtliche Schriften und Briefe.* Edited by the Preussische Akademie der Wissenschaften. Darmstadt: Reichl.

Lévy, A. 1960. "Axiom Schemata of Strong Infinity in Axiomatic Set Theory." *Pacific Journal of Mathematics*, vol. 10.

Lewis, C. I. 1918. *A Survey of Mathematical Logic.* Berkeley: University of California Press.

Lewis, C. I., and Langford, C. H. 1932. *Symbolic Logic.* New York and London: Century; 2nd ed., New York: Dover, 1959.

Lewis, D. K. 1968. "Counterpart Theory and Quantified Modal Logic." *Journal of Philosophy*, vol. 65.

1969. *Convention.* Cambridge: Harvard University Press.

Lewy, C. 1946. "Symposium: Why Are the Calculuses of Logic and Mathematics Applicable to Reality?" part 2. In *Logic and Reality. Aristotelian Society Supplementary Volume 20.*

Linsky, L. 1952. *Semantics and the Philosophy of Language.* Urbana: University of Illinois Press.

587

Lorenzen, P. 1955. *Einführung in die operative Logik und Mathematik. Die Grundlehren der Mathematischen Wissenschaften in Einzeldarstellungen*, vol. 78. Berlin: Springer.

1957. "Wie ist Philosophie der Mathematik möglich?" *Philosophia Naturalis*, vol. 4.

1962. *Metamathematik*. Mannheim: Bibliographisches Institut.

Łoś, J., Mostowski, A., and Rasiowa, H. 1956. "A Proof of Herbrand's Theorem." *Journal des mathématiques pures et appliquées*, vol. 35; correction, vol. 40 (1961).

Lucas, J. R. 1961. "Minds, Machines, and Gödel." *Philosophy: The Journal of the Royal Institute of Philosophy*, vol. 36.

Łukasiewicz, J. 1929. *Elementy logiki matematycznej*. Authorized lecture notes prepared by M. Presburger. Warsaw: Warsaw University; 2nd ed., Warsaw: Państwowe Wydawnictwo Naukowe, 1958. Translated as Łukasiewicz 1963.

1941. "Die Logik und das Grundlagenproblem." *Les Entretiens de Zurich sur les fondements et la méthode des sciences mathématiques* (1938). Edited by F. Gonseth. Zürich: Leemann.

1963. *Elements of Mathematical Logic*. A translation of Łukasiewicz 1929 by O. Wojtasiewicz. Oxford: Pergamon Press; New York: Macmillan.

Lusin, N. 1914. "Sur un problème de M. Baire." *Comptes Rendus des Séances de l'Académie des Sciences* (Paris), vol. 158.

1935. "Sur les ensembles analytiques nuls." *Fundamenta mathematicae*, vol. 25.

Lusin, N., and Sierpiński, W. 1918. "Sur quelques propriétés des ensembles." *Bulletin International de l'Académie des Sciences de Cracovie, Classe des Sciences Mathématiques et Naturelles, ser. A: Sciences Mathématiques*.

McCall, S., ed. 1967. *Polish Logic 1920-1939*. Oxford: Clarendon Press.

MacLane, S. 1961. "Locally Small Categories and the Foundations of Set Theory." In *Infinitistic Methods. Proceedings of the Symposium on Foundations of Mathematics* (Warsaw, 1959). London and New York: Pergamon Press.

McNaughton, R. 1957. "Conceptual Schemes in Set Theory." *Philosophical Review*, vol. 66.

Mahlo, P. 1911. "Über lineare transfinite Mengen." *Berichte über die Verhandlungen der Königlich Sächsischen Gesellschaft der Wissenschaften*, vol. 63. Leipzig: Teubner.

1913. "Zur Theorie und Anwendung der $\rho_0 =$ Zahlen. II." *Berichte über die Verhandlungen der Königlich Sächsischen Gesellschaft der Wissenschaften*, vol. 65. Leipzig: Teubner.

Mannoury, G. 1909. *Methodologisches und Philosophisches zur Elementar-Mathematik*. Haarlem: Visser; Assen: Van Gorcum.

Marcus, R. B. 1974. "Classes, Collections, and Individuals." *American Philosophical Quarterly*, vol. 11.

Marshall, W. 1953. "Frege's Theory of Functions and Objects." *Philosophical Review*, vol. 62.

Martin, D. A. 1968. "The Axiom of Determinateness and Reduction Principles in the Analytic Hierarchy.'' *Bulletin of the American Mathematical Society*, vol. 74.

1976. ''Hilbert's First Problem: the Continuum Hypothesis.'' In *Mathematical Developments Arising from Hilbert Problems. Proceedings of Symposia in Pure Mathematics* (Northern Illinois University, DeKalb, 1974), vol. 28. Edited by F. E. Browder. Providence, R.I.: American Mathematical Society.

Martin, R. M. 1958. *Truth and Denotation*. Chicago: University of Chicago Press.

1963. *Intension and Decision: A Philosophical Study*. Englewood Cliffs, N.J.: Prentice-Hall.

Mehlberg, H. 1962. ''The Present Situation in the Philosophy of Mathematics.'' In *Logic and Language: Studies Dedicated to Professor Rudolf Carnap on the Occasion of His Seventieth Birthday*. Edited by B. H. Kazemier and D. Vuysje. Dordrecht: Reidel.

Menger, K. 1930. ''Der Intuitionismus.'' *Blätter für deutsche Philosophie*, vol. 4.

1933. ''Die neue Logik.'' In *Krise und Neuaufbau in den exakten Wissenschaften*. Leipzig and Vienna: Deuticke.

Meyer, R. 1975. ''Weak Monadic Second Order Theory of Successor Is Not Elementary-Recursive.'' In *Logic Colloquium* (Boston, 1972–3). Edited by R. Parikh. *Lecture Notes in Mathematics*, no. 453. Edited by A. Dold and B. Eckmann. Berlin, Heidelberg, and New York: Springer.

Mill, J. S. 1843. *A System of Logic*. London: Longmans.

Mirimanoff, D. 1917a. ''Les Antinomies de Russell et de Burali-Forti et le problème fondamental de la théorie des ensembles.'' *L'Enseignement mathématique*, 1st ser. vol. 19.

1917b. ''Remarques sur la théorie des ensembles et les antinomies cantoriennes,'' part 1. *L'Enseignement mathématique*, 1st ser. vol. 19.

1920. ''Remarques sur la théorie des ensembles et les antinomies cantoriennes,'' part 2. *L'Enseignement mathématique*, 1st ser. vol. 21.

Montague, R. 1970. ''Pragmatics and Intensional Logic.'' *Dialectica*, vol. 24, and *Synthèse*, vol. 22. Reprinted in Montague 1974.

1974. *Formal Philosophy*. Edited, with an introduction, by R. H. Thomason. London and New Haven: Yale University Press.

Moore, G. E. 1954–5. ''Wittgenstein's Lectures in 1930–33.'' *Mind*, n.s. vols. 63 and 64. Reprinted in Moore 1959.

1959. *Philosophical Papers*. London: Allen and Unwin; New York: Macmillan.

Moschovakis, Y. N. 1971. ''Predicative Classes.'' In *Axiomatic Set Theory. Proceedings of Symposia in Pure Mathematics*, vol. 13, pt. 1. Edited by D. S. Scott. Providence, R.I.: American Mathematical Society.

Mostowski, A. 1957a. ''On Recursive Models of Formalised Arithmetic.'' *Bulletin de L'Académie Polonaise des Sciences, Série des Sciences Mathématiques, Astronomiques et Physiques*, vol. 5.

1957b. Review of G. Kreisel, ''A Variant to Hilbert's Theory of the Foundations of Arithmetic'' (Kreisel 1953c). *Journal of Symbolic Logic*, vol. 22.

1965. ''On Models of Zermelo–Fraenkel Set Theory Satisfying the Axiom of

Constructibility." In *Studia Logico-Mathematica et Philosophica. Acta Philosophica Fennica*, no. 18. Helsinki: Societas Philosophica Fennica.

1966. *Thirty Years of Foundational Studies. Acta Philosophica Fennica*, no. 17. New York: Barnes and Noble.

Mycielski, J. 1964. "On the Axiom of Determinateness." *Fundamenta Mathematicae*, vol. 53.

Myhill, J. 1951–2. "Two Ways of Ontology in Modern Logic." *Review of Metaphysics*, vol. 5.

1952–3. "Some Philosophical Implications of Mathematical Logic: Three Classes of Ideas." *Review of Metaphysics*, vol. 6.

Nagel, E. 1944. "Logic Without Ontology." In *Naturalism and the Human Spirit*. Edited by Y. H. Krikorian. New York: Columbia University Press. Reprinted in Feigl and Sellars 1949 and Benacerraf and Putnam 1964.

1948. "Book Review" of *Meaning and Necessity* (Carnap 1947). *Journal of Philosophy*, vol. 45.

1956. *Logic Without Metaphysics*. New York: Free Press.

Nagel, E., and Newman, J. R. 1958. *Gödel's Proof*. New York: New York University Press.

Natorp, P. G. 1910. *Die logischen Grundlagen der exakten Wissenschaften. Wissenschaft und Hypothese*, vol. 12. Lepizig and Berlin: Teubner.

Noteler, E., and Cavaillès, J., eds. 1937. *Briefwechsel Cantor–Dedekind. Actualités scientifiques et industrielles*, no. 518. Paris: Hermann.

Pap, A. 1958. *Semantics and Necessary Truth*. New Haven: Yale University Press.

1960. "Types and Meaninglessness." *Mind*, n.s. vol. 69.

Parikh, R., ed. 1975. *Logic Colloquium* (Boston, 1972–3). *Lecture Notes in Mathematics*, no. 453. Edited by A. Dold and B. Eckmann. Berlin, Heidelberg, and New York: Springer.

Paris, J., and Harrington, L. 1977. "A Mathematical Incompleteness in Peano Arithmetic." In *Handbook of Mathematical Logic*. Edited by J. Barwise. *Studies in Logic and the Foundations of Mathematics*, vol. 90. Amsterdam and New York: North-Holland.

Parsons, C. 1965. "Frege's Theory of Number." In *Philosophy in America*. Edited by M. Black. London: Allen and Unwin.

1969. "Kant's Philosophy of Arithmetic." In *Philosophy, Science, and Method: Essays in Honor of Ernest Nagel*. Edited by S. Morgenbesser, P. Suppes, and M. White. New York: St. Martin's Press.

1971. "Ontology and Mathematics." *Philosophical Review*, vol. 80.

1974a. "Informal Axiomatization, Formalization, and the Concept of Truth." *Synthèse*, vol. 27.

1974b. "The Liar Paradox." *Journal of Philosophical Logic*, vol. 3.

1974c. "Sets and Classes." *Noûs*, vol. 8.

1974d. "Sets and Possible Worlds." Abstract of a talk at Columbia University, mimeographed.

1976. "Some Remarks on Frege's Conception of Extension." In *Studien zu Frege = Studies on Frege*, 3 vols. Vol. 1, *Logik und Philosophie der Mathematik*. Edited by M. Schirn. Stuttgart: Frommann-Holzboog.

590

1977. "What Is the Iterative Conception of Set?" In *Logic, Foundations of Mathematics, and Computability Theory. Proceedings of the Fifth International Congress of Logic, Methodology and the Philosophy of Science* (London, Ontario, 1975), pt. 1. Edited by R. E. Butts and J. Hintikka. Dordrecht: Reidel. Reprinted in this volume.

1981. "Modal Set Theories." *Journal of Symbolic Logic*, vol. 46.

Forthcoming. "Quine on the Philosophy of Mathematics." In C. Parsons, *Mathematics in Philosophy*. Ithaca, N.Y.: Cornell University Press.

Peano, G. 1889. *Arithmetices principia, nova methodo exposita.* Turin: Bocca. Translated as "The Principles of Arithmetic, Presented by a New Method" by J. van Heijenoort in van Heijenoort 1967.

Peirce, C. S. 1931–58. *Collected Papers*, 8 vols. Edited by C. Hartshorne, P. Weiss, and A. W. Burks. Cambridge: Harvard University Press.

Plato. *Meno.* Translated by B. Jowett. New York: Liberal Arts Press; Indianapolis: Bobbs-Merrill, 1949.

Poincaré, H. 1894. "Sur la nature du raisonnement mathématique." *Revue de métaphysique et de morale*, vol. 2. Reprinted in Poincaré 1903. Translated as "On the Nature of Mathematical Reasoning" in Poincaré 1907. Excerpted and reprinted in Benacerraf and Putnam 1964 and in this volume.

1903. *La Science et l'Hypothèse.* Paris: Flammarion. Translated as Poincaré 1907 and in 1913b.

1905. "Les Mathématiques et la logique." *Revue de métaphysique et de morale*, vol. 13.

1907. *Science and Hypothesis.* A translation of Poincaré 1903 by "W. J. G." (W. J. Greenstreet). London: Scott; New York: Scribner; reprinted, New York: Dover, 1952.

1908a. "L'Avenir des mathématiques." *Scientia: Rivista di Scienza*, vol. 4, and *Revue générale des sciences pures et appliquées*, vol. 19. Reprinted in Poincaré 1908b. Translated in Poincaré 1913b and 1914.

1908b. *Science et Méthode.* Paris: Flammarion. Translated in Poincaré 1913b and as 1914.

1913a. *Dernières pensées. Bibliothèque de philosophie scientifique.* Paris: Flammarion.

1913b. *The Foundations of Science.* Translated by G. B. Halsted. New York: Science Press. Includes Poincaré 1903 and 1908b.

1914. *Science and Method.* A translation of Poincaré 1908b by F. Maitland. London and New York: Nelson; reprinted, New York: Dover, 1952.

Popper, K. 1946. "Symposium: Why Are the Calculuses of Logic and Mathematics Applicable to Reality?" part 3. In *Logic and Reality. Aristotelian Society Supplementary Volume 20.*

1947. "Logic Without Assumptions." *Proceedings of the Aristotelian Society* (1946–7), n.s. vol. 47.

1947–8. "New Foundations for Logic" and "Corrections and Additions to 'New Foundations for Logic.'" *Mind*, n.s. vols. 56 and 57.

1948. "What Can Logic Do for Philosophy?" In *Logical Positivism and Ethics. Aristotelian Society Supplementary Volume 22.*

Powell, W. C. 1972. "Set Theory with Predication." Ph.D. dissertation, State University of New York at Buffalo.

1975. "Extending Gödel's Negative Interpretation to ZF." *Journal of Symbolic Logic*, vol. 40.

Putnam, H. 1956. "Mathematics and the Existence of Abstract Entities." *Philosophical Studies*, vol. 7.

1962. "The Analytic and the Synthetic." In *Minnesota Studies in the Philosophy of Science*, vol. 1. Edited by H. Feigl and G. Maxwell. Minneapolis: University of Minnesota Press.

1967a. "Mathematics without Foundations." *Journal of Philosophy*, vol. 64. Reprinted in Putnam 1975, vol. 1, and in this volume.

1967b. "The Thesis that Mathematics Is Logic." In *Bertrand Russell, Philosopher of the Century*. Edited by R. Schoenman. London: Allen and Unwin.

1975. *Philosophical Papers*, 2 vols. Vol. 1: *Mathematics, Matter and Method;* vol. 2: *Mind, Language, and Reality.* Cambridge University Press.

1980. "Models and Reality." Presidential address to the annual meeting of the Association for Symbolic Logic, December 1977. *Journal of Symbolic Logic*, vol. 45. Reprinted in this volume.

Quine, W. V. 1936. "Truth by Convention." In *Philosophical Essays for Alfred North Whitehead*. Edited by O. H. Lee. New York: Longmans, Green. Reprinted in Feigl and Sellars 1949, Benacerraf and Putnam 1964, Quine 1966b, and in this volume.

1937. "New Foundations for Mathematical Logic." *American Mathematical Monthly*, vol. 44. Reprinted in Quine 1953a.

1939. "Designation and Existence." *Journal of Philosophy*, vol. 36. Reprinted in Feigl and Sellars 1949.

1940. *Mathematical Logic.* New York: Norton; rev. ed., Cambridge: Harvard University Press, 1951; reprinted, New York: Harper Torchbooks, 1962.

1941. "Whitehead and the Rise of Modern Logic." In *The Philosophy of Alfred North Whitehead.* Edited by P. A. Schilpp. Evanston and Chicago: Northwestern University Press; 2nd ed., New York: Tudor, 1951. Reprinted in Quine 1966a.

1943. "Notes on Existence and Necessity." *Journal of Philosophy*, vol. 40. Reprinted in Linsky 1952.

1947. "On Universals." *Journal of Symbolic Logic*, vol. 12.

1948. "On What There Is." *Review of Metaphysics*, vol. 2. Reprinted in "Symposium: On What There is III" in *Freedom, Language, and Reality. Aristotelian Society Supplementary Volume 25* (1951). Reprinted in Linsky 1952, Quine 1953a, Benacerraf and Putnam 1964, and Feigl, Sellars, and Lehrer 1972.

1951a. "On Carnap's Views on Ontology." *Philosophical Studies*, vol. 2. Reprinted in Quine 1966b and Feigl, Sellars, and Lehrer 1972.

1951b. "Ontology and Ideology." *Philosophical Studies*, vol. 2. Reprinted in Feigl, Sellars, and Lehrer 1972.

1951c. "Semantics and Abstract Objects." In *Contributions to the Analysis and Synthesis of Knowledge. Proceedings of the American Academy of Arts and Sciences*, vol. 80.

1951d. "Two Dogmas of Empiricism." *Philosophical Review*, vol. 60. Reprinted in Quine 1953a, Benacerraf and Putnam 1964, and Feigl, Sellars, and Lehrer 1972.

1952. "On an Application of Tarski's Theory of Truth." *Proceedings of the National Academy of Sciences*, vol. 38. Reprinted in Quine 1966a.

1953a. *From a Logical Point of View: 9 Logico-Philosophical Essays.* Cambridge: Harvard University Press; 2nd ed., 1961.

1953b. "On Mental Entities." In *Contributions to the Analysis and Synthesis of Knowledge. Proceedings of the American Academy of Arts and Sciences*, vol. 80. Reprinted in Quine 1966b.

1955. "On Frege's Way Out." *Mind*, n.s. vol. 64. Reprinted in Quine 1966a.

1958. "Speaking of Objects." *Proceedings and Addresses of the American Philosophical Association*, vol. 31.

1960. *Word and Object.* Cambridge and New York: Technology Press of MIT and Wiley; London: Chapman.

1962. "Carnap and Logical Truth." In *Logic and Language: Studies Dedicated to Professor Rudolf Carnap on the Occasion of His Seventieth Birthday.* Edited by B. H. Kazemier and D. Vuysje. Dordrecht: Reidel. Reprinted in Schilpp 1963, Quine 1966b, Feigl, Sellars, and Lehrer 1972, and in this volume.

1964. "Truth by Convention." In Benacerraf and Putnam 1964. Reprint with some changes of Quine 1936. Reprinted in this volume.

1966a. *Selected Logic Papers.* New York: Random House.

1966b. *The Ways of Paradox and Other Essays.* New York: Random House; 2nd rev. ed., Cambridge, Mass., and London: Harvard University Press, 1976.

Ramsey, F. P. 1926a. "The Foundations of Mathematics." *Proceedings of the London Mathematical Society* (1925), 2nd ser. vol. 25, pt. 5. Reprinted in Ramsey 1931.

1926b. "Mathematical Logic." *Mathematical Gazette*, vol. 13. Reprinted in Ramsey 1931.

1926c. "Universals and the 'Method of Analysis.'" In *Methods of Analysis. Aristotelian Society Supplementary Volume 6.* Reprinted in Ramsey 1931.

1931. *The Foundations of Mathematics and Other Logical Essays.* Edited by R. B. Braithwaite. New York: Harcourt, Brace; London: Routledge and Kegan Paul.

Reichenbach, H. 1944. "Bertrand Russell's Logic." In *The Philosophy of Bertrand Russell.* Edited by P. A. Schilpp. Evanston and Chicago: Northwestern University Press; 3rd ed., New York: Tudor, 1951.

Reinhardt, W. N. 1974a. "Remarks on Reflection Principles, Large Cardinals, and Elementary Embeddings." In *Axiomatic Set Theory. Proceedings of Symposia in Pure Mathematics*, vol. 13, pt. 2. Providence, R.I.: American Mathematical Society.

1974b. "Set Existence Principles of Shoenfield, Ackermann, and Powell." *Fundamenta Mathematicae*, vol. 84.

Resnik, M. D. 1966. "On Skolem's Paradox." *Journal of Philosophy*, vol. 63.

Robinson, A. 1965. "Formalism 64 [1964]." In *Logic, Methodology and Philosophy of Science: Proceedings of the 1964 International Congress.* Edited by Y. Bar-Hillel. Amsterdam: North-Holland.

    1973. "Mathematical Problems." *Journal of Symbolic Logic*, vol. 38.

Rogers, H., Jr. 1968. *Theory of Recursive Functions and Effective Computability.* New York: McGraw-Hill.

Rosser, J. B. 1936. "Constructibility as a Criterion for Existence." *Journal of Symbolic Logic*, vol. 1.

    1939. "An Informal Exposition of Proofs of Gödel's Theorems and Church's Theorem." *Journal of Symbolic Logic*, vol. 4. Reprinted in Davis 1965.

Russell, B. A. W. 1903. *The Principles of Mathematics.* Cambridge University Press; 2nd ed., London: Allen and Unwin, 1937.

    1905. "On Denoting." *Mind*, n.s. vol. 14. Reprinted in Feigl and Sellars 1949 and Russell 1956.

    1906a. "Les Paradoxes de la logique." *Revue de métaphysique et de morale*, vol. 14.

    1906b. "The Theory of Implication." *American Journal of Mathematics*, vol. 28.

    1907. "On Some Difficulties in the Theory of Transfinite Numbers and Order Types." *Proceedings of the London Mathematical Society*, 2nd ser. vol. 4.

    1908. "Mathematical Logic as Based on the Theory of Types." *American Journal of Mathematics*, vol. 30. Reprinted in Russell 1956 and van Heijenoort 1967.

    1914. *Our Knowledge of the External World as a Field for Scientific Method in Philosophy.* Chicago: Open Court; 2nd ed., Norton, 1929.

    1918–19. "The Philosophy of Logical Atomism," parts 1–7. *Monist*, vols. 28 and 29. Reprinted in Russell 1956.

    1919. *Introduction to Mathematical Philosophy.* London: Allen and Unwin; New York: Macmillan; 2nd ed., Allen and Unwin, 1920. Chapters 1, 2, and 18 reprinted in Benacerraf and Putnam 1964 and in this volume.

    1940. *An Inquiry into Meaning and Truth.* London: Allen and Unwin; New York: Norton; reprinted, Allen and Unwin, 1943 and 1951.

    1956. *Logic and Knowledge: Essays, 1901–1950.* Edited by R. C. Marsh. London: Allen and Unwin; New York: Macmillan.

    1959. *My Philosophical Development.* New York: Simon and Schuster.

Ryle, G. 1946. "Symposium: Why Are the Calculuses of Logic and Mathematics Applicable to Reality?" part 1. In *Logic and Reality. Aristotelian Society Supplementary Volume 20.*

    1949. "Discussion: *Meaning and Necessity*," a review of Carnap 1947. *Philosophy: The Journal of the Royal Institute of Philosophy*, vol. 24.

Rynin, D. 1956. "The Dogma of Logical Pragmatism." *Mind*, n.s. vol. 65.

Saaty, T. L., ed. 1963–5. *Lectures on Modern Mathematics*, 3 vols. New York: Wiley.

Scheffler, I., and Chomsky, N. 1959. "What Is Said to Be." *Proceedings of the Aristotelian Society* (1958–9), n.s. vol. 59.

Bibliography

Schilpp, P. A., ed. 1941. *The Philosophy of Alfred North Whitehead.* Evanston and Chicago: Northwestern University Press; 2nd ed., New York: Tudor, 1951.

1944. Ed. *The Philosophy of Bertrand Russell.* Evanston and Chicago: Northwestern University Press; 3rd ed., New York: Tudor, 1951.

1963. Ed. *The Philosophy of Rudolf Carnap.* La Salle, Ill.: Open Court.

Schlick, M. 1918. *Allgemeine Erkenntnislehre.* Berlin: Springer; 2nd ed., 1925.

1932. "Positivismus und Realismus." *Erkenntnis,* vol. 3. Reprinted in Schlick 1938. Translated as "Positivism and Realism" by D. Rynin in Ayer 1959.

1938. *Gesammelte Aufsätze, 1926–36.* Vienna: Gerold.

Schröder, E. 1873. *Lehrbuch der Arithmetik und Algebra für Lehrer und Studirende.* Leipzig: Teubner.

Schröter, K. 1973. "Interpretation der intuitionistischen Logik mit Hilfe des Beweisbarkeitsbegriffs." *Ajatus Suomen Filosofisen Yhdistyksen Vuosikiria,* vol. 35.

Schütte, K. 1960a. *Beweistheorie. Die Grundlehren der Mathematischen Wissenschaften in Einzeldarstellungen,* vol. 103. Berlin: Springer.

1960b. "Syntactical and Semantical Properties of Simple Type Theory." *Journal of Symbolic Logic,* vol. 25.

1968. *Vollständige Systeme modaler und intuitionistischer Logik.* Heidelberg: Springer.

Schwartz, S. P., ed. 1977. *Naming, Necessity, and Natural Kinds.* Ithaca, N.Y.: Cornell University Press.

Scott, D. 1961. "Measurable Cardinals and Constructible Sets." *Bulletin de L'Académie Polonaise des Sciences, Série des Sciences Mathématiques, Astronomiques et Physiques,* vol. 9.

1968. "Extending the Topological Interpretation to Intuitionistic Analysis." *Compositio Mathematica,* vol. 20.

1970. "Extending the Topological Interpretation to Intuitionistic Analysis II." In *Intuitionism and Proof Theory. Proceedings of the Summer Conference at Buffalo, N.Y., 1968.* Edited by A. Kino, J. Myhill, and R. Vesley. Amsterdam: North-Holland.

Searle, J. R. 1958. "Russell's Objections to Frege's Theory of Sense and Reference." *Analysis,* vol. 18.

Sellars, W. 1949. "Acquaintance and Description Again." *Journal of Philosophy,* vol. 46.

Sheffer, H. M. 1913. "A Set of Five Independent Postulates for Boolean Algebras, with Applications to Logical Constants." *Transactions of the American Mathematical Society,* vol. 14.

Shoenfield, J. R. 1967. *Mathematical Logic.* Reading, Mass.: Addison-Wesley.

Shwayder, D. S. 1961. *Modes of Referring and the Problem of Universals: An Essay in Metaphysics. University of California Publications in Philosophy,* vol. 35. Berkeley: University of California Press.

Sierpiński, W. 1934a. *Hypothèse du Continu. Monografie Matematyczne,* vol. 4. Warsaw: Z Subwencji Funduszu Kultury Narodowej: 2nd ed., New York: Chelsea, 1956.

1934b. "Sur une extension de la notion de l'homéomorphie." *Fundamenta Mathematicae*, vol. 22.

1935a. "Sur deux ensembles linéaires singuliers." *Annali della R. Scuola Normale Superiore di Pisa: Scienze Fisiche e Matematiche*, 2nd ser. vol. 4.

1935b. "Sur une hypothèse de M. Lusin." *Fundamenta Mathematicae*, vol. 25.

1951. "Sur quelques propositions concernant la puissance du continu." *Fundamenta Mathematicae*, vol. 38.

Sierpiński, W., and Tarski, A. 1930. "Sur une propriété caractéristique des nombres inaccessibles." *Fundamenta Mathematicae*, vol. 15.

Sikorski, R. 1951. "A Characterization of Alephs." *Fundamenta Mathematicae*, vol. 38.

Skolem, T. 1923. "Einige Bemerkungen zur axiomatischen Begründung der Mengenlehre." *Matematikerkongressen i Helsingfors den 4–7 Juli 1922, Den femte skandinaviska matematikerkongressen, Redogörelse.* Helsinki: Akademiska Bokhandeln. Translated as "Some Remarks on Axiomatized Set Theory" by S. Bauer-Mengelberg in van Heijenoort 1967.

1934. "Über die Nicht-charakterisierbarkeit der Zahlenreihe mittels endlich oder abzählbar unendlich vieler Aussagen mit ausschliesslich Zahlenvariablen." *Fundamenta Mathematicae*, vol. 23.

1970. *Selected Works in Logic.* Edited by J. E. Fenstad. Oslo: Universitetsforlaget.

Skyrms, B. 1967. "The Explication of 'X Knows that p'." *Journal of Philosophy*, vol. 64.

Smart, J. J. C. 1950. "Whitehead and Russell's Theory of Types." *Analysis*, vol. 10.

Smorynski, C. 1977. "The Incompleteness Theorems." In *Handbook of Mathematical Logic.* Edited by J. Barwise. *Studies in Logic and the Foundations of Mathematics*, vol. 90. Amsterdam and New York: North-Holland.

Spector, C. 1957. "Recursive Ordinals and Predicative Set Theory." In vol. 3, *Summaries of Talks Presented at the Summer Institute for Symbolic Logic (Cornell University, 1957)*, 3 vols., duplicated; 2nd ed., Princeton, N.J.: Communications Research Division, Institute for Defense Analyses, 1960.

1962. "Provably Recursive Functionals of Analysis: A Consistency Proof of Analysis by an Extension of Principles Formulated in Current Intuitionistic Mathematics." In *Recursive Function Theory. Proceedings of Symposia in Pure Mathematics*, vol. 5. Providence, R.I.: American Mathematical Society.

Steiner, M. 1973. "Platonism and the Causal Theory of Knowledge." *Journal of Philosophy*, vol. 70.

1975. *Mathematical Knowledge.* Ithaca, N.Y.: Cornell University Press.

Strawson, P. F. 1950. "On Referring." *Mind*, n.s. vol. 59. Reprinted in Feigl, Sellars, and Lehrer 1972.

1952. *Introduction to Logical Theory.* London: Methuen; New York: Wiley.

1956. "Singular Terms, Ontology, and Identity." *Mind*, n.s. vol. 65.

1959. *Individuals.* London: Methuen.

Study, E. 1914. *Die realistische Weltansicht und die Lehre vom Raume.* Brunswick: Vieweg.

Bibliography

Tait, W. W. 1965a. "Functionals Defined by Transfinite Recursion." *Journal of Symbolic Logic*, vol. 30.

1965b. "Infinitely Long Terms of Transfinite Type." In *Formal Systems and Recursive Functions: Proceedings of the Eighth Logic Colloquium* (Oxford, 1963). Edited by J. N. Crossley and M. A. E. Dummett. Amsterdam: North-Holland.

Takeuti, G. 1954. "Construction of the Set Theory from the Theory of Ordinal Numbers." *Journal of the Mathematical Society of Japan*, vol. 6.

Takeuti, G., and Zaring, W. M. 1971. *An Introduction to Axiomatic Set Theory.* New York: Springer.

Tarski, A. 1925. "Quelques théorèmes sur les alephs." *Fundamenta Mathematicae*, vol. 7.

1930–1. "O pojęciu prawdy w odniesieniu do sformalizowanych nauk dedukcyjnych." *Ruch Filozoficzny*, vol. 12. Translated as Tarski 1935b.

1934. "Z badań metodologicznych nad definiowalnością terminów." *Przegląd Filozoficzny*, vol. 37. Translated as Tarski 1935a.

1935a. "Einige methodologische Untersuchungen über die Definierbarkeit der Begriffe." *Erkenntnis*, vol. 5. Translated as "Some Methodological Investigations on the Definability of Concepts" in Tarski 1956.

1935b. *Der Wahrheitsbegriff in den formalisierten Sprachen. Studia Philosophica*, vol. 1. Translated as "The Concept of Truth in Formalized Languages" in Tarski 1956.

1936. "O pojciu wynikania logicznego." *Przegląd Filozoficzny*, vol. 39. Translated as "On the Concept of Logical Consequence" in Tarski 1956.

1938. "Über unerreichbare Kardinalzahlen." *Fundamenta Mathematicae*, vol. 30.

1939. "On Undecidable Statements in Enlarged Systems of Logic and the Concept of Truth." *Journal of Symbolic Logic*, vol. 4.

1941. *Introduction to Logic and the Methodology of the Deductive Sciences.* New York: Oxford University Press.

1944. "The Semantic Conception of Truth and the Foundations of Semantics." *Philosophy and Phenomenological Research*, vol. 4. Reprinted in Feigl and Sellars 1949 and Linsky 1952.

1956. *Logic, Semantics, Metamathematics: Papers from 1923 to 1938.* Translated by J. H. Woodger. Oxford: Clarendon Press. Includes Tarski 1930–1, 1934, and 1936.

1962. "Some Problems and Results Relevant to the Foundations of Set Theory." In *Logic, Methodology and Philosophy of Science. Proceedings of the 1960 International Congress* (Stanford, Calif.). Edited by E. Nagel, P. Suppes, and A. Tarski. Stanford, Calif.: Stanford University Press.

Tarski, A., Mostowski, A., and Robinson, R. M. 1953. *Undecidable Theories.* Amsterdam: North-Holland.

Tharp, L. 1975. "Three Theorems of Metaphysics." Unpublished.

Thomson, J. F. 1960. "What Achilles Should Have Said to the Tortoise." *Ratio*, vol. 3.

1963. "On Some Paradoxes." In *Analytical Philosophy*. Edited by R. J. Butler. Oxford: Blackwell.

597

Bibliography

Troelstra, A. S. 1969. *Principles of Intuitionism*. Berlin, Heidelberg, and New York: Springer.

——— 1973. Ed. *Metamathematical Investigation of Intuitionistic Arithmetic and Analysis. Lecture Notes in Mathematics*, no. 344. Edited by A. Dold and B. Eckmann. Berlin, Heidelberg, and New York: Springer.

——— 1977a. *Choice Sequences: A Chapter of Intuitionistic Mathematics*. Oxford: Clarendon Press.

——— 1977b. "Completeness and Validity for Intuitionistic Predicate Logic." In *Colloque International de Logique* (Clermont-Ferrand, 1975). *Colloques Internationaux du Centre National de la Recherche Scientifique*, no. 249. Paris: Éditions du Centre National de la Recherche Scientifique.

Turing, A. M. 1937. "On Computable Numbers, with an Application to the Entscheidungsproblem." *Proceedings of the London Mathematical Society*, 2nd ser. vol. 42. Reprinted in Davis 1965.

——— 1954. "Solvable and Unsolvable Problems." *Science News*, vol. 31.

Vacca, G. 1903. "La logica di Leibniz." *Rivista di matematica*, vol. 8.

van Heijenoort, J., ed. 1967. *From Frege to Gödel: A Source Book in Mathematical Logic, 1879-1931*. Cambridge: Harvard University Press.

Ville, F. 1971. "Decidabilité des formules existentielles en théorie des ensembles." *Comptes Rendus Hebdomadaires des Séances de l'Académie des Sciences*, series A (Paris), vol. 272.

von Neumann, J. 1923. "Zur Einführung der transfiniten Zahlen." *Acta Litterarum ac Scientiarum: Sectio Scientiarum Mathematicarum*, vol. 1. Translated as "On the Introduction of Transfinite Numbers" by J. van Heijenoort in van Heijenoort 1967.

——— 1925. "Eine Axiomatisierung der Mengenlehre." *Journal für die reine und angewandte Mathematik*, vol. 154. Translated as "An Axiomatization of Set Theory" by S. Bauer-Mengelberg and D. Føllesdal in van Heijenoort 1967.

——— 1927. "Zur Hilbertschen Beweistheorie." *Mathematische Zeitschrift*, vol. 26.

——— 1928. "Die Axiomatisierung der Mengenlehre." *Mathematische Zeitschrift*, vol. 27.

——— 1929. "Über eine Widerspruchsfreiheitsfrage in der axiomatischen Mengenlehre." *Journal für reine und angewandte Mathematik*, vol. 160.

——— 1931. "Die formalistische Grundlegung der Mathematik." *Erkenntnis*, vol. 2. Translated as "The Formalist Foundations of Mathematics" by E. Putnam and G. J. Massey in Benacerraf and Putnam 1964 and reprinted in this volume.

——— 1961-3. *Collected Works*, 6 vols. Edited by A. H. Taub. Oxford and New York: Pergamon Press.

Waismann, F. 1928. "Die Natur des Reduzibilitätsaxioms." *Monatshefte für Mathematik und Physik*, vol. 35.

——— 1938. "Ist die Logik eine deduktive Theorie?" *Erkenntnis*, vol. 7.

——— 1946. "Are There Alternative Logics?" *Proceedings of the Aristotelian Society* (1945-6), n.s. vol. 46.

——— 1949-53. "Analytic-Synthetic," parts 1-6. *Analysis*, vols. 10, 11, and 13.

——— 1951. *Introduction to Mathematical Thinking*. Translated by J. Benac. New York: Ungar.

Bibliography

Wang, H. 1953. "What Is an Individual?" *Philosophical Review*, vol. 62.

　1954. "Formalization of Mathematics." *Journal of Symbolic Logic*, vol. 19.

　1955. "On Formalization." *Mind*, n.s. vol. 64.

　1958. "Eighty Years of Foundational Studies." *Dialectica*, vol. 12.

　1961. "Process and Existence in Mathematics." In *Essays on the Foundations of Mathematics, Dedicated to A. A. Fraenkel on His Seventieth Anniversary*. Edited by Y. Bar-Hillel, E. I. J. Poznanski, M. O. Rabin, and A. Robinson. Jerusalem: Magnes Press, Hebrew University.

　1962. *A Survey of Mathematical Logic*. Peking: Science Press; Amsterdam: North-Holland, 1963.

　1970. "Logic, Computation, and Philosophy." *L'Âge de la Science*, vol. 3.

　1974. *From Mathematics to Philosophy*. London: Routledge and Kegan Paul. Chapter 6, "The Concept of Set," is reprinted in this volume.

　1977. "Large Sets." In *Logic, Foundations of Mathematics, and Computability Theory. Proceedings of the Fifth International Congress of Logic, Methodology and Philosophy of Science* (London, Ontario, 1975), pt. 1. Edited by R. E. Butts and J. Hintikka. Dordrecht: Reidel.

Warnock, G. J. 1956. "Metaphysics in Logic." In *Essays in Conceptual Analysis*. Edited by A. Flew. London: Macmillan.

Weyl, H. 1918. *Das Kontinuum: Kritische Untersuchungen über die Grundlagen der Analysis*. Leipzig: Veit; reprinted, Berlin and Leipzig: Gruyter, 1932.

　1921. "Über die neue Grundlagenkrise der Mathematik." *Mathematische Zeitschrift*, vol. 10.

　1927. *Philosophie der Mathematik und Naturwissenschaft*, parts 1 and 2. *Handbuch der Philosophie*, nos. 4 and 5. Munich and Berlin: Oldenbourg. Translated as Weyl 1949.

　1940. "The Mathematical Way of Thinking." *Science*, vol. 92.

　1944. "David Hilbert and His Mathematical Work." *Bulletin of the American Mathematical Society*, vol. 50.

　1946. "Mathematics and Logic: a brief survey serving as preface to a review of *The Philosophy of Bertrand Russell*" (Schilpp 1944). *American Mathematical Monthly*, vol. 53.

　1949. *Philosophy of Mathematics and Natural Science*. Weyl 1927, revised and enlarged, based on a translation by O. Helmer. Princeton: Princeton University Press; reprinted, New York: Atheneum, 1963.

White, M. G. 1950. "The Analytic and the Synthetic: An Untenable Dualism." In *John Dewey, Philosopher of Science and Freedom: A Symposium*. Edited by S. Hook. New York: Dial Press. Reprinted in Linsky 1952.

White, N. P. 1974. "What Numbers Are." *Synthèse*, vol. 27.

Whitehead, A. N., and Russell, B. A. W. 1910–13. *Principia Mathematica*, 3 vols. Cambridge University Press; 2nd ed., 1925–7.

Wiggins, D. 1977. "Truth, Invention and the Meaning of Life." In *Proceedings of the British Academy* (London, 1976), vol. 62. London: Oxford University Press.

Wilder, R. L. 1952. *Introduction to the Foundations of Mathematics*. New York: Wiley.

Bibliography

Wittgenstein, L. 1921. *Logisch-philosophische Abhandlung.: Annalen der Natur-philosophie*, vol. 14. Reprinted and translated as Wittgenstein 1922 and 1961c.

— 1922. *Tractatus Logico-Philosophicus.* A translation of Wittgenstein 1921 with German and English on opposite pages and an introduction by B. Russell. London: Routledge and Kegan Paul. Reprinted with a new translation as Wittgenstein 1961c.

— 1929. "Some Remarks on Logical Form." In *Knowledge, Experience and Realism. Aristotelian Society Supplementary Volume 9.*

— 1953. *Philosophical Investigations.* Translated by G. E. M. Anscombe. Oxford: Blackwell.

— 1956. *Remarks on the Foundations of Mathematics.* Translated by G. E. M. Anscombe. Edited by G. H. von Wright, R. Rhees, and G. E. M. Anscombe. Oxford: Blackwell; New York: Macmillan. Selections from *Remarks* were reprinted in Benacerraf and Putnam 1964.

— 1961a. *The Blue and Brown Books.* Oxford: Blackwell.

— 1961b. *Notebooks 1914–1916.* Translated by G. E. M. Anscombe. Edited by G. H. von Wright and G. E. M. Anscombe. Oxford: Blackwell.

— 1961c. *Tractatus Logico-Philosophicus.* A translation of Wittgenstein 1921 by D. F. Pears and B. F. McGuinness, with an introduction by B. Russell. London: Routledge and Kegan Paul; New York: Humanities Press.

Wood, O. P. 1961. "Symposium: On Being Forced to a Conclusion II." *Aristotelian Society Supplementary Volume 35.*

Young, J. W. 1911. *Lectures on the Fundamental Concepts of Algebra and Geometry.* New York: Macmillan.

Zermelo, E. 1904. "Beweis, dass jede Menge wohlgeordnet werden kann." *Mathematische Annalen*, vol. 59. Translated as "Proof That Every Set Can Be Well-Ordered" by S. Bauer-Mengelberg in van Heijenoort 1967.

— 1908. "Untersuchungen über die Grundlagen der Mengenlehre I." *Mathematische Annalen*, vol. 65. Translated as "Investigations in the Foundations of Set Theory I" by S. Bauer-Mengelberg in van Heijenoort 1967.

— 1909. "Sur les ensembles finis et le principe de l'induction complète." *Acta Mathematica*, vol. 32.

— 1929. "Über eine Widerspruchsfreiheitsfrage in der axiomatischen Mengenlehre." *Journal für reine und angewandte Mathematik*, vol. 160.

— 1930. "Über Grenzzahlen und Mengenbereiche." *Fundamenta Mathematicae*, vol. 16.